W9-DEO-538

8/27/92

CAMBRIDGE MONOGRAPHS ON
MECHANICS AND APPLIED MATHEMATICS

General Editor
G.K. BATCHELOR, FRS
Professor of Fluid Dynamics at the University of Cambridge

COLLOIDAL DISPERSIONS

Colloidal Dispersions

W. B. RUSSEL

D. A. SAVILLE

W. R. SCHOWALTER

Department of Chemical Engineering
Princeton University

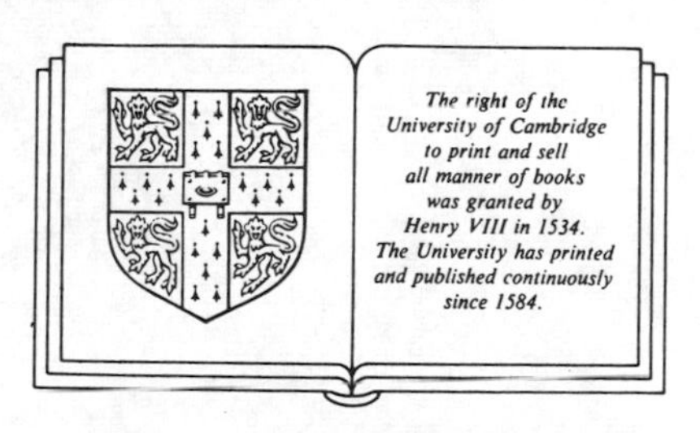
The right of the
University of Cambridge
to print and sell
all manner of books
was granted by
Henry VIII in 1534.
The University has printed
and published continuously
since 1584.

CAMBRIDGE UNIVERSITY PRESS

Cambridge

New York Port Chester

Melbourne Sydney

Published by the Press Syndicate of the University of Cambridge
The Pitt Building, Trumpington Street, Cambridge CB2 1RP
40 West 20th Street, New York NY 10011-4211, USA
10 Stamford Road, Oakleigh, Victoria 3166, Australia

First published 1989
First paperback edition (with corrections) 1991

Printed in Great Britain at the University Press, Cambridge

British Library cataloguing in publication data

Russel, W.B. (William Bailey), *1945–*
Colloidal dispersions
1. Colloids. Dispersions
I. Title II. Saville, D.A. (Dudley Albert)
1933– III. Schowalter, W.R. (William
Raymond), *1929–*
541.3′451

Library of Congress cataloguing in publication data

Russel, William B.
Colloidal dispersions / W.B. Russel, D.A. Saville, W.R.
Schowalter.
p. cm.
Bibliography: p.
Includes index.
ISBN 0 521 34188 4 (hb); ISBN 0 521 42600 6 (pb)
1. Colloids. I. Saville, D.A. II. Schowalter, William Raymond,
1929– . III. Title
QD549.R744 1989
541.3′451–dc 19 88-38589 CIP

ISBN 0 521 34188 4 hardback
ISBN 0 521 42600 6 paperback

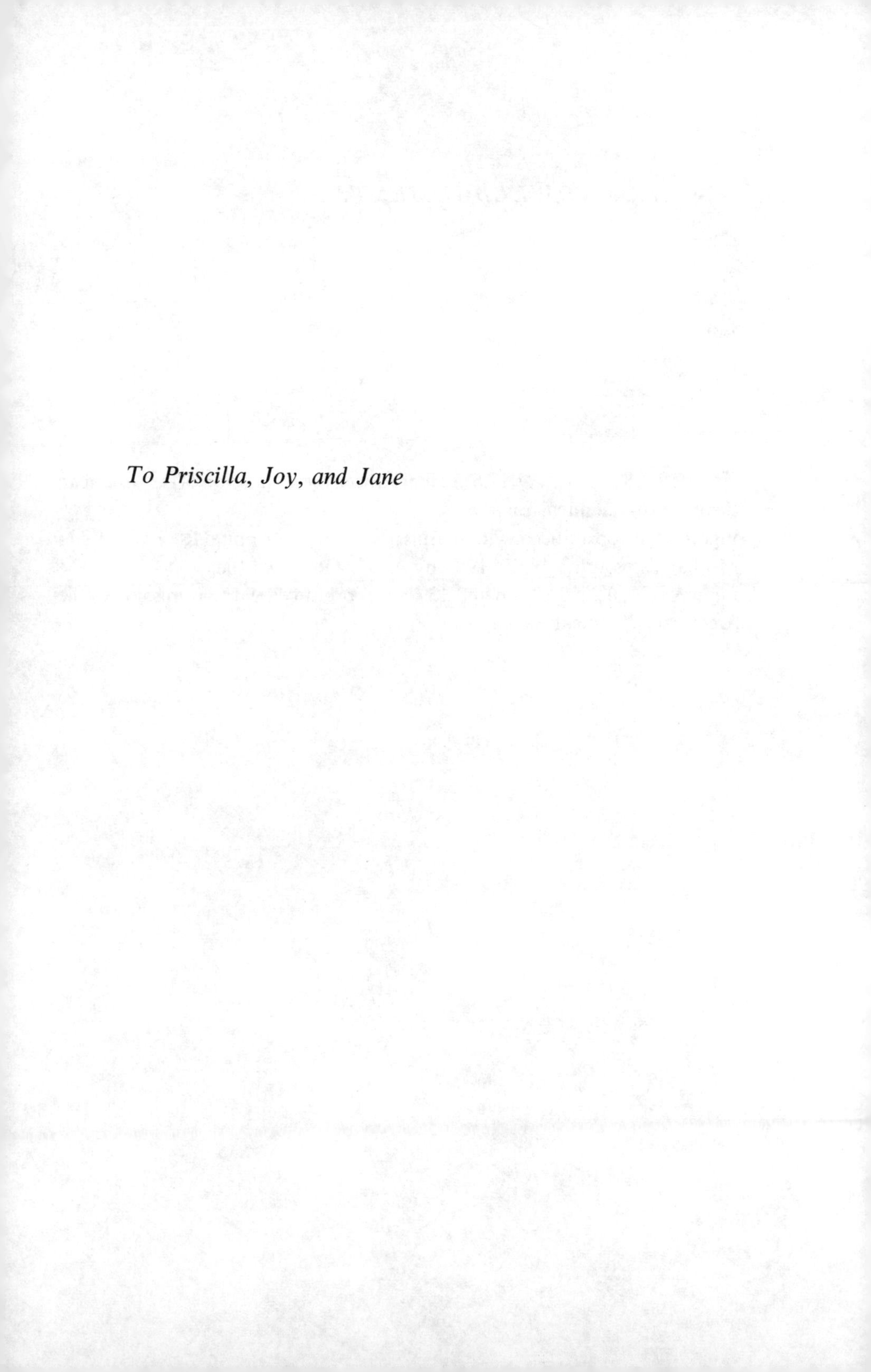

To Priscilla, Joy, and Jane

ACKNOWLEDGEMENTS

Figures 8.5, 8.6, 8.8*c*, 8.19, 12.6 are reproduced by permission of the American Institute of Chemical Engineers.
Figure 3.2 is reproduced, with permission, from the Annual Review of Fluid Mechanics, volume **13**, © 1981 by Annual Reviews Inc.
Figures 6.20, 6.22, 7.19, 9.8 and 13.4 are reproduced with permission of the American Chemical Society.

CONTENTS

PREFACE

Colloid science has its roots in nineteenth- and early twentieth-century discoveries concerning the behavior of minute particles. Its early development was stimulated by controversies regarding the very existence of molecules. Scientific interest, along with technological and biological applications, fostered several definitive monographs and textbooks in the 1930s and 1940s. However, interest in the field declined within many academic circles after the Second World War, especially in the United States, despite continued and widespread industrial applications. The resurgence of interest that began in the early 1960s arose from mutually reinforcing events. New technological problems appeared in, for example, the manufacture of synthetic dispersions for coatings, enhanced oil recovery, the development of new fuels, environmental pollution, ceramics fabrication, corrosion phenomena, biotechnology, and separations processes. In addition, monodisperse suspensions of colloidal particles of diverse sorts became readily available and advances in our understanding of fluid mechanics on the colloidal scale burgeoned almost simultaneously. Further stimuli were provided by the appreciation by colloid scientists of advances in the theory of interparticle forces coupled with the development of several new experimental techniques. Forces and particle properties have long been difficult to measure accurately on the colloidal scale and numerical values were often the result of a long uncertain chain of inference. The new techniques made possible direct, accurate measurements of size, shape, and concentration, as well as the attractive and repulsive forces between surfaces separated by a few nanometers. The advancements made over the last thirty years convinced us of the need for a broad synthesis, integrating recent discoveries with those of earlier times so as to treat dynamic as well as equilibrium properties of dispersions.

This book addresses the physical side of colloid science; the subjects range from the individual forces acting between submicron particles suspended in a liquid through the equilibrium and dynamic properties of the dispersion. The relevant forces include Brownian motion, electrostatic repulsion, attraction due to dispersion forces, attraction and repulsion caused by soluble polymers, and viscous forces arising from relative motion between the particles and the liquid. The balance between Brownian motion and the interparticle forces decides issues concerning stability and phase behavior in quiescent systems. Imposition of external fields alters the structure to produce complex effects, i.e., electrokinetic phenomena (electric field), sedimentation (gravitational field), diffusion (concentration/chemical potential gradient), and non-Newtonian rheology (shear field).

Our aim is to impart a quantitative understanding grounded in basic theory and coupled to experiments on well-characterized model systems. This provides the broad grasp of fundamentals which lends insight and helps develop the intuitive sense needed to isolate essential features of scientific and technological problems and to design critical experiments. The book is suitable both as a text for an advanced graduate course in chemical engineering, physical chemistry, physics, or applied mathematics, and as a reference for those doing industrial or academic research. Most of the material is accessible to those with a basic knowledge of mechanics and mathematics. Although exposure to fluid mechanics, statistical mechanics, and electricity and magnetism is assumed, the subjects in the book are introduced in a self-contained manner. Likewise, some facility with differential equations and vectors and tensors is required. Those interested in probing further can deepen their understanding by referring to the original works cited herein.

The book developed from complementary research interests among the authors, fostered initially by general grants from the Dreyfus and Xerox Foundations. This led to an advanced graduate course, first taught at Princeton in 1978 and now offered in alternate years. The writing began in earnest during WBR's tenure as the Olaf A. Hougen Professor in the Department of Chemical Engineering at the University of Wisconsin in 1984.

We are indebted to our students for their contributions, some of which appear explicitly in the text, as well as the interactions which advanced our understanding of the subject. In addition, we acknowledge the critical reading and constructive comments on portions of the text by Chip Zukoski and Alice Gast. Finally, we thank Elizabeth B. Bixby for her ability to deal with what appeared to be an endless sequence of revisions; her threats of divine retribution helped make the process convergent.

W.B.R., D.A.S., W.R.S.
Princeton, September 1988

UNITS AND PHYSICAL CONSTANTS

The International Metric System ('SI', from the French, *Système Internationale d'Units*) used here employs the following base units:

Quantity	Unit	Symbol
Length	meter	m
Mass	kilogram	kg
Time	second	s
Temperature	kelvin	K
Amount	mole	mol
Electric current	ampere	A

The units derived from this set are:

Quantity	Unit	Symbol	Definition
Force	newton	N	$1\,kg\,m\,s^{-2}$
Pressure	pascal	Pa	$1\,kg\,m^{-1}\,s^{-2}$
Energy	joule	J	$1\,kg\,m^2\,s^{-2}$
Electric charge	coulomb	C	$1\,s\,A$
Electric potential	volt	V	$1\,kg\,m^2\,s^{-3}\,A^{-1}$
Frequency	hertz	Hz	$2\pi\,rad\,s^{-1}$
Capacitance	farad	F	$1\,kg^{-1}\,m^{-2}\,s^4\,A^2$

Of course the units of the derived quantities can be expressed in terms of one another, e.g.,

$$1\,\mathrm{F} = 1\,\mathrm{C/V}.$$

The fundamental laws involve a number of physical constants. Those used here are:

Constant	Symbol	Numerical value
Avogadro's constant	N_A	6.02552×10^{23} molecules/mole
Boltzmann's constant	k	1.38054×10^{-23} J/K
Magnitude of charge on an electron	e	1.60210×10^{-19} C
Permittivity of the vacuum	ε_0	$8.854 \times 10^{-12}\ \mathrm{C^2/N\,m^2}$
Planck's constant	$2\pi\hbar$	6.6256×10^{-34} J s
Speed of light	c	2.9979×10^{8} m/s

Miscellaneous conversion factors

Standard acceleration due to gravity, g	$9.8066\ \mathrm{m/s^2}$
Atmospheric pressure	$1.013\,25 \times 10^{5}$ Pa
kT/e at 298.16 K	25.69×10^{-3} V
1 molar solution, M	$1\ \mathrm{mol/(dm)^3}$
1 liter	$1.000\,0028 \times 10^{-3}\ \mathrm{m^3}$

Prefixes

Meters, kilograms, seconds and the like are not always convenient scales but various multiples are. The commonly used scale factors are listed below:

kilo,k	hecto,h	deca,da	deci,d	centi,c	milli,m	micro,μ	nano,n
10^3	10^2	10	10^{-1}	10^{-2}	10^{-3}	10^{-6}	10^{-9}

MATHEMATICAL SYMBOLS

The mathematical symbols denoting constants and variables of one sort or another are defined in the text at their point of introduction. Symbols used in equations are defined as follows:

Symbol	Meaning
$\equiv$	is identical to
$=$	is equal to
$\sim$	is asymptotically equal to (in some limit)
$\approx$	is approximately equal to
$\leq$	is less than or equal to
$\ll$	is much less than
$\rightarrow$	tends to
$f=O(\varepsilon)$	limit $f/\varepsilon < \infty$ as $\varepsilon \rightarrow 0$
$f=o(\varepsilon)$	limit $f/\varepsilon = 0$ as $\varepsilon \rightarrow 0$

1

A SURVEY OF COLLOIDAL DISPERSIONS

1.1 Colloidal phenomena

Colloidal particles dispersed in liquids exhibit astonishing properties. Dispersions such as the colloidal gold sol prepared by Faraday (1791–1867) over a century ago can persist almost indefinitely, yet the addition of salt would cause rapid, irreversible flocculation. In fact, for many dispersions the physical state, i.e. the stability or phase behavior, can be altered dramatically by modest changes in composition. This complex behavior stems from the different forces that act among the particles, determining their spatial distribution and governing the dynamics. Brownian motion and dispersion forces (arising from London–van der Waals attraction) would flocculate Faraday's gold sol were it not for electrostatic repulsion between the particles. The addition of salt increases the concentration of ions screening the surface charge, suppressing repulsion and allowing flocculation. Doublets and more complicated structures formed during flocculation have long lifetimes, since Brownian motion is too weak to overcome the strong attractive force between particles near contact. Indeed, removal of the salt does not usually lead to spontaneous redispersion, so mechanical means must be used.

Another type of transformation occurs when ions are removed from electrostatically stabilized systems. Polymer latices in an electrolyte solution are milky-white fluids, but dialysis eliminates the ions and leads to iridescence owing to Bragg diffraction of visible light from an ordered structure (Fig. 1.1). Here the absence of screening allows long-range electrostatic repulsion to induce a disorder–order phase transition. Colloidal crystals contain defects and dislocations which permit flow as ordered structures at finite, but low, shear stresses. At higher stresses they melt and exhibit shear thinning (Rothen *et al.*, 1987).

Colloidal dispersions also exhibit a reversible phase transition in the presence of non-adsorbing polymer. Adding sodium polyacrylate to a suspension of polystyrene spheres, for example, causes particles to attract one another owing to osmotic pressure differences. The simplest explanation is that polymer molecules are excluded from the region between

Fig. 1.1. Micrograph of the ordered structure in a deionized polystyrene latex. The particles are 0.33 μm in diameter and the volume fraction of particles is 0.01. (From Kose *et al.*, 1973.)

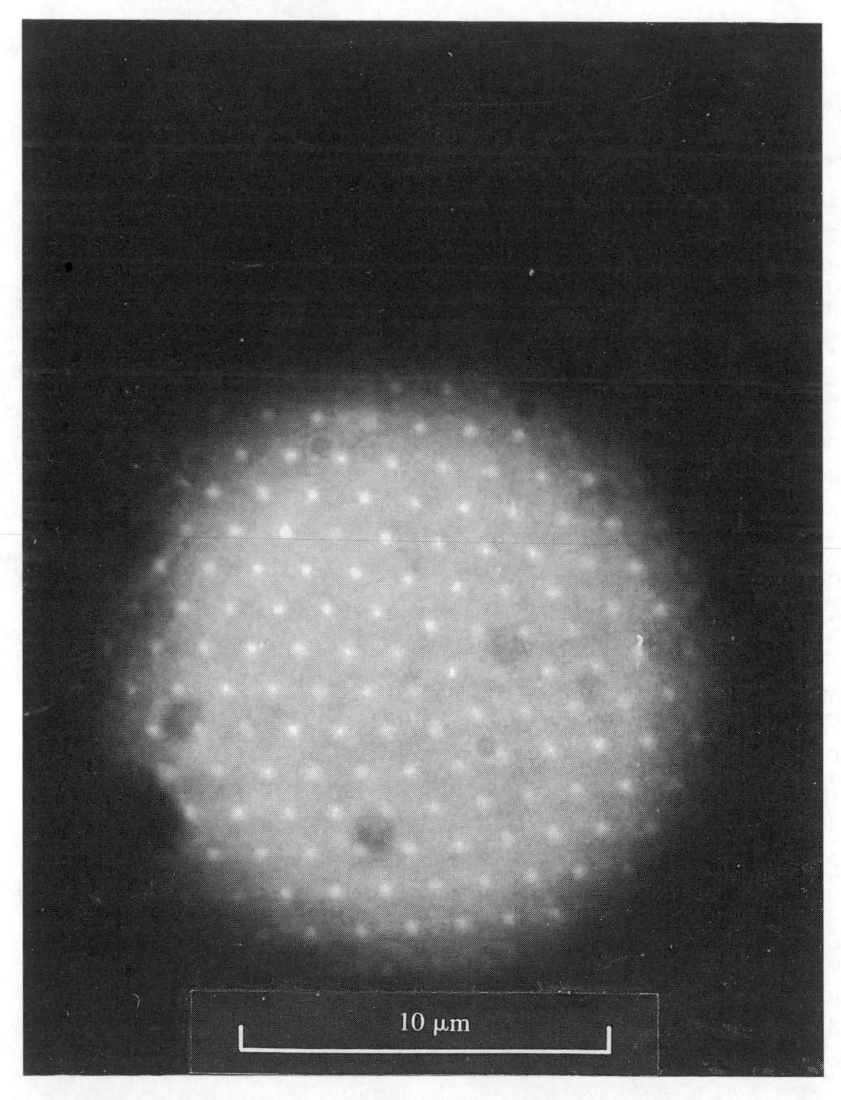

particles by steric effects, so the pressure is lower in the gap than on the portions of the particles accessible to the bulk solution. The osmotic pressure difference can cause the suspension to separate into a dense, ordered phase and a dilute, disordered phase (Fig. 1.2). Such equilibrium mixtures resemble those encountered with molecular fluids.

Colloidal systems display complex rheological behavior related to their thermodynamic non-ideality but do not show the substantial elastic recovery characteristic of polymeric liquids. Macromolecules recover from extensions several times their equilibrium dimensions, making polymer solutions very elastic. The relatively short range of interparticle forces precludes such behavior with colloidal dispersions. Dilute and moderately concentrated stable dispersions behave like low-viscosity liquids at low shear rates and may exhibit shear thinning. Changing the relative magnitudes of colloidal forces has dramatic effects. With aqueous latices, for example, lowering the ionic strength may increase the viscosity substantially as electrostatic repulsion comes into play. Alternatively, flocculated sols assume disordered structures which deform elastically under small strains, but fracture at higher strains and flow like liquids. Concentrated dispersions stabilized either electrostatically or polymerically form ordered layers under shear, and, at a critical shear stress, undergo an order–disorder transition as illustrated in Fig. 1.3. Each of these rheological properties stems from the interparticle forces responsible for the different types of phase behavior, with the added influence of hydrodynamic forces.

Sedimentation and Brownian motion of colloidal particles also reflect the balance of interparticle and hydrodynamic forces. Settling experiments produce states ranging from dilute random dispersions to dense sediments. The Frontispiece depicts the sedimentation of silica particles in cyclohexane, showing five distinct regions:

> clear fluid at the top;
> a region of particles settling at the initial, dilute concentration;
> a dense, disordered region of particles still settling freely;
> an iridescent, ordered region; and
> an opaque, disordered sediment.

This complex behavior reflects a combination of kinematic processes stemming from the concentration dependence of the settling velocity and thermodynamic factors responsible for the order–disorder transition. Brownian motion is equally sensitive to interparticle forces, but the effects are more subtle. For example, in an electrostatically stabilized system, the

Fig. 1.2. An equilibrium two-phase structure formed in a polystyrene latex with 0.000 37 per cent (by weight) sodium polyacrylate. Particle diameter is 0.33 μm. (From Kose & Hachisu, 1976.)

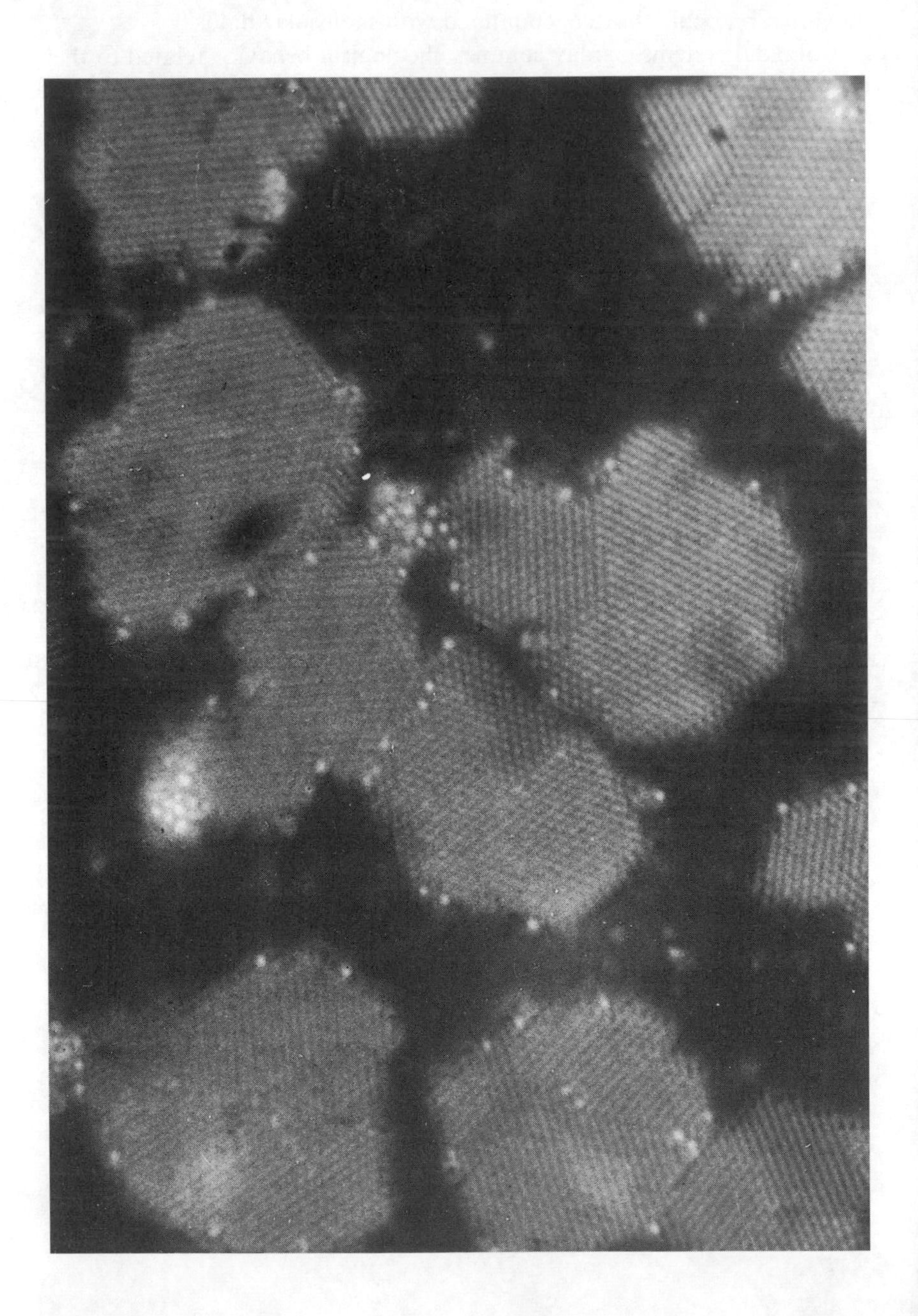

mutual-diffusion coefficient increases rapidly as the ionic strength is lowered, but the self-diffusion coefficient decreases. Both effects can be measured accurately by photon correlation spectroscopy.

Electric and magnetic fields affect dispersions in various ways. When the particles are charged, externally applied electric fields move them relative to the fluid (electrophoresis). From this motion the net electric charge on a particle can be ascertained. Electrophoresis can also be used to separate particles via the relative motion between particles of different charge. Another illustration is furnished by the behavior of ferrofluids, colloidal magnetic particles dispersed in a liquid. Magnetic fields alter the rheology of such dispersions and induce labyrinth-like patterns (Fig. 1.4).

Interactions between colloidal particles and macroscopic bodies are also governed to a large degree by colloidal forces. Small particles moving past a larger object are influenced by both electrostatic and dispersion forces, and the fate of individual particles is controlled by the delicate balance between viscous and interparticle forces. Striking configurations are possible, as illustrated in Fig. 1.5.

Our understanding of colloid dynamics and phase behavior is based largely on knowledge of the behavior of individual particles, either in

Fig. 1.3. Diffraction patterns from monodisperse suspensions in simple shear between parallel discs (Hoffman, 1972). The photograph on the left shows the diffraction of white light from an ordered structure below the critical shear stress; the other photograph shows the pattern above the critical shear stress.

Fig. 1.4. Photographs of a labyrinthine instability developed by a magnetic field acting perpendicular to a thin vertical layer of ferrofluid. The field intensity increases from left to right. (From Rosensweig, 1985.)

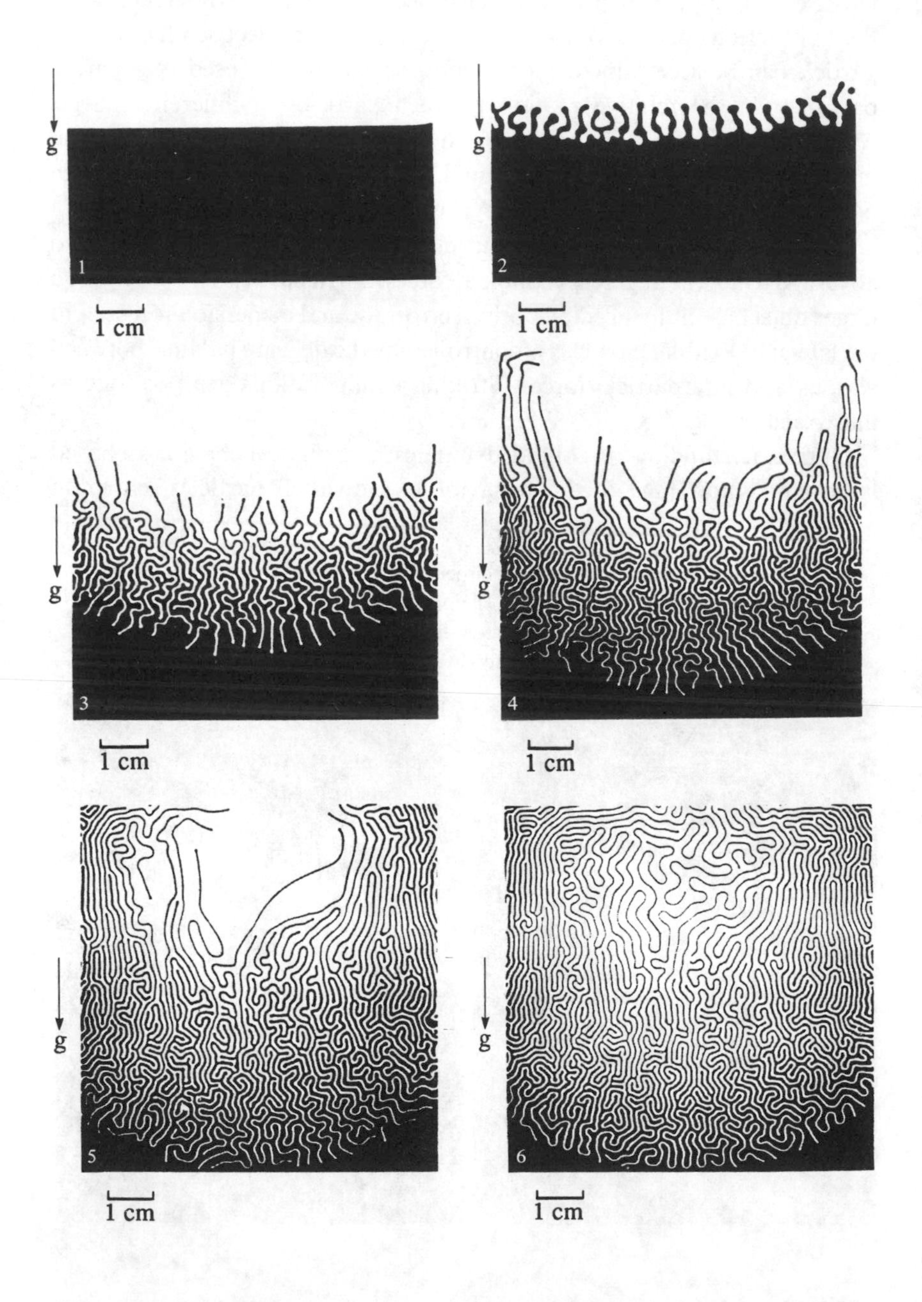

isolation or interacting with others in a pairwise manner. In the chapters that follow the salient features of those phenomena are set out and then integrated into a broad theory of colloidal behavior. To establish a context, the historical foundations are reviewed next, followed by a discussion of the general character of colloidal forces.

1.2 Historical notes

In his 1861 paper, 'Liquid Diffusion Applied to Analysis', Thomas Graham (1805–69) described two classes of matter, crystalloids and colloids. This classification differentiated between substances that would diffuse through a membrane separating water from an aqueous solution (crystalloids), and those which would not (colloids).[†] The crystalloid class included salt, sugar, and other substances that crystallize, while albumin, gum arabic, gelatin, and the like belonged to the colloid class. Graham also described how to disperse normally insoluble substances by a method called dialysis. Prussian blue, which is soluble in oxalic acid but insoluble in water, is an example of such a material. During dialysis with Prussian Blue, oxalic acid is replaced by water moving through the membrane, yet the

[†] From the Greek κόλλα, meaning glue.

Fig. 1.5. Dendritic structures built up of 1 μm polystyrene spheres stacked on a 20 μm nylon fiber. The particles were collected from an aerosol using a strong transverse electric field and are held in place by the strength of the dispersion force at close separations. (From Oak, Lamb, & Saville, 1985.)

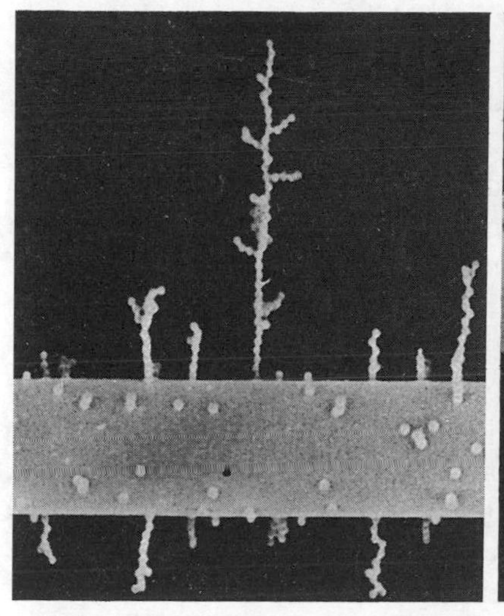

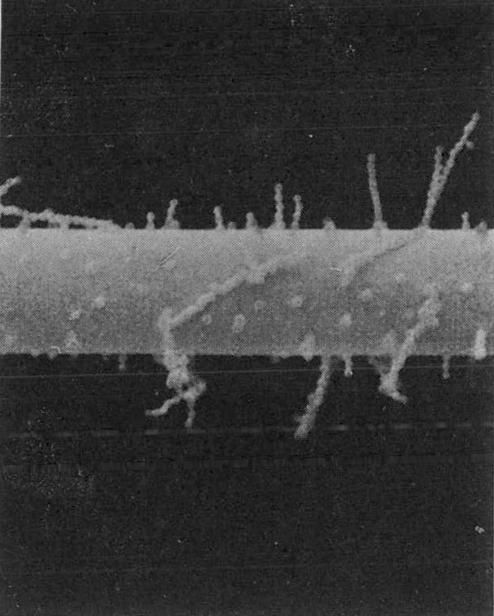

solution remains clear. Graham coined the term sol for the substance that did not dialyze.

Although Graham's techniques were original, colloids had been observed and studied much earlier. Seventeenth-century alchemists produced sols by treating gold chloride solutions with reducing agents and Berzelius (1779–1848) studied silicic acid, sulfur, and metallic sulfides. One of the most important observations was Faraday's discovery that small particles could be detected by focusing light into a conical region. This led to the development of the ultramicroscope by Zsigmondy & Siedentopf in 1903. Studies with this instrument probed the nature of the erratic motion of individual particles first observed by Robert Brown (1773–1858) and ascribed to the impact of molecules of the suspending medium. The molecular basis of the motion was settled when Perrin (1870–1942) summarized extensive observations in his book *Brownian Movement and Molecular Reality* (1910). The supporting theory was presented almost simultaneously by Einstein (1879–1955), who analyzed the sedimentation–diffusion equilibrium (Einstein, 1906) and Langevin (1872–1946), who treated the dynamics (Langevin, 1908).

The discovery that naturally occurring colloidal particles were charged dates to an 1809 study by Reuss, who noted the motion of clay particles in an electric field. Linder and Picton found in 1892 that synthetic sols of sulfur, ferrocyanides, gold, silver, or platinum are negatively charged, while oxide sols of iron, chromium, aluminium, and cerium are positive. The role of added electrolytes in suppressing the effects of charge and promoting flocculation was elucidated by Schultz (1882) and Hardy (1900). Their work provided strong evidence that the stability of aqueous dispersions derived from electrostatic repulsion. Nevertheless, no theory existed to describe the screened repulsion or relate the particle's charge to its mobility in an electric field, even though Helmholtz (1821–94) had already formulated his model of the molecular condenser. His analysis of the movement of liquid through a capillary under the action of an electric field introduced the notion of the ζ-potential to describe the electrostatic state of a surface (Helmholtz, 1879). Smoluchowski (1872–1917) derived his celebrated formula relating the ζ-potential to the electrophoretic mobility (Smoluchowski, 1903) by recognizing the similarity between the motion produced by an external electric field acting on a small particle (electrophoresis) and on a liquid in a capillary (electro-osmosis). As a result, the ζ-potential could be measured by timing the motion of a particle viewed through a microscope. Though the electric charge could be estimated, the effect of electrolyte on the double-layer thickness was still not understood.

The theory of screening of the surface charge by the diffuse charge cloud was developed by Gouy (1910) and Chapman (1913), thereby relating the thickness of the layer to the ionic strength of the solution.

After it was known that dispersions could be flocculated by screening the electrostatic repulsion with excess electrolyte, Smoluchowski (1917) deduced expressions for the rate of formation of small aggregates by Brownian and shear-induced collisions. However, the structure of the attractive interparticle potential was unknown until de Boer (1936) and Hamaker (1937) developed a theory based on pairwise summation of the intermolecular forces. Representing the total interparticle potential as the sum of the attractive and repulsive components then led to a detailed theory of colloid stability. Activity within several groups culminated in the theory published by Derjaguin & Landau (1941) in the Soviet Union and Verwey & Overbeek (1948) in the Netherlands.

1.3 Recent developments

Progress in colloid science has been stimulated by several important developments since the early 1950s, beginning with the synthesis of model colloids (Vanderhoff *et al.*, 1956).[†] Highly monodisperse latices with diameters of 0.05–3.0 μm can be formed of various polymers by emulsion polymerization. Inorganic dispersions made from hydrous metal oxides (Matijevic, 1976) and silica particles (Iler, 1979) provide a variety of particle types and shapes. Consequently, complications due to polydispersity are avoided and theories can be tested cleanly. Particle size analysis has been speeded immensely through easy access to scanning electron microscopes; see Fig. 1.6.

Another key development has been the direct measurement of interaction forces. Early efforts by Derjaguin *et al.* (1954) and Overbeek & Sparnaay (1954) were limited by the roughness of the fused quartz and polished glass surfaces employed. Tabor & Winterton (1969) recognized that cleaved muscovite mica provides a molecularly smooth surface, and this increased the accuracy of the measurement considerably. Later work by Israelachvili & Tabor (1973) produced a greatly improved instrument, leading to measurements that resolve events on the scale of a nanometer or so using crossed cylinders covered with mica sheets. The experimental results using this device and independent measurements on lipid bilayers (LeNeveu, Rand & Parsegian, 1976) have confirmed both the attractive

[†] See Vanderhoff (1964), *Preprint, Div. of Organic Coatings and Plastics Chemistry* **24** (2) 223–32, for an account of the almost accidental discovery of monodisperse latex dispersions.

and the repulsive interactions, as well as revealing complex structural forces at separations comparable to molecular dimensions (see Israelachvili, 1985).

A deeper theoretical understanding of attractive forces emerged during the same period. Early theories based on the assumption of pairwise additivity failed to account for many-body interactions important in condensed phases. Lifshitz (1955) resolved the problem by developing a continuum theory in which materials were characterized by their dielectric

Fig. 1.6. Ordered arrays formed from filtered latex suspensions. (Photo from Interfacial Dynamics Corporation, Portland, Oregon.)

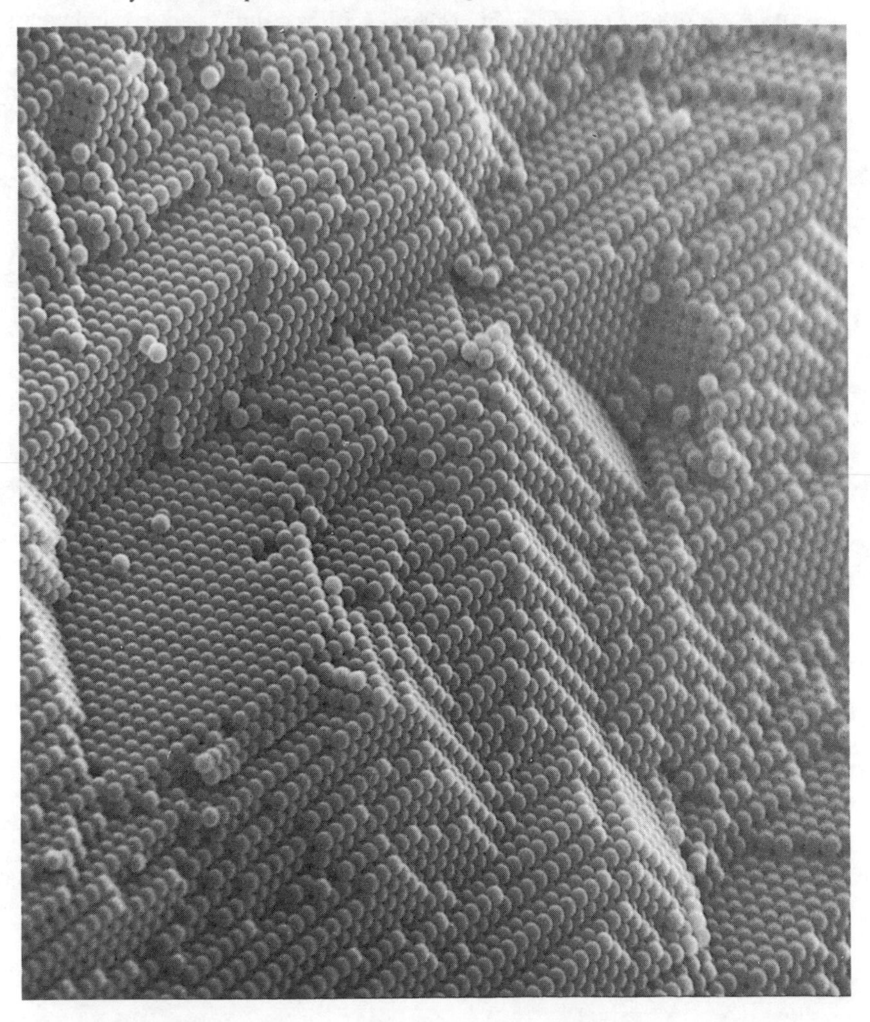

behavior. Ninham & Parsegian (1968) then devised tractable ways of applying the theory quantitatively to colloidal systems. Predictions based on measurements of the dielectric spectra for mica now agree with forces measured across a vacuum with the crossed cylinder apparatus (White *et al.*, 1976).

The need to disperse particles at high ionic strengths or in non-aqueous solvents led to improved techniques for polymeric stabilization in the 1960s. Naturally occurring macromolecules have been employed since ancient times, but the development of block and graft copolymers revolutionized the practice. These molecules combine an insoluble component for attachment to the particle surface and a soluble chain for stabilization. The pioneering work of Napper and colleagues (Napper, 1983) has revealed the close relation between the stability of the dispersion and the solution properties of the stabilizing block.

New developments in the statistical mechanics of the liquid state, coupled with synthesis of new model systems, also influenced colloid science. Silica spheres coated with octadecyl chains and dispersed in cyclohexane (Stober, Fink and Bohn, 1968; Iler, 1979) behave much like the hard spheres described by both analytical theories and computer simulations. Furthermore, perturbation theories (Barker & Henderson, 1967; McQuarrie, 1976), originally developed to deal with Lennard-Jones potentials in the liquid state, provide a convenient means for predicting phase transitions and other thermodynamic properties of concentrated colloidal dispersions.

Our detailed understanding of hydrodynamic effects on Brownian motion is another contemporary development. Although Langevin (1908) used Stokes' law to analyze the dynamics of Brownian particles, discrepancies between this approach and corresponding analyses of molecular motions were not resolved until the early 1970s. On the experimental side, rapid advancements in light-scattering techniques profoundly improved our understanding, particularly in concentrated dispersions. Photon correlation spectroscopy (dynamic light scattering) now provides a rapid, accurate method for measuring diffusion coefficients both for sizing particles and for understanding interactions between particles (see Berne & Pecora, 1976).

The full description of viscous interactions between a pair of particles is also of recent vintage; Happel & Brenner (1965) give a comprehensive review of the situation through the early 1960s. The detailed treatment necessary for understanding colloidal systems followed from work by Goldman, Cox & Brenner (1966) on a pair of spheres moving through a quiescent fluid and Lin, Lee & Sather (1970) on spheres in a linear shear flow. This knowledge of the fluid mechanics advanced our understanding of Brownian diffusion, colloid stability, sedimentation, and rheology.

All these topics will be examined in detail in subsequent chapters, but first it is useful to consider the classification of colloids and the relative magnitudes of the various forces.

1.4 The classification of colloids

Graham used the term colloid to distinguish types of matter, but it later became apparent that colloids are not separate types, but matter in a particular state of subdivision, in which effects connected with the surface are pre-eminent. The terms lyophobic and lyophilic were introduced in the early literature to distinguish sols that were extremely sensitive to small amounts of electrolyte from those that were not. Gold and Prussian blue sols, for example, were called lyophobic ('liquid hating'); sols that required a large amount of salt to produce coagulation, e.g. albumin or gum arabic, were called lyophilic. Kruyt (1949) gives an extensive survey of the historical development of the subject and Shaw (1970) describes modern aspects. A simpler classification will be used in this book, viz.: particles, macromolecules, and molecular assemblies. Silver halide and gold sols, along with clays and polymer latex particles, fall into the first category. Such particles can be treated as rigid bodies and the details of their internal structure ignored when analyzing the dynamics. Macromolecular solutions, where the conformation of the molecule reflects the nature of the solvent, exemplify the second category. Micelles and microemulsions epitomize the third.

Particle size and shape are of considerable importance in setting the properties of a colloidal system. The simplest shape, a sphere, is characteristic of emulsions, latices, some protein molecules, and gold and silica sols. Clay particles such as kaolinite are plate-like, while attapulgite is rod-like. The treatment in the following chapters is restricted to spherical particles, in part because modest departures do not alter the qualitative features of the phenomena so that non-spherical particles can be modelled in terms of an equivalent sphere. Moreover, comprehensive results are available only for spheres.

Particle sizes in colloidal systems generally range from 1 nm to 10 μm. Table 1.1 shows some representative particle dimensions.

To furnish insight into the relative importance of the forces we can look into representative orders of magnitude. The thermal energy of molecular chaos may be interpreted as a Brownian force with a magnitude of $O(kT/a)$;

Table 1.1. *Typical particle sizes*

μm			
10^2			
		Sand	Pulverized coal
10^1	↑		
		Silt	Red blood cells
10^0	Mist and fog		Paint pigment
		Clay	Latices
10^{-1}	↓		
			Coiled macromolecules
		Colloidal silica	Carbon black
10^{-2}			
		Colloid gold	Micelles
10^{-3}			

here a denotes a representative length, k stands for Boltzmann's constant (1.381×10^{-23} J/K), and T for the absolute temperature. Additivity of London– van der Waals forces on an atomic or molecular scale yields a force between macroscopic bodies, the dispersion force, which is $O(A_{\text{eff}}/a)$. The Hamaker constant, A_{eff}, depends on the nature of the particles and the intervening fluid. Electrostatic forces between two particles are moderated by ions in the intervening fluid, but an order of magnitude follows from Coulomb's law as $\varepsilon\varepsilon_0\zeta^2$; ε is the dielectric constant of the fluid, ε_0 the permittivity of free space (8.85×10^{-12} C/V m), and ζ the electrostatic potential of the particles. Viscous forces on a particle moving at a velocity U through a medium of viscosity μ are $O(\mu a U)$, according to Stokes' law, and inertial forces are $O(a^2 \rho U^2)$. Finally, if $\Delta\rho$ represents the difference between the density of the particle and the fluid, then the gravitational body force on the particle is $O(a^3 \Delta\rho g)$. To compare magnitudes we use the values listed in Table 1.2 and compute ratios of the several forces.

In situations represented by this set of parameters, repulsive forces dominate attraction, so a suspension of such particles should be stable. Viscous and Brownian forces appear to influence matters equally, but sedimentation and inertia should not be important. Nevertheless, it must be recognized that this approach overlooks an important fact: individual forces depend differently on particle separation. For example, the screened electrostatic repulsion between flat surfaces decays exponentially, whereas the dispersion force varies inversely with the cube of the separation when

Table 1.2. *Magnitudes of the characteristic forces*; $a=1\,\mu m$, $\mu=10^{-3}$ kg/m s, $U=1\,\mu/s$, $\rho=10^3$ kg/m^3, $\Delta\rho/\rho=10^{-2}$, $g=10$ m/s^2, $A_{eff}=10^{-20}$ N m, $\zeta=50$ mV, $\varepsilon=10^2$

$\dfrac{\text{electrical force}}{\text{Brownian force}}$	$\dfrac{\varepsilon\varepsilon_0\zeta^2}{kT}$	$\approx$	10^8
$\dfrac{\text{attractive force}}{\text{Brownian force}}$	$\dfrac{A_{eff}}{kT}$	$\approx$	1
$\dfrac{\text{Brownian force}}{\text{viscous force}}$	$\dfrac{kT}{\mu U a^2}$	$\approx$	1
$\dfrac{\text{gravitational force}}{\text{viscous force}}$	$\dfrac{a^3\Delta\rho g}{\mu U a}$	$\approx$	10^{-1}
$\dfrac{\text{inertial force}}{\text{viscous force}}$	$\dfrac{\rho a^2 U^2}{\mu U a}$	$\approx$	10^{-12}

the surfaces are close together. Accordingly, attraction due to the dispersion force may predominate at one separation while electrostatic repulsion dominates the interaction at another. Moreover, viscous effects are so strong a function of separation that as two surfaces are forced together, the resistance to further diminution of the gap diverges. Clearly, a more careful study is warranted.

1.5 An overview

Our presentation focuses on dynamic and thermodynamic properties of colloidal dispersions. Because the behavior is complex, study of idealized situations makes it easier to understand key factors. For example, the motion of isolated spheres provides the simplest model for particles interacting with external fields, while the study of pairs of spheres exemplifies the effects of interparticle forces. Although our approach is theoretical and relies on mathematical analyses of model problems, we endeavor to give experimental foundations full weight.

The hydrodynamic factors are set out in Chapter 2, focusing on the behavior of spheres. Solutions to complicated problems are constructed from singular solutions of the Stokes equation because this illustrates the basic physics better than, say, mathematical techniques that rely on separation of variables. Since pair interactions are so important, the motion of two spheres is studied in detail.

Brownian motion is taken up in the third chapter by using the Langevin equation to analyze the motion of an individual particle. This is followed by a complementary statistical treatment. Comparison of the two approaches establishes that the overall process can be characterized by hydrodynamic mobilities calculated from the Stokes equation, despite the unsteady nature of the detailed motion. An analysis of pairs of spheres shows how pair mobilities describe the Brownian motion of interacting particles as well as the translation and rotation of a doublet consisting of two touching spheres. Brownian motion can be studied experimentally using light-scattering techniques as well as macroscopic observations. A sampling of experimental results demonstrates the agreement between experiment and theory when electrostatic and dispersion forces are negligible.

The treatment of interparticle forces begins with electrostatics in Chapter 4. Maxwell's equations are taken as axioms and magnetic effects ignored. A brief description of homogeneous dielectrics sets the stage for a detailed study of situations where fully ionized electrolytes are present. The structure of the diffuse layer adjacent to a charged surface is analyzed first, followed by an inquiry into the electrostatic forces between plates and spheres. Ion densities adjacent to a charged surface follow the Boltzmann distribution, so the equations for the electrostatics are non-linear. Thus, approximations must be used to obtain explicit expressions for the electrostatic part of the interparticle potential. Comparison with results from exact numerical solutions and experimental data helps define the range of validity of the various approximations.

Dispersion forces are less familiar, so Chapter 5 begins by describing interactions between two oscillating dipoles. This shows how the interaction energy varies with separation and forms the basis for a microscopic treatment of condensed phases involving a pairwise summing of dipole interactions. The result is an analytical form for the interaction potential which characterizes the material properties in terms of the Hamaker constant. A different methodology is required to account for the many body interactions neglected in this microscopic theory. The continuum theory based on the dielectric response of the different phases rectifies the shortcomings of the microscopic approach and provides the interaction potential between macroscopic bodies, including the effect of electromagnetic retardation and the intervening electrolyte. The chapter closes with a discussion of the experimental test of the theory using the crossed-cylinder apparatus with mica surfaces and simplified forms for the interaction potential.

Soluble polymers, either free in solution, attached to the colloid surface at one end, or adsorbed reversibly at a number of points along the polymer backbone, also mediate particle–particle interactions. Forces due to polymer–polymer interactions are analyzed in Chapter 6 by combining thermodynamic and statistical mechanical methods. The task is to describe how the interactions produce a non-uniform polymer distribution in the solution around each particle. Alteration of this distribution by particle–particle interactions produces an interparticle force. Our understanding of such forces is less developed than for electrostatic or dispersion forces. Thus, although the theory can explain some features of the phenomena, many important details remain to be worked out.

Electrokinetic phenomena derived from the relative motion between a charged surface and an ionic solution reflect the electrical characteristics of the diffuse charge cloud and the particle surface. Chapter 7 presents a detailed survey of the behavior of individual particles and suspensions. Electrophoresis is taken up first to explain how static and dynamic situations differ. Though measurements of electrophoretic mobility traditionally serve to characterize the charge on colloidal particles, measurements of the electrical conductivity or the complex impedence can serve the same purpose. The former characterizes the suspension in terms of its response to a steady field, whereas the real and imaginary parts of the impedence reflect a frequency-dependent conductivity and dielectric constant. Dielectric relaxation arises from polarization and relaxation processes in the diffuse layer and on the particle surface. Attention is focused on dilute systems, since theories for concentrated systems are not highly developed. Detailed comparisons between theory and experiment are limited to the static conductivity, because measurements of the complex impedence are clouded by interpretation problems at present.

Electrostatic forces, dispersion forces, and polymer-induced forces are combined into an interparticle potential in Chapters 8 and 9 to assess colloid stability. Electrostatic stabilization refers to a metastable condition where kinetic factors control the persistence of the system. Rates of doublet formation due to Brownian motion and shear are analyzed and compared to experimental results. Now that it is possible to evaluate the electrostatic and dispersion contributions to the interaction potential independently, measurements of doublet formation rates test the theory unambiguously instead of simply serving to estimate the Hamaker constant. For polymerically stabilized systems we concentrate on incipient Brownian flocculation, using the theory developed in Chapter 7 to rationalize the qualitative features of the experiments.

Colloidal systems also undergo reversible phase transformations, which are addressed in Chapter 10 using a statistical mechanical approach to calculate thermodynamic properties. A perturbation theory based on the behavior of hard spheres explains the order–disorder transition for charged spheres that produces the iridescence noted earlier. Phase transformations involving two fluid phases or a fluid and a solid phase are produced by dissolved polymer for both electrostatically and polymerically stabilized dispersions. These can be predicted from the weak attraction induced by the presence of non-adsorbing polymer.

The last four chapters describe dynamic processes where colloidal forces are central features: particle capture on stationary objects (Chapter 11), sedimentation (Chapter 12), diffusion (Chapter 13), and rheology (Chapter 14). In each instance the emphasis is on ways interparticle forces affect dynamics. For example, it is very difficult to bring two surfaces into contact in the presence of viscous forces. Particle inertia can produce capture, but the speeds of particles in liquids rarely exceed the threshold where capture begins. Thus, without an attractive interaction most filtration processes would be very inefficient. Chapter 11 deals with the ways interparticle forces alter particle paths so as to promote or inhibit capture.

Sedimentation in a closed container produces complex structures, as the Frontispiece amply demonstrates. Here consideration must be given to the effect of particle number density, the external force field, and interparticle potential on the settling velocity, Brownian motion, and the microstructure of the sediment. For example, a purely repulsive interaction hinders settling whereas attraction promotes it. Moreover, the microstructure of the suspension is set by Brownian motion and the interparticle force in the absence of the external field, but this structure is altered by the settling process. Thus, the speed of the process and the evolution of sediment structure depend crucially on colloidal forces. These issues are taken up in Chapter 12.

As noted earlier, the diffusion process in concentrated dispersions is very sensitive to interparticle forces. Chapter 13 first addresses gradient diffusion by distinguishing hydrodynamic and thermodynamic effects through a generalized Stokes–Einstein equation. Then self-diffusion coefficients are defined through the mean-square displacement of a Brownian particle in the presence of neighbors. Theoretical expressions relate the 'short-time' and 'long-time' behavior to the interparticle forces and hydrodynamic interactions. At finite particle concentrations, the self-diffusion and gradient diffusion coefficients differ because of different weightings of the various interactions. These diffusion coefficients can be measured by a combination of photon correlation spectroscopy and other techniques.

The rheology of colloidal suspensions is discussed in the final chapter. The presentation begins with a description of the phenomena and their classification using dimensional analysis. The utility of this approach is then demonstrated by correlating sets of experimental data for dispersions of hard, charged, polymerically stabilized, and weakly flocculated spheres. Then pair-interaction theories are developed that describe the non-equilibrium microstructure and the bulk stress in dilute systems. The good agreement with experimental results confirms the general applicability of the approach.

Our treatment centers on simple systems, primarily those involving monodisperse spherical particles with well-defined interactions at dilute concentrations. Dispersions of technological importance are often concentrated and polydisperse, with oddly shaped particles and a variety of uncharacterized components. Much of the difficulty encountered with these systems lies in simply identifying the components present and sorting out the important forces. These problems are formidable, and theoretical descriptions of the type developed here will rarely apply quantitatively. Nevertheless, it is clear that the relation between fundamental knowledge and technology is synergistic; each profits from the stimulus of the other. A broad understanding of the fundamentals does provide insight and helps develop the intuitive sense needed to isolate essential features of technological problems and design critical experiments. Providing that fundamental knowledge is the purpose of this book.

References

Barker, J. A. & Henderson, D. (1967). Perturbation theory and equation of state for fluids: I, The square well potential. *J. Chem. Physics* **47**, 2856–61. II, A successful theory of liquids. *Ibid.* **47**, 4714–21.

Berne, B. J. & Pecora, R. (1976). *Dynamic Light Scattering.* Wiley.

Chapman, D. L. (1913). A contribution to the theory of electroencapillarity. *Phil. Mag.* **25(6)**, 475–81.

de Boer, J. H. (1936). The influence of van der Waals forces and primary bonds on binding energy, strength and orientation, with special reference to some artificial resins. *Trans. Faraday Soc.* **32**, 10–38.

Derjaguin, B. V., Titijevskaia, A. S., Abrikossova, I. I. & Malikina, A. D. (1954). Investigations of the forces of interaction of surfaces in different media and their application to the problem of colloid stability. *Disc. Faraday Soc.* **18**, 24–41.

Derjaguin, B. V. & Landau, L. D. (1941). Theory of the stability of strongly charged lyophobic sols and the adhesion of strongly charged particles in solutions of electrolytes. *Acta Physicochim. URSS* **14**, 633–62.

Einstein, A. (1906). On the theory of Brownian movement, in *Investigations on the Theory of Brownian Movement* (ed. R. Furth). Dover, 1956. A translation of a paper appearing in *Ann. d. Phys.* **19**, 371–81.

Goldman, A. J., Cox, R. G. and Brenner, H. (1966). The slow motion of two identical arbitrarily oriented spheres through a viscous fluid. *Chem. Eng. Sci.* **21**, 1151–70.

Gouy, G. (1910). Sur la constitution de la électrique à la surface d'un electrolyte. *J. Phys. Radium* **9**, 457–68.

Hamaker, H. C. (1937). London–van der Waals attraction between spherical particles. *Physica* **4**, 1058–72.

Hardy, W. B. (1900). A preliminary investigation of the conditions which determine the stability of irreversible hydrosols. *Proc. Roy. Soc. Lond.* **66**, 110–25.

Happel, J. & Brenner, H. (1965). *Low Reynolds Number Hydrodynamics.* Prentice-Hall.

Helmholtz, H. von (1879). Stüdien über electrische grenschichten. *Ann. der Physik und Chemie* **7**, 337–87.

Hoffman, R. L. (1972). Discontinuous and dilatant viscosity behavior in concentrated suspensions. I. Observations of flow instability. *Soc. of Rheology, Transactions* **16**, 155–73.

Iler, R. K. (1979). *The Chemistry of Silica.* Wiley.

Israelachvili, J. N. (1985). *Intermolecular and Surface Forces.* Academic Press.

Israelachvili, J. N. & Tabor, D. (1972). The measurement of van der Waals dispersion forces in the range 1.5 to 130 nm. *Proc. Roy. Soc. Lond. A***331**, 19–38.

Kose, A., & Hachisu, S. (1976). Ordered structure in weakly flocculated monodisperse latex. *J. Colloid Interface Sci.* **55**, 487–98.

Kose, A., Ozka, M., Takano, K., Kobayashi, Y. & Hachisu, S. (1973). Direct observation of ordered latex suspension by metallurgical microscope. *J. Colloid Interface Sci.* **44**, 330–8.

Kruyt, J. R. (1949). *Colloid Science*, vols. I & II. Elsevier.

Langevin, P. (1908). Theory of Brownian motion. *C. R. Acad. Sci.* **146**, 530–3.

LeNeveu, D. M., Rand, R. P. & Parsegian, V. A. (1976). Measurement of forces between lecithin bilayers. *Nature* **259**, 601–3.

Lifshitz, E. M. (1955). The theory of molecular attractive forces between solids. *Soviet Physics JETP* **2**, 73–83.

Lin, C. J., Lee, K. J. & Sather, N. F. (1970). Slow motion of two spheres in a shear field. *J. Fluid Mech.* **43**, 35–47.

Matijevic, E. (1976). Preparation and characterization of monodisperse metal hydrous oxide sols. *Prog. Colloid & Polymer Sci.* **61**, 24–35.

McQuarrie, D. A. (1976). *Statistical Mechanics.* Harper & Row.

Napper, D. H. (1983). *Polymeric Stabilization of Colloidal Dispersions.* Academic Press.

Ninham, B. W. & Parsegian, V. A. (1970). van der Waals forces. Special characteristics in lipid–water systems and a general method of calculation based on Lifshitz theory. *Biophysical J.* **10**, 646–63.

Oak, M.-J., Lamb, G. E. R. & Saville, D. A. (1985). Particle capture on fibers in strong electric fields. *J. Colloid Interface Sci.* **106**, 490–501.

Overbeek, J. Th. G. & Sparnaay, M. J. (1954). Coagulation and flocculation, II. Classical coagulation. London–van der Waals attraction between macroscopic objects. *Disc. Faraday Soc.* **18**, 12–24.

Perrin, J. (1910). *Brownian Motion and Molecular Reality*. Taylor & Francis, London.

Rosensweig, R. (1985). *Ferrohydrodynamics.* Cambridge University Press.

Rothen, F., Jorand, M., Koch, A.-J., Dubois-Violette, E. & Pansu, B. (1987). Dislocations in Colloidal Crystals in *Physics of Complex and Supermolecular Fluids* (ed. S. Safran & N. A. Clark), pp. 413–45. Wiley.

Schulze, H. (1882). Schwefelarsen im wässeriger Losung. *J. Prakt. Chem.* **25**, 431–52.

Shaw, D. J. (1970). *Introduction to Colloid and Surface Chemistry*, 2nd edn. Butterworths.

Smoluchowski, M. von (1903). Contribution à la théorie de l'endosmose électrique et de quelques phenomènes corrélatifs. *Bulletin International de l'Academie des Sciences de Cracovie* **8**, 182–200.

Smoluchowski, M. von (1917). Versuch einer mathematischen Theorie der Koagulationkinetik kollider lösungen. *Z. Phys. Chem.* **92**, 129–68.

Stöber, W., Fink, A. & Bohn, E. (1968). Controlled growth of monodisperse silica spheres in the micron size range. *J. Colloid Interface Sci.* **26**, 62–9.

Tabor, D. & Winterton, R. H. S. (1969). Direct measurements of normal and retarded van der Waals forces. *Proc. Roy. Soc. Lond.* *A***312**, 435–50.

Vanderhoff, J. W., Vitkuske, J. F., Bradford, E. B. & Alfrey, T. (1956). Some factors involved in the preparation of uniform particle latexes. *J. Polymer Science* **20**, 225–34.

Verwey, E. J. W. & Overbeek, J. Th. G. (1948). *Theory of the Stability of Lyophobic Colloids.* Elsevier.

White, L. R., Israelachvili, J. N. & Ninham, B. W. (1976). Dispersion interaction of crossed mica cylinders: a reanalysis of the Israelachvili–Tabor experiments. *J. Chem. Soc. Faraday Trans. I* **72**, 2526–36.

2

HYDRODYNAMICS

2.1 Introduction

Because colloidal particles generally reside in a viscous fluid, the behavior of a dispersion is strongly influenced by hydrodynamic forces generated by the relative particle–fluid motion. Although many hydrodynamic effects can be deduced from the behavior of an isolated particle, the disturbance it causes decays so slowly with distance that interparticle effects are seldom negligible. Consequently, hydrodynamic forces transmitted from one particle to another through a viscous fluid must be understood. Interactions, as well as the behavior of isolated particles, are discussed here. The presentation is not meant to be a scaled-down text on hydrodynamics, but is intended to provide tools to deal with phenomena encountered in colloidal systems.

The next section presents the basic differential equations governing the behavior of an incompressible Newtonian fluid and an analysis of the relative importance of viscous and inertial effects. The analysis of two simple flows illustrates some basic principles about the kinematics of fluid motion. Then we turn our attention to flows for which inertial effects are negligible, Stokes flows. Special emphasis is given to singular solutions resulting from forces applied at points in the fluid. Subsequent sections deal with isolated spheres and two interacting spheres, first in a quiescent fluid and then in fluid undergoing laminar shear flow.

Throughout this chapter, most of our attention will be fixed on situations where spatial accelerations are unimportant, and conditions under which this is valid will be set forth. Time-dependent situations are analyzed for small-amplitude motions to illustrate the effect of particle accelerations. This serves to establish the validity of quasi-static approximations where temporal effects enter only insofar as they refer to changes in particle configurations.

2.2 Description of the motion of continuous media

For the applications of interest here, the classical laws of Newtonian mechanics suffice. However, the fluid between colloidal particles is a continuum and the mass-point mechanics of elementary physics is inappropriate. Instead, the momentum conservation principle is applied to a stationary control volume through which fluid may enter or leave.

It is convenient to divide the forces acting in the fluid into body forces and contact forces. Gravitational and electrostatic body forces are relevant to colloidal dispersions. For the present we include only that represented by $\mathbf{g}$, the gravitational force per unit mass, and leave electrostatic forces to Chapter 4. Thus the body force $\mathbf{F}_g$ acting on a volume V is

$$\mathbf{F}_g = \int_V \rho \mathbf{g}\, \mathrm{d}V, \tag{2.2.1}$$

where ρ is the density of the fluid. Contact forces act at the bounding surface of a control volume (Fig. 2.1). If $\mathbf{t}$ is the stress, i.e. force per unit area acting over an infinitesimal portion $\mathrm{d}S$ of the surface S enclosing V, then the contact force is

$$\mathbf{F}_c = \int_S \mathbf{t}\, \mathrm{d}S. \tag{2.2.2}$$

$\mathbf{F}_c + \mathbf{F}_g$ is equated with the rate of change in momentum associated with the volume V.

The stress vector $\mathbf{t}$ depends on the orientation of the area over which it acts. For example, in a fluid at rest $\mathbf{t} = -p\mathbf{n}$ since the only contact force is a pressure, acting normal to any surface. For this reason $\mathbf{t}$ is associated with the outer surface normal $\mathbf{n}$ shown in Fig. 2.1. For a fluid in motion, $\mathbf{t}$ will

Fig. 2.1. Stress vector acting as a contact force on a surface enclosing V.

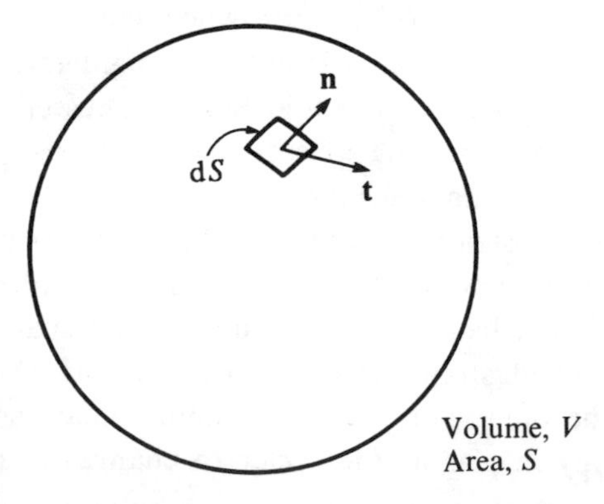

not be aligned with $\mathbf{n}$, but can be shown to be a linear vector function of $\mathbf{n}$ (Batchelor, 1967, Ch. 1).

The dependence of the stress vector on $\mathbf{n}$ complicates derivation of an equation for momentum conservation, but the difficulty can be overcome through the linear relationship between $\mathbf{n}$ and $\mathbf{t}$. Defining a stress tensor, $\boldsymbol{\sigma}$, as a linear operator that transforms the unit normal into the stress vector $\mathbf{t}$ yields (Batchelor, 1967, Ch. 1)

$$\mathbf{t}=\boldsymbol{\sigma}\cdot\mathbf{n}. \tag{2.2.3}$$

For the fluids dealt with here, the stress tensor is symmetric.

Conservation of momentum dictates that the rate of change of momentum in an arbitrary volume V equals the net force acting on the volume. With (2.2.1–3) and the divergence theorem, it is straightforward to show that

$$\frac{\partial(\rho\mathbf{u})}{\partial t}+\nabla\cdot(\rho\mathbf{u}\mathbf{u})=\rho\mathbf{g}+\nabla\cdot\boldsymbol{\sigma}; \tag{2.2.4}$$

$\mathbf{u}$ is the local fluid velocity.

Similarly, one can derive a differential statement of mass conservation,

$$\frac{\partial\rho}{\partial t}+\nabla\cdot(\rho\mathbf{u})=0. \tag{2.2.5}$$

However, until further stipulations are made regarding the stress tensor and the density the model is incomplete.

We will be concerned solely with fluids of constant density, for which (2.2.5) reduces to

$$\nabla\cdot\mathbf{u}=0. \tag{2.2.6}$$

This restriction (usually expressed by describing the fluid as incompressible) creates a problem because the equation of state for an incompressible fluid is indeterminate, and a unique definition of pressure is not possible. Thus only pressure differences or pressure gradients are meaningful.

For a Newtonian fluid of constant density,

$$\boldsymbol{\sigma}=-P\boldsymbol{\delta}+\mu[\nabla\mathbf{u}+(\nabla\mathbf{u})^{\mathrm{T}}], \tag{2.2.7}$$

where μ denotes the shear viscosity of the fluid. Thus the stress is equal to the isotropic pressure plus a symmetric linear tensor function of the velocity gradient. Combining (2.2.7) with (2.2.4) and (2.2.6) results in the Navier–Stokes equation,

$$\rho\left(\frac{\partial \mathbf{u}}{\partial t}+\mathbf{u}\cdot\nabla\mathbf{u}\right)=-\nabla p+\mu\nabla^2\mathbf{u}. \tag{2.2.8}$$

This form of the equation assumes a constant viscosity and absorbs the effect of gravity into a dynamic pressure, p,

$$p=P+\Phi, \tag{2.2.9}$$

with Φ being a scalar potential given by $\rho\mathbf{g}=-\nabla\Phi$. Equations (2.2.6) and (2.2.8) are the basic equations governing the motion of an incompressible Newtonian fluid.

A complete description requires boundary conditions. Application of the conservation principles shows that the normal component of the velocity, and, in the absence of interfacial tension, the stress vector, are continuous across interfaces. Experience shows that the tangential velocity is also continuous for the situations under study.

Before attempting solutions, it is instructive to write the equations in dimensionless form. Defining characteristic quantities as velocity, U, length, L, time, L/U, and denoting dimensionless quantities by overbars, e.g., as $\bar{x}=x/L$, gives

$$Re\left(\frac{\partial \bar{\mathbf{u}}}{\partial \bar{t}}+\bar{\mathbf{u}}\cdot\bar{\nabla}\bar{\mathbf{u}}\right)=-\bar{\nabla}\bar{p}+\bar{\nabla}^2\bar{\mathbf{u}}, \tag{2.2.10}$$

where

$$\bar{p}\equiv\frac{pL}{\mu U},$$

$$Re\equiv\frac{\rho UL}{\mu}.$$

It is clear that a single dimensionless group, the Reynolds number, Re, indicates the importance of inertia relative to viscous forces.

Integral properties of the flow can also be converted to dimensionless forms. For example, the force acting on a suspended particle becomes

$$\begin{aligned}\mathbf{F}&=\int_S \boldsymbol{\sigma}\cdot\mathbf{n}\,\mathrm{d}S\\ &=\mu UL\int_{S_0}\bar{\boldsymbol{\sigma}}\cdot\mathbf{n}\,\mathrm{d}S,\end{aligned} \tag{2.2.11}$$

identifying the appropriate dimensionless force as $\mathbf{F}/\mu UL$.

There are several reasons for writing hydrodynamic equations in dimensionless form. As we have just seen, it makes possible, via inspection, an assessment of regimes in which the physics is dominated by one or another force. For example, for $Re \ll 1$, viscous forces dominate the effects of inertia, while for $Re \gg 1$ inertia is predominant, although subtle viscous effects often play significant roles. Second, it is the basis for similarity analysis, which capitalizes on the fact that the dimensionless velocities and pressure, as well as integral quantities such as forces, depend only on the dimensionless independent variables and groups appearing in the governing equations. The latter includes the boundary conditions, i.e. continuity of velocity and stress vector across interfaces. Here additional groups can enter, e.g., a viscosity ratio for two fluid phases in contact, the Froude number U^2/Lg for a free surface, or the aspect ratio of an anisotropic body. However, for the rigid spheres of primary interest here the dimensionless force defined above depends only on $\bar{t}$, Re, and the geometrical factors. In later chapters similarity analysis is used repeatedly to present experimental data efficiently for situations in which non-hydrodynamic forces introduce additional dimensionless groups.

For low-viscosity liquids such as water, inertial effects are negligible only if the flow is slow or the dimensions of interest are small. For example, water flowing at 0.1 m/s through an array of 1 mm diameter fibers produces $Re \approx 100$. In cases such as this, inertia cannot be ignored. However, for colloidal particles suspended in such a flow, the appropriate Reynolds number depends on the particle size and the relative velocity between the particle and the fluid, which is generally much less than the fluid velocity. Even for 1 μm spheres moving at 0.1 m/s, the Reynolds number is small, i.e. $Re \approx 10^{-1}$. Hence inertia is generally negligible for flows on the scale of colloidal particles.

To set the stage for later descriptions of the dynamics of colloidal particles, we next examine two characteristic macroscopic flows. The first, steady laminar shear, illustrates the kinematics to which colloidal particles can be subjected. The second, potential flow past a sphere, introduces the concepts of streamlines and stream functions and represents a prototype flow for discussions of particle collection in Chapter 11.

2.3 Two simple flow fields

Steady laminar shear

One of the simplest flow fields is generated in the fluid between parallel infinite planes moved parallel to each other, as shown in Fig. 2.2.

For a constant velocity U, the Navier–Stokes equations reduce to

$$0=\mu\frac{\partial^2 u}{\partial y^2}. \tag{2.3.1}$$

Thus, from (2.3.1) and the no-slip boundary conditions on the planes,

$$u=U\frac{y}{d}. \tag{2.3.2}$$

Although (2.3.2) was obtained for a Newtonian fluid, the result is valid for any fluid for which the stress is determined by the history of the velocity gradient.

The idealization shown in Fig. 2.2 is a prototype for certain viscometers. The relationship between shear stress and velocity gradient is readily found from measurements of the wall shear stress, or the drag, on one of the planes at various values of U. For a Newtonian fluid, defined by (2.2.7), it follows from (2.3.2) that

$$\sigma_{xy}(y)=\mu\frac{du}{dy}=\mu\frac{U}{d}. \tag{2.3.3}$$

Hence the (constant) slope of the relation between shear stress and velocity gradient equals the viscosity of the fluid.

The velocity profile (2.3.2) is a special case of the general linear velocity field,

$$\mathbf{u}=\mathbf{\Gamma}\cdot\mathbf{x}, \tag{2.3.4}$$

where $\mathbf{\Gamma}$ is the constant velocity gradient tensor $\nabla\mathbf{u}^T$. Indeed, the velocity near a point $\mathbf{x}_0$ can be expressed through a Taylor series expansion as

$$\mathbf{u}=\mathbf{u}(\mathbf{x}_0)+(\mathbf{x}-\mathbf{x}_0)\cdot\nabla\mathbf{u}(\mathbf{x}_0)+\ldots. \tag{2.3.5}$$

Fig. 2.2. Laminar shear flow.

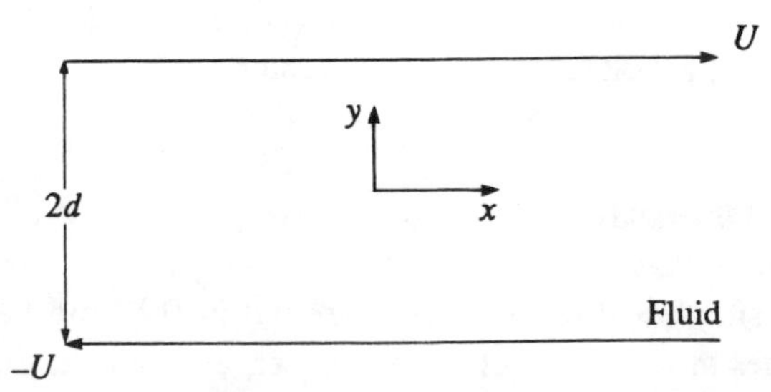

The first two terms suffice to describe the velocity field experienced by a colloidal particle with dimension $a^2 \ll |\mathbf{u}(\mathbf{x}_0)|/|\nabla^2\mathbf{u}(\mathbf{x}_0)|$.

Later we will see that the motion of a particle depends on the nature of $\mathbf{\Gamma}$. For the present it is instructive to express $\mathbf{\Gamma}$ for a two-dimensional flow in terms of a shear rate γ and a parameter α as

$$\mathbf{\Gamma} = \tfrac{1}{2}\gamma \begin{bmatrix} 1+\alpha & 1-\alpha & 0 \\ -1+\alpha & -(1+\alpha) & 0 \\ 0 & 0 & 0 \end{bmatrix} \tag{2.3.6}$$

in a Cartesian coordinate system. The effect of α on the flow structure is clearly evident from Fig. 2.3. Negative values of α denote an increasing amount of rotation, with a purely rotational flow for $\alpha = -1$. Simple shear flow corresponds to $\alpha = 0$ with the axes rotated by 45° relative to (2.3.2). The case of $\alpha = 1$ is a two-dimensional extension where a small particle placed in the flow experiences no rotation but is subjected to tensile and compressive forces. This contrasts with simple shear, for which stretching, compression, and rotation all take place.

Fig. 2.3. Streamlines for various values of $0 \leq \alpha \leq 1.0$ (Stone, Bentley, & Leal, 1986).

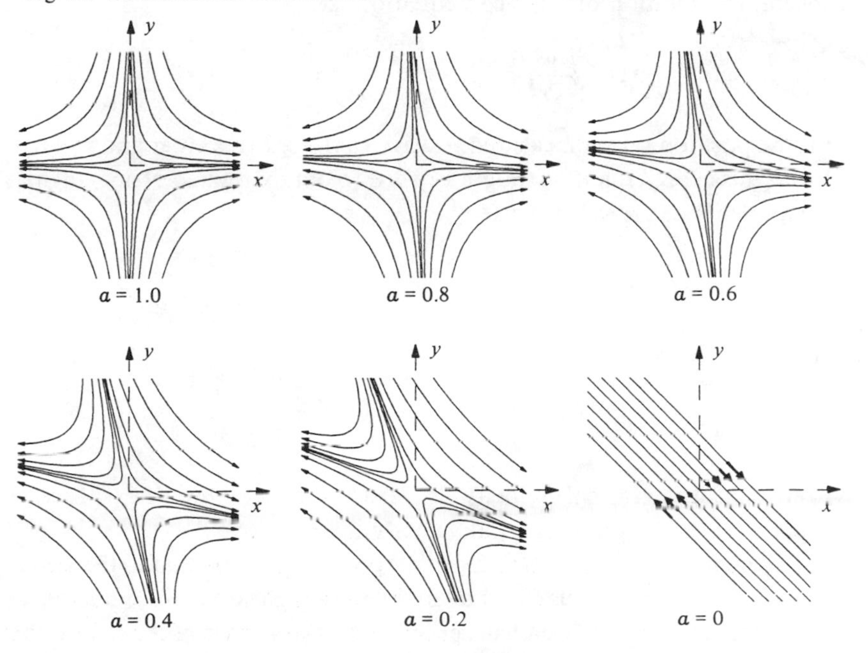

Potential flow past a sphere

Another interesting flow is that past a stationary sphere. Just such situations occur when particles in suspension encounter a stationary obstacle. Here the large-scale motion must be known before the paths of individual particles can be ascertained. Individual particle trajectories are analyzed in Chapter 11; here we delve into the structure of the motion and illustrate the concepts of streamlines and stream functions.

Imagine the situation when a sphere is placed in a flow with a uniform velocity **U**. If the Reynolds number is large and the flow steady, temporal accelerations and viscous effects can be ignored over most of the region around the sphere. The uniform flow is irrotational, i.e., $\nabla \times \mathbf{u} = \mathbf{0}$, and, in the absence of viscous effects, this characteristic is preserved (see, for example, Batchelor, 1967, §§2.7 and 6.8). Hence the velocity can be written as the gradient of a scalar potential, ϕ, as

$$\mathbf{u} = \nabla\phi. \tag{2.3.7}$$

From (2.2.6) and (2.3.7) it follows that

$$\nabla^2\phi = 0. \tag{2.3.8}$$

Situations in which (2.3.8) is valid form a class of flows known as potential flows. For a rigid impermeable sphere of radius a with $\mathbf{u} \to \mathbf{U}$ far from the sphere, the solution of Laplace's equation is

$$\phi(\mathbf{r}) = \mathbf{U} \cdot \mathbf{r}\left(1 + \frac{a^3}{2r^3}\right); \tag{2.3.9}$$

$\mathbf{r}$ is the position vector measured relative to the sphere center and $r^2 = \mathbf{r} \cdot \mathbf{r}$. From this expression it is straightforward to obtain the radial and angular components of the velocity as

$$u_r = U\cos\theta\left(1 - \frac{a^3}{r^3}\right)$$

and (2.3.10)

$$u_\theta = -U\sin\theta\left(1 + \frac{a^3}{2r^3}\right).$$

Note that the normal velocity is zero at $r = a$, since the sphere is impermeable, but the tangential velocity is non-zero.

One might expect (2.3.10) to furnish a good approximation to the steady flow of a real (i.e. viscous) fluid at sufficiently high Reynolds numbers, as noted earlier. This is indeed true, except near the surface of the sphere and downstream. Viscous effects alter the structure in a thin boundary layer near the surface, so as to satisfy the no-slip condition, and play a role in separation of the flow, which leads to formation of a wake.

Equations (2.3.10) can be used to introduce streamlines and stream functions. For an incompressible fluid, the continuity equation for two-dimensional or axisymmetric flows can define a scalar function, ψ, whose derivatives yield the velocity. In an axisymmetric flow, the relation between the stream function and velocity is expressed as (Batchelor, 1967, §2.3)

$$u_r = \frac{1}{r^2 \sin\theta}\frac{\partial\psi}{\partial\theta}, \qquad u_\theta = -\frac{1}{r\sin\theta}\frac{\partial\psi}{\partial r} \tag{2.3.11}$$

in spherical polar coordinates. Thus, the continuity equation,

$$\frac{1}{r^2}\frac{\partial}{\partial r}r^2 u_r + \frac{1}{r\sin\theta}\frac{\partial}{\partial\theta}(u_\theta \sin\theta) = 0, \tag{2.3.12}$$

is satisfied identically by (2.3.11). Integrating the expressions given in (2.3.11) with the velocity from the potential flow solution shows that the stream function for this flow is

$$\psi = \tfrac{1}{2}Ur^2\sin^2\theta\left(1 - \frac{a^3}{r^3}\right). \tag{2.3.13}$$

Figure 2.4 depicts streamlines in a plane through the axis of the sphere aligned with the flow direction, the z-axis. Note that the lines $\theta = 0$ and π coincide with the streamline $\psi = 0$ for $r > a$ and that the stream function is zero on the surface of the sphere. The points on the surface $r = a$ at $\theta = 0$ and

Fig. 2.4. Streamlines and equipotential lines for potential flow past a sphere at rest.

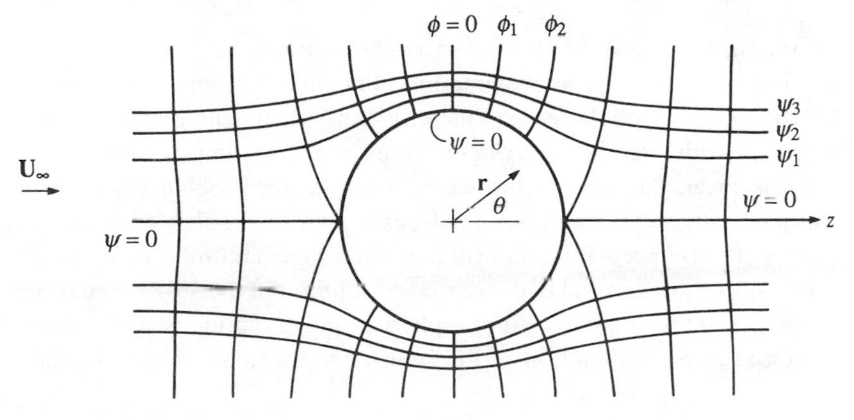

π are stagnation points, where the velocity is zero and the flow divides and then converges as it passes around the sphere.

It is also easy to show that the lines of constant ϕ are perpendicular to the lines of constant ψ. Accordingly, the velocity is tangent to the streamlines and no fluid flows across streamlines. To illustrate the connection between the stream function and the volumetric flow rate, imagine a surface formed by rotating a line of constant ψ around the z-axis to form a stream tube. Two such stream tubes delineate an annular region within which the volumetric flow rate must be constant. A plane intersecting the center of the sphere perpendicular to the flow, i.e., the plane $\theta = \pi/2$, intersects the two stream tubes on circles with radii r_1 and r_2. Since the local normal to the plane, **n**, is in the direction of the incident flow, **U**, the flow rate through the annular ring is

$$\int_{r_1}^{r_2} \mathbf{u} \cdot \mathbf{n} 2\pi r \, dr = 2\pi \int_{r_1}^{r_2} (-\sin\theta) u_\theta r \, dr = 2\pi \int_{r_1}^{r_2} \frac{\partial \psi}{\partial r} dr = 2\pi(\psi_2 - \psi_1).$$

Hence, the flow rate through a stream tube depends only on the difference of the stream function on the two stream surfaces.

In this section viscous effects were neglected on the grounds that $Re \gg 1$; slow flows will be analyzed next to show how viscous effects modify the flow structure when $Re \ll 1$.

2.4 Characteristics of Stokes flow

In §2.2 we noted that for many problems relevant to colloid science the Reynolds number is small. There, to a good approximation, the flow is governed by the Stokes equations,

$$\begin{aligned} \nabla \cdot \mathbf{u} &= 0, \\ \nabla p &= \mu \nabla^2 \mathbf{u}. \end{aligned} \tag{2.4.1}$$

Two features of the Stokes equations deserve particular attention. Since the low Reynolds number approximation eliminates the time derivative in (2.2.8), solutions of the equations with time-dependent boundary conditions provide pseudo-steady descriptions of the motion. Second, the set (2.4.1) is linear. Therefore, solutions to a complicated problem can often be synthesized by superposition of solutions to simpler problems. Indeed, the superposition concept is fundamental to some important solution methods.

Consider, now, several issues central to solution of the Stokes equations for motion involving one or more spheres. Sphere motion is conveniently expressed as a combination of the translational velocity $\mathbf{U}_0$ of a sphere center and the rotational velocity $\mathbf{\Omega}$ of the sphere about its center. Then the velocity of the surface of a sphere centered at $\mathbf{x}_0$ is

$$\mathbf{U} = \mathbf{U}_0 + \mathbf{\Omega} \times (\mathbf{x} - \mathbf{x}_0). \tag{2.4.2}$$

The viscous forces over the surface of a sphere give rise to a net force $\mathbf{F}$ on the sphere

$$\mathbf{F} = \int_{S_0} \boldsymbol{\sigma} \cdot \mathbf{n} \, dS, \tag{2.4.3}$$

and a torque $\mathbf{L}$ about the sphere center

$$\mathbf{L} = \int_{S_0} (\mathbf{x} - \mathbf{x}_0) \times \boldsymbol{\sigma} \cdot \mathbf{n} \, dS. \tag{2.4.4}$$

The integration is over the sphere surface S_0 and $\mathbf{n}$ is the unit normal, taken as positive outward. In some situations the value of $\mathbf{U}$ for one or more spheres with specified forces and torques is needed. This is known as the *mobility* problem. However, it is also possible to seek the force (or torque) on the spheres when the motion is specified. This is the *resistance* problem.

There are numerous methods for solving the Stokes equations for spheres moving in a Newtonian fluid. Separation of variables leads to Lamb's general solution (Lamb, 1932) with spherical harmonics used to represent the velocity and pressure fields. Happel & Brenner (1965, pp. 62–78) describe the methodology in some detail.

A second approach draws upon solutions to the Stokes equations generated by a point force, a force dipole, etc. These solutions are known as fundamental or singular solutions. To solve a particular problem, the singular solutions are combined so as to meet the boundary conditions in question. This approach is well suited to problems associated with colloid dynamics and will be used extensively, e.g. in Brownian motion (Chapter 3), electrokinetics (Chapter 7), and sedimentation (Chapter 12).

2.5 Singular solutions to the Stokes equations

Certainly the most important solution of the Stokes equations is that which describes the flow caused by a sphere moving due to an applied force $\mathbf{F}$ through an unbounded quiescent fluid. Sufficiently far from the sphere, one expects the same velocity and pressure fields as those due to a concentrated force of equal magnitude. Thus the point force solution represents a 'far-field' approximation and provides a basic building block for the superposition method mentioned above.

Oseen (1927) employed Green's functions to obtain the point force solution. Although that approach permits an appreciation of the physics of the problem, the derivation is less direct than an alternative route, using Fourier transforms. We follow the latter approach as outlined by Ladyzhenskaya (1969, pp. 50–1).

To determine the velocity and pressure fields induced in an infinite expanse of fluid we use the Dirac delta function (Stakgold, 1968, p. 5) to represent a force $\mathbf{f}$ imposed at the origin. Two properties of the delta function are

$$\int \delta(\mathbf{x})\,\mathrm{d}V = 1, \tag{2.5.1}$$

and the convenient sifting property for some function $g(\mathbf{x})$,

$$\int g(\mathbf{x})\,\delta(\mathbf{x})\,\mathrm{d}V = g(\mathbf{0}). \tag{2.5.2}$$

Then the Stokes equations can be written as

$$\nabla\cdot\mathbf{u} = 0; \qquad \nabla p = \mu\nabla^2\mathbf{u} + \delta(\mathbf{x})\mathbf{f}. \tag{2.5.3}$$

The solution to (2.5.3) is one of the fundamental solutions to the Stokes equations mentioned above.

The key to solving (2.5.3) is to recognize some special relations between the Dirac delta function and Fourier transforms. Recall that a function $\mathbf{f}(\mathbf{x})$ and its Fourier transform $\hat{\mathbf{f}}(\mathbf{q})$ are related by (Stakgold, 1968, pp. 36ff.)

$$\hat{\mathbf{f}}(\mathbf{q}) = \int \mathbf{f}(\mathbf{x})\,\mathrm{e}^{i\mathbf{q}\cdot\mathbf{x}}\,\mathrm{d}\mathbf{x} \tag{2.5.4}$$

and

$$\mathbf{f}(\mathbf{x}) = \frac{1}{(2\pi)^3}\int \hat{\mathbf{f}}(\mathbf{q})\,\mathrm{e}^{-i\mathbf{q}\cdot\mathbf{x}}\,\mathrm{d}\mathbf{q}. \tag{2.5.5}$$

Integrations are over the full three-dimensional space of the integration variable. From these relations and the properties of $\delta(\mathbf{x})$ one can show that

$$\delta(\mathbf{x}) = \frac{1}{(2\pi)^3}\int \mathrm{e}^{-i\mathbf{q}\cdot\mathbf{x}}\,\mathrm{d}\mathbf{q}. \tag{2.5.6}$$

$$\nabla^2\frac{1}{4\pi x} = \nabla^4\frac{x}{8\pi} = -\delta(\mathbf{x}), \tag{2.5.7}$$

and hence

$$\frac{1}{4\pi x}=\frac{1}{(2\pi)^3}\int\frac{1}{q^2}\mathrm{e}^{-i\mathbf{q}\cdot\mathbf{x}}\,\mathrm{d}\mathbf{q} \tag{2.5.8}$$

and

$$\frac{x}{8\pi}=-\frac{1}{(2\pi)^3}\int\frac{1}{q^4}\mathrm{e}^{-i\mathbf{q}\cdot\mathbf{x}}\,\mathrm{d}\mathbf{q}, \tag{2.5.9}$$

where $q^2=\mathbf{q}\cdot\mathbf{q}$, $x^2=\mathbf{x}\cdot\mathbf{x}$. Taking the Fourier transform of (2.5.3) yields

$$\mathbf{q}\cdot\hat{\mathbf{u}}=0;\ -\mu q^2\hat{\mathbf{u}}+i\mathbf{q}\hat{p}=-\mathbf{f}. \tag{2.5.10}$$

Solving separately for $\hat{\mathbf{u}}$ and $\hat{p}$ gives

$$\hat{\mathbf{u}}(x)=\frac{1}{\mu q^2}\left(\mathbf{f}-\frac{(\mathbf{q}\cdot\mathbf{f})\mathbf{q}}{q^2}\right)$$

and (2.5.11)

$$\hat{p}(\mathbf{x})=i\frac{\mathbf{q}\cdot\mathbf{f}}{q^2}.$$

To invert the transformed solution we use (2.5.8) and (2.5.9) to obtain

$$\mathbf{f}\cdot\nabla\frac{1}{4\pi x}=-\frac{i}{(2\pi)^3}\int\frac{\mathbf{f}\cdot\mathbf{q}}{q^2}\mathrm{e}^{-i\mathbf{q}\cdot\mathbf{x}}\,\mathrm{d}\mathbf{q}$$

and (2.5.12)

$$\mathbf{f}\cdot\nabla\nabla\frac{x}{8\pi}=\frac{1}{(2\pi)^3}\int\frac{\mathbf{f}\cdot\mathbf{q}\mathbf{q}}{q^4}\mathrm{e}^{-i\mathbf{q}\cdot\mathbf{x}}\,\mathrm{d}\mathbf{q}.$$

Then, from (2.5.11) and (2.5.12) we find

$$\mathbf{u}(\mathbf{x})=\frac{1}{8\pi\mu}\left(\frac{\boldsymbol{\delta}}{x}+\frac{\mathbf{x}\mathbf{x}}{x^3}\right)\cdot\mathbf{f}$$

and (2.5.13)

$$p(\mathbf{x})=\frac{1}{4\pi x^3}\mathbf{x}\cdot\mathbf{f}.$$

The $\boldsymbol{\delta}$ symbol refers to the unit tensor. In rectangular Cartesian coordinates with the basis $\mathbf{i}$, $\mathbf{j}$, $\mathbf{k}$,

$$\boldsymbol{\delta}=\mathbf{ii}+\mathbf{jj}+\mathbf{kk}. \tag{2.5.14}$$

The multiplier of $\mathbf{f}$ in (2.5.13) is called the Oseen tensor, $\mathbf{I}$,

$$\mathbf{I}=\frac{1}{8\pi\mu x}\left(\boldsymbol{\delta}+\frac{\mathbf{xx}}{x^2}\right), \tag{2.5.15}$$

and $\mathbf{f}$ is often expressed in terms of a strength $\boldsymbol{\alpha}$ as

$$\mathbf{f}=8\pi\mu\boldsymbol{\alpha}. \tag{2.5.16}$$

The force $\mathbf{f}$ (and often the velocity field (2.5.13) due to $\mathbf{f}$) is called a Stokeslet of strength $\boldsymbol{\alpha}$.

The Stokeslet is not the only singular solution of the Stokes equations; derivatives of the Stokeslet are also fundamental solutions. For example, if $\mathbf{d}$ is an arbitrary constant vector then $(\mathbf{df}):\nabla\mathbf{I}$ represents the velocity field of a Stokes doublet (Chwang & Wu, 1975). The doublet also provides far-field behavior for flows involving small force-free particles. These fields can be separated into two parts, viz:

$$(\mathbf{df}):\nabla(\mathbf{I},\mathbf{x}/4\pi x^3)=(\mathbf{u},p)_s+(\mathbf{u},p)_c. \tag{2.5.17}$$

The first set of velocity and pressure fields is known as the stresslet, i.e.,

$$\mathbf{u}_s=\frac{1}{8\pi\mu}\left(\frac{1}{x^3}\mathbf{S}:\boldsymbol{\delta}\mathbf{x}-\frac{3}{x^5}\mathbf{x}\cdot\mathbf{S}\cdot\mathbf{xx}\right),$$

$$p_s=\frac{1}{4\pi}\left(\frac{1}{x^3}\mathbf{S}:\boldsymbol{\delta}-\frac{3}{x^5}\mathbf{x}\cdot\mathbf{S}\cdot\mathbf{x}\right), \tag{2.5.18}$$

where

$$\mathbf{S}=\tfrac{1}{2}[(\mathbf{df})+(\mathbf{fd})].$$

Note that this definition of the stresslet corresponds to that of Batchelor (1970) for situations where $\mathbf{S}$ is traceless. The stresslet represents a straining motion with principal axes in the $\mathbf{d}+\mathbf{f}$, $\mathbf{d}-\mathbf{f}$, and $\mathbf{d}\times\mathbf{f}$ directions. An integral of the stress vector over a surface encompassing the singularity vanishes, showing that the stresslet exerts no force on the fluid. However, the integral of the moment of the stress vector does not vanish and

$$\int_S[\boldsymbol{\sigma}\cdot\mathbf{nx}-\mu(\mathbf{un}+\mathbf{nu})]\,\mathrm{d}S=\mathbf{S}, \tag{2.5.19}$$

when $\mathbf{S}$ is traceless. This result will prove useful shortly in calculating the bulk stress in a suspension of colloidal particles.

The other part of the doublet is called the couplet and represents a torque applied to the fluid by the singularity. Here

$$\mathbf{u}_c = \frac{\mathbf{L}\times\mathbf{x}}{8\pi\mu x^3}, \qquad p_c = 0. \tag{2.5.20}$$

where

$$\mathbf{L} = \mathbf{f}\times\mathbf{d}. \tag{2.5.21}$$

The moment applied by the couplet is

$$\int_S \mathbf{x}\times\boldsymbol{\sigma}\cdot\mathbf{n}\,\mathrm{d}S = -\mathbf{L}. \tag{2.5.22}$$

2.6 Dynamics of isolated spheres

In many applications the full velocity field is not needed, but, as we have already noted, only the force on a particle or some other integrated characteristic of the flow. For these purposes a set of expressions known as Faxen's laws can be useful. To derive these we follow the approach of Batchelor (1972), based on the velocity field due to a point force, (2.5.13).

Consider a rigid sphere of radius a and surface S_0 immersed in a fluid with velocity field $\mathbf{u}_\infty(\mathbf{x})$ in the absence of the sphere. The sphere is centered at $\mathbf{x}_0$, translates with velocity $\mathbf{U}_0$ and rotates about its center with angular velocity $\boldsymbol{\Omega}$. Owing to its motion the sphere affects the fluid through forces distributed over its surface, $\mathbf{t}(\mathbf{x})$ per unit area. From (2.5.13) the velocity field at $\mathbf{x}$ is now $\mathbf{u}_\infty(\mathbf{x})$ plus that generated by the superposition of forces $\mathbf{t}(\mathbf{x}')\,\mathrm{d}\mathbf{x}'$, or

$$\mathbf{u}(\mathbf{x}) = \mathbf{u}_\infty(\mathbf{x}) - \int_{S_0} \mathbf{I}(\mathbf{x}-\mathbf{x}')\cdot\mathbf{t}(\mathbf{x}')\,\mathrm{d}\mathbf{x}'. \tag{2.6.1}$$

At the surface of the sphere this must match the sphere velocity given by (2.4.2). Thus,

$$\mathbf{U}_0 + \boldsymbol{\Omega}\times(\mathbf{x}-\mathbf{x}_0) = \mathbf{u}_\infty(\mathbf{x}) - \int_{S_0} \mathbf{I}(\mathbf{x}-\mathbf{x}')\cdot\mathbf{t}(\mathbf{x}')\,\mathrm{d}\mathbf{x}'. \tag{2.6.2}$$

Integration of (2.6.2) over the sphere surface results in

$$\int_{S_0} \mathbf{t}(\mathbf{x}')\,\mathrm{d}\mathbf{x}' = \mathbf{F} = \frac{3}{2}\frac{\mu}{a}\int_{S_0} [\mathbf{u}_\infty(x) - \mathbf{U}_0]\,\mathrm{d}\mathbf{x}. \tag{2.6.3}$$

Next, expanding $\mathbf{u}_\infty(\mathbf{x})$ about the sphere center and noting that $\nabla^2\nabla^2\mathbf{u}=0$ for creeping flow yields Faxen's first law,

$$\mathbf{F}=6\pi\mu a[(\mathbf{u}_\infty+\tfrac{1}{6}a^2\nabla^2\mathbf{u}_\infty)_0-\mathbf{U}_0] \tag{2.6.4}$$

(Batchelor, 1972). The subscript zero indicates evaluation at $\mathbf{x}_0$.

A striking example of the utility of Faxen's law can be obtained immediately from inspection of (2.6.4). Stokes' law for the drag exerted on a sphere held at rest in a fluid moving with velocity $\mathbf{u}_\infty$ (or, alternatively, the force on a sphere moving at velocity $\mathbf{U}_0$ through a fluid otherwise at rest) is traditionally obtained by solving the Stokes equations, computing the stress tensor and integrating over the sphere surface. From (2.6.4) the drag force is obtained directly by setting $\mathbf{U}_0=\mathbf{0}$ or $\mathbf{u}_\infty=\mathbf{0}$ as

$$\mathbf{F}=6\pi\mu a\mathbf{u}_\infty \qquad \text{or} \qquad =-6\pi\mu a\mathbf{U}_0. \tag{2.6.5}$$

Writing (2.6.5) in the reciprocal form $\mathbf{U}_0=-\mathbf{F}/6\pi\mu a\equiv-\boldsymbol{\omega}\cdot\mathbf{F}$ identifies the mobility of a sphere as $\boldsymbol{\omega}=\boldsymbol{\delta}/6\pi\mu a$.

Faxen's second law can be derived by taking the vector product of (2.6.2) with $(\mathbf{x}-\mathbf{x}_0)$ before carrying out the second integration and then separating the equation into antisymmetric and symmetric parts, which must be satisfied individually. In this way Batchelor & Green (1972) and Rallison (1978) obtain the torque on a sphere as

$$\mathbf{L}=8\pi\mu a^3[\tfrac{1}{2}(\nabla\times\mathbf{u}_\infty)_0-\boldsymbol{\Omega}] \tag{2.6.6}$$

and the force dipole (stresslet) as

$$\mathbf{S}=\tfrac{10}{3}\pi\mu a^3(1+\tfrac{1}{10}a^2\nabla^2)(\nabla\mathbf{u}_\infty+(\nabla\mathbf{u}_\infty)^{\mathrm{T}}). \tag{2.6.7}$$

Equation (2.6.6) illustrates that a freely suspended sphere, i.e. where the force and torque are zero, rotates at the local angular velocity of the fluid, $\tfrac{1}{2}(\nabla\times\mathbf{u}_\infty(\mathbf{x}))_0$. The connection between the stresslet defined in §2.5 and the velocity field around a freely suspended sphere is provided by (2.6.7).

Equation (2.6.7), which represents the symmetric part of the dipole that the sphere exerts on the fluid, can also be used to derive the bulk stress in a suspension of freely suspended spheres undergoing a simple shearing deformation (Batchelor, 1970). First, note that the average strain rate $\langle\mathbf{e}\rangle$ is calculated by averaging the local strain rate over the volume V as

$$\langle\mathbf{e}\rangle=\frac{1}{V}\int\mathbf{e}\,dV,$$

where (2.6.8)

$$2\mathbf{e}=\nabla\mathbf{u}+(\nabla\mathbf{u})^{\mathrm{T}}.$$

This expression can be decomposed into integrals over the fluid volume, V_f, and the volumes of suspended particles, V_i. The average stress can be written in a similar fashion as (Batchelor, 1970)

$$\langle \boldsymbol{\sigma} \rangle = \frac{1}{V}\int_{V_f} \boldsymbol{\sigma}\, dV + \frac{1}{V}\sum_i \int_{V_i} \boldsymbol{\sigma}\, dV. \tag{2.6.9}$$

Now, by using (2.2.7) to represent the stress in the fluid, (2.6.9) can be rewritten as

$$\langle \boldsymbol{\sigma} \rangle = -\langle P \rangle \boldsymbol{\delta} + 2\mu \langle \mathbf{e} \rangle + \frac{1}{V}\sum_i \int_{S_i} \{\boldsymbol{\sigma} \cdot \mathbf{nx} - \mu(\mathbf{un} + \mathbf{nu})\}\, dS. \tag{2.6.10}$$

Since particles in a dilute suspension do not interact, the integrals can be evaluated using the results for an isolated sphere. Furthermore, the particles are force-free, so (2.5.19) shows that the integral is the dipole, **S**, once the contribution from the trace of **S** is assigned to the isotropic pressure. Then, for a linear shear flow, (2.6.7) relates the dipole for an isolated sphere $\mathbf{S}_0$ to the strain rate as

$$\mathbf{S}_0 = \frac{20\pi\mu a^3}{3}\langle \mathbf{e} \rangle \tag{2.6.11}$$

and the average stress is

$$\langle \boldsymbol{\sigma} \rangle = -\langle P \rangle \boldsymbol{\delta} + 2\mu(1 + \tfrac{5}{2}\phi)\langle \mathbf{e} \rangle, \tag{2.6.12}$$

where ϕ denotes the volume fraction of particles. Thus the celebrated Einstein relation (Einstein, 1906) for the effective viscosity of a suspension, η, is obtained, viz.,

$$\eta = \mu(1 + \tfrac{5}{2}\phi). \tag{2.6.13}$$

To obtain the full velocity and pressure fields, (2.4.1) must be solved or the integral form (2.6.1) evaluated. For example, the velocity field generated by the steady translation of a sphere is obtained from (2.6.1) by setting $\mathbf{t} = \mathbf{F}/4\pi a^2 = -3\mu\mathbf{U}_0/2a$ and expanding the Oseen tensor about the center of the sphere, leaving

$$\begin{aligned}\mathbf{u}(\mathbf{x}) &= \frac{3\mu}{2a}\mathbf{U}_0 \cdot \int_{S_0} \{\mathbf{I}(\mathbf{r}) + (\mathbf{x}_0 - \mathbf{x}') \cdot \nabla \mathbf{I}(\mathbf{r}) + \tfrac{1}{2}(\mathbf{x}_0 - \mathbf{x}')(\mathbf{x}_0 - \mathbf{x}') \\ &\quad : \nabla\nabla \mathbf{I}(\mathbf{r}) + \ldots\}\, d\mathbf{x}' \\ &= 6\pi\mu a \mathbf{U}_0 \cdot \left(1 + \tfrac{1}{6}a^2\nabla^2\right)\mathbf{I}(\mathbf{r})\end{aligned} \tag{2.6.14}$$

with $\mathbf{r}=\mathbf{x}-\mathbf{x}_0$. The terms shown derive from the identities

$$\int_{S_0} \mathrm{d}\mathbf{x}' = 4\pi a^2$$
$$\int_{S_0} (\mathbf{x}'-\mathbf{x}_0)(\mathbf{x}'-\mathbf{x}_0)\,\mathrm{d}\mathbf{x}' = \frac{4\pi a^4}{3}\boldsymbol{\delta} \tag{2.6.15}$$

and the remaining terms integrate to zero, since

$$\int_{S_0} (\mathbf{x}'-\mathbf{x}_0)^n \mathrm{d}\mathbf{x}' = 0 \qquad \text{for } n \text{ odd}$$

and $\nabla\cdot\mathbf{I}=0$, $\nabla^2\nabla^2\mathbf{I}=0$.

Substitution of (2.5.15) into (2.6.14) provides the velocity field

$$\mathbf{u}(\mathbf{x}) = \frac{3a}{4r}\left(1+\frac{a^2}{3r^2}\right)\mathbf{U}_0 + \frac{3a}{4r}\left(1-\frac{a^2}{r^2}\right)\frac{\mathbf{r}\cdot\mathbf{U}_0\mathbf{r}}{r^2} \tag{2.6.16}$$

which satisfies the boundary conditions

$$\mathbf{u}=\mathbf{U}_0, \qquad r=a$$
$$\mathbf{u}\to\mathbf{0}, \qquad r\to\infty. \tag{2.6.17}$$

The associated pressure field from (2.5.13) is

$$p = \frac{3\mu a}{2r^3}\mathbf{U}_0\cdot\mathbf{r}. \tag{2.6.18}$$

Note that in the far-field $\mathbf{u}\sim r^{-1}$ and $p\sim r^{-2}$, as expected from the point force solution.

For a sphere in a linear velocity field

$$\mathbf{u}\to\boldsymbol{\Gamma}\cdot\mathbf{x}=(\mathbf{E}+\boldsymbol{\Omega})\cdot\mathbf{x}, \qquad r\to\infty,$$
$$\mathbf{u}=\boldsymbol{\Omega}\cdot\mathbf{r}+(\mathbf{E}+\boldsymbol{\Omega})\cdot\mathbf{x}_0, \qquad r=a \tag{2.6.19}$$

with $\mathbf{E}=(\boldsymbol{\Gamma}+\boldsymbol{\Gamma}^{\mathrm{T}})/2$ and $\boldsymbol{\Omega}=(\boldsymbol{\Gamma}-\boldsymbol{\Gamma}^{\mathrm{T}})/2$. The form of the dipole extracted from Faxen's second law (2.6.7) suggests setting

$$\mathbf{t} = \frac{3}{4\pi a^3}\left(\frac{\mathbf{x}-\mathbf{x}_0}{a}\right)\cdot\mathbf{S}$$
$$= 5\mu\mathbf{E}\cdot\left(\frac{\mathbf{x}-\mathbf{x}_0}{a}\right). \tag{2.6.20}$$

Expansion and integration as above produces a velocity field,

$$\mathbf{u}(\mathbf{x})=\mathbf{\Gamma}\cdot\mathbf{x}+\frac{20\pi\mu a^3}{3}\mathbf{E}:\left(1+\frac{a^2}{10}\nabla^2\right)\nabla\mathbf{I}(\mathbf{r}), \tag{2.6.21}$$

which conforms to (2.6.19) at $r = a$. The explicit velocity and pressure fields are

$$\begin{aligned}\mathbf{u}(\mathbf{x})&=(\mathbf{E}+\mathbf{\Omega})\cdot\mathbf{x}-\mathbf{E}\cdot\mathbf{r}\left(\frac{a}{r}\right)^5-\frac{5}{2}\frac{\mathbf{rr}\cdot\mathbf{E}\cdot\mathbf{r}}{r^2}\left(\frac{a}{r}\right)^3\left(1-\frac{a^2}{r^2}\right),\\ p(\mathbf{x})&=-\frac{5\mu a^3}{r^3}\frac{\mathbf{r}\cdot\mathbf{E}\cdot\mathbf{r}}{r^2}.\end{aligned} \tag{2.6.22}$$

In this case the disturbance velocity decays as r^{-2} as $r\to\infty$, as for the dipole in (2.5.18).

Cox, Zia & Mason (1968) noted an important feature of this solution. In the case of simple shear flow with $\Gamma_{ij}=\gamma\delta_{i2}\delta_{j1}$, there exists a region surrounding the sphere within which the streamlines are closed. Representative streamlines in the xy-plane shown in Fig. 2.5(a) demonstrate that a fluid element within this region undergoes orbital motion. The full surface, shown in Fig. 2.5(b), is given implicitly by

$$y=\pm rf(r)[g(r)]^{1/2}, \tag{2.6.23}$$

where

$$\begin{aligned}f(r)&=(r^3-\tfrac{5}{2}+\tfrac{3}{2}r^{-2})^{-1/3},\\ g(r)&=\int_r^{\infty}\zeta^{-3}f(\zeta)\,\mathrm{d}\zeta,\end{aligned} \tag{2.6.24}$$

and

$$r=(x^2+y^2+z^2)^{1/2}.$$

Note that the enclosed volume is infinite because, as $r\to\infty$, $f(r)\sim r^{-1}$ and $g(r)\sim r^{-3}$.

The existence of closed streamlines depends on the nature of $\mathbf{\Gamma}$, or, equivalently, the value of α in (2.3.6). In the uniaxial extension with $\alpha=1$, there is no rotation and therefore no vorticity; streamlines in the xy-plane have the shape shown in Fig. 2.6 and are all open.

Fig. 2.5. Laminar shear flow past a sphere: (*a*) Streamlines in the *xy*-plane (Cox, Zia, & Mason, 1968). (*b*) The limiting surface enclosing the closed streamlines. Note that the surface is symmetric about the *y*-axis.

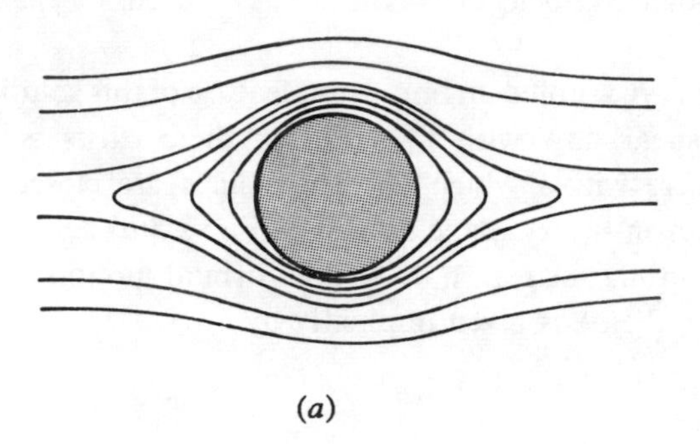

(*a*)

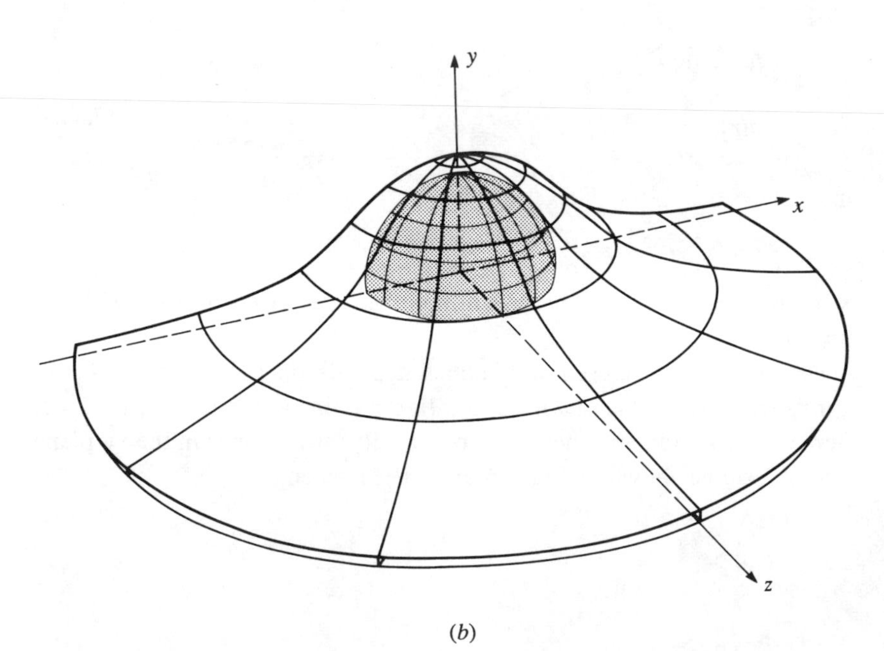

(*b*)

Fig. 2.6. Streamlines about a sphere in the extensional flow described by (2.3.6) with $\alpha = 1$.

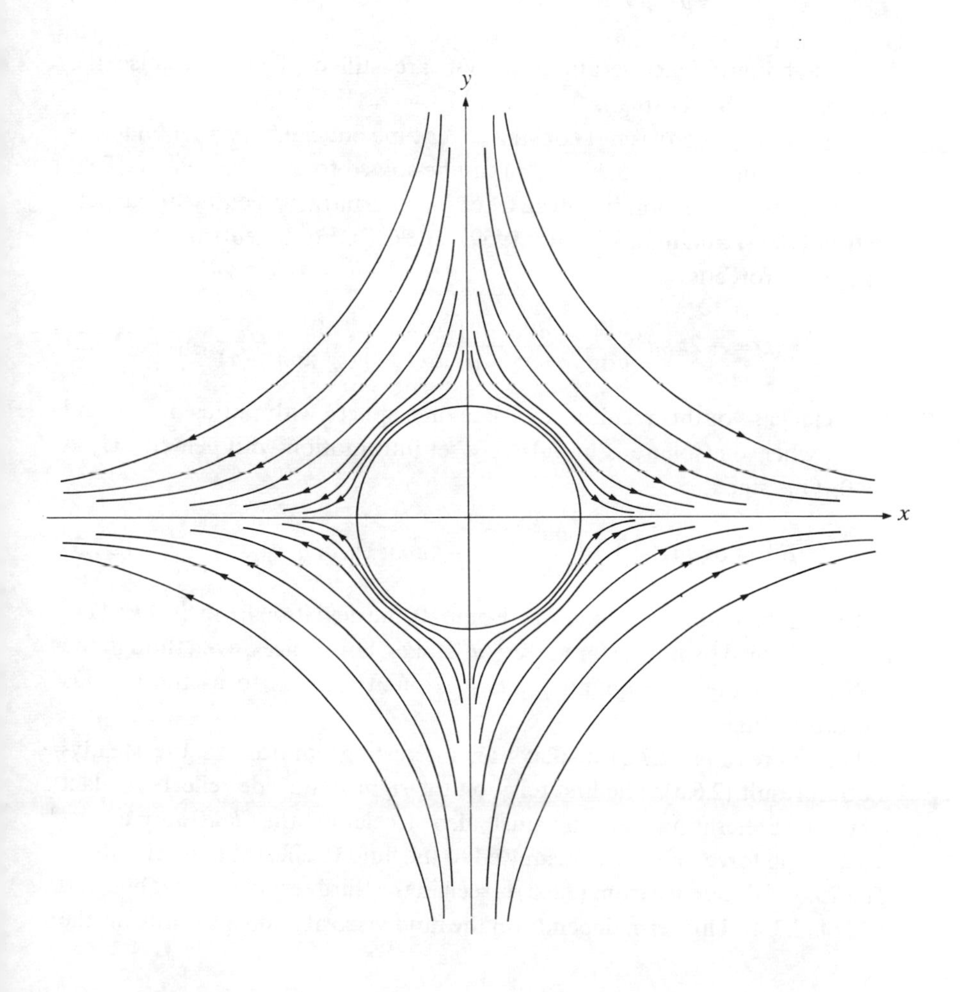

2.7 Unsteady translation of spheres

Time-dependent processes enter the description of particle motions in low Reynolds number flows in two ways: through time-dependent changes in the particle configuration and temporal accelerations. If the characteristic time scale associated with particle motion, t_p, is much longer than $\rho L^2/\mu$, then the Stokes equations are applicable in a pseudo-steady sense. In other words, the solution for any instant in time corresponds to that for the instantaneous particle configuration and time dependence appears only through the boundary conditions applied to the particles. On the other hand, if t_p is comparable to or shorter than $\rho L^2/\mu$ then the temporal acceleration must be retained as

$$\rho\frac{\partial \mathbf{u}}{\partial t}=-\nabla p+\mu\nabla^2\mathbf{u} \qquad \text{and} \qquad \nabla\cdot\mathbf{u}=0. \tag{2.7.1}$$

Note that inertial accelerations, $\mathbf{u}\cdot\nabla\mathbf{u}$, are still negligible because the Reynolds number is small.

If a particle velocity is not constant over time but can be written in terms of Fourier components then (2.7.1) can be solved to give the velocity field $\mathbf{u}(\mathbf{x}, t)$ and the corresponding drag force for an arbitrary, rectilinear particle motion $\mathbf{U}(t)$ (Landau & Lifshitz, 1959, pp. 96–7). For a neutrally buoyant sphere the force is

$$\mathbf{F}(t)=-2\pi\rho a^3\left\{\frac{\mathrm{d}\mathbf{U}}{\mathrm{d}t}+\frac{3\mu\mathbf{U}}{\rho a^2}+\frac{3}{a}\left(\frac{\mu}{\pi\rho}\right)^{1/2}\int_{-\infty}^{t}\frac{\mathrm{d}\mathbf{U}}{\mathrm{d}\tau}\frac{\mathrm{d}\tau}{(t-\tau)^{1/2}}\right\}. \tag{2.7.2}$$

A special case of this result is the force on a sphere with radius a in a fluid at rest when the sphere is impulsively set into motion with velocity $\mathbf{U}_0$ at $t > 0$. One finds

$$\mathbf{F}=-6\pi\mu a\mathbf{U}_0\left[1+\left(\frac{\rho a^2}{\pi\mu t}\right)^{1/2}\right]-\tfrac{2}{3}\pi\rho a^3\mathbf{U}_0\,\delta(t), \tag{2.7.3}$$

$\delta(t)$ being the Dirac delta function. For $t>0$, the deviation from Stokes' law, (2.6.5), is scaled by $\rho a^2/\mu$; for $t \gg \rho a^2/\mu$, Stokes' law suffices, even though the flow is unsteady. The quantity $\rho a^2/\mu$ is often referred to as the viscous relaxation time.

The expression (2.7.2) displays an interesting contrast to the steady-motion result (2.6.5). The first term on the right-hand side reflects the fact that an accelerating particle must also accelerate the *fluid* and a corresponding force must be transmitted to the fluid to effect this acceleration. The second difference from (2.6.5) is seen in the third term on the right-hand side of (2.7.2). This term depends on the fluid viscosity and accounts for the force associated with an unsteady diffusion of vorticity into the fluid. The details of this process at some past time affect the force at the present time t. For all but simple particle motions the integral can be difficult to evaluate.

In the terminology of §2.4, (2.7.3) is a solution to the *resistance* problem. In some applications the force on a particle is specified, so one wishes to solve the inverse or *mobility* problem. We look at two special cases. In the first a sphere, initially motionless in a fluid at rest, is subjected to a constant force beginning at time $t=0$, so that

$$\mathbf{F}(t)=\mathbf{F}_0 H(t). \tag{2.7.4}$$

The Heaviside function has values $H(t)=0$ for $t\leq 0$ and $H(t)=1$ for $t>0$. Taking the Laplace transform of (2.7.2) yields the transformed velocity

$$\hat{\mathbf{U}}=\frac{\mathbf{F}_0}{6\pi\mu a}\frac{1}{s}\left[\frac{\rho a^2 s}{3\mu}+\left(\frac{\rho a^2 s}{\mu}\right)^{1/2}+1\right]^{-1} \tag{2.7.5}$$

in terms of the transform variables.

The full expression for the time dependence of the velocity follows from a partial fraction expansion and inversion, but short and long time limits suffice for the present purposes. Expanding (2.7.5) for $\rho a^2 s/\mu \gg 1$, i.e. for short times, and then inverting leads to

$$\mathbf{U}(t)=\frac{1}{2\pi a^3\rho}\mathbf{F}_0 t+\ldots, \tag{2.7.6}$$

indicating an initially constant acceleration of the particle plus the 'added' mass of fluid, $2\pi a^3\rho/3$. For $\rho a^2 s/\mu \ll 1$, or long times, expansion and inversion gives the behavior as

$$\mathbf{U}(t)=\frac{\mathbf{F}_0}{6\pi\mu a}\left[1-\left(\frac{\rho a^2}{\pi\mu t}\right)^{1/2}+\ldots\right]. \tag{2.7.7}$$

Thus steady state is approached slowly, exactly as indicated by (2.7.3) for the inverse problem.

The second example is the reaction of a sphere to a force impulse,

$$\mathbf{F}(t)=\mathbf{M}_0\delta(t). \tag{2.7.8}$$

In this case the velocity is the derivative of (2.7.6) and (2.7.7). The impulse accelerates the sphere to

$$\mathbf{U}(0)=\frac{1}{2\pi a^3\rho}\mathbf{M}_0 \tag{2.7.9}$$

and the velocity decays as

$$\mathbf{U}(t) = \frac{1}{9\sqrt{\pi}} \frac{3\mathbf{M}_0}{4\pi a^3 \rho} \left(\frac{\rho a^2}{\mu t} \right)^{3/2} + \ldots \tag{2.7.10}$$

at long times. Now the impulse propels the sphere a finite distance

$$\int_0^\infty \mathbf{U}(t)\,\mathrm{d}t = \lim_{s \to 0} \hat{\mathbf{U}} = \frac{\mathbf{M}_0}{6\pi\mu a}. \tag{2.7.11}$$

This result is central to the discussion of Brownian motion in Chapter 3.

These results indicate that the velocity field and integral properties such as the drag assume steady-state values for $t \gg \rho a^2/\mu$. For colloidal particles, e.g. spheres of 1 µm radius in water at room temperature for which $\rho a^2/\mu \approx 10^{-6}$ s, the time scale is sufficiently short to neglect the transient for most situations. Hence we almost invariably invoke the pseudo-steady-state approximation in the dynamical treatments.

2.8 Two spheres translating through a quiescent fluid

Many colloidal phenomena involve collaborative motion of several particles. Because of the linearity of the Stokes equations, the solution methods described previously are applicable. Multipole expansions can be constructed for any number of particles by superimposing the velocity fields generated by point forces and successively higher-order singularities with the strengths of the singularities, i.e. the coefficients in the expansion, chosen to satisfy the boundary conditions on the surfaces of the spheres (e.g. Kim & Mifflin, 1985). For two spheres, eigenfunction expansions can be derived by separation of variables (e.g., Stimson & Jeffery, 1926). In either approach, the series converges slowly for small separations between the particles and must be summed using a digital computer. Nevertheless, complete solutions are available for interactions between two equal spheres and extensive results can be found for unequal spheres.

Rigid particles will translate and rotate in response to external forces and torques. For anisotropic particles or interacting spheres, the rotational and translational motions are coupled. However, we limit our attention to pair interactions between torque-free spheres (Fig. 2.7), for which the linearity of the Stokes equations allows the velocities to be written as

$$\mathbf{U}_i = \sum_{j=1}^{2} \boldsymbol{\omega}_{ij} \cdot \mathbf{F}_j, \tag{2.8.1}$$

with $i=1,2$. The mobility tensors $\boldsymbol{\omega}_{ij}$ express the response of the ith sphere to a force acting on the jth and depend on the separation. The dynamics of the interacting spheres can also be represented in terms of the relative and center of mass motion by defining $d\mathbf{r}/dt=\mathbf{U}_2-\mathbf{U}_1$ and $d\mathbf{x}/dt=(\mathbf{U}_1+\mathbf{U}_2)/2$, so that

$$\frac{d\mathbf{r}}{dt}=(\boldsymbol{\omega}_{21}-\boldsymbol{\omega}_{11})\cdot\mathbf{F}_1+(\boldsymbol{\omega}_{22}-\boldsymbol{\omega}_{12})\cdot\mathbf{F}_2$$

and (2.8.2)

$$\frac{d\mathbf{x}}{dt}=\tfrac{1}{2}(\boldsymbol{\omega}_{21}+\boldsymbol{\omega}_{11})\cdot\mathbf{F}_1+\tfrac{1}{2}(\boldsymbol{\omega}_{22}+\boldsymbol{\omega}_{12})\cdot\mathbf{F}_2.$$

Two spheres at sufficiently large separations behave as if isolated, so

$$\lim_{r\to\infty}\boldsymbol{\omega}_{ii}=\frac{\boldsymbol{\delta}}{6\pi\mu a_i} \tag{2.8.3}$$

and, for $i\neq j$,

$$\lim_{r\to\infty}\boldsymbol{\omega}_{ij}=\mathbf{0}.$$

Fig. 2.7. Two interacting spheres.

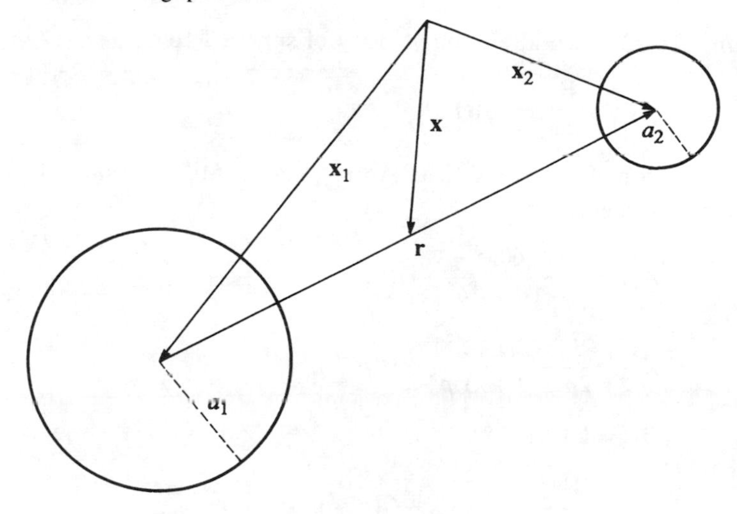

At finite separations, the dependence of the mobilities on the relative position of the spheres, $\mathbf{r}=\mathbf{x}_2-\mathbf{x}_1$, can be represented as (Batchelor, 1976)

$$\boldsymbol{\omega}_{ij}=\frac{1}{3\pi\mu(a_i+a_j)}\left[A_{ij}(r)\frac{\mathbf{rr}}{r^2}+B_{ij}(r)\left(\boldsymbol{\delta}-\frac{\mathbf{rr}}{r^2}\right)\right], \qquad (2.8.4)$$

with the scalar coefficients A_{ij} and B_{ij} characterizing mobility along and perpendicular to the line of centers, respectively. Thus, from (2.8.3), $\lim_{r\to\infty} A_{ii}=B_{ii}=1$ and $\lim_{r\to\infty} A_{ij}=B_{ij}=0$ for $i\neq j$. For equal spheres, symmetry arguments show that $\boldsymbol{\omega}_{12}=\boldsymbol{\omega}_{21}$ and $\boldsymbol{\omega}_{11}=\boldsymbol{\omega}_{22}$.

Interactions can be accounted for quite simply if the particles are far apart (e.g. Batchelor, 1976). Consider, for example, the situation with $\mathbf{F}_1=\mathbf{F}$ and $\mathbf{F}_2=\mathbf{0}$. Sphere 1 moves approximately at velocity $\mathbf{F}/6\pi\mu a_1$ and generates a disturbance velocity $\mathbf{u}_1$ corresponding to (2.6.16). This velocity field, however, violates the no-slip boundary conditions on sphere 2, which responds by translating at a velocity determined by Faxen's law (2.6.4),

$$\mathbf{U}_2=\mathbf{u}_1(\mathbf{r})+\tfrac{1}{6}a_2^2\nabla^2\mathbf{u}_1(\mathbf{r}), \qquad (2.8.5)$$

and acting as a force dipole with strength set by (2.6.7),

$$\mathbf{S}_2=\tfrac{20}{3}\pi\mu a_2^3\left[\mathbf{e}_1(\mathbf{r})+\tfrac{1}{10}a_2^2\nabla^2\mathbf{e}_1(\mathbf{r})\right]. \qquad (2.8.6)$$

Here $\mathbf{e}_1$ is the rate of strain corresponding to the velocity field $\mathbf{u}_1$. The velocity field generated by sphere 2,

$$\mathbf{u}_2=\mathbf{S}_2:\nabla\mathbf{I}, \qquad (2.8.7)$$

from (2.6.22), then alters the velocity of sphere 1 to

$$\mathbf{U}_1=\frac{\mathbf{F}}{6\pi\mu a_1}+\mathbf{u}_2(\mathbf{r}). \qquad (2.8.8)$$

Comparison of (2.8.5) and (2.8.8) with (2.8.4) identifies the far-field forms of the scalar coefficients,

$$\begin{aligned}
A_{11}&=1-\frac{60\lambda^3}{(1+\lambda)^4\rho^4}+O(\rho^{-6}),\\
A_{12}&=\frac{3}{2\rho}-\frac{2(1+\lambda^2)}{(1+\lambda)^2\rho^3}+O(\rho^{-7}),\\
B_{11}&=1+O(\rho^{-6}),\\
B_{12}&=\frac{3}{4\rho}+\frac{1+\lambda^2}{(1+\lambda)^2\rho^3}+O(\rho^{-7}),
\end{aligned} \qquad (2.8.9)$$

with $\lambda = a_2/a_1$ and $\rho = 2r/(a_1 + a_2)$. The interaction decays slowly, as ρ^{-1}, for the velocity induced by the force on the opposite sphere, but more rapidly for the effect of the passive second sphere on the velocity of the first. Indeed the latter disappears entirely as $\lambda \to 0$.

Some information about the mobilities near contact, i.e. as $\rho \to 2$, can be obtained from lubrication theory. For example, the squeeze flow between two spheres with $\mathbf{U}_2 = \mathbf{0}$ and $\mathbf{U}_1 = -U\mathbf{e}_z$, with $\rho - 2 \ll \min(1, \lambda)$, is governed in cylindrical coordinates r and z by (Fig. 2.8)

$$0 = -\frac{\partial p}{\partial r} + \mu \frac{\partial^2}{\partial z^2} u_r$$

$$0 = -\frac{\partial p}{\partial z} \tag{2.8.10}$$

and

$$\frac{1}{r}\frac{\partial}{\partial r} r u_r + \frac{\partial}{\partial z} u_z = 0,$$

Fig. 2.8. Gap between two spheres with surface-to-surface separation h_0.

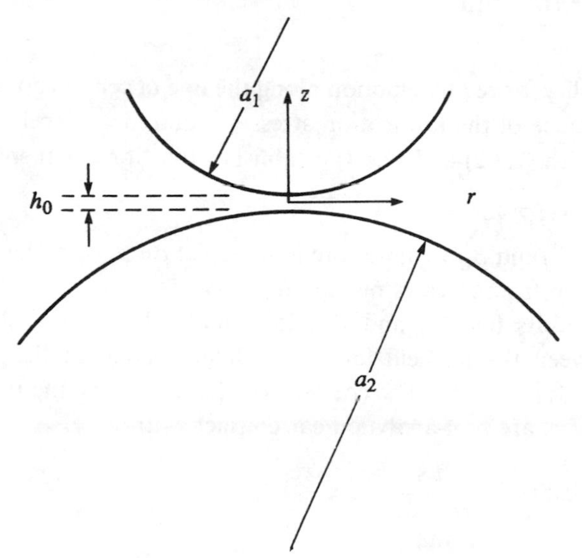

with

$$u_r = 0, \qquad u_z = -U, \qquad z = \frac{1}{2}\left(h_0 + \frac{r^2}{a_1}\right) \equiv h_1(r)$$

$$u_r = u_z = 0, \qquad z = -\frac{1}{2}\left(h_0 + \frac{r^2}{a_2}\right) \equiv -h_2(r)$$

$$\frac{\partial p}{\partial r} = 0, \qquad r = 0$$

$$p \to p_0, \qquad r \to \infty.$$

Integrating the r-momentum equation and applying the boundary conditions yields

$$u_r = \frac{1}{2\mu}\frac{\partial p}{\partial r}[z^2 + (h_2 - h_1)z - h_1 h_2]. \tag{2.8.11}$$

Substitution into the continuity equation, integration once in the z-direction and twice in the r-direction, and application of the boundary conditions on the pressure leaves

$$p = p_0 + \frac{3\mu U a_1 a_2}{h_0^2(a_1 + a_2)}\left(1 + \frac{(a_1 + a_2)}{2h_0 a_1 a_2} r^2\right)^{-2}. \tag{2.8.12}$$

Integrating the pressure over sphere 1 then fixes the force as

$$F_z = \frac{6\pi\mu U}{h_0}\frac{a_1^2 a_2^2}{(a_1 + a_2)^2}. \tag{2.8.13}$$

Thus the mobility for relative motion along the line of centers goes to zero at contact because of the lubrication stresses required to expel the fluid. Comparison with (2.8.2) and (2.8.4) establishes that for equal spheres

$$A_{11} - A_{12} \sim 2(\rho - 2). \tag{2.8.14}$$

The individual A_{ij} and B_{ij} are finite and non-zero at $\rho = 2$, since the leading-order terms stem from stresses outside the gap.

The exact results for A_{11} and A_{12} for equal spheres (Fig. 2.9) vary smoothly between the far-field forms, valid for $r/a \gg 2$, to the value at contact, $A_{11} = A_{12} = 0.7750$. The coefficients characterizing the transverse motion, however, are non-analytic near contact, with

$$B_{11} = 0.891 - \frac{0.388}{\ln(\rho - 2)} + \ldots,$$
$$B_{12} = 0.490 + \frac{0.144}{\ln(\rho - 2)} + \ldots. \tag{2.8.15}$$

This behavior apparently arises because the lubrication stresses within the gap suppress the individual rotation of the spheres at small separations. Jeffrey & Onishi (1984) have extracted the asymptotic forms from the solutions in each limit, leaving series which converge rapidly to the exact values.

The results of Jeffrey & Onishi (1984) also apply to unequal spheres. Here we consider only the limit $\lambda \to \infty$, corresponding to a sphere of radius a near a stationary plane wall, treated by Brenner (1961) and Goldman, Cox & Brenner (1966). The $\mathbf{U}_2 = \mathbf{0}$ and

$$\mathbf{U}_1 = \boldsymbol{\omega} \cdot \mathbf{F}_1,$$

with

$$\boldsymbol{\omega} = \boldsymbol{\omega}_{11} - \boldsymbol{\omega}_{12} \cdot \boldsymbol{\omega}_{22}^{-1} \cdot \boldsymbol{\omega}_{21} = \frac{1}{6\pi\mu a}\left[A_1\left(\frac{x}{a}\right)\frac{\mathbf{xx}}{x^2} + B_1\left(\frac{x}{a}\right)\left(\boldsymbol{\delta} - \frac{\mathbf{xx}}{x^2}\right)\right], \tag{2.8.16}$$

Fig. 2.9. Coefficients for the mobility function (2.8.4) for two equal spheres (Batchelor, 1976).

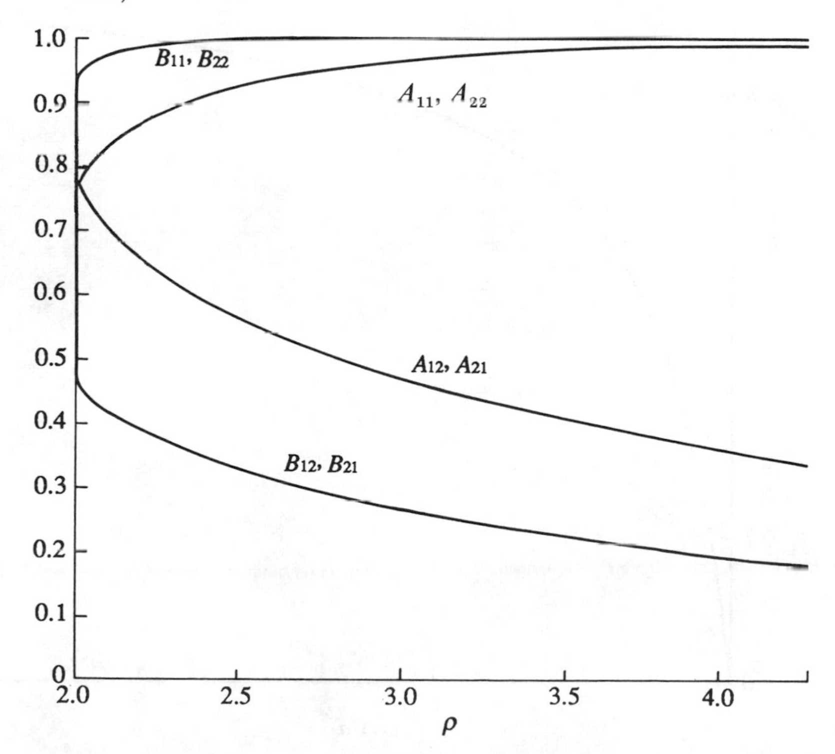

with $\mathbf{x}$ the vector from the surface of the plate to the center of the sphere and $\mathbf{x}\cdot\mathbf{x}=x^2$. The asymptotic forms for A_1 and B_1 can be obtained by techniques similar to those described above:

$$x\to\infty \qquad A_1=1-\frac{9a}{8x}+\ldots,$$

$$B_1=1-\frac{9a}{16x}+\ldots; \tag{2.8.17}$$

$$x\to a \qquad A_1=\frac{x}{a}-1+\ldots,$$

$$B_1=-\frac{1}{\ln(x/a-1)}+\ldots.$$

Note that the mobilities both along and transverse to the line of centers vanish at contact. The full form of A_1 is plotted in Fig. 2.10; B_1 can be extracted from results in Goldman, Cox, & Brenner (1966).

Fig. 2.10. Mobility of sphere normal to a wall (Brenner, 1961).

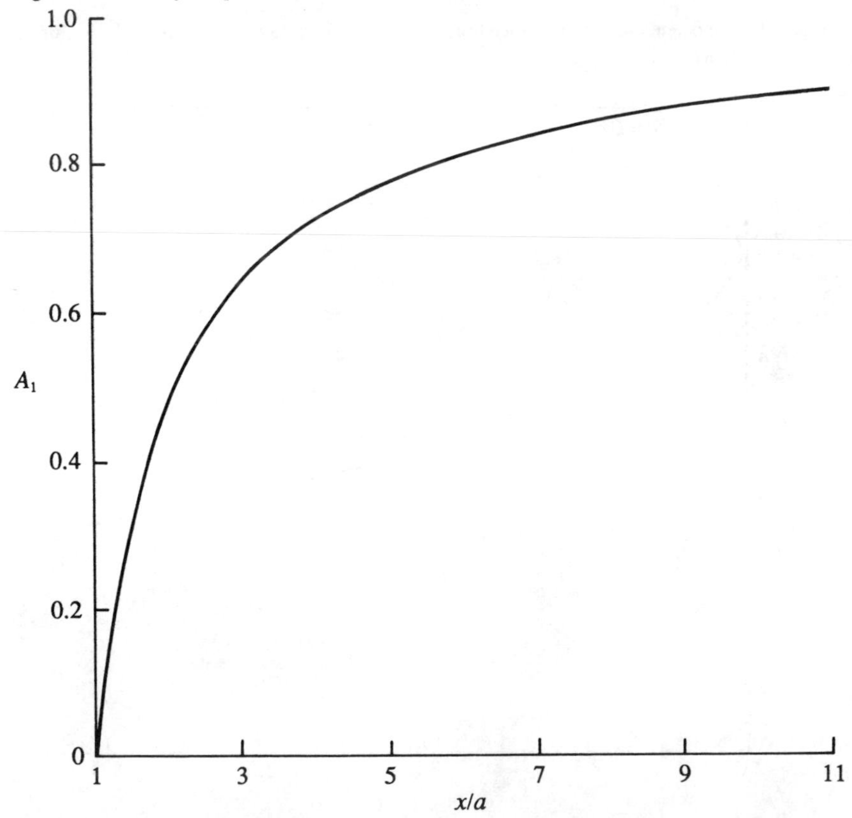

Though analyses of the dynamics of interacting particles in subsequent chapters primarily use this mobility formulation, the inverse resistance representation is sometimes needed. This means relating the forces and velocities as

$$\mathbf{F}_i = \sum_{j=1}^{2} \mathbf{f}_{ij} \cdot \mathbf{U}_j, \tag{2.8.18}$$

with the $\mathbf{f}_{ij}$ comprising the friction tensors for a pair of interacting particles. Comparison with (2.8.1) identifies the relationship between the friction and mobility tensors as

$$\begin{aligned} \mathbf{f}_{ii} &= (\boldsymbol{\omega}_{ii} - \boldsymbol{\omega}_{ij} \cdot \boldsymbol{\omega}_{jj}^{-1} \cdot \boldsymbol{\omega}_{ji})^{-1}, \\ \mathbf{f}_{ij} &= (\boldsymbol{\omega}_{ji} - \boldsymbol{\omega}_{jj} \cdot \boldsymbol{\omega}_{ij}^{-1} \cdot \boldsymbol{\omega}_{ii})^{-1} \text{ for } i \neq j. \end{aligned} \tag{2.8.19}$$

From these one can work out the scalar coefficients in the corresponding forms,

$$\mathbf{f}_{ij} = 3\pi\mu(a_1 + a_2)\left[X_{ij}(\rho)\frac{\boldsymbol{\rho\rho}}{\rho^2} + Y_{ij}(\rho)\left(\boldsymbol{\delta} - \frac{\boldsymbol{\rho\rho}}{\rho^2}\right)\right], \tag{2.8.20}$$

in terms of the A_{ij} and B_{ij} in the mobility tensors.

We close this section by illustrating the consequences of these results with four examples (Fig. 2.11(i)–(iv)):

Fig. 2.11. Examples of interactions: (i) motion along a line of centers; (ii) translation due to equal forces; (iii) the rotation of a doublet; (iv) a sphere moving relative to a wall.

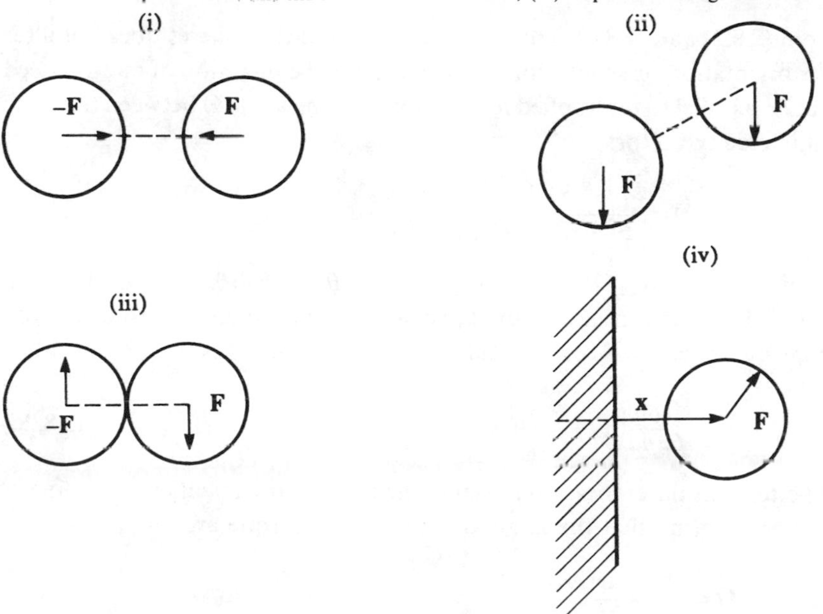

(i) Two equal spheres with $\mathbf{F}_1 = -\mathbf{F}_2 = F\mathbf{r}/r$ and F = constant approach along their line of centers with relative velocity

$$\frac{\mathrm{d}r}{\mathrm{d}t} = -\frac{F}{3\pi\mu a}(A_{11} - A_{12}). \tag{2.8.21}$$

Integrating this as

$$\frac{F}{3\pi\mu a}t = \int_r^{r_0} \frac{\mathrm{d}r}{A_{11} - A_{12}} \tag{2.8.22}$$

implicitly determines the separation as a function of time with $r = r_0$ at $t = 0$. However, as a consequence of (2.8.14),

$$\frac{Ft}{6\pi\mu a^2} \sim \ln \frac{1}{\rho - 2} \tag{2.8.23}$$

as $t \to \infty$. Thus perfectly smooth spheres approaching under a constant force never achieve a true mathematical contact.

(ii) Two identical spheres acted upon by an external force $\mathbf{F}_1 = \mathbf{F}_2 = \mathbf{F}$ have relative and center-of-mass velocities

$$\begin{aligned} 6\pi\mu a \frac{\mathrm{d}\mathbf{x}}{\mathrm{d}t} &= (B_{11} + B_{12})\mathbf{F} + (A_{11} + A_{12} - B_{11} - B_{12}) \frac{\mathbf{rr} \cdot \mathbf{F}}{r^2}, \\ \frac{\mathrm{d}\mathbf{r}}{\mathrm{d}t} &= \mathbf{0}. \end{aligned} \tag{2.8.24}$$

from (2.8.2) and (2.8.4). Thus the translation of the spheres does not alter the orientation or separation of the pair, but the direction of motion need not be parallel to the applied force. However, the angle Θ between the force and velocity vectors,

$$\cos \Theta = \frac{B + (A - B)\cos\theta}{[B^2 + (A^2 - B^2)\cos^2\theta]^{1/2}}, \tag{2.8.25}$$

with $A = A_{11} + A_{12}$, $B = B_{11} + B_{12}$, and $\cos\theta = \mathbf{r} \cdot \mathbf{F}/|\mathbf{F}|r$, is never large.

(iii) The rotation of a doublet, i.e. a pair with $r = 2a$, due to equal and opposite forces, $\pm\mathbf{F}$, on the individual spheres satisfies

$$\frac{\mathrm{d}\mathbf{r}}{\mathrm{d}t} = \frac{0.80}{6\pi\mu a}\left(\mathbf{F} - \frac{\mathbf{rr}}{r^2} \cdot \mathbf{F}\right). \tag{2.8.26}$$

The term in parentheses represents the force perpendicular to the line of centers. Noting that the angular velocity and torque are

$$\mathbf{\Omega} = \frac{\mathbf{r}}{4a^2} \times \frac{\mathrm{d}\mathbf{r}}{\mathrm{d}t}$$

and (2.8.27)

$$\mathbf{T}=\mathbf{r}\times\mathbf{F}$$

identifies the rotational mobility as $\omega_r = 1/30\pi\mu a^3$.

(iv) For a sphere moving relative to a plane wall the equation of motion is

$$6\pi\mu a\frac{d\mathbf{x}}{dt}=B_1\mathbf{F}+(A_1-B_1)\frac{\mathbf{xx}}{x^2}\cdot\mathbf{F}. \tag{2.8.28}$$

A force acting parallel to the wall, i.e. with $\mathbf{x}\cdot\mathbf{F}=0$, induces a velocity which is retarded by the wall, since $B_1\leq 1$, but is still parallel to it. The velocity due to a force normal to the wall depends only on A_1. Since $A_1\to 0$ as $x\to a$, a finite force requires an infinite time to bring the sphere into contact with the wall.

2.9 Two spheres in a shear flow

Interactions between spheres embedded in a fluid subjected to a linearly varying velocity field provide the basis for analyzing shear-induced flocculation and the rheology of concentrated dispersions. For two spheres the solution techniques described in the previous section suffice. The results summarized here derive from Lin, Lee, & Sather (1970), Batchelor & Green (1972), and Kim & Mifflin (1985).

Since the effects of external and interparticle forces can be superimposed, we concentrate here on force-free spheres. In the absence of inertia, the velocity of each sphere varies linearly with the imposed rate of strain $\mathbf{E}$ and vorticity $\mathbf{\Omega}$ according to

$$\mathbf{U}_i=(\mathbf{E}+\mathbf{\Omega})\cdot\mathbf{x}_i+\mathbf{C}_{ij}\cdot\mathbf{E}\cdot(\mathbf{x}_j-\mathbf{x}_i); \tag{2.9.1}$$

see §2.6 for the definitions of $\mathbf{E}$ and $\mathbf{\Omega}$.

Decomposing the motion of two spheres into the relative and center-of-mass velocities leads to

$$\begin{aligned}\frac{d\mathbf{r}}{dt}&=(\mathbf{E}+\mathbf{\Omega})\cdot\mathbf{r}+(\mathbf{C}_{12}+\mathbf{C}_{21})\cdot\mathbf{E}\cdot\mathbf{r},\\ \frac{d\mathbf{x}}{dt}&=(\mathbf{E}+\mathbf{\Omega})\cdot\mathbf{x}+\tfrac{1}{2}(\mathbf{C}_{21}-\mathbf{C}_{12})\cdot\mathbf{E}\cdot\mathbf{r}.\end{aligned} \tag{2.9.2}$$

The tensor $\mathbf{C}=-(\mathbf{C}_{12}+\mathbf{C}_{21})$ accounts for the effect of interactions on the relative velocity and has the general form

$$\mathbf{C}=A(r)\frac{\mathbf{rr}}{r^2}+B(r)\left(\boldsymbol{\delta}-\frac{\mathbf{rr}}{r^2}\right), \tag{2.9.3}$$

The scalar functions A and B characterize motion along and perpendicular to the line of centers, respectively.

The interaction also causes the dipole induced in each sphere by the flow to vary with the separation and orientation of the pair according to (Batchelor & Green, 1972):

$$\mathbf{S}_i = \frac{20\pi\mu a_i^3}{3}\left[\left(1+K(r)\right)\mathbf{E} + \left(\frac{\mathbf{rr}\cdot\mathbf{E}+\mathbf{E}\cdot\mathbf{rr}}{r^2} - \frac{2\mathbf{r}\cdot\mathbf{E}\cdot\mathbf{r}}{3r^2}\boldsymbol{\delta}\right)L(r) + \frac{\mathbf{r}\cdot\mathbf{E}\cdot\mathbf{r}}{r^2}\left(\frac{\mathbf{rr}}{r^2} - \tfrac{1}{3}\boldsymbol{\delta}\right)M(r)\right]. \tag{2.9.4}$$

Since isolated spheres are convected by the flow, but cannot deform with the fluid, they acquire dipoles given by (2.6.11); therefore, the functions A, B, K, L, and M, which depend on $a_j/a_i(\equiv\lambda)$, must all decay to zero as $r\to\infty$.

The far-field form of the scalar functions can be obtained with the Faxen laws (2.6.4) and (2.6.7). In the absence of interactions, the shear flow generates dipoles $\mathbf{S}_{i0}$ (2.6.11) in each sphere, producing the velocity disturbances (2.5.17)

$$\mathbf{u}_i = \mathbf{S}_{i0} : \nabla\mathbf{I}(\mathbf{x}-\mathbf{x}_i). \tag{2.9.5}$$

However, each sphere also experiences the velocity field generated by the other, which alters the velocity to

$$\mathbf{U}_j = (\mathbf{E}+\boldsymbol{\Omega})\cdot\mathbf{x}_j + (1+\tfrac{1}{6}a_j^2\nabla^2)\mathbf{u}_i(\mathbf{r}), \tag{2.9.6}$$

and augments the dipole according to (2.6.7) to

$$\mathbf{S}_j = \tfrac{20}{3}\pi\mu a_j^3[\mathbf{E} + (1+\tfrac{1}{10}a_j^2\nabla^2)\mathbf{e}_i(\mathbf{r})] \tag{2.9.7}$$

with $2\mathbf{e}_i = (\nabla\mathbf{u}_i + (\nabla\mathbf{u}_i)^{\mathrm{T}})$. Comparison with the expressions (2.9.2) and (2.9.4) identifies the far-field forms

$$\begin{aligned}
A &= \frac{20(1+\lambda^3)}{(1+\lambda)^3\rho^3} - \frac{48(1+\lambda^5)+80\lambda^2(1+\lambda)}{(1+\lambda)^5\rho^5} + \ldots,\\
B &= 32\frac{1+\lambda^5+\frac{5}{3}\lambda^2(1+\lambda)}{(1+\lambda)^5\rho^5} + \ldots,\\
K &= -32\frac{\lambda^3(1+\lambda^2)}{(1+\lambda)^5\rho^5} + \ldots,\\
L &= -\frac{20\lambda^3}{(1+\lambda)^3\rho^3} + \frac{160\lambda^3(1+\lambda^2)}{(1+\lambda)^5\rho^5} + \ldots,\\
M &= \frac{100\lambda^3}{(1+\lambda)^3\rho^3} - \frac{560\lambda^3(1+\lambda^2)}{(1+\lambda)^5\rho^5} + \ldots,
\end{aligned} \tag{2.9.8}$$

with $\rho = 2r/(a_i + a_j)$. Note that the velocity induced along the line of centers decays as ρ^{-2}, i.e. as the velocity field generated by the dipoles. Similarly, the dipole strength falls as ρ^{-3}, i.e. as the velocity gradient.

If $\lambda = a_2/a_1 \to 0$, i.e., the second sphere is much smaller, $K = L = M = 0$ and

$$A = \frac{5}{2}\left(\frac{a_1}{r}\right)^3 - \frac{3}{2}\left(\frac{a_1}{r}\right)^5,$$
$$B = \left(\frac{a_1}{r}\right)^5. \tag{2.9.9}$$

Thus the smaller sphere is simply swept along with the flow around the first. The dipole induced in the smaller sphere, from (2.9.8) with now, $\lambda = a_1/a_2 \to \infty$, is simply determined by the rate of strain produced by the flow around the larger sphere,

$$K = -\left(\frac{a_1}{r}\right)^5,$$
$$L = -\frac{5}{2}\left(\frac{a_1}{r}\right)^3 + 5\left(\frac{a_1}{r}\right)^5, \tag{2.9.10}$$
$$M = \frac{25}{2}\left(\frac{a_1}{r}\right)^3 - \frac{35}{2}\left(\frac{a_1}{r}\right)^5.$$

The behavior of each of the five functions is depicted in Figs 2.12(a) and (b). The far-field forms prove accurate at separations of one radius or more in each case; at contact the functions assume the following values for equal spheres (Batchelor & Green, 1972):

$$\begin{aligned} A(2) &= 1.0 \\ B(2) &= 0.4060 \\ K(2) &= -0.0472 \\ L(2) &= 0.1928 \\ M(2) &= 1.0508. \end{aligned} \tag{2.9.11}$$

All but L vary monotonically from zero at infinite separation to the finite value at contact.

The asymptotic forms near contact may be extracted from analyses of the viscous stresses in the gap between spheres. In this case, though, the spheres are driven together by viscous stresses acting on the surfaces outside the gap exposed to the ambient rate of strain. Hence the analysis requires the opposing forces and couples on the individual spheres of a rigid, force-free and couple-free doublet embedded in the shear flow, i.e.

$$\mathbf{F}_i = 3\pi\mu a_i(1+\lambda)\left[\frac{\mathbf{rr}}{r^2}P_0 + \left(\boldsymbol{\delta} - \frac{\mathbf{rr}}{r^2}\right)Q_0\right]\cdot\mathbf{E}\cdot\mathbf{r},$$

$$\mathbf{L}_i = 3\pi\mu a_i^2(1+\lambda)^2\frac{\mathbf{r}\times\mathbf{E}\cdot\mathbf{r}}{r^2}R_0, \tag{2.9.12}$$

Fig. 2.12. Hydrodynamic functions for two equal spheres in a shear flow, with dashed lines indicating the far-field forms: (*a*) for the relative velocity (Batchelor & Green, 1972); (*b*) for the dipole strength (Kim & Mifflin, 1985).

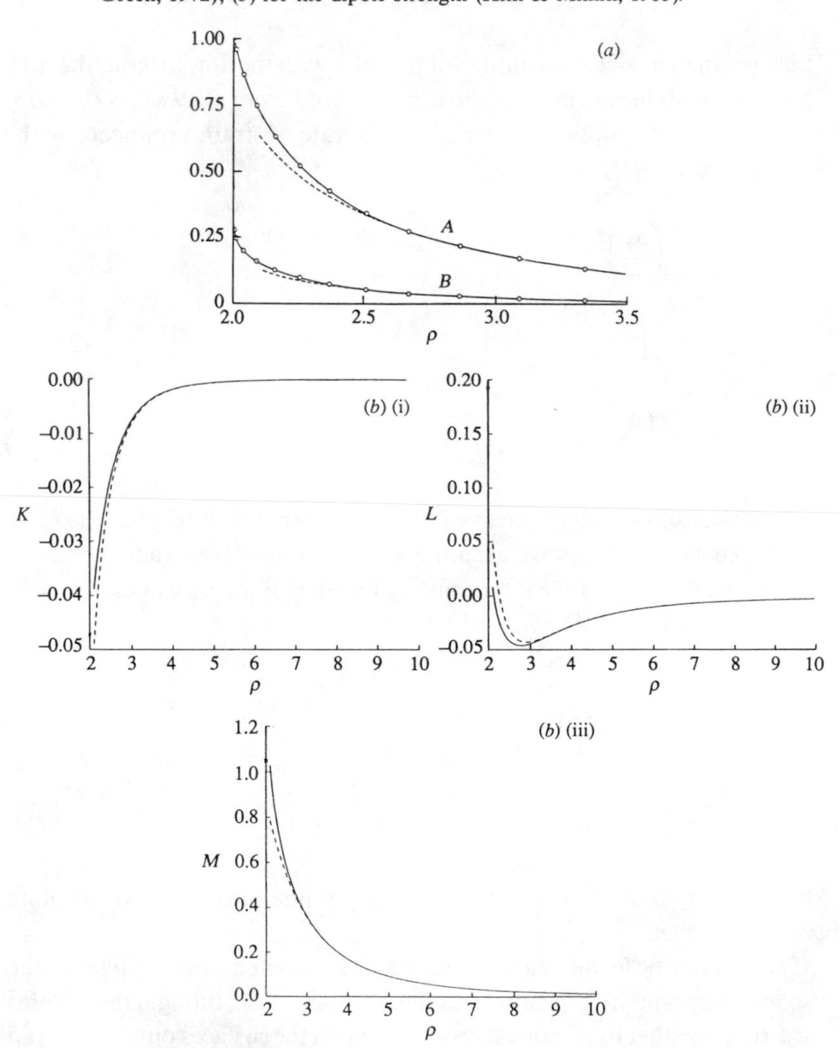

with P_0, Q_0, and R_0 functions of λ alone. The lubrication analysis for motion along the line of centers in §2.8 yielded the relation (2.8.13) between force and velocity. The total force on the sphere must be zero, so setting $\mathbf{F}_1 \cdot \mathbf{r}/r = -F_z$ determines the relative velocity. Comparison with the corresponding velocity from (2.9.1),

$$U_z = \frac{\mathbf{r}}{r} \cdot \mathbf{U} = (1-A)\frac{\mathbf{r} \cdot \mathbf{E} \cdot \mathbf{r}}{r}, \tag{2.9.13}$$

identifies the asymptotic form of A, e.g. for equal spheres with $P_0 = 1.019$ (Batchelor & Green, 1972)

$$A(r) = 1 - 4.077\frac{h_0}{a} + \ldots. \tag{2.9.14}$$

Thus lubrication stresses within the gap cause the velocity along the line of centers to go to zero, preventing the spheres from coming into contact in a finite time, regardless of the rate of strain.

The velocity perpendicular to the line of centers remains finite at contact since the doublet still rotates, but approaches the value in a singular fashion,

$$B(r) = 0.4060 - \frac{0.78}{\ln a/h_0} + \ldots. \tag{2.9.15}$$

This resembles the forms for B_{11} and B_{12}, reflecting the suppression of individual rotation of the spheres near contact.

Two examples illustrate the consequences of these results for the dynamics of interacting spheres:

(i) The equations governing the trajectories of two force-free spheres interacting in a simple shear flow with

$$\mathbf{E} = \frac{\gamma}{2}\begin{bmatrix} 0 & 1 & 0 \\ 1 & 0 & 0 \\ 0 & 0 & 0 \end{bmatrix}, \qquad \mathbf{\Omega} = \frac{\gamma}{2}\begin{bmatrix} 0 & -1 & 0 \\ 1 & 0 & 0 \\ 0 & 0 & 0 \end{bmatrix} \tag{2.9.16}$$

are found from the general expression (2.9.1) as

$$\begin{aligned} U_r &= \gamma(1-A) r \sin^2\theta \sin\phi \cos\phi \\ U_\theta &= \gamma(1-B) r \sin\theta \cos\theta \sin\phi \cos\phi \\ U_\phi &= -\gamma r \sin\theta[\sin^2\phi + \tfrac{1}{2}B(\cos^2\phi - \sin^2\phi)]. \end{aligned} \tag{2.9.17}$$

The trajectory of the second sphere relative to the first is determined by

$$\frac{1}{r_3}\frac{\partial r}{\partial \theta}=\frac{U_r}{U_\theta}=\frac{1-A}{1-B}\tan\theta$$
$$\sin\theta\frac{\partial\phi}{\partial\theta}=\frac{U_\phi}{U_\theta}=-\frac{B\cos^2\phi+(2-B)\sin^2\phi}{2(1-B)\cos\theta\sin\phi\cos\phi}. \tag{2.9.18}$$

With $r_3=r\cos\theta$ and $r_2=r\sin\theta\sin\phi$, these may be rewritten as

$$\frac{1}{r_3}\frac{\partial r_3}{\partial r}=\left(\frac{B-A}{1-A}\right)\frac{1}{r}$$
$$\frac{\partial}{\partial r}\left(\frac{r_2}{r_3}\right)^2=-\left(\frac{B}{1-A}\right)\frac{r}{r_3^2}. \tag{2.9.19}$$

Integrating the first gives $r_3(r)$, while the second provides the corresponding r_2.

Each trajectory is characterized by initial values of $(r_2, r_3)=(R_2, R_3)$ for $r\to\infty$. Real values of (R_2, R_3) lead to symmetric open trajectories which begin far upstream and finish far downstream as shown in Fig. 2.13 for the plane $R_3=0$. Clearly, the hydrodynamic interactions oppose the relative velocity due to the imposed shear, pushing the spheres apart during the

Fig. 2.13. Trajectories in the r_1r_2-plane for two equal force-free spheres in a simple shear flow (Batchelor & Green, 1972).

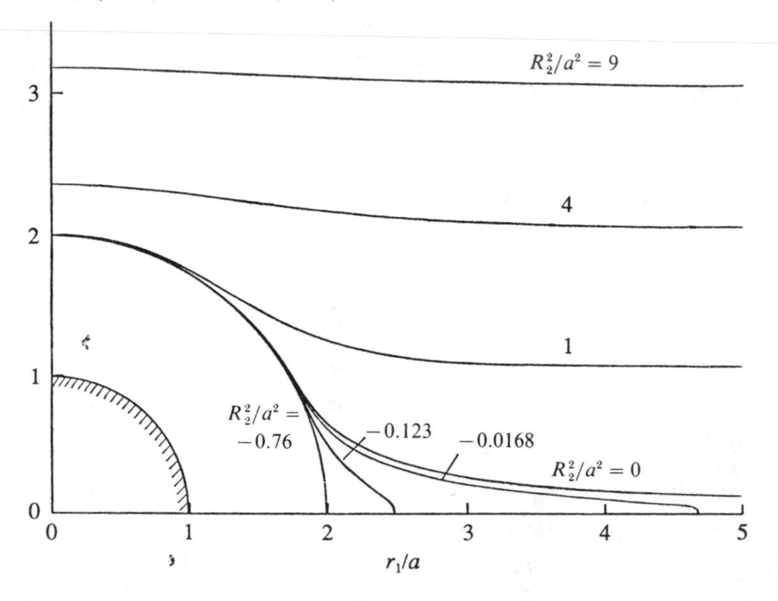

approach but pulling them together during the separation. A sphere beginning near $R_2=0$ is displaced upward and retarded by the hydrodynamic interaction, thereby leaving a region of space inaccessible to open trajectories.

Batchelor & Green (1972) discovered that closed trajectories with $R_2^2<0$ occupy the volume between the excluded sphere $r=2a$ and the limiting trajectory with $R_2=0$. The surface formed by the limiting trajectories resembles that enclosing the closed streamlines for flow about a single sphere (Fig. 2.5(b)) and the enclosed volume is infinite. However, spheres of equal size approach to exceedingly small separations, suggesting that very smooth spheres are required to observe closed trajectories.

Trajectories of spheres interacting in a simple shear flow have been observed with a travelling microtube apparatus (e.g. Takamura, Goldsmith & Mason, 1981). In this experiment a fixed movie camera records the interaction between spheres embedded in flow through a small tube. Translation of the tube relative to the camera keeps the center of mass of the pair fixed. Figure 2.14 depicts the trajectory for two 2.6 μm diameter polystyrene latex spheres at conditions chosen to minimize non-hydrodynamic forces. The interaction is symmetric with an indetectably small minimum separation as expected from the analysis.

Fig. 2.14. Trajectory for two polystyrene latex spheres ($a=1.3$ μm) in a simple shear flow (Takamura, Goldsmith, & Mason, 1981).

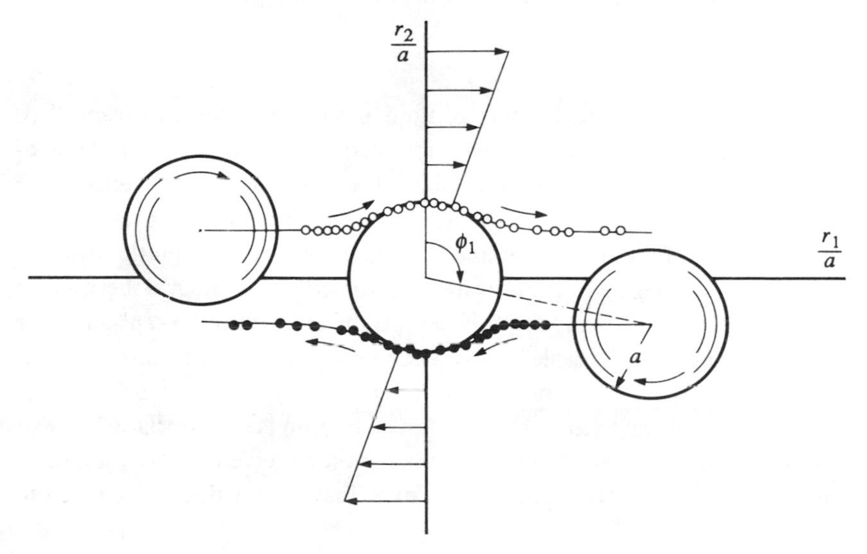

(ii) Permanent doublets exposed to a shear flow rotate at a rate given by

$$\frac{\mathbf{r}}{r^2}\times\frac{\mathrm{d}\mathbf{r}}{\mathrm{d}t}=(1-B)\frac{\mathbf{r}\times\mathbf{E}\cdot\mathbf{r}}{r^2}+\frac{\mathbf{r}\times\boldsymbol{\Omega}\cdot\mathbf{r}}{r^2}=\boldsymbol{\omega}. \tag{2.9.20}$$

For a doublet in the r_1-r_2 plane in a simple shear $\boldsymbol{\omega}=(0,0,\mathrm{d}\phi/\mathrm{d}t)$ and

$$\frac{\mathrm{d}\phi}{\mathrm{d}t}=\frac{\gamma}{2}[1+(1-B)\cos 2\phi]. \tag{2.9.21}$$

Then, for fixed r, the period of rotation is

$$T=\frac{2}{\gamma}\int_0^{2\pi}\frac{\mathrm{d}\phi}{1+(1-B)\cos 2\phi},$$

so that (van de Ven & Mason, 1976)

$$T\gamma=\frac{4\pi}{[B(2-B)]^{1/2}}. \tag{2.9.22}$$

Since B varies rapidly near contact as lubrication stresses suppress free rotation of the individual spheres, the period is a sensitive function of the separation (Fig. 2.15). In fact, van de Ven & Mason (1976) used this measurement to distinguish between doublets of particles in contact, with $T\gamma\approx 15.6$, and doublets held in potential energy minima at small but finite separations, with $T\gamma\approx 17$–18. The data of Koerner (1988) in Fig. 2.15, for doublets of polystyrene latices with radius of 1.09 μm at separations controlled by the ionic strength, conform with the prediction.

2.10 Summary

In this chapter, aspects of fluid mechanics deemed essential to understanding the dynamics of colloidal dispersions were set out. Hydrodynamic interactions were emphasized because interparticle effects are often important. Fortunately, the Reynolds number based on the characteristic length of the particle is small so the linear Stokes equations apply. Solutions for specific situations were constructed using integral methods to superpose singular solutions of the equations of motion. Faxen's laws for the force, torque, and dipole on suspended particles followed directly from the integral representations. An analysis of temporal accelerations demonstrated that the pseudo-steady approximation is appropriate for most situations to be treated. Finally, expressions for the velocities of a pair of spheres moving due to applied forces or a linear shear flow were given in detail.

Before an attempt is made to synthesize a comprehensive picture of colloidal phenomena, knowledge of several non-hydrodynamic forces is needed. The next four chapters deal with Brownian motion, electrostatic forces, attraction due to dispersion forces, and forces due to soluble polymer. In subsequent chapters this knowledge is combined with the hydrodynamics set forth here to describe the behavior of colloidal dispersions in detail.

Fig. 2.15. Period of doublet rotating in simple shear flow (Koerner, 1988): ——, prediction from (2.9.22); □, observations with polystyrene spheres ($a = 1.09\,\mu m$) in glycerine–water mixtures of varying ionic strength.

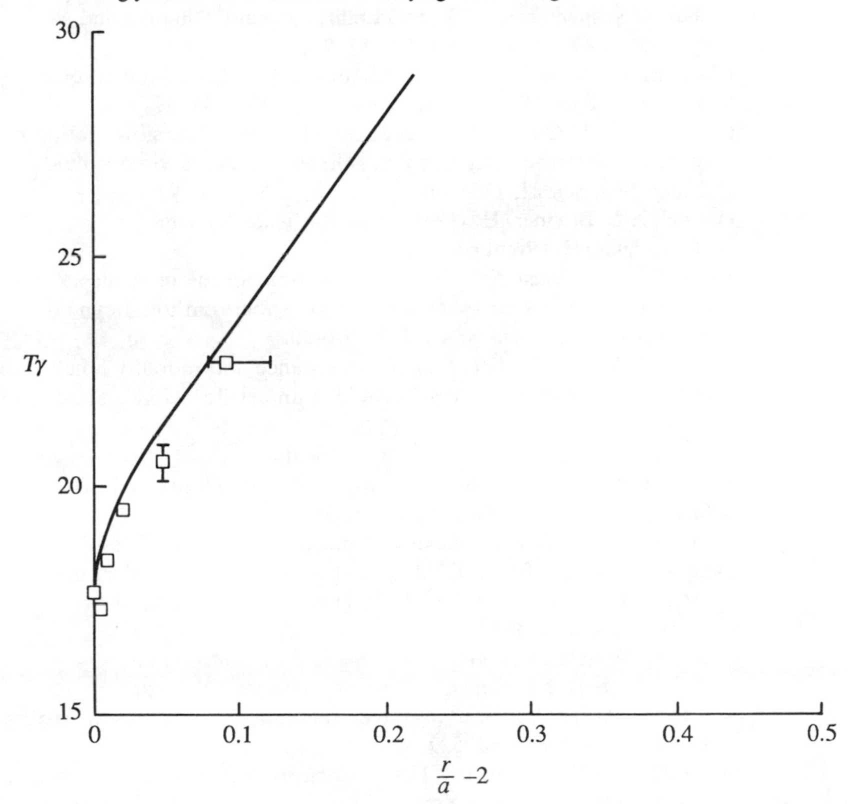

References

Batchelor, G. K. (1967). *An Introduction to Fluid Dynamics.* Cambridge University Press.

Batchelor, G. K. (1970). The stress system in a suspension of force-free particles. *J. Fluid Mech.* **41**, 545–70.

Batchelor, G. K. (1972). Sedimentation in a dilute dispersion of spheres. *J. Fluid Mech.* **52**, 245–68.

Batchelor, G. K. (1976). Brownian diffusion of particles with hydrodynamic interaction. *J. Fluid Mech.* **74**, 1–29.

Batchelor, G. K. & Green, J. T. (1972). The hydrodynamic interaction of two small freely-moving spheres in a linear flow field. *J. Fluid Mech.* **56**, 375–400.

Brenner, H. (1961). The slow motion of a sphere through a viscous fluid towards a plane surface. *Chem. Eng. Sci.* **16**, 242–51.

Chwang, A. T. & Wu, T. Y. (1975). Hydromechanics of low-Reynolds-number flow. Part 2. Singularity method for Stokes flow. *J. Fluid Mech.* **67**, 787–816.

Cox, R. G., Zia, I. Y. & Mason, S. G. (1968). Particle motions in sheared suspensions. XXV. Streamlines around cylinders and spheres. *J. Colloid Interface Sci.* **27**, 7–18.

Einstein, A. (1906). Eine neue Bestimmung der Moleküdimensionen. *Ann. Physik* **19**, 289–306. (See also (1911): *ibid.* **34**, 591–2.)

Goldman, A. J., Cox, R. G. & Brenner, H. (1966). The slow motion of two identical arbitrarily oriented spheres through a viscous fluid. *Chem. Eng. Sci.* **21**, 1151–70.

Happel, J. & Brenner, H. (1965). *Low Reynolds Number Hydrodynamics.* Prentice-Hall.

Jeffrey, D. J. & Onishi, Y. (1984). Calculations of the resistance and mobility functions for two unequal rigid spheres in low-Reynolds-number flow. *J. Fluid Mech.* **139**, 261–90.

Kim, S. & Mifflin, R. T. (1985). The resistance and mobility functions of two equal spheres in low-Reynolds-number flow. *Phys. Fluids* **28**, 2033–45.

Koerner, M. R. (1988). Private communication.

Ladyzhenskaya, O. A. (1969). *The Mathematical Theory of Viscous Incompressible Flow.* Gordon & Breach.

Lamb, H. (1932). *Hydrodynamics.* Cambridge University Press.

Landau, L. D. & Lifshitz, E. M. (1959). *Fluid Mechanics.* Pergamon.

Lin, C. J., Lee, K. J. & Sather, N. F. (1970). Slow motion of two spheres in a shear field. *J. Fluid Mech.* **43**, 35–47.

Oseen, C. W. (1927). *Neuere Methoden und Ergebnisse in der Hydrodynamik.* Akademische Verlagsgesellschaft, Leipzig.

Rallison, J. M. (1978). Note on Faxen relations for a particle in Stokes flow. *J. Fluid Mech.* **88**, 529–34.

Stakgold, I. (1968). *Boundary Value Problems of Mathematical Physics, vol. II.* Macmillan, New York.

Stimson, M. & Jeffery, G. B. (1926). The motion of two spheres in a viscous fluid. *Proc. Roy. Soc. London A* **111**, 110–26.
Stone, H. A., Bentley, B. J. & Leal, L. G. (1986). An experimental study of transient effects in the breakup of viscous drops. *J. Fluid Mech.* **173**, 131–58.
Takamura, K., Goldsmith, H. L. & Mason, S. G. (1981). The microrheology of colloidal dispersions. XII. Trajectories of orthokinetic pair collisions of latex spheres in simple electrolyte. *J. Colloid. Interface Sci.* **82**, 175–89.
van de Ven, T. G. M. & Mason, S. G. (1976). The microrheology of colloidal dispersions. V. Primary and secondary doublets of spheres in shear flow. *J. Colloid Interface Sci.* **57**, 517–34.

Problems

1 The rate of sedimentation of small particles is a strong function of their size. For a density difference of Δ the balance among gravity, buoyancy, and the viscous force (2.6.5) leads to the terminal settling velocity

$$U = 2a^2 \Delta g / 9\mu$$

for a sphere of radius a. To illustrate the time scales involved, calculate t_0, the time required to accelerate from rest to $0.9U$, and t_S, the time required to settle 0.1 m, for

(i) a 0.1 μm diameter silica sphere in cyclohexane;
(ii) a 1.0 μm diameter polystyrene sphere in water;
(iii) a 10.0 μm diameter glass sphere in decane;
(iv) a 50.0 μm diameter water droplet in air.

2 A force- and torque-free sphere being convected in a slit of width $2d$ by a Poiseuille flow experiences the quadratically varying velocity field

$$\mathbf{u}_\infty = (0, 0, U_\mathrm{m}(1 - x^2/d^2).$$

Use Faxen's laws to determine whether the translational and angular velocities of the sphere, $\mathbf{U}$ and $\mathbf{\Omega}$, deviate from the local velocity and vorticity of the fluid. Neglect hydrodynamic interactions with the walls.

3 A doublet formed from spheres of unequal densities and radii may rotate while sedimenting. Derive an expression for the angular velocity of such a doublet in terms of the hydrodynamic mobilities. Then evaluate the orientation angle as a function of time from an initially horizontal position for

(i) $a_2/a_1 = 0.25$ and $\Delta\rho_2/\Delta\rho_1 = 1$;
(ii) $a_2/a_1 = 0.50$ and $\Delta\rho_2/\Delta\rho_1 = 5$.

Jeffrey & Onishi (1984) provide the requisite values for the mobilities as in the table:

a_2/a_1	B_{11}	$B_{12}=B_{21}$	B_{22}
0.50	0.927	0.535	0.764
0.25	0.973	0.571	0.473

4 The drag coefficient is often measured by observing motion of an isolated particle in a container of fluid. Use the results of §2.8 to assess the error induced by wall effects for a sphere of radius a falling along the centerline of a container of rectangular cross-section $(L \times W)$ with $L \gg W \gg a$.

5 The trajectories for two equal spheres interacting in a simple shear flow in Fig. 2.13 pertain to calculations with exact hydrodynamics. To assess the effect of detailed near-field interactions integrate the trajectory equations (2.9.18) with the far-field forms (2.9.8) for A and B. Since $A \neq 1$ at contact, explicitly exclude overlap by maintaining $r=2a$ as long as $r\,\mathrm{d}r/\mathrm{d}t<0$. Confine the calculation to pairs in the plane of shear, $R_3=0$, and compare the results for $R_2=0, 1, 2, 3$ with those in Fig. 2.13. Note that (2.9.18) are best integrated as $\mathrm{d}r_1/\mathrm{d}t$, $\mathrm{d}r_2/\mathrm{d}t$, and $\mathrm{d}r_3/\mathrm{d}t$.

3

BROWNIAN MOTION

3.1 Introduction

Optical microscopic observations of small particles dispersed in water reveal a constant state of random motion. The discovery of this phenomenon is now attributed to Robert Brown, a botanist, although other publications predate his descriptions of 1828 and 1829. While Brown correctly attributed the motion to the molecular nature of matter, controversy persisted until the experiments of Gouy in 1888 ruled out extraneous causes such as mechanical vibrations, convection currents, and illumination and focused attention on molecular agitation. As Perrin (1910) concluded, the particles seem to move independently with no effect of density or composition, although the amplitude of the motion is greater for smaller particles, with less viscous fluids, and at higher temperatures. The displacements are significant; for example, 0.2-μm spheres in water wander 10 μm from their starting point in a bit over 30 seconds. Gouy and Perrin both attributed the motion of the particle to incessant impacts of fluid molecules which impart kinetic energy equal to $\frac{3}{2}kT$, partitioned equally among the three translational degrees of freedom. The irregularity of the translational motion and the rapid damping of the random fluctuations by the viscous fluid, however, confounded early attempts to measure this kinetic energy by calculating the instantaneous velocity from the observed trajectory. This failure to verify directly the origin of Brownian motion led to theoretical treatments appropriate for the longer diffusion time scale.

The literature on Brownian motion extends from early descriptions of microscopic observations through elegant analyses of stochastic equations. Perrin (1910) and Nelson (1967) recount the history of the subject. Einstein's work (1906) illustrates the early approach, while the papers assembled by Wax (1954) along with the book by Nelson (1967) and the review of Pomeau & Resibois (1975) contain more rigorous mathematical analyses. Here we attempt to develop an appreciation for the physics by following a relatively simple mathematical path.

In this chapter we focus on events on the diffusion time scale, first for isolated spheres and then with pair interactions. Two parallel paths are necessary. In one approach the mean square displacements of the individual particles are calculated as functions of time by balancing the random Brownian forces against the viscous drag. The second approach leads to a diffusion equation governing the probability of finding the particles in a particular configuration. Comparison of the two relates the diffusion coefficients in the latter to the hydrodynamic mobilities of the individual particles in the former. Supplementary sections describe the use of photon correlation spectroscopy to measure diffusion coefficients in dilute dispersions and the generalization of the basic treatment to simulate the dynamics of Brownian particles subject to interparticle forces in various (applied) flow fields. The analyses pertain to homogeneous suspensions, i.e. without macroscopic concentration gradients. Chapter 13 addresses the related but more difficult problem of gradient diffusion.

3.2 The Langevin equation

The motion of a neutrally buoyant sphere of radius a, in a fluid of viscosity μ and density ρ, and subject to Brownian motion as well as viscous and inertial forces, is governed by a momentum balance. For motions of small amplitude ($\leqslant a$) on time scales long relative to viscous relaxations ($\rho a^2/\mu$), the Stokes equations apply. Consequently, the pseudo-steady Stokes drag characterizes the viscous response of the fluid as shown in §2.7. Then the viscous drag and particle inertia must balance the fluctuating Brownian force $\mathbf{f}(t)$ as

$$m\frac{\mathrm{d}^2\mathbf{x}}{\mathrm{d}t^2}+6\pi\mu a\frac{\mathrm{d}\mathbf{x}}{\mathrm{d}t}=\mathbf{f}(t), \qquad (3.2.1)$$

with $\mathbf{x}(t)$ the position of the sphere center and $m=4\pi a^3\rho/3$. The condition

$$\mathbf{x}=0 \qquad \text{at} \qquad t=0$$

fixes the starting point for the trajectory; setting

$$\frac{\mathrm{d}\mathbf{x}}{\mathrm{d}t}=0 \qquad \text{at} \qquad t=-\infty \qquad (3.2.2)$$

assures time to reach thermodynamic equilibrium by $t=0$ and thereby eliminates any effect of the initial velocity. This formulation implicitly separates the forces that the fluid molecules exert on the particle into rapid fluctuations $\mathbf{f}(t)$, on time scales corresponding to molecular motions ($\approx 10^{-13}$ s for water), and a much slower viscous drag.

Two assumptions suffice to characterize $\mathbf{f}(t)$ (e.g., Brenner, 1967; Hinch, 1975; McQuarrie, 1976, pp. 452–6). First, the Brownian forces are taken to be random in direction and magnitude and uncorrelated on the time scale of particle motion. These conditions can be expressed mathematically through the ensemble average defined by

$$\langle \mathbf{f}(t) \rangle = \frac{1}{N} \sum_{i=1}^{N} \mathbf{f}^{(i)}(t) \tag{3.2.3}$$

with $\mathbf{f}^{(i)}$ the value in the ith realization of a process having identical initial conditions. Then random, uncorrelated forces satisfy

$$\begin{aligned} &\langle \mathbf{f}(t) \rangle = \mathbf{0}, \\ &\langle \mathbf{f}(t)\mathbf{f}(t+\tau) \rangle = \mathbf{F}\delta(\tau). \end{aligned} \tag{3.2.4}$$

Here $\delta(\tau)$ is the Dirac delta function with the properties $\delta(\tau)=0, \tau \neq 0$, and $\int_{-\infty}^{\infty} \delta(\tau)\,\mathrm{d}\tau = 1$ (Lighthill, 1968, pp. 10–14). In this approach, the unknown constant tensor $\mathbf{F}$ must be determined from the second assumption that, at equilibrium, kinetic energy is partitioned equally among the three translational modes of the particle so that

$$\tfrac{1}{2} m \left\langle \frac{\mathrm{d}\mathbf{x}}{\mathrm{d}t} \frac{\mathrm{d}\mathbf{x}}{\mathrm{d}t} \right\rangle = \tfrac{1}{2} kT \boldsymbol{\delta}. \tag{3.2.5}$$

Integration of (3.2.1) and application of the initial condition (3.2.2) leads to

$$\frac{\mathrm{d}\mathbf{x}}{\mathrm{d}t} = \frac{1}{m} \int_{-\infty}^{t} \mathbf{f}(t') \exp(-6\pi\mu a(t-t')/m)\,\mathrm{d}t'. \tag{3.2.6}$$

This solution demonstrates that the particle has a short memory. The velocity autocorrelation function

$$\begin{aligned} \mathbf{R}(\tau) &= \left\langle \frac{\mathrm{d}\mathbf{x}}{\mathrm{d}t}(t) \frac{\mathrm{d}\mathbf{x}}{\mathrm{d}t}(t+\tau) \right\rangle \\ &= \frac{\mathbf{F}}{12\pi\mu a m} \exp\left(-\frac{6\pi\mu a}{m} \tau \right) \end{aligned} \tag{3.2.7}$$

makes this point even more clearly. The simple exponential form corresponds to that for a sphere released with a specified velocity at $t=0$ and no subsequent forcing. Thus the energy imparted to a particle by each thermal impulse decays on the viscous time scale $m/6\pi\mu a = 2\rho a^2/9\mu$ ($\approx 10^{-9}$ s for a neutrally buoyant 0.1 μm sphere in water) independently of the random forcing during the decay process.

Comparison of (3.2.7) for $\tau = 0$ and (3.2.5) establishes

$$\mathbf{F} = 12\pi\mu akT\boldsymbol{\delta}, \tag{3.2.8}$$

relating the strengths of the random Brownian fluctuations to the steady frictional forces that dissipate the energy. Both originate with interaction between the particles and the solvent molecules, but differ substantially in time scales. Clearly, higher viscosities, implying more efficient momentum transfer between molecules of the fluid, correlate with stronger Brownian forces. This consequence is obvious from the kinetic energy alone, since maintaining the magnitude of the fluctuating velocities requires stronger forces in a more viscous fluid.

Since detection of processes on the nanosecond time scale is difficult, the principal observable consequence of Brownian motion is displacement rather than velocity. The time rate of change of the variance in position follows from the velocity autocorrelation function as

$$\tfrac{1}{2}\frac{\mathrm{d}}{\mathrm{d}t}\langle \mathbf{x}(t)\mathbf{x}(t)\rangle = \int_0^t \mathbf{R}(\tau)\,\mathrm{d}\tau. \tag{3.2.9}$$

At times long with respect to the viscous relaxations, i.e., $t \gg \rho a^2/\mu$,

$$\frac{1}{2}\langle \mathbf{x}(t)\mathbf{x}(t)\rangle = \frac{kT}{6\pi\mu a}\boldsymbol{\delta}t, \tag{3.2.10}$$

indicating that the root-mean-square displacement $\langle \mathbf{x}\cdot\mathbf{x}\rangle^{1/2}$ becomes comparable to the sphere radius at times on the order of $6\pi\mu a^3/kT$ ($\approx 10^{-3}$ s for a 0.1 μm sphere in water).

3.3 Brownian motion and diffusion

The preceding analysis involved integration of the equation of motion for a single particle to obtain its position as a function of time. In the end, however, only statistical information, e.g. the mean-square displacement, survived because of the stochastic nature of the Brownian forces. Alternatively, we can couch the treatment in statistical terms from the outset, in essence viewing the displacements of Brownian particles as a diffusion process. This complementary approach serves two purposes: (i) to demonstrate explicitly the link between Brownian motion and diffusion and (ii) to provide the basis for subsequent analyses of a variety of dynamical processes.

The first step is to represent the results described above in terms of the transition probability $p(\Delta\mathbf{x}, \Delta t)$ that a particle is displaced $\Delta\mathbf{x}$ in a time interval $\Delta\mathbf{t}$. If the transition probability is normalized such that

$$\int p(\Delta\mathbf{x}, \Delta t)\,\mathrm{d}\Delta\mathbf{x} = 1, \tag{3.3.1}$$

and Δt is long relative to the viscous relaxation time, then (3.2.10) establishes

$$\langle \Delta\mathbf{x}\Delta\mathbf{x} \rangle = \int p(\Delta\mathbf{x}, \Delta t)\,\Delta\mathbf{x}\Delta\mathbf{x}\,\mathrm{d}\Delta\mathbf{x} = \frac{kT}{3\pi\mu a}\boldsymbol{\delta}\,\Delta t. \tag{3.3.2}$$

If $n(\mathbf{x}, t)\,\mathrm{d}\mathbf{x}$ is the probability of finding a sphere within $\mathbf{x}$ to $\mathbf{x}+\mathrm{d}\mathbf{x}$ at time t, then (McQuarrie, 1976, §§20–2)

$$n(\mathbf{x}, t+\Delta t) = \int n(\mathbf{x}-\Delta\mathbf{x}, t)\,p(\Delta\mathbf{x}, \Delta\mathbf{t})\,\mathrm{d}\Delta\mathbf{x}, \tag{3.3.3}$$

and, from Taylor series expansions,

$$\begin{aligned} n(\mathbf{x}, t+\Delta t) &= n(\mathbf{x}, t) + \Delta t\frac{\partial n}{\partial t} + \ldots, \\ n(\mathbf{x}+\Delta\mathbf{x}, t) &= n(\mathbf{x}, t) + \Delta\mathbf{x}\cdot\nabla n + \tfrac{1}{2}\Delta\mathbf{x}\Delta\mathbf{x} : \nabla\nabla n + \ldots. \end{aligned} \tag{3.3.4}$$

Combination of these yields

$$\Delta t\frac{\partial n}{\partial t} = \int \Delta\mathbf{x}\,p(\Delta\mathbf{x}, \Delta t)\,\mathrm{d}\Delta\mathbf{x}\cdot\nabla n + \tfrac{1}{2}\int \Delta\mathbf{x}\,\Delta\mathbf{x}\,p(\Delta\mathbf{x}, \Delta t)\,\mathrm{d}\Delta\mathbf{x} : \nabla\nabla n + \ldots. \tag{3.3.5}$$

The first integral vanishes on symmetry grounds, since $p(-\Delta\mathbf{x}, \Delta t) = p(\Delta\mathbf{x}, \Delta t)$; with (3.3.2) the remainder establishes

$$\frac{\partial n}{\partial t} = \frac{kT}{6\pi\mu a}\nabla^2 n. \tag{3.3.6}$$

Since (3.3.6) describes transient diffusion, we have the Stokes–Einstein result

$$D_0 = \frac{kT}{6\pi\mu a} \tag{3.3.7}$$

for the diffusivity of an isolated sphere.

If one solves (3.3.6) for a particle initially at the origin, i.e.,

$$n(\mathbf{x},0)=\delta(\mathbf{x}),$$

then the solution

$$n(\mathbf{x},t)=\frac{\exp(-x^2/4D_0t)}{(4\pi D_0 t)^{3/2}}, \tag{3.3.8}$$

with $x^2=\mathbf{x}\cdot\mathbf{x}$, predicts a Gaussian distribution of displacements about the initial position, with variance

$$\langle\mathbf{xx}\rangle=2D_0t\boldsymbol{\delta} \tag{3.3.9}$$

in accord with (3.2.10). Thus, for this situation, $n(\mathbf{x},t)=p(\mathbf{x},t)$.

The time dependence of the variance in position (3.2.10) formed the principal result of the original Langevin analysis. Experiments with the ultramicroscope reported by Perrin (1910) confirmed these predictions over a wide range of conditions. Such microscopic observations measure the probability of a displacement x_i along a particular coordinate axis,

$$\begin{aligned} f(x_i,t)&=\int_{-\infty}^{\infty}\int_{-\infty}^{\infty} n(\mathbf{x},t)\,\mathrm{d}x_j\,\mathrm{d}x_k \qquad j\neq k\neq i \\ &=\frac{\exp(-x_i^2/4D_0t)}{(4\pi D_0 t)^{1/2}}. \end{aligned} \tag{3.3.10}$$

Results for polystyrene latices in water (Fig. 3.1) demonstrate the displacements to be Gaussian distributed as expected.

Despite evidence for the validity of this description of the diffusion process, the theory in fact errs at short time scales. As discussed in §2.7, the pseudo-steady form for the viscous drag requires the vorticity of the fluid to diffuse faster than the particle loses inertia. However, for neutrally buoyant particles the time scale for both processes is $\rho a^2/\mu$; so the theory should account for the unsteadiness of the fluid motion. In addition, at very short times, $\approx a/c \ll \rho a^2/\mu$, compressibility redistributes the initial thermal impulse received by the particle over the particle plus the fluid added mass of $2\pi a^3\rho/3$. Inclusion of these two effects (Hauge & Martin-Löf, 1973; Chow & Hermans, 1973; Hinch, 1975; Zwanzig & Bixon, 1975) yields a velocity autocorrelation function with the correct short-time behavior and a long tail decaying as $\tau^{-3/2}$ in accord with numerical simulations, rather than exponentially as predicted above.

Fortunately, the low-frequency components, representing almost steady motion, provide the largest net displacements and dominate the diffusion process so (3.2.7) and (3.2.10) give the right result. The physical explanation stems from the relation

$$\lim_{t\to\infty} \tfrac{1}{2}\frac{\mathrm{d}}{\mathrm{d}t}\langle \mathbf{x}(t)\mathbf{x}(t)\rangle = \int_0^\infty \mathbf{R}(\tau)\,\mathrm{d}\tau = 2\pi\hat{\mathbf{R}}(0), \tag{3.3.11}$$

where $\hat{\mathbf{R}}(\omega)$ is the Fourier transform of $\mathbf{R}(\tau)$ (Batchelor, 1976). The Fourier transforms of the correct autocorrelation function and the pseudosteady approximation (3.3.7) share a common zero-frequency limit (Fig. 3.2), so substitution of the former into (3.2.9) generates the same result for the diffusion limit as does the exponential form. The remainder of the spectrum, while interesting, is superfluous. This fact can be demonstrated simply by

Fig. 3.1. Distribution of displacements due to translational Brownian motion of an isolated 1.19 μm diameter sphere: ● ▼, observed by Vadas *et al.* (1976; ——, predicted from (3.3.10), with $D_0 = 4.12 \times 10^{-13}\ \mathrm{m^2/s}$ and $t = 2.5\ \mathrm{s}$.

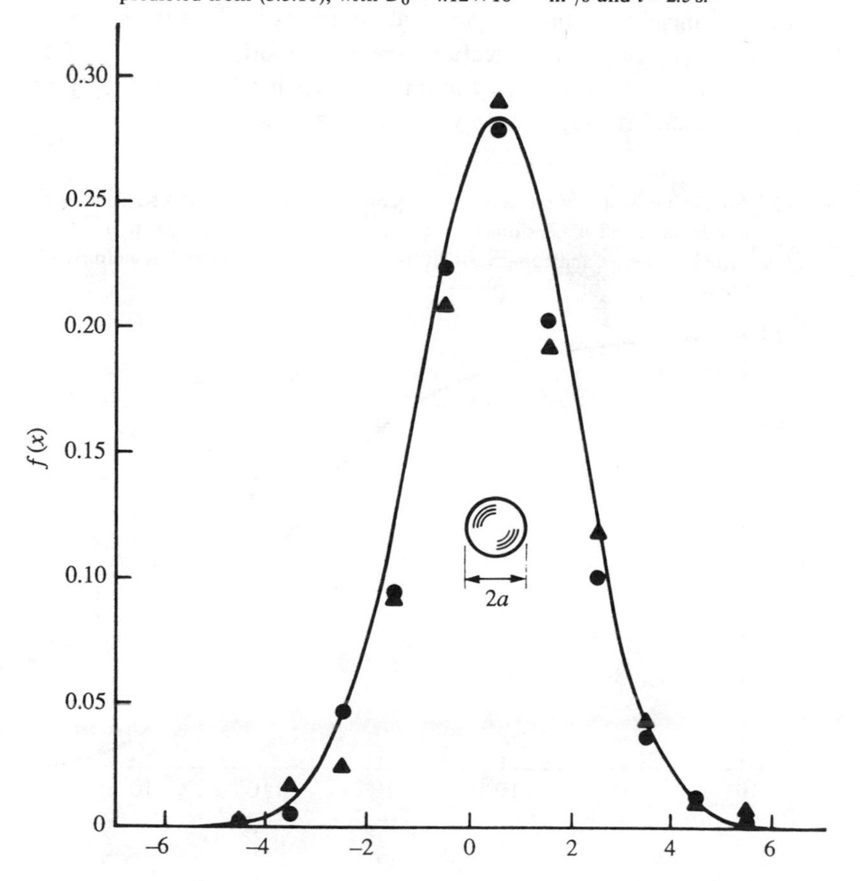

comparing the total displacement expected from an impulsive force accounting for both fluid and particle inertia through (2.7.11) with that obtained from (3.2.6), which recognizes particle inertia alone (Problem 3.3).

These results indicate clearly that the pseudo-steady hydrodynamic mobilities determine the consequences of the Brownian motion of colloidal particles on the diffusion time scale. Hence, for interacting particles or bodies of arbitrary shape with characteristic size l, predictions of the increase in the mean-square displacement or orientation based on these mobilities should be accurate for $t \gg \rho l^2/\mu$ (Brenner, 1967). We capitalize on this result in later sections without further proof.

3.4 Measurement by photon correlation spectroscopy

The Brownian motion of colloidal particles causes light scattered from a dispersion to fluctuate with time. Detection and autocorrelation of these fluctuating intensities is a rapid and accurate method for measuring diffusion coefficients in suspensions and solutions (e.g. Berne & Pecora, 1976; Dahneke, 1983). This section presents a brief treatment of the application to dilute dispersions. Later (§13.4–6) we delve more deeply into the technique and its application to concentrated systems.

Fig.3.2. Fourier transform of the velocity autocorrelation function for a single sphere as a function of the dimensionless frequency (Russel, 1981): ——, exact result including inertia and compressibility; - - -, approximation with pseudo-steady Stokes drag.

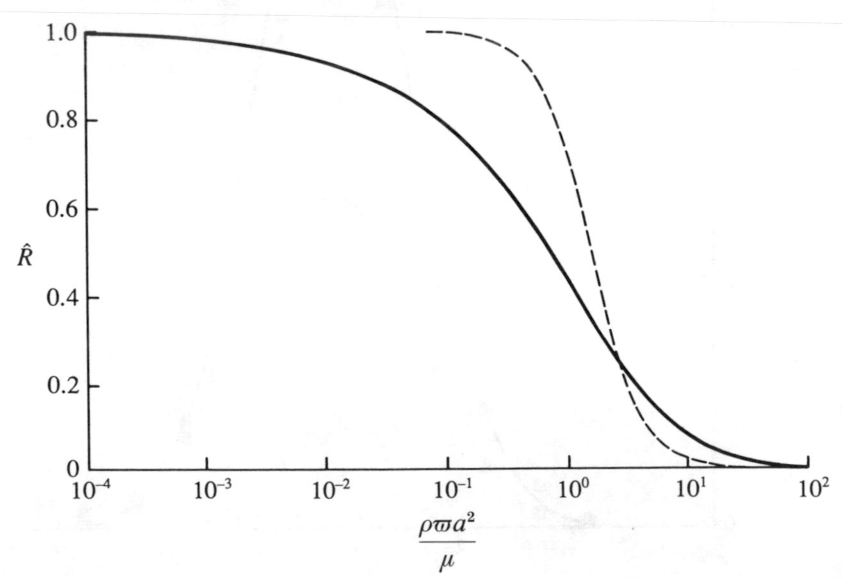

In the experiment a volume containing N particles is illuminated by laser light of wavelength λ (Fig. 3.3). The electric field induces a dipole in each particle with magnitude proportional to the refractive index difference between the fluid and the particle and with frequency and phase corresponding to the local field, e.g.

$$\mathbf{p}_j = 4\pi n^2 a^3 \frac{n^2 - \bar{n}^2}{n^2 + 2\bar{n}^2} \mathbf{E}_0 \exp[i(\omega t - \mathbf{q}_i \cdot \mathbf{x}_j)] \tag{3.4.1}$$

for a sphere of radius $a \ll \lambda$ and refractive index $\bar{n}$ at position $\mathbf{x}_j$ in a fluid with refractive index n. Here $\mathbf{E}_0$ indicates the magnitude and polarization, ω the frequency, and $\mathbf{q}_i$ the wave vector (with magnitude $4\pi/\lambda$) for the incident field. The scattering consists of the superposition of the fields generated by these oscillating dipoles. The fields interfere constructively or destructively depending on the phases of the dipoles, $-\mathbf{q}_i \cdot \mathbf{x}_j$, and the additional phase shifts introduced by propagation to the detector, $-\mathbf{q}_s \cdot (\mathbf{r} - \mathbf{x}_j)$. Hence the magnitude of the scattered field, expressed as

$$E(q,t) = 4\pi n^2 a^3 \frac{n^2 - \bar{n}^2}{n^2 + 2\bar{n}^2} E_0 \exp[i(\omega t - \mathbf{q}_s \cdot \mathbf{r})] \sum_{j=1}^{N} \exp[i\mathbf{q} \cdot \mathbf{x}_j(t)], \tag{3.4.2}$$

with $\mathbf{q} = \mathbf{q}_s - \mathbf{q}_i$ and $q = (4\pi n/\lambda)\sin(\theta/2)$, depends on the relative positions of the particles. The coherence of laser light preserves phase relations between the light scattered by different particles, so the scattered beam resembles a random diffraction or speckle pattern which fluctuates with time as the particles execute Brownian motion.

Fig. 3.3. Schematic of light-scattering experiment used for photon correlation spectroscopy.

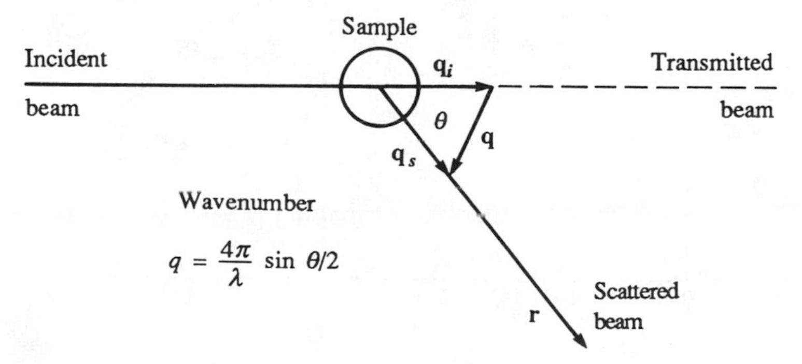

The autocorrelation of these amplitudes represents the simplest characterization of such a randomly fluctuating quantity. For a dilute suspension this function is directly related to the positions and scattering intensities $I_j = 4\pi a^3(n^2 - \bar{n}^2)/(n^2 + 2\bar{n}^2)$ of the individual particles through

$$F_s(q,\tau) = \frac{\sum_{i=1}^{N} I_i^2 \langle \exp[i\mathbf{q}\cdot(\mathbf{x}_i(\tau) - \mathbf{x}_i(0))]\rangle}{\sum_{i=1}^{N} I_i^2}. \tag{3.4.3}$$

Fig. 3.4. (*a*) Autocorrelation function for dilute polystyrene latex spheres with a = 0.30 μm showing experimental points (□) and single exponential decay (—) expected from (3.4.6).
(*b*) Confirmation of the q^2 dependence of the decay constant for $2a = 0.357$ μm (●) and 0.091 μm (■) polystyrene latices (Lee, Tscharnuter, & Chu, 1972).

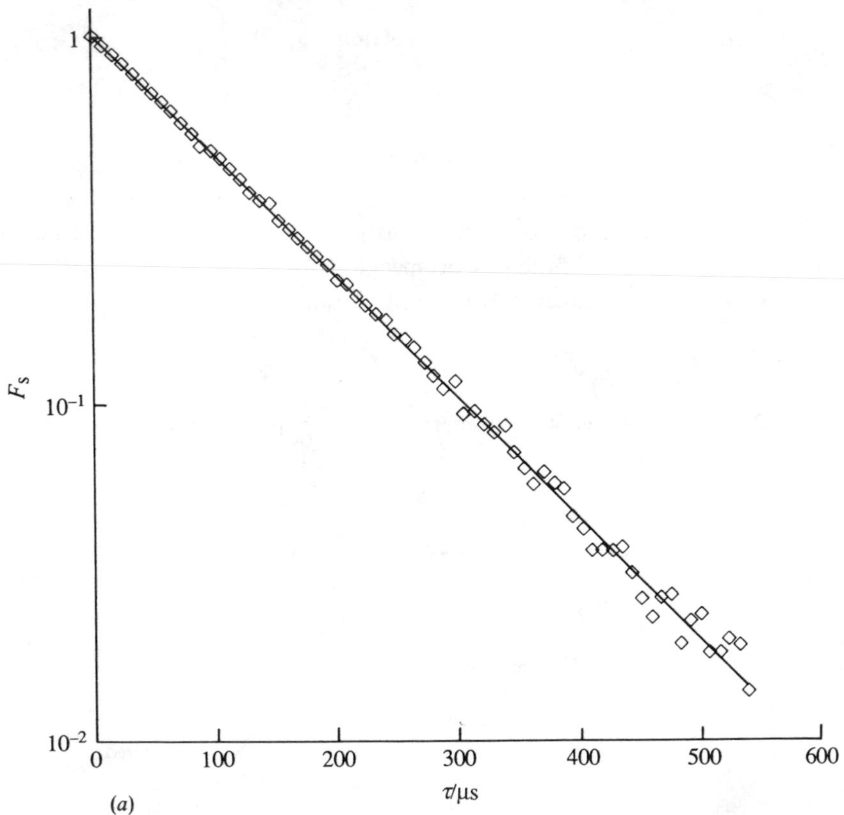

For identical spheres the summation can be converted to an ensemble average, i.e.,

$$\frac{1}{N}\sum_{j=1}^{N}(\quad)=\int p(\mathbf{x},\tau)(\quad)d\mathbf{x}, \tag{3.4.4}$$

with the probability of a particle changing its position by $\mathbf{x}=\mathbf{x}(\tau)-\mathbf{x}(0)$ during time τ specified by (3.3.8) as

$$p(\mathbf{x},\tau)=(4\pi D_0\tau)^{-3/2}\exp\left(-\frac{x^2}{4D_0\tau}\right). \tag{3.4.5}$$

Substitution and integration yields

$$F_s(q,\tau)=\exp(-\Gamma\tau), \qquad \text{with} \qquad \Gamma=q^2D_0, \tag{3.4.6}$$

so the exponential decay of the autocorrelation function yields the diffusion coefficient of the particles.

Numerous experiments with a wide range of suspended particles have demonstrated the validity of this result. The example depicted in Fig. 3.4 illustrates, for a dilute polystyrene latex, the exponential decay of the autocorrelation function at a single angle, compared with that expected from independent measurement of the size, and the expected q^2 dependence

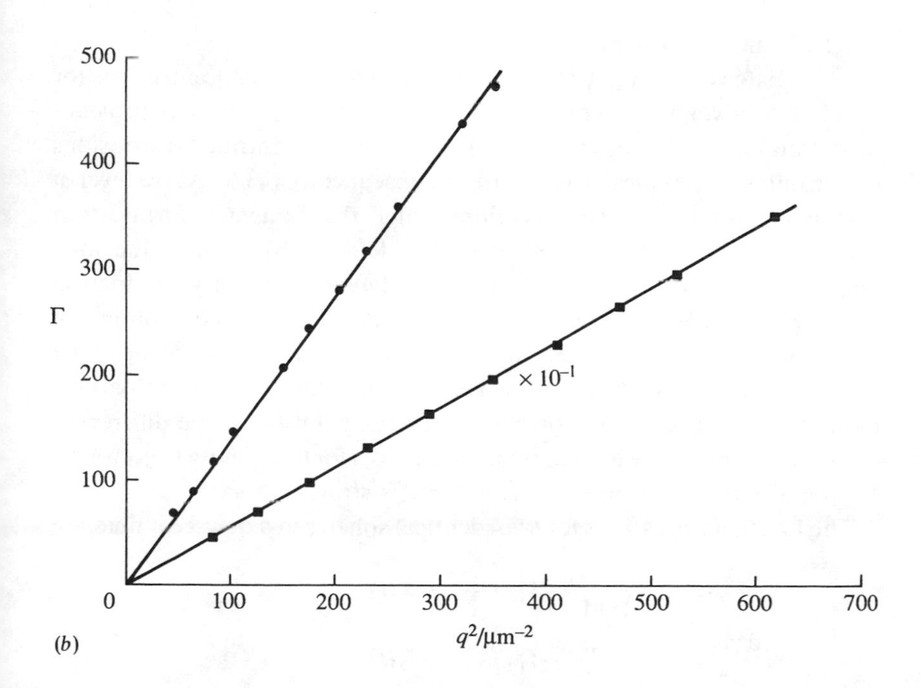

of the slope. For monodisperse spheres in the range of 1 nm to 1 μm, accurate diffusivities can be obtained within minutes.

For polydisperse suspensions, the autocorrelation function comprises a sum of exponentials weighted by the characteristic intensity and the number density n_i of the individual sizes as

$$F_s = \frac{\Sigma n_i I_i^2 \exp(-q^2 D_i \tau)}{\Sigma n_i I_i^2}. \tag{3.4.7}$$

The most common means of analyzing the data is the method of moments or cumulants (Koppel, 1972), which amounts to expanding as

$$\ln F_s = -\Gamma\tau + \ldots \tag{3.4.8}$$

Then the first cumulant, Γ, provides a weighted average of the diffusion coefficient as

$$\Gamma = \frac{\Sigma n_i I_i^2 D_i}{\Sigma n_i I_i^2} q^2. \tag{3.4.9}$$

The ratio of higher-order cumulants to Γ indicates the polydispersity, but quantitative interpretation is difficult except for relatively monodisperse samples. We will not pursue the question further here, but will use the first cumulant in the later discussion of flocculation kinetics (§8.7).

3.5 Pair interactions

Extension of the treatment described above to concentrations for which hydrodynamic interactions become significant eventually provides the means for simulating a variety of processes in concentrated suspensions on the diffusion time scale, as described subsequently (§3.6). At the level of pair interactions, two complications enter the Langevin formulation (Ermak & McCammon, 1978; Hess & Klein, 1978), even in the low-frequency limit with pseudo-steady hydrodynamics. Clearly the friction coefficients – now configuration dependent – couple the motions of interacting particles. Less obvious is the coupling that arises between the fluctuating Brownian forces at separations on the order of the particle radius. The analysis below for pair interactions illustrates the differences from the single particle limit and forms the basis for calculating trajectories. A generalization to N interacting spheres is straightforward.

The Langevin equations for two identical spheres in a quiescent fluid are

$$\begin{aligned} m\frac{d^2\mathbf{x}_1}{dt^2} &= -\mathbf{f}_{11}\cdot\frac{d\mathbf{x}_1}{dt} - \mathbf{f}_{12}\cdot\frac{d\mathbf{x}_2}{dt} + \mathbf{f}_1(t) \\ m\frac{d^2\mathbf{x}_2}{dt^2} &= -\mathbf{f}_{12}\cdot\frac{d\mathbf{x}_1}{dt} - \mathbf{f}_{11}\cdot\frac{d\mathbf{x}_2}{dt} + \mathbf{f}_2(t). \end{aligned} \tag{3.5.1}$$

The friction tensors, which are related to the mobilities defined in §2.8 through

$$\mathbf{f}_{11}=(\boldsymbol{\omega}_{11}-\boldsymbol{\omega}_{12}\cdot\boldsymbol{\omega}_{11}^{-1}\cdot\boldsymbol{\omega}_{12})^{-1}$$

$$\mathbf{f}_{12}=(\boldsymbol{\omega}_{12}-\boldsymbol{\omega}_{11}\cdot\boldsymbol{\omega}_{12}^{-1}\cdot\boldsymbol{\omega}_{11})^{-1}, \tag{3.5.2}$$

depend on the relative position $\mathbf{r}$ and thereby couple the motions. Only for well-separated spheres, where

$$\mathbf{f}_{11}\sim 6\pi\mu a\boldsymbol{\delta}$$
$$\mathbf{f}_{12}\sim \mathbf{0},$$

do the particles move independently.

Conversion to center of mass coordinates, $\mathbf{x}=\frac{1}{2}(\mathbf{x}_1+\mathbf{x}_2)$ and $\mathbf{r}=(\mathbf{x}_2-\mathbf{x}_1)$, yields

$$m\frac{\mathrm{d}^2\mathbf{x}}{\mathrm{d}t^2}=-(\mathbf{f}_{11}+\mathbf{f}_{12})\cdot\frac{\mathrm{d}\mathbf{x}}{\mathrm{d}t}+\tfrac{1}{2}(\mathbf{f}_1+\mathbf{f}_2)$$

$$m\frac{\mathrm{d}^2\mathbf{r}}{\mathrm{d}t^2}=-(\mathbf{f}_{11}-\mathbf{f}_{12})\cdot\frac{\mathrm{d}\mathbf{r}}{\mathrm{d}t}+\mathbf{f}_2-\mathbf{f}_1, \tag{3.5.3}$$

with

$$\mathbf{x}=\mathbf{0}$$
$$\mathbf{r}=\mathbf{r}_0 \qquad \text{at} \qquad t=0$$

and

$$\frac{\mathrm{d}\mathbf{x}}{\mathrm{d}t}=\frac{\mathrm{d}\mathbf{r}}{\mathrm{d}t}=\mathbf{0} \qquad \text{at} \qquad t=-\infty.$$

As with the single sphere, the Brownian forces remain random,

$$\langle\mathbf{f}_i\rangle=\mathbf{0}, \tag{3.5.4}$$

and uncorrelated on the time scales of interest,

$$\langle\mathbf{f}_i(t)\mathbf{f}_j(t+\tau)\rangle=\mathbf{F}_{ij}\delta(\tau). \tag{3.5.5}$$

The equipartition of kinetic energy among the translational modes for each particle, i.e.

$$\tfrac{1}{2}m\left\langle\frac{\mathrm{d}\mathbf{x}_i}{\mathrm{d}t}\frac{\mathrm{d}\mathbf{x}_j}{\mathrm{d}t}\right\rangle=\tfrac{1}{2}kT\,\boldsymbol{\delta}\,\delta_{ij}, \tag{3.5.6}$$

leads to

$$\tfrac{1}{2}(2m)\left\langle\frac{\mathbf{dx}}{\mathrm{d}t}\frac{\mathbf{dx}}{\mathrm{d}t}\right\rangle=\tfrac{1}{2}kT\,\boldsymbol{\delta},$$

$$\tfrac{1}{2}m\left\langle\frac{\mathbf{dr}}{\mathrm{d}t}\frac{\mathbf{dr}}{\mathrm{d}t}\right\rangle=kT\,\boldsymbol{\delta}, \tag{3.5.7}$$

$$\left\langle\frac{\mathbf{dx}}{\mathrm{d}t}\frac{\mathbf{dr}}{\mathrm{d}t}\right\rangle=\mathbf{0}.$$

These reflect the translational kinetic energy of a pair of mass $2m$, the vibrational and rotational energy of the pair, and the decoupling of translational and relative motions, respectively.

Equations (3.5.3) can be integrated analytically provided the relative position $\mathbf{r}$ does not change significantly on the viscous time scale, so that the friction tensors are constant. Convenient representation of the solutions is possible with the matrix exponential (Amundson, 1966, §7.10)

$$\exp(\mathbf{A}t)\equiv\sum_{n=0}^{\infty}\frac{\mathbf{A}^n t^n}{n!}, \tag{3.5.8}$$

which has properties similar to the scalar analog. For example,

$$\frac{\mathrm{d}}{\mathrm{d}t}\exp(\mathbf{A}t)=\exp(\mathbf{A}t)\cdot\mathbf{A}. \tag{3.5.9}$$

Explicit evaluation through summation of the infinite series would be tedious but proves unnecessary here. With the matrix exponential one can construct integrating factors and express the solutions in a form analogous to (3.2.6).

Evaluation of the kinetic energies and velocity autocorrelation functions from the formal solution establishes

$$\mathbf{F}_{ij}=2kT\mathbf{f}_{ij}(\mathbf{r}_0),$$

$$\mathbf{R}_x(\tau)=\frac{kT}{m}\exp\left(-\mathbf{f}_x\frac{\tau}{m}\right), \tag{3.5.10}$$

$$\mathbf{R}_r(\tau)=\frac{4kT}{m}\exp\left(-\mathbf{f}_r\frac{\tau}{m}\right),$$

with

$$\mathbf{f}_x=\mathbf{f}_{11}(\mathbf{r}_0)+\mathbf{f}_{12}(\mathbf{r}_0)$$
$$\mathbf{f}_r=\mathbf{f}_{11}(\mathbf{r}_0)-\mathbf{f}_{12}(\mathbf{r}_0).$$

The displacements at times long with respect to the viscous relaxation times then follow from (3.2.9) and (3.5.10) as

$$\begin{aligned} \tfrac{1}{2}\langle \mathbf{x}(t)\mathbf{x}(t)\rangle &= 2kT\mathbf{f}_x^{-1}t, \\ \tfrac{1}{2}\langle (\mathbf{r}(t)-\mathbf{r}_0)(\mathbf{r}(t)-\mathbf{r}_0)\rangle &= 2kT\mathbf{f}_r^{-1}t. \end{aligned} \tag{3.5.11}$$

From (3.5.11), one can readily verify that the change in the relative position during the viscous relaxation time, $2\rho a^2/9\mu$, is

$$\frac{\langle (\mathbf{r}-\mathbf{r}_0)\cdot(\mathbf{r}-\mathbf{r}_0)\rangle^{1/2}}{r_0-2a} \approx \left(\frac{2}{27\pi}\frac{\rho kT}{\mu^2 a}\right)^{1/2},$$

which remains small ($\leqslant 10^{-2}$ for $a > 1$ nm) for conditions of interest.

The combination of (3.5.5) and (3.5.10) represents a generalized fluctuation–dissipation theorem, revealing an interesting facet of the Brownian forces in a coupled system. Since

$$\langle \mathbf{f}_i(t)\mathbf{f}_j(t+\tau)\rangle = 2kT\mathbf{f}_{ij}(\mathbf{r}_0)\,\delta(\tau), \tag{3.5.12}$$

the forces on two interacting particles are coupled and their magnitudes depend on the relative position. The physical interpretation is straightforward. As the particles approach one another, the viscous resistance to motion, particularly along the line of centers, increases; hence, the magnitudes of the Brownian forces must increase to maintain the prescribed kinetic energy.

These results for the mean-square displacements, together with the arguments in §3.3, identify the diffusion coefficients for interacting spheres as

$$\begin{aligned} \mathbf{D} &= 2kT\mathbf{f}_r^{-1} \\ &= 2kT(\boldsymbol{\omega}_{11}-\boldsymbol{\omega}_{12}) \end{aligned} \tag{3.5.13}$$

for relative motion, and

$$\begin{aligned} \mathbf{D}_c &= 2kT\mathbf{f}_x^{-1} \\ &= 2kT(\boldsymbol{\omega}_{11}+\boldsymbol{\omega}_{12}) \end{aligned} \tag{3.5.14}$$

for motion of the center of mass. For two spheres in contact (Fig. 3.5), the relative diffusion tensor reduces to

$$\mathbf{D} = 0.401\frac{kT}{3\pi\mu a}\left(\boldsymbol{\delta}-\frac{\mathbf{rr}}{r^2}\right), \tag{3.5.15}$$

corresponding to rotational diffusion of the doublet. Since

$$\mathbf{D}_c = \frac{kT}{3\pi\mu a}\left[1.55\frac{\mathbf{rr}}{r^2}+1.38\left(\boldsymbol{\delta}-\frac{\mathbf{rr}}{r^2}\right)\right], \tag{3.5.16}$$

translational diffusion along the axis is slightly faster than the transverse process.

Microscopic observations of the dynamics of interacting latex spheres confirm to some extent the predictions of (3.5.11). For example, Table 3.1 compares the mean-square displacements, parallel,

$$\langle x_{\|}^2 \rangle = \left\langle \frac{\mathbf{r}}{r} \cdot \mathbf{xx} \cdot \frac{\mathbf{r}}{r} \right\rangle$$

and perpendicular,

$$\langle x_{\perp}^2 \rangle = \langle x^2 \rangle - \langle x_{\|}^2 \rangle,$$

to the line of centers observed for polystyrene latex doublets with those expected from (3.5.11) and (3.5.16). Figure 3.6 shows corresponding results for the rotational Brownian motion of a doublet. The prediction represented by the solid curve requires a modified form of (3.5.11) that constrains $\mathbf{r} \cdot \mathbf{r} = 4a^2$ (Problem 3.2).

This section addresses the Brownian motion of two interacting spheres on time scales long with respect to the viscous time $2\rho a^2/9\mu$ but too short for the relative separation to change significantly. The mean-square displacements (3.5.11) increase linearly with time as for isolated spheres, but hydrodynamic interactions cause the corresponding diffusion coefficients (3.5.13, 3.5.14) and Brownian forces (3.5.12) to vary with the relative separation. The limited observations available are in complete accord with the predictions.

Fig. 3.5. Coordinate system for doublet.

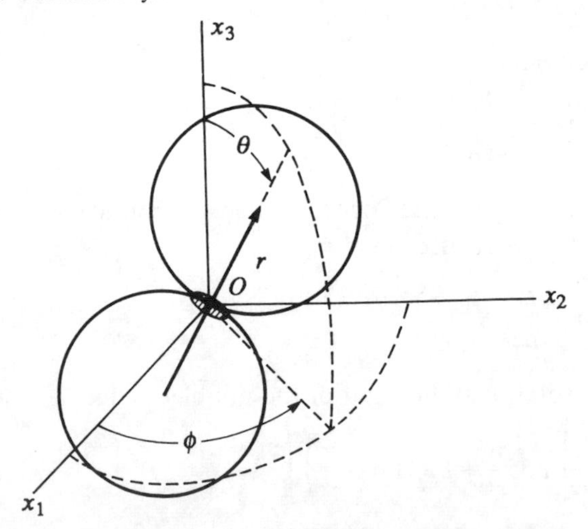

Table 3.1. *Mean-square displacements of the center of mass of doublets of latex spheres parallel* (| |) *and perpendicular* (⊥) *to the line of centers without flow* (*Vadas* et al., *1976*)

		$\langle x_{\parallel}^2\rangle/10^{-2}\,\mu m^2$		$\langle x_{\perp}^2\rangle/10^{-2}\,\mu m^2$	
$2a/\mu m$	t/s	Meas.	Calc.	Meas.	Calc.
1.19	0.060	3.81	3.80	3.57	3.40
2.07	0.10	3.82	3.66	3.21	3.26
	0.20	8.52	7.31	7.08	6.53

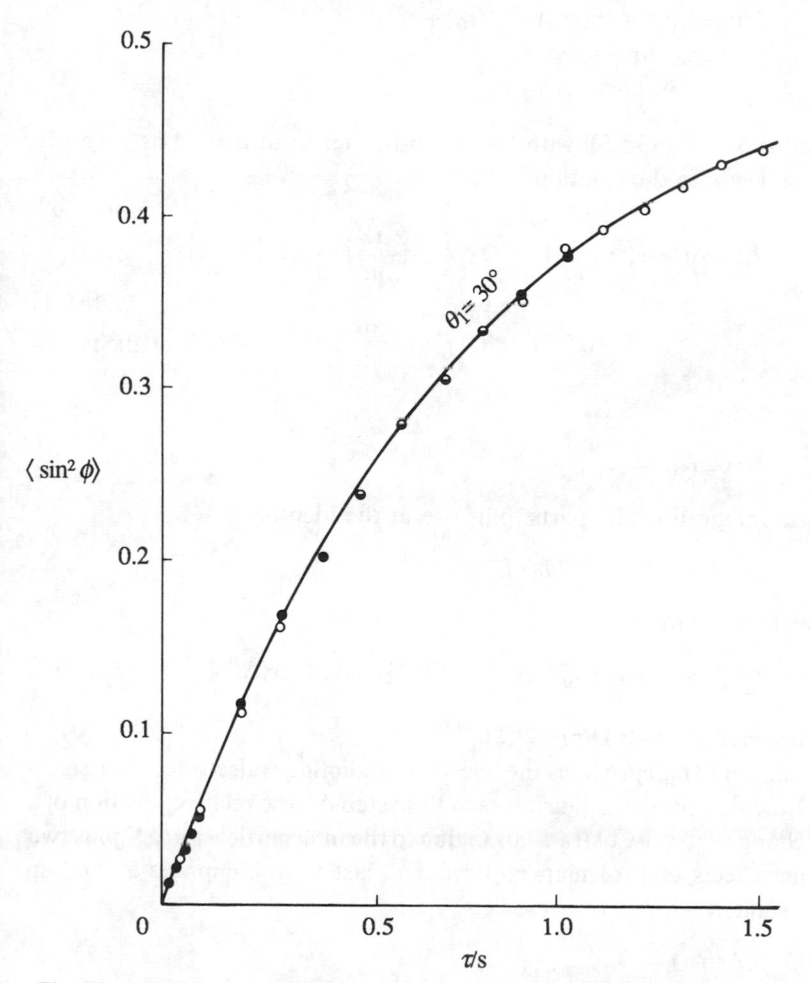

Fig. 3.6. Rotary Brownian motion of a doublet of 1.19 μm diameter spheres: ○ ●, observations of Vadas *et al.* (1976); ——, predictions from the analogue of (3.3.9).

3.6 Brownian dynamics

The preceding analysis of pair interactions is restricted to infinitesimal displacements by the assumption of constant friction tensors and includes neither interparticle forces nor macroscopic flow fields. Simulation of the motion of two or more spheres for longer times subject to these influences as well as Brownian motion, requires numerical integration of the equations. In the following we illustrate the approach for the relative motion of two spheres acted upon by forces $\mathbf{F}=\mathbf{F}_2-\mathbf{F}_1$.

Ermak & McCammon (1978) extended the time interval in the integration by correcting for the **r**-dependence of $\mathbf{f}_r$ via the Taylor series expansions

$$\mathbf{f}_r(\mathbf{r})=\mathbf{f}_r(\mathbf{r}_0)+(\mathbf{r}-\mathbf{r}_0)\cdot\nabla_r\mathbf{f}(\mathbf{r}_0)+\ldots$$
$$\equiv\mathbf{f}_0+(\mathbf{r}-\mathbf{r}_0)\cdot\nabla_r\mathbf{f}_0+\ldots$$
$$\mathbf{F}(\mathbf{r})=\mathbf{F}(\mathbf{r}_0)+\ldots.$$

Substitution into (3.5.3), with the second-order term treated as a known forcing, leads to the solution

$$\mathbf{r}(t+\Delta t)=\mathbf{r}(t)+\frac{1}{m}\int_t^{t+\Delta t}\exp\left(-\mathbf{f}_0\frac{t'}{m}\right)\cdot\int_t^{t'}\exp\left(\mathbf{f}_0\frac{t''}{m}\right)$$
$$\cdot\left[-(\mathbf{r}(t'')-\mathbf{r}(t))\cdot\nabla_r\mathbf{f}_0\cdot\frac{\mathrm{d}\mathbf{r}}{\mathrm{d}t}+\mathbf{F}(t'')+\mathbf{f}(t'')\right]\mathrm{d}t'\,\mathrm{d}t'', \tag{3.6.1}$$

with

$$\mathbf{f}(t)=\mathbf{f}_2(t)-\mathbf{f}_1(t).$$

Several integrations by parts, plus use of the identity

$$\nabla_r\cdot\mathbf{f}_0^{-1}=-\mathbf{f}_0^{-1}:\nabla_r\mathbf{f}_0\cdot\mathbf{f}_0^{-1},$$

convert (3.6.1) to

$$\mathbf{r}(t+\Delta t)=\mathbf{r}(t)+\mathbf{f}_0^{-1}\cdot\mathbf{F}\,\Delta t+\nabla_r\cdot\mathbf{D}(\mathbf{r})\,\Delta t+\Delta\mathbf{r}(\Delta t) \tag{3.6.2}$$

for $\Delta t\gg m\,|\mathbf{f}_0^{-1}|$ with $\mathbf{D}(\mathbf{r})=2kT\mathbf{f}_0^{-1}$.

Equation (3.6.2) provides the basis for predicting trajectories of interacting Brownian particles. During each time step Δt, the relative position of a pair changes because of translation due to the interparticle force **F** plus two distinct effects of Brownian motion. The last term comprises a random displacement with zero mean, i.e.

$$\langle\Delta\mathbf{r}(\Delta t)\rangle=\mathbf{0}, \tag{3.6.3}$$

but with a variance proportional to the time interval and the pair mobility as

$$\langle \Delta \mathbf{r} \, \Delta \mathbf{r} \rangle = 2\mathbf{D}(\mathbf{r}) \, \Delta t. \tag{3.6.4}$$

The third term corrects for the spatial dependence of the diffusivity, introducing a drift velocity that causes the spheres to move apart, i.e. towards the region of higher mobility. Thus a simulation simply marches forward in time, recalculating the interparticle force, diffusion coefficient, and random displacement for each new configuration.

A calculation must preserve small relative changes in the pair configuration, i.e.

$$\frac{\delta r}{r} \equiv \frac{[\mathbf{r}(t+\Delta t) - \mathbf{r}(t)] \cdot \mathbf{r}(t)}{\mathbf{r}(t) \cdot \mathbf{r}(t)} \leqslant 10^{-1} \frac{r-2a}{a}, \tag{3.6.5}$$

by adjusting the time step Δt. When $r - 2a \geqslant a$, the drift term becomes negligible, leaving from (3.6.2) with $\mathbf{F} = \mathbf{0}$

$$\frac{\delta r}{r} \approx \left(\frac{kT}{3\pi\mu a^3} \Delta t \right)^{1/2},$$

so that

$$\Delta t \leqslant 10^{-2} \frac{3\pi\mu a^3}{kT}$$

suffices. At small separations, however,

$$\mathbf{D} \sim \frac{kT}{3\pi\mu a} \left[2\left(\frac{r-2a}{a} \right) \frac{\mathbf{rr}}{r^2} + 0.401 \left(\boldsymbol{\delta} - \frac{\mathbf{rr}}{r^2} \right) \right],$$

so that

$$\frac{\mathbf{r}}{r} \cdot (\nabla \cdot \mathbf{D}) \sim 1.504 \frac{kT}{3\pi\mu a^2},$$

necessitating, according to (3.6.5), a smaller time step,

$$\Delta t \leqslant 10^{-2} \frac{3\pi\mu a^3}{kT} \left(\frac{r-2a}{a} \right).$$

Then the displacement due to the gradient in mobility compares with the random Brownian displacement. For convection-dominated flows, adjustment of the time step, of course, would be based on the shear rate.

Figure 3.7 illustrates a trajectory observed in shearing flows with negligible external or interparticle forces. Although Brownian motion is weak ($a^2\gamma/D_0 \approx 5.5$), the random motion can displace the pair from the initial trajectory into the region of closed orbits (Fig. 2.13). Then, after one or more revolutions, Brownian motion can separate the particles sufficiently for the shear to complete the process. Thus the situation $a^2\gamma/D_0 \gg 1$ is singular in the sense that very weak Brownian motion makes a qualitative difference in the behavior. This interesting phenomenon could, in principle, be simulated through (3.6.2).

3.7 Summary

This chapter examines the motion of particles subject to Brownian motion. Two important points emerge:

(i) Pseudo-steady hydrodynamic mobilities, as calculated from the Stokes equations, lead to incorrect predictions for the instantaneous velocities, but suffice to characterize the observable displacements.

(ii) Langevin equations incorporating pseudo-steady hydrodynamics accounting for pair interactions accurately determine the detailed dynamics on the diffusion time scale for both quiescent and flowing systems.

Together, these form the foundation for understanding the dynamics of dispersions. In later chapters we frequently describe suspension microstructure through conservation equations that treat Brownian motion as a

Fig. 3.7. Trajectory observed for the interaction between polystyrene latices ($a = 2.0\,\mu$m) in water in a linear shear flow with $\mathbf{u}_\infty = (0, 0, \gamma x_2)$ and $a^2\gamma/D_0 = 5.5$ (Takamura, Goldsmith & Mason, 1981). The circles represent projections of the relative center-to-center position onto the $x_2 - x_3$ plane. Since $x_1 > 0$, the projections can lie within the excluded sphere $x_2^2 + x_3^2 = 4a^2$.

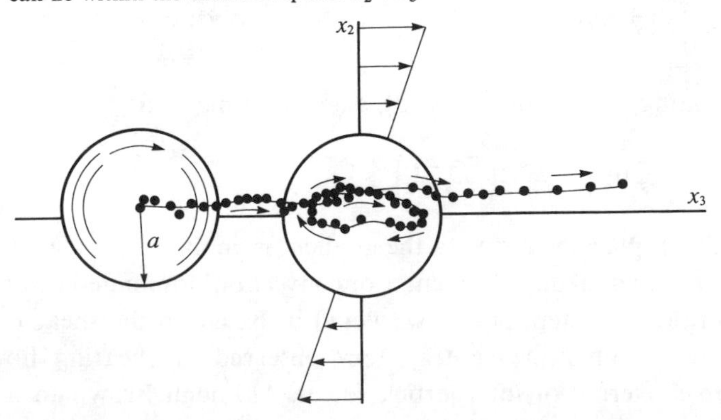

diffusional process characterized by diffusivities obtained directly from the appropriate hydrodynamic mobilities. However, the simulation of the dynamics through the integrated Langevin equations presents an equally powerful technique for dealing with many-body interactions in concentrated systems.

References

Amundson, N. R. (1966). *Mathematical Methods in Chemical Engineering: Matrices and Their Applications.* Prentice-Hall.

Batchelor, G. K. (1976). Developments in microhydrodynamics. In *Theoretical and Applied Mechanics* (ed. W. T. Koiter). North Holland, 33–55.

Berne, B. J. & Pecora, R. (1976). *Dynamic Light Scattering.* Wiley.

Brenner, H. (1967). Coupling between translational and rotational Brownian motions of rigid particles of arbitrary shape. Part II. General Theory. *J. Colloid Interface Sci.* **23**, 407–36.

Chow, T. S. & Hermans, J. J. (1973). Brownian motion of a spherical particle in a compressible fluid. *Physica* **65A**, 156–62.

Dahneke, B. E. (ed.) (1983). *Measurement of Suspended Particles by Quasi-Elastic Light Scattering.* Wiley–Interscience.

Einstein, A. (1906). On the theory of Brownian movement, in *Investigations on the Theory of Brownian Movement* (ed. R. Furth), Dover, 1956. A translation of a paper appearing in *Ann. d. Phys.* **19**, 371–81.

Ermak, D. L. & McCammon, J. A. (1978). Brownian dynamics with hydrodynamic interactions. *J. Chem. Phys.* **69**, 1352–60.

Hauge, E. H. & Martin-Löf, A. (1973). Fluctuating hydrodynamics and Brownian motion. *J. Stat. Phys.* **7**, 259–81.

Hess, W. & Klein, R. (1978). Dynamic properties of colloidal systems. 1. Derivation of stochastic transport equations. *Physica* **94A**, 71–90.

Hinch, E. J. (1975). Application of the Langevin equation to fluid suspensions. *J. Fluid Mech.* **72**, 499–511.

Koppel, D. E. (1972). Analysis of macromolecular polydispersity in intensity correlation spectroscopy. Method of cumulants. *J. Chem. Phys.* **57**, 4814–20.

Lee, S. P., Tscharnuter, W. & Chu, B. (1972). Calibration of an optical self-beating spectrometer by polystyrene latex spheres and confirmation of the Stokes–Einstein formula. *J. Poly. Sci.: Poly. Phys. Ed.* **10**, 2453–9.

Lighthill, M. J. (1958). *Fourier Analysis and Generalised Functions.* Cambridge University Press.

McQuarrie, D. A. (1976). *Statistical Mechanics.* Harper & Row.

Nelson, E. (1967). *Dynamical Theories of Brownian Motion.* Princeton University Press.

Perrin, J. (1910). *Brownian Motion and Molecular Reality*. Taylor & Francis, London.
Pomeau, Y. & Resibois, P. (1975). Time dependent correlation functions and mode–mode coupling theories. *Phys. Rep.* **19C**, 64–139.
Russel, W. B. (1981). Brownian motion of small particles suspended in liquids. *Ann. Rev. Fluid Mech.* **13**, 425–55.
Takamura, K., Goldsmith, H. L. & Mason, S. G. (1981). The microrheology of colloidal dispersions. XII. Trajectories of orthokinetic pair collisions of latex spheres in a simple electrolyte. *J. Colloid Interface Sci.* **82**, 175–89.
Vadas, E. B., Cox, R. G., Goldsmith, H. L. & Mason, S. G. (1976). The microrheology of colloidal dispersions. II. Brownian diffusion of doublets of spheres. *J. Colloid Interface Sci.* **57**, 308–26.
Wax, N. (ed.) (1954). *Noise and Stochastic Processes*. Dover.
Zwanzig, R. & Bixon, M. (1975). Compressibility effects in the hydrodynamic theory of Brownian motion. *J. Fluid Mech.* **69**, 21–5.

Problems

1 Show that Brownian motion of a sphere in a fluid with a spatially varying viscosity of the form

$$\mu(\mathbf{x}) = \mu(\mathbf{x}_0) + (\mathbf{x} - \mathbf{x}_0) \cdot \nabla\mu(\mathbf{x}_0)$$

produces a steady drift velocity. Note that for $\varepsilon \equiv |a\nabla\mu(\mathbf{x}_0)/\mu(\mathbf{x}_0)| \ll 1$ the displacement can be expanded as

$$\mathbf{x}(t) = \mathbf{x}_0 + \mathbf{x}_1(t) + \varepsilon\mathbf{x}_2(t) + \ldots$$

and substituted into the Langevin equation to determine

$$\lim_{t\to\infty} \left\langle \frac{\mathrm{d}\mathbf{x}_2}{\mathrm{d}t} \right\rangle.$$

State the direction of particle drift and assess whether the magnitude of the velocity could be significant with respect to sedimentation.

2 Develop the equation of motion for a permanent doublet, i.e. two equal spheres in contact, and integrate to derive $\langle \sin^2 \phi \rangle$ as a function of time normalized by the rotational diffusion coefficient. Verify that the result conforms to the theoretical curve in Fig. 3.6.

3 A sphere subjected to an impulsive force $MT\delta(t)$ accelerates to a maximum velocity determined by inertial effects and then decelerates owing to viscous drag. Compare the maximum velocities and the total displacements predicted from (2.7.9) to (2.7.11), which accounts for both fluid and particle inertia, and (3.2.6), which includes only the latter.

4 For anisotropic particles such as doublets the coupling of Brownian rotation with sideways drift during sedimentation results in vertical and lateral diffusion in excess of that due to translational Brownian motion alone. The effect can be calculated by adapting the treatment of pair diffusion in §3.5 to determine the dispersion coefficient D_{eff} for the lateral migration process through

$$\lim_{t\to\infty}[x_1^2(t)+x_2^2(t)]=2D_{eff}t.$$

(i) Derive the expression for the lateral drift velocity from (2.8.2) as a function of $\theta(t)$, the angle between the center-to-center vector and the x_3-axis.
(ii) Extract from (3.5.3) the equation governing the orientation and integrate to obtain $\theta(t)$.
(iii) The evaluation of D_{eff} from the expression for the lateral drift velocity and $\theta(t)$ requires some tricky integrations described by S. Goren (1979): Effective diffusivity of nonspherical, sedimenting particles, *J. Colloid Interface Sci.* **71**, 209–15. Complete the problem by specializing his final result to the case of doublets.

4

ELECTROSTATICS

4.1 Introduction

Everyone has empirical knowledge of electrostatic and electromagnetic phenomena based on experiences such as the buildup of static charge on a comb or nature's grand displays of lightning and the auroras borealis and australis. Less obvious but no less familiar are the stabilizing effects of electrostatic forces in colloidal suspensions. Clay particles and silt carried in suspensions by rivers coagulate upon encountering the higher salt concentration of the sea to form huge deltas. Electrostatic stabilization is also responsible for the long shelf-life of certain latex paints. Needless to say, electrostatic forces play central roles in the behaviour of biological systems. Despite such diversity, electromagnetic and electrostatic phenomena can be understood in terms of the elegant theory embodied in Maxwell's equations. Here we take these equations as axioms and proceed deductively.

The presentation is organized as follows. First the equations governing quasi-static electric fields are set out. Starting with the balance laws and conditions prescribed at boundaries where electrical properties change abruptly, we are led to discuss dielectrics, polarization, free charge, and the electrical stress embodied in Maxwell's stress tensor. Then emphasis shifts to the electrical double layer and mathematical models describing its behavior. Here layers of charge are immobilized on a surface, while ions in the adjacent solution move freely under the influence of electrical forces and Brownian motion. After studying matters near a single surface, we turn our attention to the region between two surfaces and electrostatic forces between macroscopic particles in solutions containing dissolved ions. The theory of the electrical double layer abounds with approximate analytical formulas which must be used judiciously. Accordingly, some of the numerical computations which establish the validity of these approximations are described before recent measurements testing the theory are reviewed. Magnetic effects are small in most situations of interest, so our discussion focuses on electrostatics, ignoring electromagnetic phenomena. They receive special attention later in the section dealing with dispersion forces.

4.2 Electrostatic fields

We begin with Maxwell's equations, simplified for electrostatics (see Feynman, Leighton, & Sands, 1964):

$$\varepsilon_0 \nabla \cdot \mathbf{E} = \rho^{(e)} \tag{4.2.1}$$

and

$$\nabla \times \mathbf{E} = \mathbf{0}. \tag{4.2.2}$$

The fundamental entities introduced here are the electric field, $\mathbf{E}$, (the force per unit charge), the total electric charge per unit volume, $\rho^{(e)}$, and ε_0, the permittivity of the vacuum. Combining (4.2.1) with the divergence theorem relates the electric field on a closed surface, S, to the charge, Q, enclosed in the volume, V, as

$$\int_S \mathbf{E} \cdot \mathbf{n}\, \mathrm{d}S = \frac{1}{\varepsilon_0} \int_V \rho^{(e)}\, \mathrm{d}V = \frac{Q}{\varepsilon_0}. \tag{4.2.3}$$

Here $\mathbf{n}$ is the outer unit normal.

Consequently, for a spherical surface of radius r centered on a point charge in a vacuum,

$$\mathbf{E} \cdot \mathbf{n} = \frac{Q}{4\pi\varepsilon_0 r^2}, \tag{4.2.4}$$

indicating that

$$\mathbf{E} = \frac{Q}{4\pi\varepsilon_0 r^3}\mathbf{r}, \tag{4.2.5}$$

where $\mathbf{r}$ is centered on the point charge. Since the electric field is defined as the force per unit charge, the force exerted by one point charge on another at relative position $\mathbf{r}_{12}$ (Coulomb's law) is

$$\mathbf{F}_{12} = \mathbf{E}_1 Q_2 = \frac{Q_1 Q_2}{4\pi\varepsilon_0 r_{12}^3}\mathbf{r}_{12}. \tag{4.2.6}$$

Note that the rationalized SI system used here measures the charge in coulombs and the electric field strength in volts per meter; the permittivity of free space, ε_0, is $8.854 \times 10^{-12}\ \mathrm{C}^2/\mathrm{N\,m}^2$.

According to (4.2.2), the electric field is conservative, i.e., the line integral of $\mathbf{E}\cdot\mathbf{t}$ (where $\mathbf{t}$ is tangent to a closed curve) is zero. Thus, there exists a potential function such that

$$\mathbf{E} = -\nabla\psi. \tag{4.2.7}$$

The potential must satisfy (see (4.2.1))

$$\varepsilon_0 \nabla^2 \psi = -\rho^{(e)}. \tag{4.2.8}$$

For example, around an isolated point charge, Q,

$$\frac{1}{r^2}\frac{\partial}{\partial r} r^2 \frac{\partial \psi}{\partial r} = 0, \tag{4.2.9}$$

with

$$\psi \to 0 \quad \text{as} \quad r \to \infty$$

and (4.2.10)

$$\lim_{r\to 0} \int \frac{\partial \psi}{\partial r} dS = -\frac{Q}{\varepsilon_0}.$$

The solution for the potential,

$$\psi = \frac{Q}{4\pi\varepsilon_0} r^{-1}, \tag{4.2.11}$$

corresponds to the electric field of (4.2.4).

The effects of matter on electrostatic fields are empirical facts. Familiar examples are the diminution of the force between two charges in a dielectric and the increased charge carrying ability of a capacitor containing a dielectric medium. The latter can be illustrated by considering a charged spherical conductor of radius a. Equation (4.2.11) connects charge and potential in a vacuum so the potential of the sphere, ψ_s (the work required to bring in a unit charge from infinity), and the sphere's charge, Q, are related by

$$Q = 4\pi\varepsilon_0 a \psi_s. \tag{4.2.12}$$

The factor $4\pi\varepsilon_0 a$ is called the capacity of the sphere.

If the vacuum is replaced by a dielectric medium, the capacity of the sphere increases because of polarization of the dielectric. Electric fields polarize matter in two ways: by orienting molecules with permanent dipoles and by deforming electron clouds within individual atoms and molecules. The polarization vector, **P**, is related to the characteristics of individual dipoles by

$$\mathbf{P} = NQ\mathbf{d} \tag{4.2.13}$$

(Feynman, Leighton & Sands, 1964). Here N represents the number of dipoles per unit volume, Q is the magnitude of the charge separated to produce the dipole, and **d** is a vector describing the average orientation of the dipole and the charge separation distance. In linear materials, polarization from the incident field is expressed as

$$\mathbf{P} = N\alpha\varepsilon_0 \mathbf{E} \tag{4.2.14}$$

where α is the polarizability of the atom or molecule (with dimension L^3). The product $N\alpha$ is called the electric susceptibility of the material, χ.

The polarization vector is then used to define the volumetric polarization charge density $\rho^{(p)}$ as

$$\nabla \cdot \mathbf{P} = -\rho^{(p)}. \tag{4.2.15}$$

Combining (4.2.1), (4.2.14), and (4.2.15) yields

$$\varepsilon_0 \nabla \cdot (1+\chi)\mathbf{E} = \rho^{(e)} - \rho^{(p)} \equiv \rho^{(f)}. \tag{4.2.16}$$

If the free charge density, $\rho^{(f)}$, is zero and the dielectric homogeneous, then (4.2.1) and (4.2.16) together indicate that the volumetric polarization charge is also zero. Note, however, that polarization charge will appear at a surface (see (4.2.19)).

Now we examine a spherical capacitor immersed in a homogeneous dielectric. If the dielectric has no free charge then

$$\varepsilon_0 \nabla \cdot (1+\chi)\mathbf{E} = 0 \tag{4.2.17}$$

and, since the field is still irrotational, a potential exists, viz.

$$\psi = Ar^{-1}. \tag{4.2.18}$$

However, the situation at the surface differs from that in a vacuum, owing to the polarization charge. From (4.2.15), the divergence theorem, and the Gaussian† surface outlined in Fig. 4.1, we have

$$\mathbf{P} \cdot \mathbf{n} = -\frac{Q^{(p)}}{4\pi a^2} \tag{4.2.19}$$

† The Gaussian surface is defined as the surface enclosing the free charge plus the polarization charge at the surface of the dielectric.

where $Q^{(p)}$ is the polarization charge on the surface of the dielectric. Applying (4.2.3) to the Gaussian surface yields

$$\mathbf{E}\cdot\mathbf{n}=\frac{Q+Q^{(p)}}{4\pi\varepsilon_0 a^2} \tag{4.2.20}$$

since the total electric charge consists of the free charge on the capacitor, Q, plus the polarization charge on the surface of the dielectric, $Q^{(p)}$. Thus

$$\mathbf{E}\cdot\mathbf{n}=\frac{Q}{4\pi\varepsilon_0 a^2}-\chi\mathbf{E}\cdot\mathbf{n}. \tag{4.2.21}$$

This determines the integration constant, A, and

$$Q=4\pi\varepsilon_0(1+\chi)a\psi_s. \tag{4.2.22}$$

Hence the capacity increases by the factor $(1+\chi)$ compared with the situation in a vacuum; $(1+\chi)$ is often designated the dielectric constant, ε. Similarly, we can show that the electric field at a distance r from a point charge in a uniform dielectric, (4.2.5), and the Coulomb force, (4.2.6), are each diminished by the factor ε.

4.3 Boundary conditions

At this point we have a mathematical structure that describes fields in bulk matter. A complementary description of conditions prevailing at interfaces between two materials can be derived by applying the same balance equations to a disc-shaped volume and the simple closed curve depicted in Fig. 4.2. Using the divergence theorem and the disc-shaped volume with (4.2.1) we have,† in the limit as $h\to 0$,

$$\varepsilon_0[(\mathbf{E}\cdot\mathbf{n})_1+(\mathbf{E}\cdot\mathbf{n})_2]=q \tag{4.3.1}$$

† The subscripts indicate the side of the interface on which the quantity in parenthesis is evaluated.

Fig. 4.1. Spherical capacitor in a dielectric.

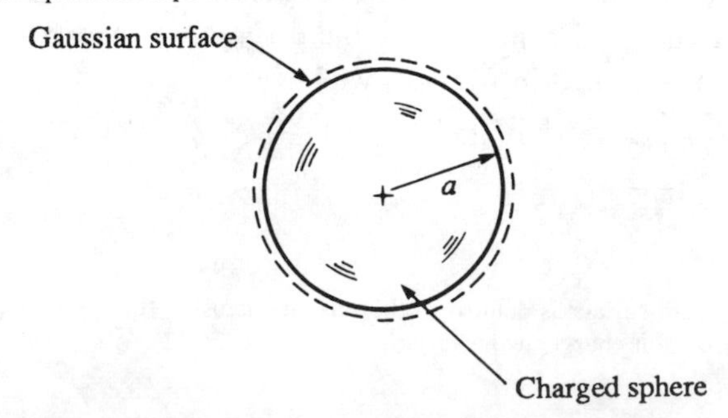

where q stands for the total charge per unit area. From the expression for the polarization, (4.2.15),

$$(\mathbf{P}\cdot\mathbf{n})_1+(\mathbf{P}\cdot\mathbf{n})_2=-q^{(\mathrm{p})} \tag{4.3.2}$$

and so

$$[(\varepsilon_0\mathbf{E}+\mathbf{P})\cdot\mathbf{n}]_1+[(\varepsilon_0\mathbf{E}+\mathbf{P})\cdot\mathbf{n}]_2=q^{(\mathrm{f})}. \tag{4.3.3}$$

Here $q^{(\mathrm{f})}$ stands for the extra or 'free' charge, $q-q^{(\mathrm{p})}$, beyond that due to polarization, i.e. charge positioned at the interface by means other than polarization. Thus, for linear dielectrics,

$$[\varepsilon_0(1+\chi)\mathbf{E}\cdot\mathbf{n}]_1+[\varepsilon_0(1+\chi)\mathbf{E}\cdot\mathbf{n}]_2=q^{(\mathrm{f})}. \tag{4.3.4}$$

If only one of the materials is a conductor then the current normal to the surface must be zero so the corresponding electrostatic field vanishes.

A condition on the tangential components of the field at the interface is obtained by evaluating the line integral of $\mathbf{E}\cdot\mathbf{t}$ around a rectangular path in a plane perpendicular to the surface (cf. Figure 4.2). Taking the limiting circuit where the path length perpendicular to the surface shrinks to zero shows, using (4.2.2), that

$$(\mathbf{E}\cdot\mathbf{t})_1+(\mathbf{E}\cdot\mathbf{t})_2=0. \tag{4.3.5}$$

Fig. 4.2. Definition sketch to illustrate the boundary conditions for electrostatic fields.

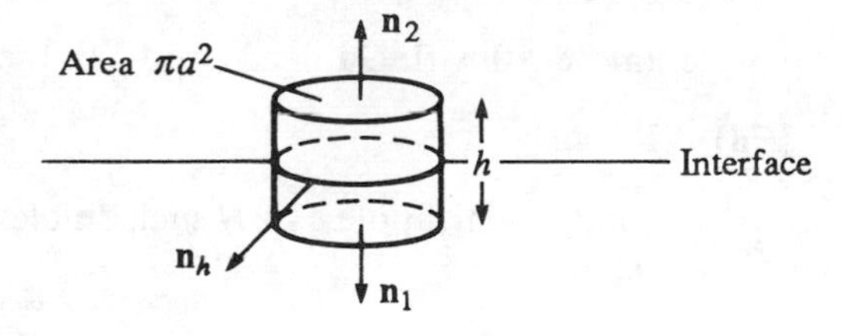

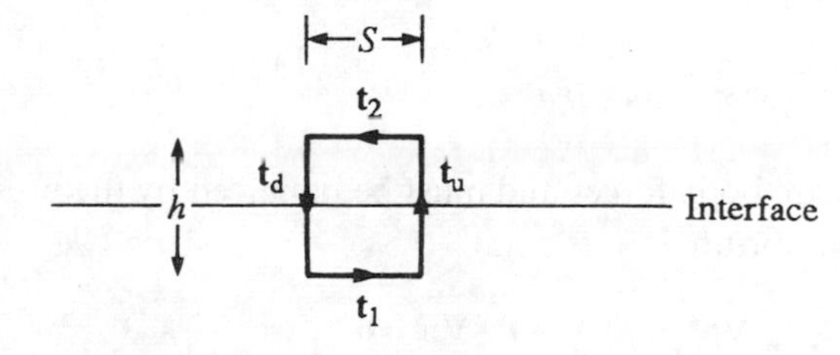

Here $\mathbf{t}$ is any vector tangent to the interface and it follows that the potentials on either side of the interface differ by at most a constant. If no work is done in transferring charge across the interface then the constant is zero.

4.4 The electric stress tensor

The study of interactions between macroscopic bodies requires the force per unit area, i.e. the stress that an external field exerts on a surface. First we derive an expression for the stress tensor in a homogeneous fluid, then illustrate its application to a sphere immersed in an uncharged dielectric.

The derivation of the stress in a fluid dielectric containing free charge is anything but straightforward, in part because of the difficulty in establishing how the presence of the electric field contributes to the pressure and thereby alters the stress. Our derivation is based on those of Landau & Lifshitz (1960) and Sanfeld (1968). Melcher (1981) arrives at the same result by a somewhat different approach.

First, the force due to the electric field acting on an isolated dipole is derived. Consider a pair of charges, Q and $-Q$, at relative position $\mathbf{d}$. The electrical force on the pair is

$$-Q\mathbf{E}(\mathbf{x})+Q\mathbf{E}(\mathbf{x}+\mathbf{d}).$$

Expanding the second term yields

$$-Q\mathbf{E}(\mathbf{x})+Q\mathbf{E}(\mathbf{x})+Q\mathbf{d}\cdot\nabla\mathbf{E}+\ldots,$$

and taking the limit $\mathbf{d}\to 0$ with $Q\mathbf{d}$ fixed produces the expression

$$(Q\mathbf{d})\cdot\nabla\mathbf{E}$$

for the force on an individual dipole. For N such dipoles per unit volume the force will be (cf. (4.2.13))

$$\mathbf{P}\cdot\nabla\mathbf{E}.$$

Accordingly, the force per unit volume acting on the free charge and dipoles is

$$\rho^{(f)}\mathbf{E}+\mathbf{P}\cdot\nabla\mathbf{E}.$$

These are body forces and must be balanced by the pressure gradient, ∇p^*. At equilibrium

$$-\nabla p^*+\rho^{(f)}\mathbf{E}+\mathbf{P}\cdot\nabla\mathbf{E}=\mathbf{0}. \tag{4.4.1}$$

With the expressions relating charge and dipole density to field strength, (4.2.14) and (4.2.15), we can transform this expression into one involving the divergence of a tensor as

$$-\nabla p^* + \nabla \cdot (\varepsilon\varepsilon_0 \mathbf{E}\mathbf{E} - \tfrac{1}{2}\varepsilon_0 \mathbf{E}\cdot\mathbf{E}\boldsymbol{\delta}) = \mathbf{0}. \qquad (4.4.2)$$

The pressure p^* differs from that in the absence of an electric field owing to electrical modifications to the short-range intermolecular forces. Accordingly, we identify the 'pressure' due to kinetic energy and short-range intermolecular forces without electrical effects as p and write

$$p^* = p + \tfrac{1}{2}\varepsilon_0\left[\varepsilon - 1 - \rho\left(\frac{\partial\varepsilon}{\partial\rho}\right)_{\mathrm{T}}\right]\mathbf{E}\cdot\mathbf{E} \qquad (4.4.3)$$

(see Landau & Lifshitz, 1960, Ch. II, or Sanfeld, 1968, Ch. 6 and 7). Here ρ denotes the density of the material and the derivative $(\partial\varepsilon/\partial\rho)$ is taken at constant temperature, T. Now, the divergence of the total stress yields

$$-\nabla p + \nabla \cdot \left\{\varepsilon\varepsilon_0 \mathbf{E}\mathbf{E} - \tfrac{1}{2}\varepsilon_0\varepsilon\left[1 - \frac{\rho}{\varepsilon}\left(\frac{\partial\varepsilon}{\partial\rho}\right)_{\mathrm{T}}\right]\mathbf{E}\cdot\mathbf{E}\boldsymbol{\delta}\right\} = \mathbf{0}. \qquad (4.4.4)$$

The electric stress tensor reduces to what is known as the Maxwell form for the vacuum where $\varepsilon = 1$.

Equation (4.4.4) also can be written in the form

$$-\nabla\left[p - \tfrac{1}{2}\varepsilon_0\rho\left(\frac{\partial\varepsilon}{\partial\rho}\right)_{\mathrm{T}}\mathbf{E}\cdot\mathbf{E}\right] - \tfrac{1}{2}\varepsilon_0\mathbf{E}\cdot\mathbf{E}\nabla\varepsilon + \rho^{(\mathrm{f})}\mathbf{E} = \mathbf{0} \qquad (4.4.5)$$

to recover the force arising from the action of the field on the local free charge.

To illustrate use of the electric stress tensor, a familiar result is derived: the force on a conducting sphere. Figure 4.3 depicts the situation: a conducting sphere immersed in a uniform (non-conducting) dielectric in the presence of a uniform field, $\mathbf{E}_\infty$. The free charge in the dielectric is nil,

Fig. 4.3. Conducting sphere in a dielectric.

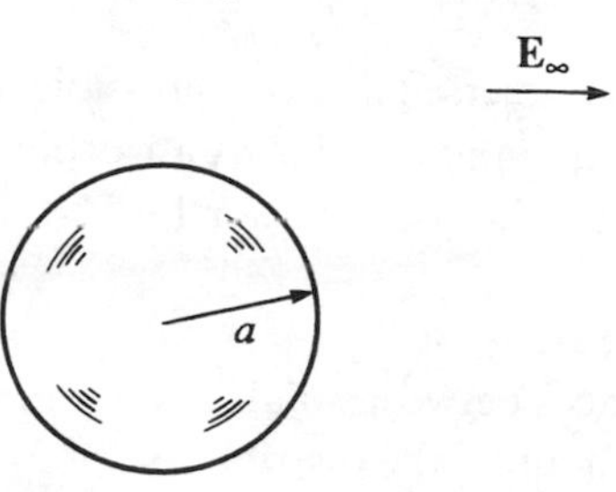

while the surface charge found on the conductor is Q; cf. (4.2.22). Clearly the force must emerge as $Q\mathbf{E}_\infty$, since the sphere appears as a point when viewed on a large length scale and Coulomb's law must apply.

From (4.2.17) and the uniformity of the dielectric, we deduce the form of the potential as

$$\psi = -(1 + Ar^{-3})\mathbf{E}_\infty \cdot \mathbf{x} + Br^{-1}, \tag{4.4.6}$$

if the field is to have the requisite behavior far from the sphere. Since the sphere is a conductor with charge Q, the boundary conditions show that $A = -a^3$ and $B = Q/4\pi\varepsilon\varepsilon_0$.

The electric stress is calculated from the potential but the force on the sphere also includes a contribution from the inhomogeneous pressure generated by the electric field. Integrating (4.4.5) shows that

$$p - \tfrac{1}{2}\varepsilon_0\rho\left(\frac{\partial\varepsilon}{\partial\rho}\right)_T \mathbf{E}\cdot\mathbf{E} = \text{constant}. \tag{4.4.7}$$

Thus the pressure variation due to the field is cancelled by electrical effects and the net force is

$$\mathbf{F} = \int_S \varepsilon\varepsilon_0(\mathbf{EE} - \tfrac{1}{2}\mathbf{E}\cdot\mathbf{E}\boldsymbol{\delta})\cdot\mathbf{n}r^2\,\mathrm{d}\Omega, \tag{4.4.8}$$

with $\mathrm{d}\Omega = \sin\theta\,\mathrm{d}\theta\,\mathrm{d}\phi$. Use of the divergence theorem converts the integral to one over a spherical surface with an infinitely large radius. Now we need only those parts of the integral that survive the limiting process and the force on the sphere is

$$\mathbf{F} = Q\mathbf{E}_\infty. \tag{4.4.9}$$

As expected, the force is the same as that on a concentrated point charge placed in an undisturbed field.

A final point worth noting concerns the term

$$\tfrac{1}{2}\varepsilon_0\rho\left(\frac{\partial\varepsilon}{\partial\rho}\right)_T \mathbf{E}\cdot\mathbf{E}.$$

For problems involving rigid bodies immersed in incompressible, homogeneous materials, reference to this term can be avoided by absorbing it into a modified pressure. This procedure is adopted henceforth.

4.5 The origins of interfacial charge

In colloidal systems, interfaces between solids and ionic solutions almost always acquire charge which alters the distribution of free ions in solution. Two such surfaces interact differently from how they would if uncharged or the ions were absent. In this section charged interfaces in fully ionized electrolyte solutions are discussed. The brief treatment establishes a context for treating the mechanics, rather than reviewing the chemistry and physics of the interfacial charging processes. More detailed presentations are given by Haydon (1964), Sparnaay (1972), Dukhin & Derjaguin (1974), and Hunter (1981).

Several mechanisms have been identified as responsible for interfacial charge. Hunter (1981) divides them into four classes:

(i) Differences between the electron affinities of two phases. Examples are the contact potentials that develop between dissimilar metals or between liquid mercury and an aqueous electrolyte solution, a situation that has been studied extensively (cf. Sparnaay (1972)).
(ii) Differences between the affinities of the two phases for ions (or ionizable species), exemplified by the interface between silver iodide crystals and aqueous electrolyte solutions.
(iii) Ionization of surface moieties such as sialic acid groups on the surface of the red blood cell, carboxylic acid or sulfonic acid groups on synthetic latex particles, and amphoteric hydroxyl groups on oxide surfaces such as silica.
(iv) Entrapment of ions in, for example, clay minerals where substitution of one ion for another in the lattice can confer a charge (usually negative); e.g. the isomorphous substitution of Si^{+4} by Al^{+3}.

In each scenario, the surface charge depends on the compositions of the two phases. In some instances, the state of the interface can be related to the conditions in the bulk phases by simple models, as illustrated next.

Consider, first, silver iodide particles suspended in an aqueous solution containing AgI as the major constituent. Differences between the electrochemical potentials of the kth ion in the solid phase and in the solution, $\Delta\mu^k$, can be separated into chemical and electrical contributions, albeit arbitrarily as (Guggenheim, 1967)

$$\Delta\mu^k = \Delta\mu_0^k + ez^k \Delta\psi; \tag{4.5.1}$$

z^k stands for the valence of the kth ionic species and e for the fundamental charge, 1.6×10^{-19} C. The electrical potential difference between the solid and the fluid, $\Delta\psi$, includes both polarization of the interface and free charge. At equilibrium, $\Delta\mu^k = 0$, so changes in the electric potential across the interface are related to changes in the chemical potential as

$$\Delta\psi = -\Delta\mu_0^k / ez^k. \tag{4.5.2}$$

With silver iodide systems, changes in the composition of the solution affect the chemical potentials of the Ag^+ or I^- ions in the solid very little since all other species are present in minor amounts. If we treat the solution as ideal, with

$$d\mu_0^k = -kT\, d\ln n^k, \tag{4.5.3}$$

then the change in the electrical potential difference will be proportional to the change in the logarithm of the concentration, n^k. Thus,

$$d(\Delta\psi) = \frac{kT}{ez^k} d(\ln n^k). \tag{4.5.4}$$

The ions responsible for the change in surface potential, called potential determining ions (p.d.i.), are Ag^+ and I^- for AgI sols. According to (4.5.4) (the Nernst equation), the interphase potential difference (between the solid and the bulk of the liquid) changes by about 60 millivolts when the concentration of potential determining ions changes by a factor of ten. More extensive discussions, including the subtleties connected with other potentials, e.g. the so-called Volta and Galvani potentials and Lange's χ-potential, are given by Kruyt (1952) and Sparnaay (1972).

A simple model of a surface with ionizable groups affords a second example of how changes in the composition of the bulk solution alter the surface potential. Suppose acidic groups embedded in the surface ionize according to†

$$AH \rightleftarrows A^- + H^+. \tag{4.5.5}$$

At equilibrium the concentrations are related as

$$K = [A^-][H^+]_0/[AH], \tag{4.5.6}$$

where $[A^-]$ and $[AH]$ represent surface concentrations (moles/area) and $[H^+]_0$ stands for the hydronium concentration in the solution at the surface. If $[n]$ designates the number of ionizable groups per unit area then the charge density follows as

$$-e[A^-] = \frac{-eK[n]}{K + [H^+]_0}. \tag{4.5.7}$$

† The proton in solution is associated with one water molecule to form the hydronium ion, H_3O^+, and the hydronium ion is solvated with 3 additional water molecules. The symbol H^+ will be used with the understanding that it represents the hydrated form.

Table 4.1. *Surface potential due to ionization of surface groups*

$[H^+]_b$	ψ_s/mV
10^{-4} M	−93
10^{-3} M	−32
10^{-2} M	−3

Next we need relations between the surface charge and potential and between the hydronium ion concentrations adjacent to the surface and the bulk solution. It will be shown (see (4.6.3)) that ion distributions follow the Boltzmann expression

$$[H^+] = [H^+]_b \exp(-e\psi/kT); \tag{4.5.8}$$

$[H^+]_b$ denotes the hydronium ion concentration in the bulk of the solution where $\psi = 0$. For a low potential surface we have, from (4.7.5),

$$-e[A^-] = \left(\frac{2\varepsilon\varepsilon_0 e^2 [H^+]_b}{kT}\right)^{1/2} \psi_s \tag{4.5.9}$$

where ψ_s denotes the potential at the surface. Combining (4.5.7) to (4.5.9) enables us to calculate the potential at the surface. Table 4.1 shows results of calculations for $K = 10^{-4}$ M, $[n] = 5 \times 10^{17}/\text{m}^2$. Note that the magnitude of the change resembles that found from the Nernst equation for the AgI surface, viz. an order of magnitude change in the ionic strength alters the potential by roughly 60–70 mV.

4.6 The Gouy–Chapman model of the diffuse layer

Having seen how the electrostatic potential at an interface responds to changes in the electrolyte composition, we proceed to investigate the detailed structure of the electrostatic field. This requires knowledge of ion distributions because the field produces a net charge in the electrolyte adjacent to the interface. Ions whose charge is opposite the sign of the charge on the interface will be attracted and the others will be repulsed. At the same time, each ion participates in the randomizing thermal motion of the solution. It follows that the fluid adjacent to the charged interface contains a charge which balances the surface charge, making the combination of surface and solution electrically neutral. The region containing the surface charge is often called the compact or Stern layer, while the region where ions move freely under the influence of electrical and thermal forces is termed the diffuse layer. Together, these make up the electric double layer. Our task is to describe the structure of this double layer, especially the diffuse region.

According to the earlier development (cf. (4.2.16)), the electrostatic potential must satisfy

$$\varepsilon\varepsilon_0 \nabla^2 \psi = -\rho^{(f)} \tag{4.6.1}$$

for a homogeneous, linear dielectric, so the first task is to describe the ionic concentrations that produce the free charge density $\rho^{(f)}$. Because the ions in the diffuse region are in equilibrium, the force, which equals the gradient of the electrochemical potential, must vanish as

$$kT\nabla \ln n^k + ez^k \nabla \psi = \mathbf{0}. \tag{4.6.2}$$

Thus the ions follow the Boltzmann distribution

$$n^k = n_b^k \exp(-ez^k \psi / kT). \tag{4.6.3}$$

It should be noted that the potentials appearing in (4.6.1) and (4.6.3) are, strictly speaking, different. The potential in the Boltzmann equation represents the potential of mean force, whereas that in the Poisson equation is the local average potential.† Nevertheless, to simplify matters any differences are ignored.

Recognizing that the free-charge density equals the local excess of ionic charge arising from N ionic species, i.e.

$$\rho^{(f)} = \sum_1^N ez^k n^k, \tag{4.6.4}$$

and combining the various expressions leads to the Poisson–Boltzmann equation describing the electrostatic potential in ionic solutions,

$$\varepsilon\varepsilon_0 \nabla^2 \psi = -e\sum_1^N z^k n_b^k \exp(-ez^k \psi / kT). \tag{4.6.5}$$

This equation is the basis of the Gouy–Chapman model of the diffuse charge cloud adjacent to a charged surface (Gouy, 1910; Chapman, 1913).

† The potential of mean force is the potential whose gradient gives the average force acting on an ion whereas the local average electrostatic potential is the cannonical ensemble average of the electrostatic potential. (Cf. McQuarrie (1976), Ch. 15).

The principal assumptions thus far are that the electrolyte is an ideal solution with uniform dielectric properties, the ions are point charges, and the potential of mean force and the average electrostatic potential are identical. A substantial amount of work has been done to identify limitations on the equation. However, more detailed formulations, ranging from analytical modifications to Monte-Carlo calculations, lead to the conclusion that the Poisson–Boltzmann equation provides very accurate results for the conditions of interest here, viz., electrolyte concentrations that do not exceed 1 M and surface potentials less than 200 mV. In this regard, see Haydon (1964), Sparnaay (1972), and Hunter (1981) for summaries of other investigations.

Another condition is required to complete the specification of the potential. At the interface, the relation between the charge and the potential can be established from (4.2.16) as

$$-\int_S \varepsilon\varepsilon_0 \nabla\psi \cdot \mathbf{n}\, \mathrm{d}S = Q^{(\mathrm{f})} \qquad (4.6.6)$$

where $Q^{(\mathrm{f})}$ is the free charge enclosed by the surface. If all the charge is on the interface, $Q^{(\mathrm{f})}$ is equal to the surface charge, Q, and for a uniform surface

$$-\varepsilon\varepsilon_0 \nabla\psi \cdot \mathbf{n} = q, \qquad (4.6.7)$$

where q is the surface charge per unit area.

4.7 The diffuse layer near a flat plate

Next we investigate some solutions of the equations embodied in the Gouy–Chapman model. For a flat interface and a $z-z$ electrolyte,† e.g. KCl, (4.6.5) becomes

$$\frac{\mathrm{d}^2\psi}{\mathrm{d}x^2} = \frac{2ez}{\varepsilon\varepsilon_0} n_{\mathrm{b}} \sinh(ez\psi/kT) \qquad (4.7.1)$$

where z is the valence of the cationic species. Linearization of the sinh-function for small dimensionless potentials, $e\psi/kT$, leads to the Debye–Hückel approximation, viz.

$$\frac{\mathrm{d}^2\psi}{\mathrm{d}x^2} = \frac{2e^2z^2n_{\mathrm{b}}}{\varepsilon\varepsilon_0 kT}\psi. \qquad (4.7.2)$$

† A $z-z$ electrolyte is a symmetrical, binary electrolyte wherein the cation valence is z and the anion valence is $-z$; NaCl is a $1-1$ electrolyte.

The expression $(\varepsilon\varepsilon_0 kT/2e^2z^2n_b)^{1/2}$ represents the Debye decay length,† often symbolized by κ^{-1}. Solving (4.7.2) shows that

$$\psi = \psi_s \exp(-\kappa x) \tag{4.7.3}$$

where ψ_s denotes the potential at the surface $x=0$. Hence the potential and space charge are non-zero in a region of thickness κ^{-1} adjacent to the interface.

The exact solution to (4.7.1) follows upon multiplying both sides by $d\psi/dx$ to obtain exact differentials which, after integrating twice and applying the boundary conditions, yields

$$\Psi = 2\ln\left(\frac{1+\exp(-\kappa x)\tanh(\frac{1}{4}\Psi_s)}{1-\exp(-\kappa x)\tanh(\frac{1}{4}\Psi_s)}\right) \tag{4.7.4}$$

where $\Psi = ez\psi/kT$. The surface charge follows from (4.6.7), as

$$q = 2(2\varepsilon\varepsilon_0 kTn_b)^{1/2}\sinh(\tfrac{1}{2}\Psi_s). \tag{4.7.5}$$

A numerical example given shortly illustrates the distributions of potential, co-ions, and counterions adjacent to a charged surface. However, certain limiting forms provide additional insight into the way the ionic atmosphere screens the surface charge. From (4.7.4), we recover the Debye–Hückel approximation for $\Psi_s/4 \ll 1$. The factor of $\frac{1}{4}$ accounts for the numerical accuracy of the Debye–Hückel formula for dimensionless potentials somewhat larger than unity. For large positive surface potentials, $\Psi_s \gg 1$, and $x>0$,

$$\Psi \sim 2\ln\frac{1+\exp(-\kappa x)}{1-\exp(-\kappa x)}, \tag{4.7.6}$$

leaving the local potential independent of the surface potential away from the interface. For negative potentials, the positive and negative signs in front of the exponentials are reversed. Similarly, for $\kappa x \gg 1$, $\exp(-\kappa x)$ is small and (4.7.4) shows that

$$\Psi \sim 4\tanh(\tfrac{1}{4}\Psi_s)\exp(-\kappa x), \tag{4.7.7}$$

and so the decay is always exponential far from the surface. Thus, for large surface potentials, there is a saturation effect, and, viewed from a distance, the surface potential is 4.

† For a general electrolyte the Debye thickness is defined as

$$\kappa^{-1} = [\varepsilon\varepsilon_0 kT/\Sigma e^2(z^k)^2 n_b^k]^{1/2}.$$

To help fix matters more concretely we turn to numerical examples for conditions set out in Table 4.2.

Table 4.2. *Conditions prescribed for the numerical examples*

Density of fully ionized negative charge groups on the interface	$5 \times 10^{17}/\mathrm{m}^2$
Electrolyte concentration	0.01 M
Temperature	298 K
Solvent dielectric constant (H_2O)	80
Double layer thickness, κ^{-1}	3.08×10^{-9} m

Accordingly, the area per surface charge group is $2\,\mathrm{nm}^2$ ($200\,\text{Å}^2$) and the dimensionless potential is -5.21 (-133.8 mV). Figure 4.4 compares potentials calculated from the exact solution (4.7.4) and the Debye–Hückel approximation, (4.7.3) using the same surface potential. Given the large surface potential, it is not surprising that the approximate solution errs substantially. Figure 4.5 shows the asymptotic formulas, (4.7.6) and (4.7.7).

Figure 4.6, the counterion (cation) and co-ion (anion) distributions, indicates a large concentration of counterions adjacent to the surface. It should be noted, however, that the assumption of 'point ions' is dubious at such small distances, so the actual ion densities will be different. The Debye–Hückel approximation gives a poor picture of the ion distributions and the charge density. For example, close to the surface, co-ion concentrations less than zero are predicted and the charge density is almost two orders of magnitude below the value from the exact solution.

The numerical results discussed thus far pertain to a surface with a given charge density. To see how the electrolyte composition affects matters for a surface with ionogenic groups, we use the model presented earlier (cf. (4.5.5) to (4.5.7)). Figures 4.7 and 4.8 depict situations where the charge groups (cf. Table 4.1) have an equilibrium constant of 10^{-4} M. In one case, a single $z-z$ electrolyte is present (HCl, say) and the cations bind with surface groups to set the surface charge; in the other situation an indifferent electrolyte (NaCl, say) is added.

With the indifferent electrolyte present, the surface potential and the fraction of ionized groups increase as the bulk HCl concentration is lowered (Figs 4.7 and 4.8). Here the Debye thickness is almost constant, owing to the excess indifferent electrolyte, NaCl. Since the potential is roughly proportional to the product of the surface charge and the Debye

Fig. 4.4. The decay of the electrostatic potential near a flat surface in an ionic solution as predicted using the exact solution of the Poisson–Boltzmann equation and the Debye–Hückel approximation.

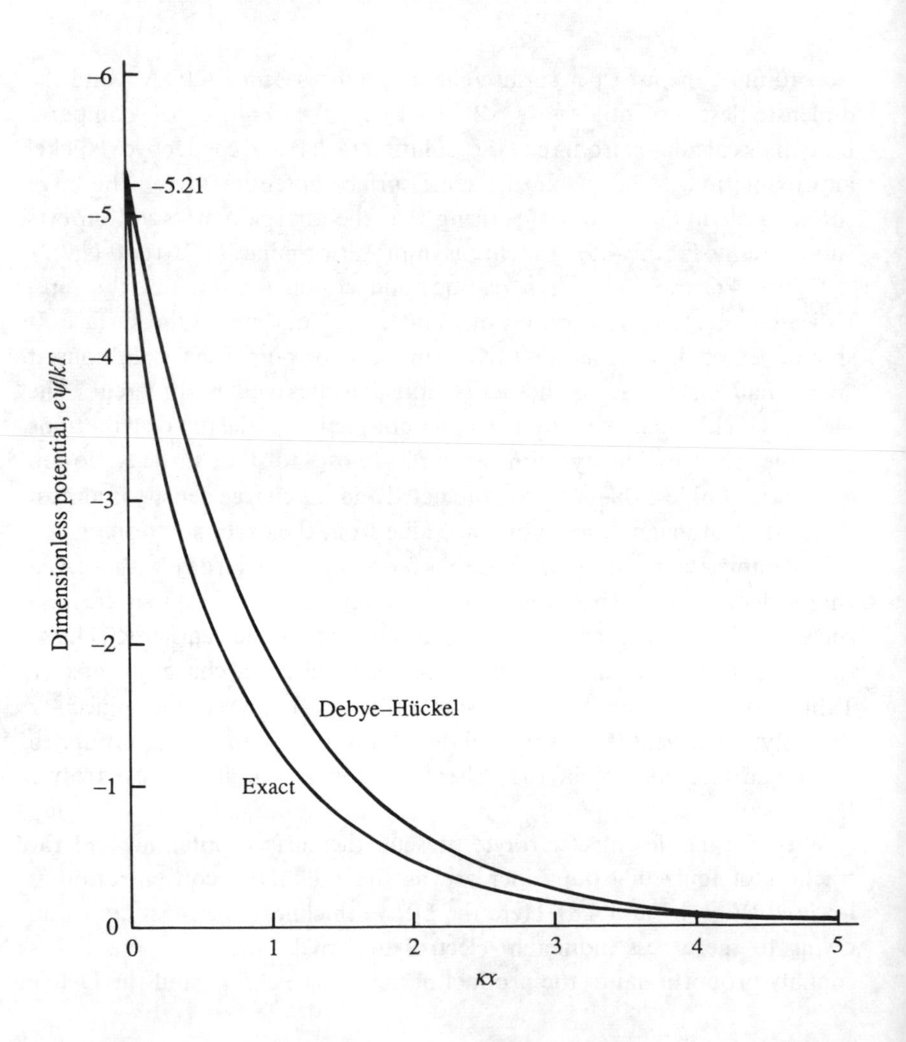

Fig. 4.5. The decay of the electrostatic potential near a flat surface according to the exact solution of the Poisson–Boltzmann equation and two approximate solutions.

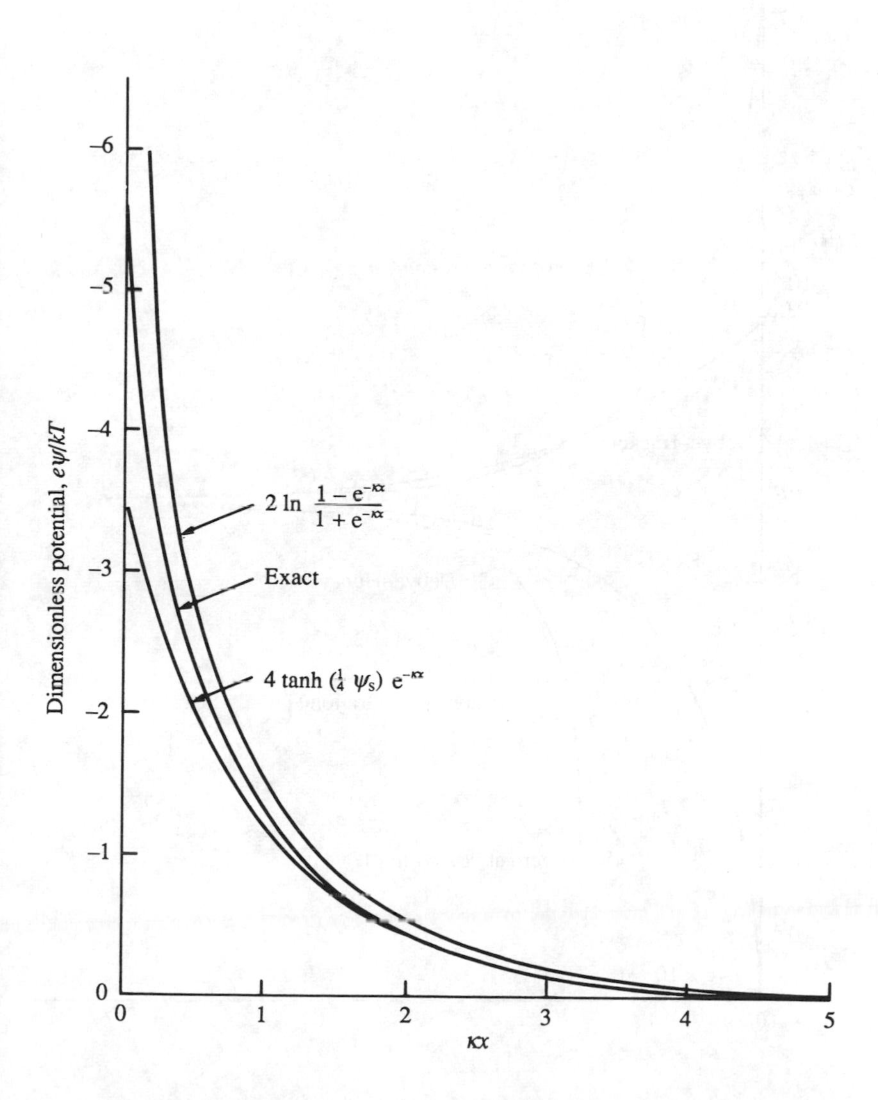

Fig. 4.6. Concentrations of ions near a charged, flat surface. The surface charge is $-8\,\mu C/cm^2$ (5×10^{17} charges/m^2), which yields a dimensionless surface potential of -5.21 (-133.9 mV).

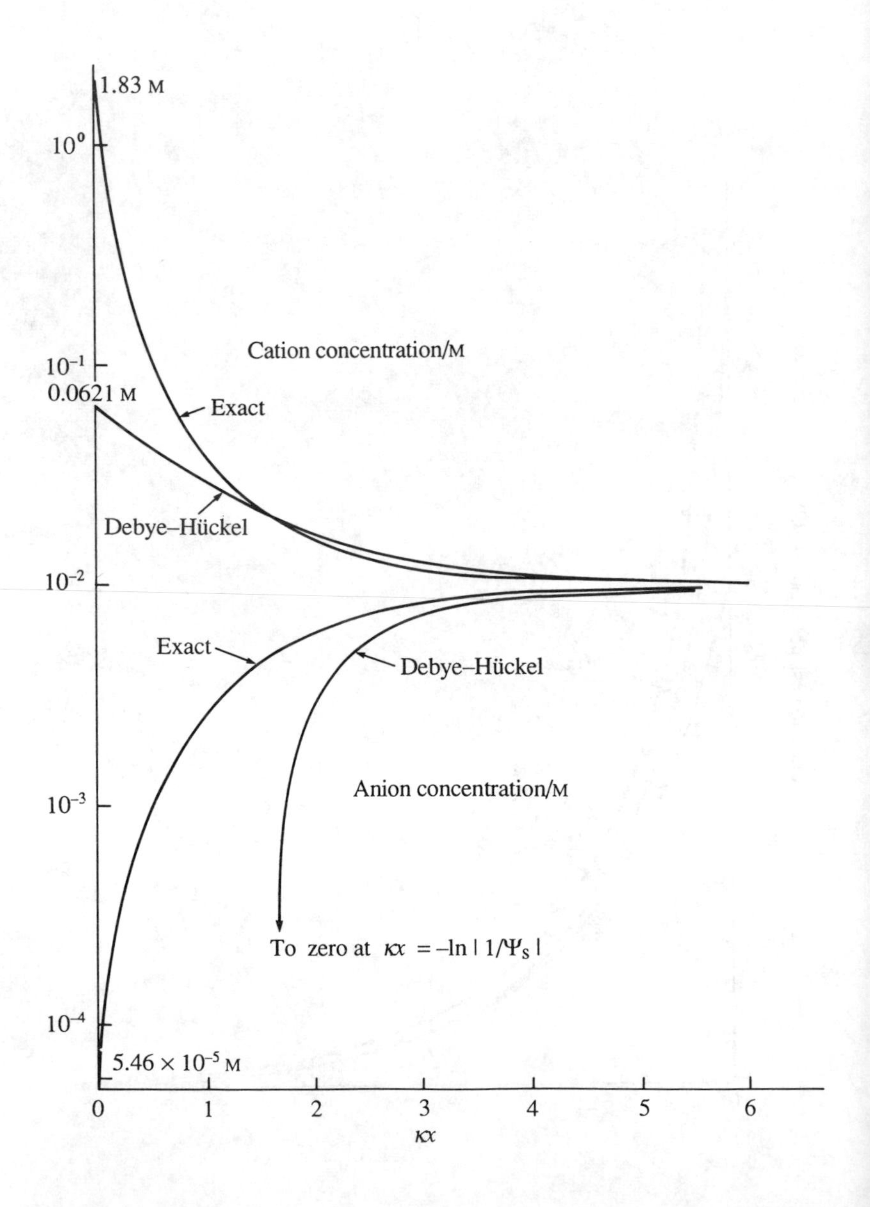

Fig. 4.7. The dimensionless surface potentials corresponding to the situations depicted in Fig. 4.8.

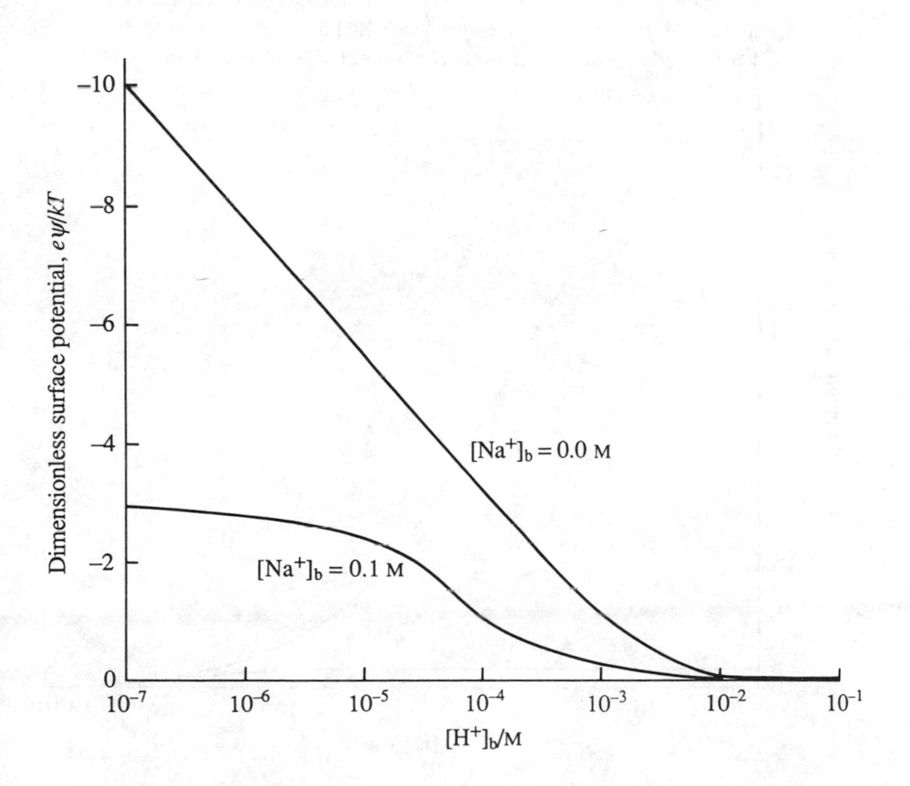

length, the surface potential and surface charge change in concert. In the example, the surface charge and potential roughly double as the bulk hydronium ion concentration decreases from 10^{-4} M to 10^{-5} M.

This behavior contrasts sharply with that without indifferent electrolyte. Here surface groups are only slightly ionized irrespective of the hydronium ion concentration (cf. Fig. 4.8). Now the Debye thickness increases as the hydronium ion concentration diminishes. This tends to increase the surface potential (Fig. 4.7) which, in turn, attracts more counterions to the surface suppressing ionization and the countervailing effects keep the degree of ionization almost constant.

These examples illustrate that the surface potential and degree of ionization of surface groups can depend strongly on the electrolyte type and composition.

Fig. 4.8. Dissociation of ionogenic groups attached to a surface immersed in an ionic solution. The number of surface groups per square meter is 5×10^{17}; the dissociation constant, K, is 10^{-4} M. The lower curve shows the fraction of dissociated groups when the electrolyte is HCl in water; the upper curve depicts matters when the solution contains 0.1 M NaCl in addition to HCl.

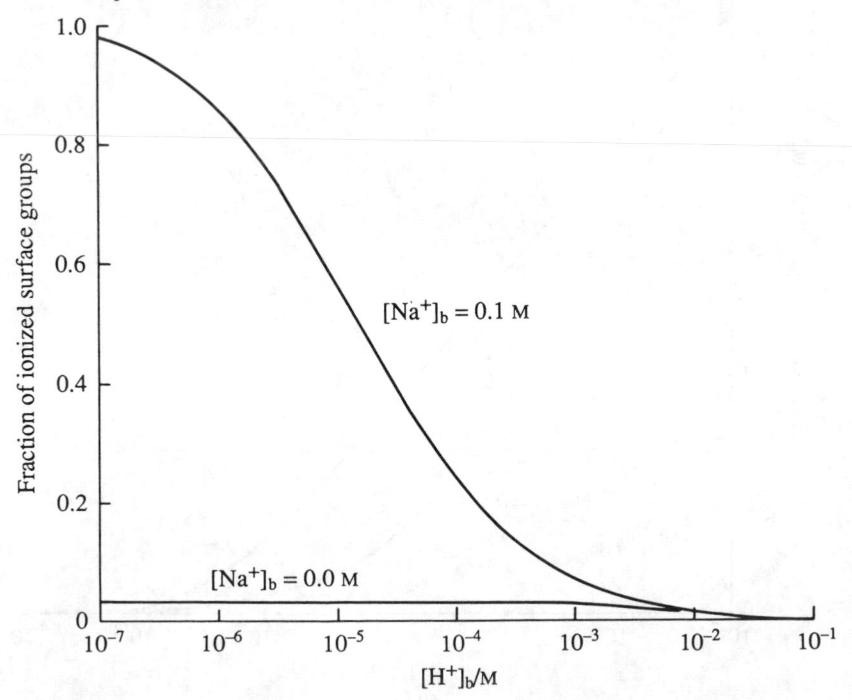

4.8 The diffuse layer around a sphere

We turn now to a spherical geometry where curvature complicates matters and precludes analytical solution. However, approximate solutions depict qualitative features accurately. For a $z-z$ electrolyte we must solve

$$\frac{1}{r^2}\frac{\partial}{\partial r}r^2\frac{\partial\psi}{\partial r}=\frac{2ezn_{\rm b}}{\varepsilon\varepsilon_0}\sinh(ez\psi/kT), \tag{4.8.1}$$

subject to

$$\begin{aligned}\psi&=\psi_{\rm s} \quad \text{at} \quad r=a\\ \psi&\to 0 \quad \text{as} \quad r\to\infty.\end{aligned} \tag{4.8.2}$$

No analytical solutions exist, leaving numerical calculations for specific situations or asymptotic solutions to describe limiting behavior.

For thin double layers, i.e. $\kappa a \gg 1$, we can rescale the equations with

$$\begin{aligned}r&=a[1+y/(\kappa a)]\\ \Psi&=ez\psi/kT\end{aligned} \tag{4.8.3}$$

to obtain

$$\frac{{\rm d}^2\Psi}{{\rm d}y^2}+\frac{2}{(\kappa a)[1+y/(\kappa a)]}\frac{{\rm d}\Psi}{{\rm d}y}=\sinh\Psi, \tag{4.8.4}$$

with

$$\begin{aligned}\Psi&=\Psi_s \quad \text{at} \quad y=0,\\ \Psi&\to 0 \quad \text{as} \quad y\to\infty.\end{aligned} \tag{4.8.5}$$

For $\kappa a\to\infty$, we recover a description pertinent to a flat interface. Thin double layers, therefore, exhibit a saturation effect similar to that for planar layers, viz. for $\Psi_{\rm s}\gg 1$,

$$\Psi\sim 4\exp(-y), \tag{4.8.6}$$

so the potential decays as though the surface potential is $4kT/ez$.

Loeb, Overbeek, & Wiersema (1961) gave the first comprehensive numerical treatment of spherical systems and tabulated extensive results for both symmetrical and unsymmetrical electrolytes (the LOW Tables). The LOW Tables also include numerous comparisons between exact (numerical) results and approximate formulas of one sort or another. For example, the formula

$$q=\frac{Q}{4\pi a^2}=\frac{\varepsilon\varepsilon_0 kT}{ez}\kappa\left[2\sinh(\tfrac{1}{2}\Psi_{\rm s})+\frac{4}{\kappa a}\tanh(\tfrac{1}{4}\Psi_{\rm s})\right] \tag{4.8.7}$$

(adapted from equation 4.50, p. 37, of the LOW Tables) gives the surface charge density to within 5 per cent for $\kappa a > 0.5$ for any surface potential.

The numerical results also demonstrate that the saturation effect at high potentials is a general feature of spherical double layers. Because the potential diminishes with distance from the surface, linearization becomes appropriate and the decay is exponential, viz.

$$\Psi \sim \Psi_A \frac{a}{r} \exp(-\kappa(r-a)). \tag{4.8.8}$$

Numerical results from the LOW Tables (cf. Table 20, p. 30) show how the saturation potential, Ψ_A, depends on κa (see Fig. 4.9).

It is possible to construct useful asymptotic series of the form

$$\Psi \sim \Psi_s f_0(\kappa a, R) + \Psi_s^2 f_1(\kappa a, R) + \ldots, \tag{4.8.9}$$

with $R = r/a$ if the surface potential is not too large (Gronwall, LaMer, & Sandved, 1928). The function

$$f_0(\kappa a, R) = R^{-1} \exp[-\kappa a(R-1)] \tag{4.8.10}$$

satisfies the Debye–Hückel equation for spherical geometry. The second term in the expansion vanishes for symmetrical electrolytes, e.g. $BaSO_4$ or NaCl, and the third term is a complicated expression involving exponential integrals (Booth, 1950). Many analytical approximations have been

Fig. 4.9. The relation between the saturation potential defined in (4.8.8) and the double-layer thickness. Sherwood (1980).

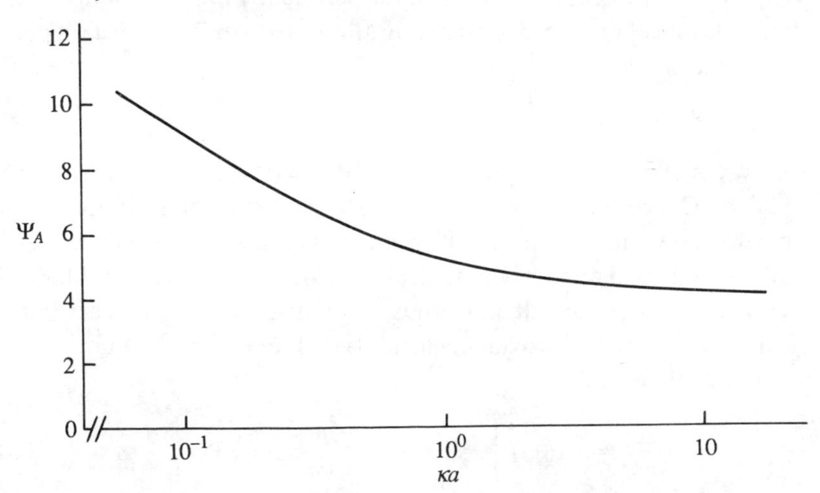

developed for the spherical diffuse layer (see, e.g., White, 1977) but the advent of fast, efficient numerical schemes has diminished their utility.

4.9 Repulsion between charged plates

To calculate the repulsive force, we need the local electrostatic potential and the local stress. Combining the equilibrium expression, (4.4.5), (modified as noted in §4.4) with the Gouy–Chapman model of the diffuse layer and integrating yields the general relation

$$p+kT\sum_{1}^{N} n_{\mathrm{b}}^{k}[1-\exp(-ez^{k}\psi/kT)]=p_{\mathrm{b}}. \tag{4.9.1}$$

From this expression and one for the total stress, cf. (4.4.4), we calculate the repulsive force per unit area, F, on either of the two identical parallel plates (Fig. 4.10) as follows. A force balance is constructed on a system bounded by the midplane and a parallel surface far away. The forces on the system consist of the force on the plate, F, the pressure on the system boundary at infinity, and the force on the midplane, where, because of symmetry, the electric field vanishes. Since the electric stress on the midplane is zero, the force per unit area follows as

$$F=kT\sum_{1}^{N} n_{\mathrm{b}}^{k}[\exp(-ez^{k}\psi_{0}/kT)-1]. \tag{4.9.2}$$

Fig. 4.10. Definition sketch for the calculation of the electrostatic force exerted on charged plates immersed in ionic solutions.

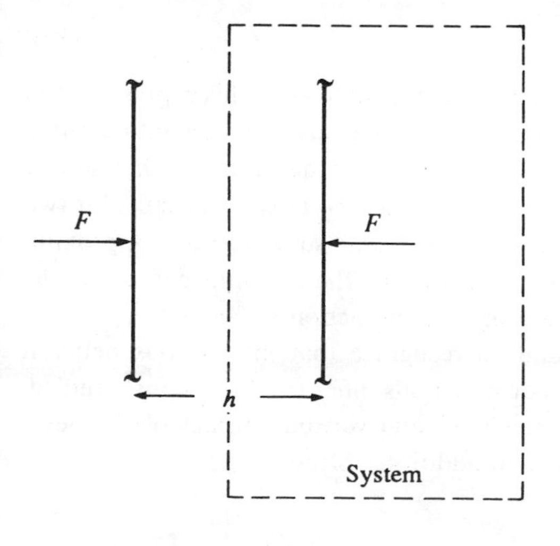

Here the potential, ψ_0, is evaluated at the midplane. The expression in brackets is simply the excess ionic concentration at the midplane, so the repulsive force per unit area is equal to the osmotic pressure.

For low potentials the force is proportional to the square of the potential, viz.

$$F = \frac{e^2 z^2 n_b}{kT} \psi_0^2. \tag{4.9.3}$$

Then, ignoring interactions between the plates and simply adding potentials from two isolated plates (4.7.7) yields

$$F = 64kTn_b \tanh^2(\tfrac{1}{4}\Psi_s)\exp(-\kappa h) \tag{4.9.4}$$

This non-linear superposition approximation clearly requires that the plate separation, h, is large compared with the Debye length.

To obtain accurate values for the variation of the repulsive force with separation, (4.7.1) must be solved. Constant-charge and constant-potential boundary conditions furnish bounds on the force–distance relation for surfaces that regulate their charge according to the mass action equilibria studied in connection with the isolated plate (Ninham & Parsegian, 1971; Chan, Perram, White, & Healy, 1975; Chan, Healy, & White, 1976). For a $z-z$ electrolyte, the Poisson–Boltzmann equation can be integrated once analytically. A second integration yields

$$\int_{\zeta_0}^{\zeta} (\beta^2 - 1)^{-1/2} (\beta - \zeta_0)^{-1/2}\, d\beta = \sqrt{2}\,\kappa x, \tag{4.9.5}$$

where

$$\zeta = \cosh ez\psi/kT \tag{4.9.6}$$

and the subscript indicates conditions at the midplane. This expression can be evaluated numerically with either constant charge or constant potential at the surface $x = h/2$. Then, using the potential at the midplane, the osmotic pressure or repulsive force is calculated. Figure 4.11 shows results for two different potentials, along with the non-linear superposition approximation (4.9.4). Note that at small separations the constant charge results begin to diverge, reflecting the singular interaction at contact.

At this point it is important to recognize that interactions between surfaces with unequal charges or potentials differ from those just studied. The force can be calculated in much the same way but the lack of symmetry about the midplane introduces an additional stress,

$$-\tfrac{1}{2}\varepsilon\varepsilon_0 \mathbf{E} \cdot \mathbf{E}.$$

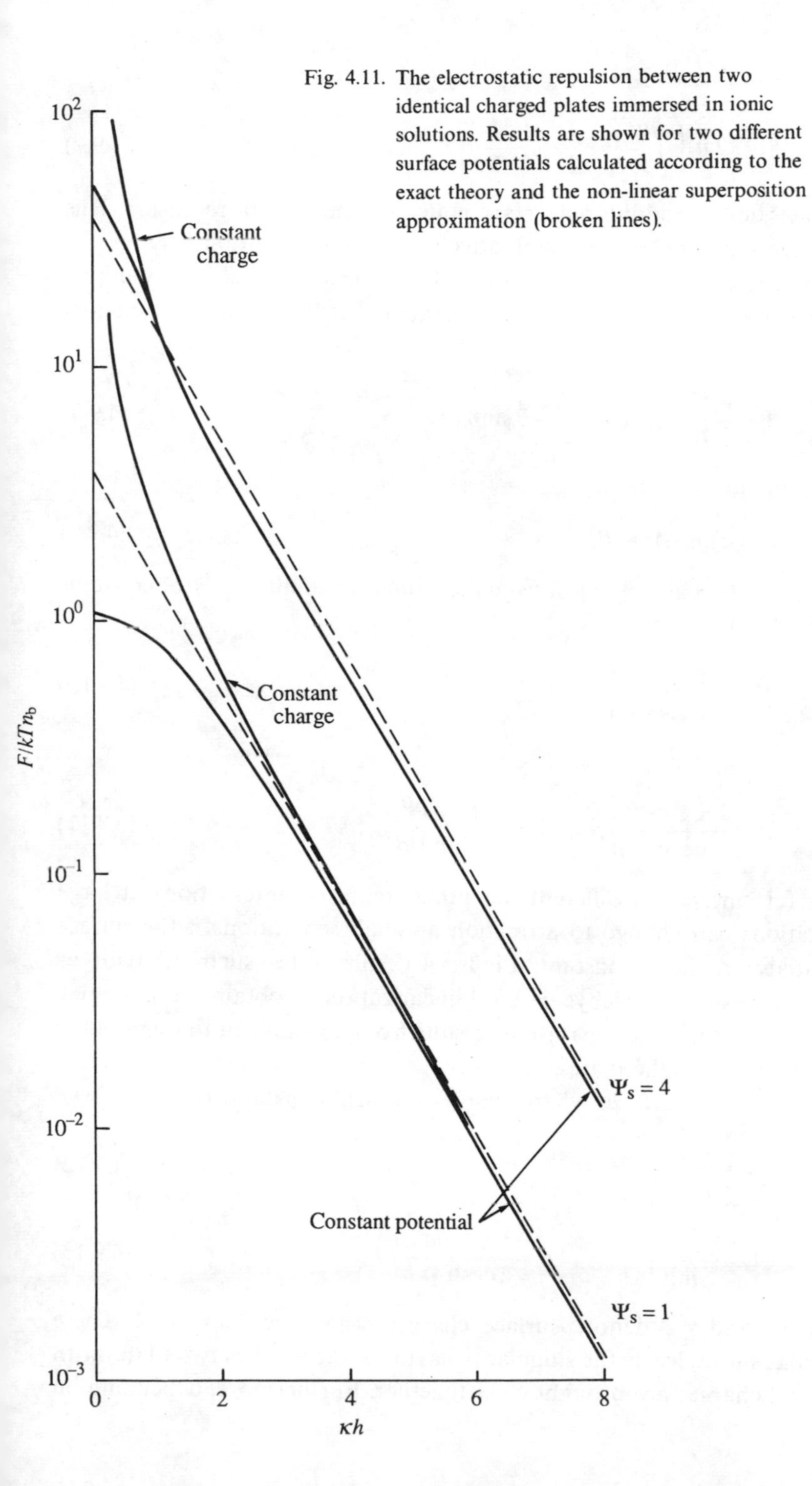

Fig. 4.11. The electrostatic repulsion between two identical charged plates immersed in ionic solutions. Results are shown for two different surface potentials calculated according to the exact theory and the non-linear superposition approximation (broken lines).

The linearized form of the force

$$F = kTn_b\left(\frac{ez\psi}{kT}\right)_0^2 - \frac{1}{2}\varepsilon\varepsilon_0\left(\frac{\partial\psi}{\partial x}\right)_0^2 \tag{4.9.7}$$

illustrates how the additional stress counteracts the osmotic repulsion. This stress alters the interaction qualitatively as well as quantitatively.

All this is illustrated by the solution of the Debye–Hückel equation for a $z-z$ electrolyte between two plates separated by a distance h. The potential is

$$\Psi = \frac{ez\psi}{kT} = A\cosh\kappa y + B\sinh\kappa y, \tag{4.9.8}$$

and so the force is simply

$$F = kTn_b(A^2 - B^2). \tag{4.9.9}$$

The constants A and B follow from the boundary conditions. For constant potentials, i.e.,

$$\Psi = \begin{cases} \Psi_+, & y = h/2 \\ \Psi_-, & y = -h/2 \end{cases} \tag{4.9.10}$$

we obtain

$$A = \frac{1}{2}\frac{\Psi_+ + \Psi_-}{\cosh\frac{1}{2}\kappa h}, \qquad B = \frac{1}{2}\frac{\Psi_+ - \Psi_-}{\sinh\frac{1}{2}\kappa h}. \tag{4.9.11}$$

Thus, for surfaces at different potentials, repulsive interactions at large separations can change to attraction at small separations if the surface potentials are held constant. Figure 4.12 illustrates such behavior as calculated using the Debye–Hückel linearization to obtain the potential. The behavior at close separations results from a change in the sign of the charge on one of the plates.

Similarly, for surfaces where charges are held constant, i.e.,

$$q = \begin{cases} q_+, & y = h/2, \\ q_-, & y = -h/2, \end{cases} \tag{4.9.12}$$

$$A = \frac{1}{2}\frac{q_+ + q_-}{\sinh\frac{1}{2}\kappa h}, \qquad B = \frac{1}{2}\frac{q_+ - q_-}{\cosh\frac{1}{2}\kappa h}. \tag{4.9.13}$$

Here q_+ and q_- denote surface charges scaled on $\varepsilon\varepsilon_0 kT\kappa/ez$. These formulas show clearly the singular behavior of the force as two plates with identical charges are brought close together. Furthermore, depending on the relative magnitude of the two charges, the force can change from attraction to repulsion as the plates are brought together. Interactions between particles with dissimilar charges are at the core of the subject of heterocoagulation (see Hogg, Healy, & Fuerstenau, 1966; Gregory, 1975; Ohshima, Healy, & White, 1982).

4.10 Repulsion between charged spheres

Given our experience with single spheres, the absence of closed-form solutions for the two-sphere problem comes as no surprise. Moreover, because of curvature, the repulsive force derives from both osmotic pressure and an electric stress. We can use the general relation, (4.9.1), along with the equilibrium condition, to show that the repulsive force can be obtained from an integration over the central plane (cf. Fig. 4.13) as

$$\mathbf{F} = \int_S kT \sum_1^N n_b^k [\exp(-ez^k\psi_0/kT) - 1]\mathbf{n}\, dS + \int_S \varepsilon\varepsilon_0 [\mathbf{EE} - \tfrac{1}{2}\mathbf{E}\cdot\mathbf{E}\boldsymbol{\delta}]\cdot\mathbf{n}\, dS. \tag{4.10.1}$$

Fig. 4.12. The electrostatic force on plates charged to different potentials, as calculated according to the Debye–Hückel theory. One plate has a dimensionless surface potential of 1.0; the other is 0.5.

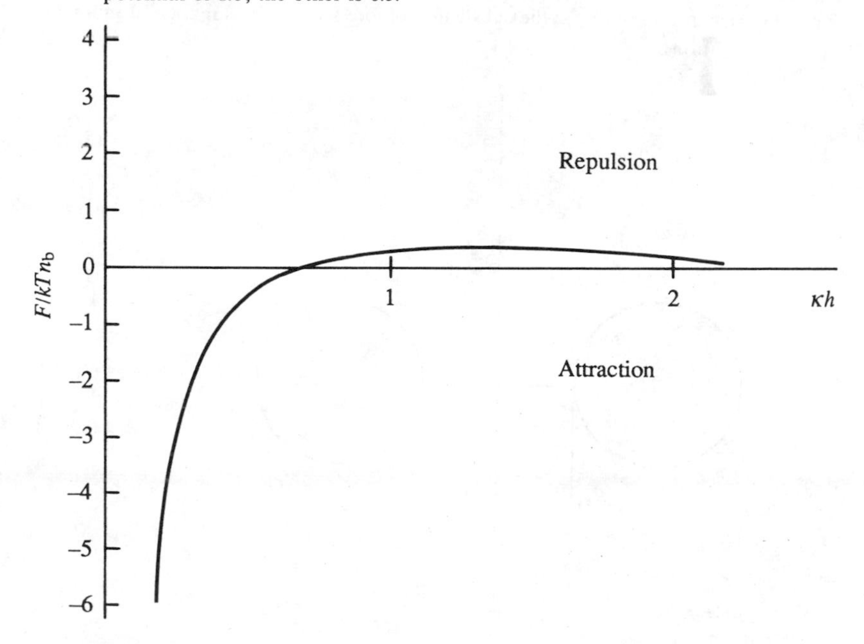

A variety of approximations provide insight into the qualitative and quantitative behavior. The Derjagiun approximation, for example, is applicable for separations small compared with the radius of the spheres (Derjaguin, 1934). Under such conditions, elements on each sphere interact as parallel plane elements at the same separation; the total interaction is a sum over the infinitesimal elements. To proceed formally, we adopt a polar cylindrical coordinate system with its axis joining the centers of the spheres and centered at the midpoint (Fig. 4.13). A sphere surface is defined by

$$z_* = \tfrac{1}{2}h + a[1-(1-r_*^2/a^2)^{1/2}]. \tag{4.10.2}$$

Scaling distances as

$$z = \kappa z_*,\ r = (\kappa a)^{1/2} r_*/a \tag{4.10.3}$$

and expanding the potential as

$$\Psi = \Psi_1(r,z) + (\kappa a)^{-1}\Psi_2(r,z) + \ldots, \tag{4.10.4}$$

leads to

$$\frac{\partial^2 \Psi_1}{\partial z^2} = \sinh \Psi_1 \tag{4.10.5}$$

Fig. 4.13. Definition sketch for the calculation of forces on spheres immersed in ionic solutions.

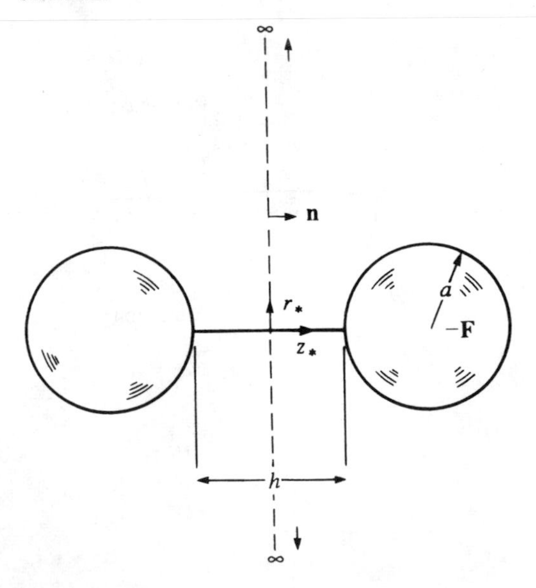

for a $z-z$ electrolyte, with

$$\frac{\partial \Psi_1}{\partial z}=0 \tag{4.10.6}$$

at the midplane where $z=0$. On the surface of the sphere, $z=g(r)$, the condition is

$$\Psi_1=\Psi_s \qquad \text{or} \qquad \frac{\partial \Psi_1}{\partial z}=q, \tag{4.10.7}$$

where

$$g(r)=\tfrac{1}{2}\kappa h+\tfrac{1}{2}r^2, \tag{4.10.8}$$

and the charge is scaled on $\varepsilon\varepsilon_0 kT\kappa/ez$.

The problem has been reduced to one dimension but further analytical progress requires linearization of the differential equation, i.e. small potentials (see Glendinning & Russel, 1983). The force derived from (4.10.1),

$$F\approx\pi\varepsilon\varepsilon_0\left(\frac{kT}{ze}\right)^2\int_0^\infty\left[(\kappa a)\Psi_1^2+\left(\frac{\partial\Psi_1}{\partial r}\right)^2\right]r\,\mathrm{d}r, \tag{4.10.9}$$

with potentials scaled on kT/ez, yields the following expressions for the linearized Derjaguin approximation:

Constant potential: $$F\approx 2\pi\varepsilon\varepsilon_0\left(\frac{kT}{ze}\right)^2\kappa a\,\Psi_s^2\frac{\exp(-\kappa h)}{1+\exp(-\kappa h)}, \tag{4.10.10}$$

Constant change: $$F\approx 2\pi\varepsilon\varepsilon_0\left(\frac{kT}{ze}\right)^2\kappa a\,q^2\frac{\exp(-\kappa h)}{1-\exp(-\kappa h)}. \tag{4.10.11}$$

Extensions of the analytical solution show that the error for constant potential boundary conditions remains small as the gap is diminished. Conversely, the terms neglected in the constant charge calculation grow without bound, showing that this approximation is invalid when the gap is much smaller than the Debye thickness. This problem stems from the radial gradients in the potential neglected in (4.10.5).

Linear superposition of single sphere potentials also provides a useful approximation for the repulsive force (see Bell, Levine, & McCartney, 1970, for a discussion of techniques). From the Debye–Hückel solution around a single sphere, (4.8.8), we find

$$F\approx\pi\varepsilon\varepsilon_0\left(\frac{kT}{ze}\right)^2\Psi_s^2\frac{1+\kappa(h+2a)}{(h/2a+1)^2}\exp(-\kappa h) \tag{4.10.12}$$

for the force. For spheres with thin double-layers, (4.8.1) yields

$$F \approx 32\pi\varepsilon\varepsilon_0\left(\frac{kT}{ze}\right)^2 \kappa a \tanh^2(\tfrac{1}{4}\Psi_s)\exp(-\kappa h). \tag{4.10.13}$$

By solving the linearized Poisson–Boltzmann equation through a multipole expansion, Glendinning & Russel (1983) mapped out regions

Fig. 4.14. Comparison between the linearized Derjaguin approximation and linear superposition for the electrostatic force between two spheres. Labels in each domain indicate which of the approximations is within 10 per cent of the exact force when the potential is small. Shaded domains show where neither approximation is appropriate (Glendinning & Russel, 1983). (*a*) Constant-potential boundary condition. (*b*) Constant-charge boundary condition.

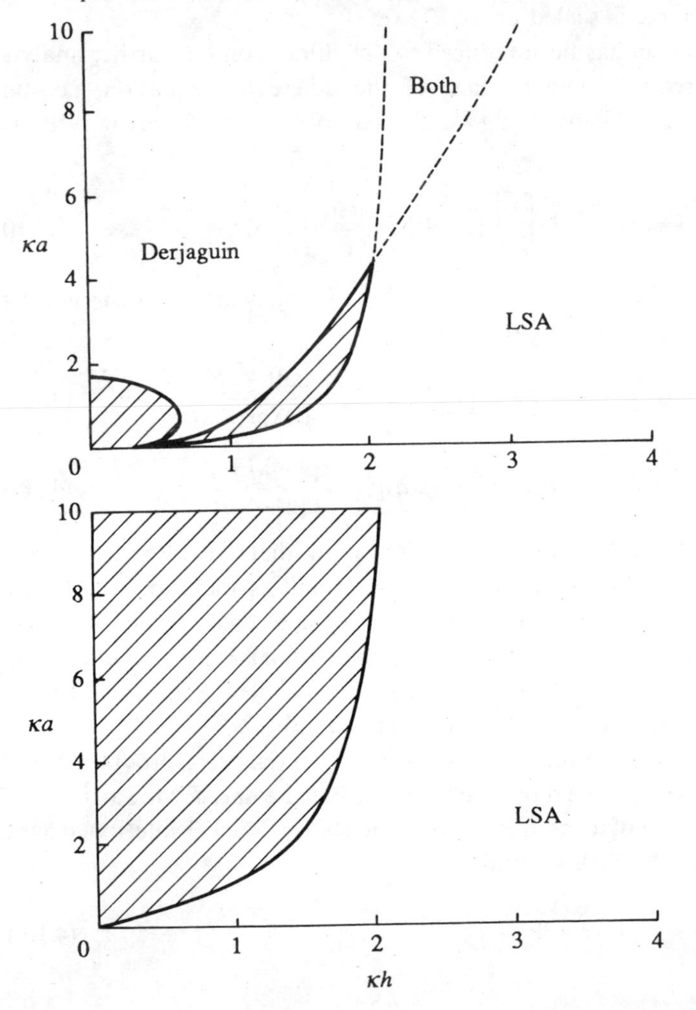

where the Derjaguin and linear superposition approximations are valid with small potentials. Figure 4.14 shows results for constant charge and constant potential. Note especially the limitation on the linearized Derjaguin approximation for constant-charge boundary conditions.

For thin double-layers, the situation for which it is intended, the Derjaguin approximation produces very accurate results for constant-potential boundary conditions when the potential is derived from the one-dimensional, non-linear Poisson–Boltzmann equation. A convincing set of examples was given by Chan & Chan (1983), who used exact, numerical solutions for reference. Figure 4.15 compares the repulsive force calculated from a finite-element solution with that from the non-linear Derjaguin approximation and linear superposition, all for the constant-potential boundary condition.

Attention has centered on the repulsive force due to electrostatic interactions since this is measured directly in many experiments. However, it is also necessary to know the electrostatic interaction energy, Φ, defined as

$$F = -\frac{\partial \Phi}{\partial h}. \tag{4.10.14}$$

Fig. 4.15. The repulsive force (scaled on $\varepsilon\varepsilon_0(kT/e)^2$) between two identical spheres for $\kappa a = 10$ and various potentials. The different symbols represent: (*a*) the non-linear Derjaguin approximation, ——; (*b*) finite-element solutions of the exact equations, +; the linear superposition approximation, ●; and numerical, finite-difference solutions by McCartney & Levine (1969), ■; (Chan & Chan, 1983).

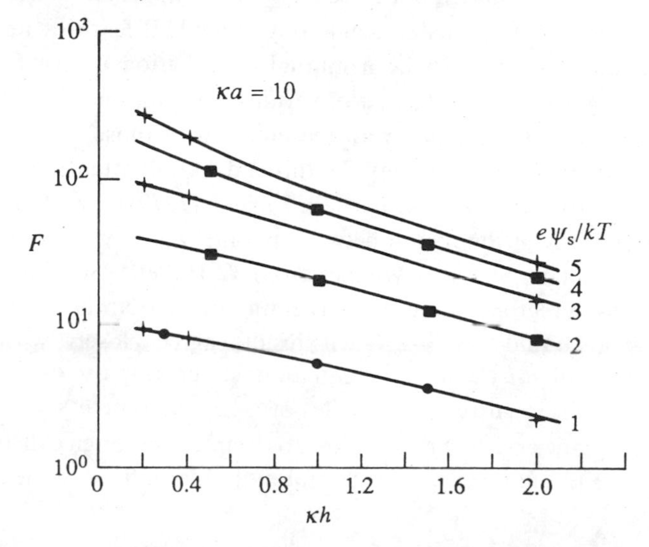

Table 4.3. *Interaction energies*

Geometry	Constraint	Force expression	Φ
Two flat plates	Superposition	(4.9.4)	$64kTn_b\kappa^{-1}\tanh^2(\frac{1}{4}\Psi_s)\exp(-\kappa h)$
Two spheres	Constant potential	(4.10.10)	$2\pi\varepsilon\varepsilon_0\left(\frac{kT}{ze}\right)^2 a\Psi_s^2 \ln(1+e^{-\kappa h})$
Two spheres	Constant charge	(4.10.11)	$-2\pi\varepsilon\varepsilon_0\left(\frac{kT}{ze}\right)^2 aq^2 \ln(1-e^{-\kappa h})$
Two spheres	Linear superposition	(4.10.12)	$4\pi\varepsilon\varepsilon_0\left(\frac{kT}{ze}\right)^2 \frac{a^2}{h+2a}\Psi_s^2 \exp(-\kappa h)$
Two spheres	Superposition	(4.10.13)	$32\pi\varepsilon\varepsilon_0\left(\frac{kT}{ze}\right)^2 a\tanh^2(\frac{1}{4}\Psi_s)\exp(-\kappa h)$

This energy can be calculated from the repulsive force by integration. Interaction energies (or potentials) will be used extensively in the sections dealing with colloid stability. Table 4.3 lists approximate forms.

4.11 Tests of the Gouy–Chapman theory

Before we discuss the measurements themselves, some related matters deserve mention. The Poisson–Boltzmann equation stands at the core of the Gouy–Chapman model, but there are so many approximations involved that it would be perilous to single out individual features for criticism. Indeed, there appears to be a mutual cancellation of errors. Instead of attempting a piecemeal testing of various features, one should compare results of calculations made with the model with those derived from a completely different methodology. To this end we call attention to results from Monte Carlo simulations. Torrie & Valleau (1979) carried out Monte Carlo simulations of the region between parallel plates with ionic strengths of 0.1 M and 1 M; Jonsson, Wennerstrom, & Halle (1980) did a similar calculation for a region containing only counterions to simulate the behavior of lamellar liquid crystals. Their results agree closely with predictions from the Poisson–Boltzmann equation, supporting use of the Gouy–Chapman model down to length scales of a few nanometers.

Another matter of concern is the region close to the interface, often called the Stern layer. Here the finite sizes of ions and molecules influence their arrangement, leading to localized perturbations in the electric field which are ignored in the Gouy–Chapman model. Evidence based on approximate mathematical models leads us to believe that 'discrete charge effects' are short-range and can be ignored, as long as the separation between two opposed surfaces is more than the lateral distance between the charges on the surface (see Israelachvili, 1985). Accordingly, the Gouy–Chapman model appears to be on firm ground insofar as the theory is concerned. It remains to establish that the theory is applicable to real systems.

It is difficult to construct an experimental test of the theory uninfluenced by other phenomena. Measurements of the diffuse-layer capacitance with silver iodide or mercury surfaces in contact with ionic solutions, for example, are confounded by charge stored in the compact (Stern) layer, necessitating a reliable model of adsorption in the surface region. Haydon (1964) reviews the data on AgI in detail; Sparnaay (1972) summarizes capacitance measurements on both types of interfaces. Although capacitance studies are extremely valuable in their own right, they do not test the Gouy Chapman theory, *per se*. Recent developments in techniques for direct measurement of forces between opposed objects, on the other hand, offer convincing evidence that the theory gives a correct representation of the electrostatic repulsion.

The technique currently employed for measuring forces directly involves positioning two objects close together and then measuring both the force and the separation distance. Early measurements made with glass bodies separated by an air gap were carried out in the USSR by Derjaguin's group and in the Netherlands by Overbeek's group. Recently the technique has been improved by J. N. Israelachvili, and it is now possible to make accurate measurements at separations down to 0.1 nm in aqueous solutions with molecularly smooth micra surfaces. Further details are given in Chapter 5 where we discuss dispersion forces.

The force between surfaces immersed in an ionic solution consists of attractive (van der Waals) and repulsive (electrostatic) parts. Figures 4.16 to 4.18 compare the experimental measurements with calculations from the Gouy–Chapman theory as outlined in §4.10, combined with the theory of attractive forces discussed in Chapter 5. Note that 1-1, 2-1, and 3-1 electrolytes are included, and the measurements cover a range of ionic strengths. Since the surface charge cannot be measured directly, the electrostatic potential is chosen to give the best 'fit' to the data. Accordingly the theory should be judged on the accuracy of the force–distance relation. The good agreement implies that the Gouy–Chapman model provides an accurate representation of electrostatic repulsion for these surfaces.

In §4.10, differences between electrostatic interactions at constant charge and constant potential were discussed. Figure 4.16 indicates that the two are indistinguishable down to separations of roughly 10 nm in a 10^{-4} M KNO_3 solution. This is about one-third the thickness of the diffuse layer (29 nm); cf. Fig. 4.11. At smaller separations the interaction appears to be nearer that expected for constant charge, until the surfaces jump into contact as the attractive component becomes dominant.

Other measurements by Pashley (1984) and Pashley & Israelachvili (1984) show good agreement with the Gouy–Chapman theory for divalent and trivalent ions; cf. Figs 4.17 and 4.18. Pashley and Israelachvili (1984) also showed that a charge regulation model of the sort discussed in §§4.5 and 4.7 could represent the relation between surface potential and concentration for mica surfaces in solutions with divalent metal ions. These experimental results provide unambiguous evidence that the Gouy–Chapman model is an accurate representation of electrostatic interactions between charged surfaces.

Fig. 4.16. Repulsive force measured between two curved mica surfaces in 10^{-4} M KNO_3 solutions, ●. The lower curve is calculated on the basis of interactions at constant potential, the upper one for constant charge; each is based on a surface potential of -130 mV at infinite separation. Note that the force, F is divided by the radius of curvature of the surfaces, R, so as to have the units of energy per unit area. Redrawn from Israelachvili & Adams (1978).

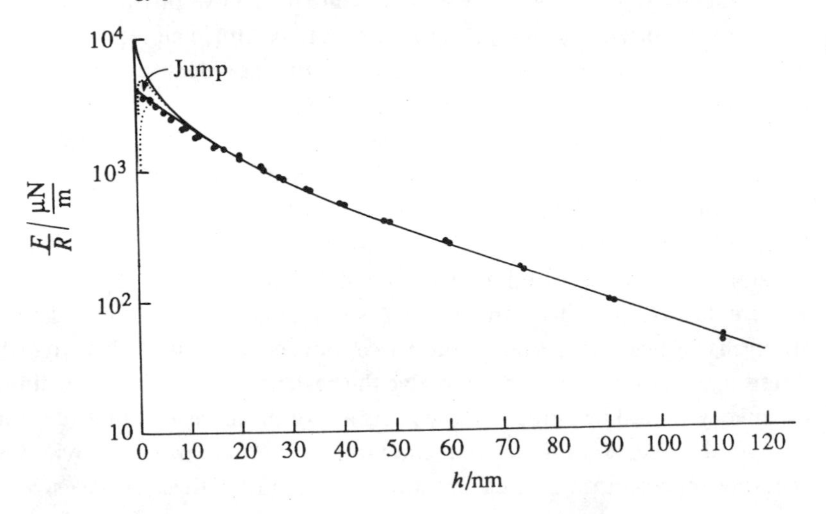

4.12 Summary

In the model of the electrostatic behavior presented here, the suspending fluid is envisioned as a uniform, structureless dielectric material containing ions derived from the complete dissociation of one or more electrolytes. The ions move under the influence of electrostatic and Brownian forces, distributing themselves to produce an equilibrium state. Macroscopic bodies carry a charge on their surfaces and this produces non-uniform ion distributions. The balance laws used to describe the equilibrium state derive from Maxwell's equations, simplified by the suppression of magnetic effects. Effects due to the finite size of the ions and the non-uniform packing of ions and molecules close to the interface (the Stern layer) are also neglected. Nevertheless, the Gouy–Chapman model yields results consistent with those from more detailed treatments.

Fig. 4.17. The force between two curved mica surfaces in solutions of 2–1 electrolytes (Mg^{+2}, Ca^{+2}, Sr^{+2}, and Ba^{+2} at pH = 5.8). The solid lines are based on the theory for potentials and concentrations shown along with van der Waals attraction corresponding to a Hamaker constant of 2.2×10^{-20} J (see §5.5). Pashley & Israelachvili (1984).

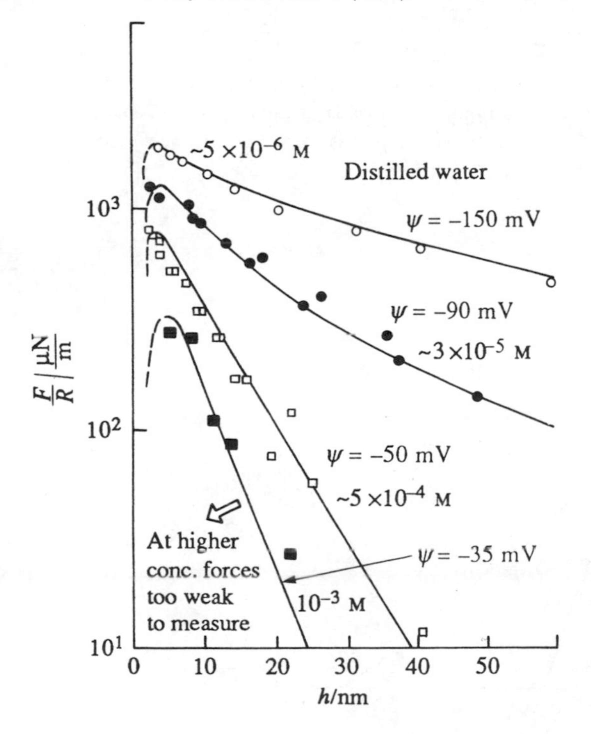

The results presented here are used in two ways in later sections. First, the electrostatic forces are incorporated into a description of equilibrium properties of colloidal suspensions, principally concerning questions of stability. Then they are used to describe the behavior of colloidal particles under the influence of various external forces so as to understand non-equilibrium aspects of colloidal behavior.

References

Bell, G. M., Levine, S. & McCartney, L. N. (1970). Approximate methods of determining the double-layer free energy of interaction between two charged colloidal spheres. *J. Colloid Interface Sci.* **33**, 335–59.

Booth, F. (1950). The cataphoresis of spherical, solid non-conducting particles in a symmetrical electrolyte. *Proc. Roy. Soc. Lond. A* **203**, 514–33.

Chan, B. K. C. & Chan, D. Y. C. (1983). Electrical double-layer interaction between spherical colloid particles: An exact solution. *J. Colloid Interface Sci.* **92**, 281–3.

Fig. 4.18. The force between two curved mica surfaces in solutions of a 3–1 electrolyte, $LaCl_3$. The attractive part of the interaction is calculated for a Hamaker constant of 2.2×10^{-20} J. The repulsive part is calculated for the surface potentials shown. Pashley (1984).

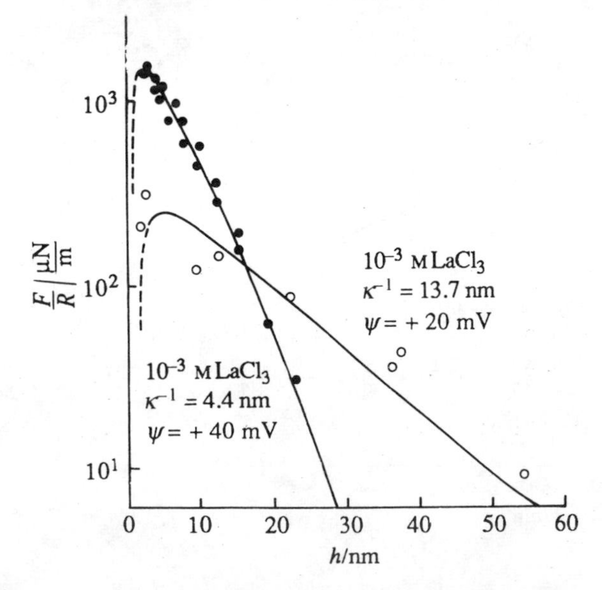

Chan, D. Y. C., Healy, T. W. & White, L. R. (1976). Electrical double-layer interactions under regulation by surface ionization equilibria Dissimilar amphoteric surfaces. *J. Chem. Soc. Faraday Trans. I.* **72**, 2844–65.

Chan, D. Y. C., Perram, J. W., White, L. R. & Healy, T. W. (1975). Regulation of surface potential at amphoteric surfaces during particle–particle interaction. *J. Chem. Soc. Faraday Trans. I.* **71**, 1046–57.

Chapman, D. L. (1913). A contribution to the theory of electroencapillarity. *Phil. Mag.* **25** (6), 475–81.

Derjaguin, B. V. (1934). Friction and adhesion. IV: The theory of adhesion of small particles. *Kolloid Z.* **69**, 155–64.

Dukhin, S. S. & Derjaguin, B. V. (1974). Electrokinetic phenomena, in *Surface and Colloid Science* (ed. E. Matjevic), vol. 7. Wiley.

Feynman, R. P., Leighton, R. B. & Sands, M. (1964). *The Feynman Lectures on Physics*, vol. II. Addison-Wesley.

Glendinning, A. B. & Russel, W. B. (1983). The electrostatic repulsion between charged spheres from exact solutions to the linearized Poisson–Boltzmann equation. *J. Colloid Interface Sci.* **93**, 95–104.

Gronwall, T. H., LaMer, V. K. & Sandved, R. (1928). Über den Einfluss der sogenanten hoheren Glieder in der Debye–Hückelschen Theorie der Losungen starker Electrolyte. *Phys. Z.* **29**, 358–93.

Gouy, G. (1910). Sur la constitution de la charge électrique à la surface d'un électrolyte. *J. Phys. Radium* **9**, 457–68.

Gregory, J. (1970). Interaction of unequal double layers at constant charge. *J. Colloid Interface Sci.* **51**, 44–51.

Guggenheim, E. A. (1967). *Thermodynamics.* North Holland.

Haydon, D. A. (1964). The electrical double-layer and electrokinetic phenomena, in *Progress in Surface Science*, vol. I (ed. J. F. Danelli, K. G. A. Pankhurst and A. C. Riddiford), pp. 94–158. Academic Press.

Hogg, R., Healy, T. W. & Fuerstenau, D. W. (1966). Mutual coagulation of colloidal dispersions. *Trans. Faraday Soc.* **62**, 1638–51.

Hunter, R. J. (1981). *Zeta Potential in Colloid Science.* Academic Press.

Israelachvili, J. N. (1985). *Intermolecular and Surface Forces.* Academic Press.

Israelachvili, J. N. & Adams, G. E. (1978). Measurement of forces between two mica surfaces in aqueous electrolyte solutions in the range 0–100 nm. *J. Chem. Soc. Faraday Trans. I.* **74**, 975–1001.

Jonsson, B., Wennerstrom, H. & Halle, B. (1980). Ion distribution in lamellar liquid crystals. A comparison between Monte Carlo simulations and solutions of the Poisson–Boltzmann equation. *J. Phys. Chem.* **84**, 2179–85.

Kruyt, J. R. (1949). *Colloid Science*, vol. I. Elsevier.

Landau, L. D. & Lifshitz, E. M. (1960). *Electrodynamics of Continuous Media.* Pergamon.

Loeb, A. L., Overbeek, J. Th. G. & Wiersema, P. H. (1961). *The Electrical Double-Layer Around a Spherical Colloid Particle.* MIT Press.

McCartney, L. N. & Levine, S. (1969). An improvement on Derjaguin's expression at small potentials for the double-layer interaction energy of two spherical colloidal particles. *J. Colloid Interface Science* **30**, 345–54.

McQuarrie, D. A. (1976). *Statistical Mechanics.* Harper & Row.

Melcher, J. R. (1981). *Continuum Electromechanics.* MIT Press.

Ninham, B. W. & Parsegian, V. A. (1971). Electrostatic potential between surfaces bearing ionizable groups in ionic equilibrium with physiologic saline solution. *J. Theor. Biol.* **31**, 405–28.

Ohshima, H., Healy, T. W. & White, L. R. (1982). Improvement on the Hogg–Healy–Fuerstenau formulas for the interaction of dissimilar double layers. *J. Colloid Interface Sci.* **89**, 484–93.

Pashley, R. M. (1984). Forces between mica surfaces in La^{3+} and Cr^{3+} electrolyte solutions. *J. Colloid Interface Sci.* **102**, 23–35.

Pashley, R. M. & Israelachvili, J. N. (1984). DLVO and hydration forces between mica surfaces in Mg^{2+}, Ca^{2+}, Sr^{2+}, and Ba^{2+} chloride solutions. *J. Colloid Interface Sci.* **97**, 446–55.

Sanfeld, A. (1968). *Thermodynamics of Charged and Polarized Layers.* Wiley.

Sherwood, J. D. (1980). The primary electroviscous effect. *J. Fluid Mech.* **101**, 609–29.

Sparnaay, M. J. (1972). *The Electrical Double Layer.* Pergamon.

Torrie, G. M. & Valleau, J. P. (1979). A Monte Carlo study of an electrical double layer. *Chem. Phys. Lett.* **65**, 343–6.

White, L. R. (1977). Approximate analytic solution of the Poisson–Boltzmann equation for a spherical colloid particle. *J. Chem. Soc. Faraday II* **73**, 377–96.

Problems

1 Electrostatic fields cause effects similar to those of surface tension. Consider the effect when two large parallel plates are dipped into a dielectric liquid. Imagine that one plate is held at a potential ψ_1, while the other is at ψ_0, and calculate the height to which the liquid will rise against the force of gravity. All the fluid interfaces are assumed flat, so surface tension effects are negligible; the liquid is a perfect dielectric and no ions are present.

2 A liquid droplet immersed in a second liquid assumes a spherical shape under the influence of interfacial tension. Suppose a uniform field is imposed and calculate the deformation of the droplet. Both liquids are perfect dielectrics and the free charge density is zero.

3 Show that the second integration involved in deriving (4.7.4) can be expressed in terms of an integral of the form

$$2\int \frac{dX}{X^2-1} \qquad \text{where } X=\exp(\tfrac{1}{2}\Psi).$$

Use this result to derive (4.7.4).

4 Derive an analytical expression relating potential and position for a 2–1 electrolyte adjacent to a flat plate.

Graph the potential as a function of position in 0.1 mM solutions for 1–1 and 2–1 electrolytes using surface potentials of 50 mV and 150 mV. Compare the exact results with those obtained with the Debye–Hückel approximation.

5 Show that the charge in the diffuse region around a charged particle balances the surface charge so that the particle and its diffuse charge comprise an electrically neutral system.

6 Derive the equations that must be solved to obtain the $O(\Psi^2)$ and $O(\Psi^3)$ terms in the asymptotic expansion (4.8.9) for a symmetrical electrolyte.

7 Use the linearized Derjaguin approximation to calculate the electrostatic potential between two crossed cylinders. Does the torque tend to align the cylinders?

8 Particles in a shear flow can be brought into close proximity if they lie on different streamlines. Calculate the separation at which electrostatic repulsion balances the viscous force tending to bring two particles together using the near-field forms of the hydrodynamic functions described in Chapter 2.

9 The non-linear Poisson–Boltzmann equation must be integrated numerically to calculate the electrostatic repulsion between flat plates. As noted in §4.9 , this can be reduced to a simple quadrature, i.e. (4.9.5), and an integration by parts will remove the singularity from the denominator. Then, given the potential on the plates, separations can be calculated for selected values of the midplane potential.

Derive the relation between the midplane potential and the potential on either of identical flat plates when the potential on the plates is set by a charge regulation mechanism. The ionization leaves covalently bound, negatively charged groups according to the scheme

$$AH \rightleftharpoons A^- + H^+.$$

Develop a computer algorithm to carry out the quadrature necessary to calculate the surface potential and spacing for a given value of the midplane potential. Calculate the midplane potential and spacing for selected values of the relevant parameters.

10 The electrostatic repulsion between identical parallel plates in a $z-z$ electrolyte is given by

$$F = 2kTn_b[\cosh(ez\psi_0/kT) - 1],$$

where ψ_0 denotes the midplane potential and F denotes the force per unit area. Show that the potential energy of the interaction per unit area, Φ_{fp}, is given by

$$\int_0^{F(h)} (x-h)\mathrm{d}F, \qquad (**)$$

where x is the distance measured from the midplane.

To calculate the potential as a function of spacing, it is convenient to use equally spaced values of F; (**) can be used to select ψ_0 then the spacing can be calculated from the algorithm developed in problem 9.

The interaction potential for spherical particles with thin double-layers can be calculated from the flat-plate solution using what is known as the Derjaguin approximation; see §5.7. Accordingly, for identical spheres

$$\Phi = \pi a \int_h^{\infty} \Phi_{fp}\, \mathrm{d}x.$$

Show that the integral can be transformed into

$$\tfrac{1}{2}\int_0^h (x-h)^2\, \mathrm{d}F.$$

Develop a computer algorithm to calculate the electrostatic interaction potential between identical spheres by evaluating the integral using results from problem 9. Calculate the interaction potential as a function of spacing for selected values of the relevant parameters.

5

DISPERSION FORCES

5.1 Introduction

Microscopic observations of colloidal particles in the nineteenth century revealed their tendency to form persistent aggregates through collisions induced by Brownian motion, clearly indicating an attractive interparticle force. Identification of its origin, however, awaited the quantitative descriptions of van der Waals forces between molecules developed in the 1920s (Israelachvili, 1985). This development prompted Kallman & Willstätter (1932) and Bradley (1932) to realize the summation over pairs of molecules in interacting particles would yield a long-range attraction.

Subsequently, de Boer (1936) and Hamaker (1937) performed explicit calculations of dispersion forces between colloidal particles by assuming the intermolecular forces to be strictly pairwise additive. Although approximate, this theory captures the essence of the phenomenon. The attraction arises because local fluctuations in the polarization within one particle induce, via the propagation of electromagnetic waves, a correlated response in the other. The associated free energy decreases with decreasing separation. Phase shifts introduced at large separations by the finite velocity of propagation reduce the degree of correlation, and, therefore, the magnitude of the attraction. Although the intermolecular potential decays rapidly on the molecular scale, the cumulative effect is a long-range interparticle potential that scales on the particle size.

Several difficulties with the Hamaker theory were recognized from the outset (e.g., Mahanty & Ninham, 1976, §1.7). One stems from the importance of many-body interactions in condensed media that render suspect the assumption of pairwise additivity for exactly the materials of interest. The theory also fails at separations on the order of molecular dimensions, predicting an infinite attraction at contact.

The continuum theory (Lifshitz, 1956; Dzaloshinskii, Lifshitz, & Pitaevski, 1961) accounts naturally for many-body effects by treating the particles and the intervening fluid as individual macroscopic phases characterized by dielectric permittivities. The full treatment employed the methods of quantum electrodynamics but was reformulated in semi-classical terms by van Kampen, Nijboer, & Schram (1968). Quantitative implementation of the theory for colloidal problems began with Parsegian & Ninham (1969, 1970). The results provide means for quantitative predictions of dispersion forces for many systems of interest.

In this chapter we strive to present the continuum theory in a straightforward manner proceeding from the basic formulation through calculations for specific materials. The treatment begins with a discussion of intermolecular forces and their summation according to the Hamaker approach. Then we delve into the continuum theory, addressing in turn the dielectric properties of materials, the form for the interaction energy, and the application to flat plates. Subsequent sections describe the implementation of the results for specific materials, experimental tests for interactions between macroscopic surfaces, and suitable approximations for interactions between colloidal spheres. The physical arguments and discussions of implementation draw on Parsegian (1975), Hough & White (1980), and Israelachvili (1985), while the mathematical derivations parallel those found in Langbein (1974) and Mahanty & Ninham (1976).

5.2 Intermolecular forces and the microscopic theory

A discussion of intermolecular forces and their pairwise summation to predict forces between colloidal particles serves as a prelude to the continuum theory, illustrating the physical origin of the attraction and the qualitative form of the potential. Our treatment follows that of Langbein (1974, §3.1).

We begin with the solution to Laplace's equation for the electrostatic potential due to a point dipole $\mathbf{p}_1$ in a vacuum,

$$\psi_1(\mathbf{r}) = -\frac{1}{4\pi\varepsilon_0}\mathbf{p}_1 \cdot \nabla r^{-1}. \tag{5.2.1}$$

A second dipole $\mathbf{p}_2$, placed in this field at relative position $\mathbf{r}$, acquires the interaction energy

$$U = \mathbf{p}_2 \cdot \nabla \psi_1, \tag{5.2.2}$$

as constructed from the energy of the individual point charges. Thus the intermolecular potential follows once the dipole moments are known.

Most molecules do not possess permanent dipoles, however. Instead, the absorption of photons from the background radiation field causes random fluctuations with instantaneous values $\mathbf{p}_{\text{inst}}$. The resulting field due to an individual molecule induces a dipole $\mathbf{p}_{\text{ind}}$ in a second molecule, with the strength depending on the molecular polarizability α which can be expressed as

$$\alpha(\omega) = \int_0^\infty f(\tau) e^{i\omega\tau}\, d\tau$$
$$\equiv \alpha'(\omega) + i\alpha''(\omega) \tag{5.2.3}$$

with $f(\tau)$ a memory function. The total interaction between the two molecules then reflects an average over many random fluctuations.

The frequency dependence of α and the time dependence of $\mathbf{p}_{\text{inst}}$ suggest a Fourier representation† with

$$\mathbf{p}_{\text{inst}} = \frac{1}{2\pi}\int_{-\infty}^{\infty} \mathbf{p}(\omega) \exp(-i\omega t)\, d\omega. \tag{5.2.4}$$

The induced dipole follows from (4.2.13), (4.2.14) and (5.2.4) as

$$\mathbf{p}_{\text{ind}} = -\frac{1}{8\pi^2}\int_{-\infty}^{\infty} \alpha(\omega)\mathbf{p}(\omega)\cdot\nabla\nabla r^{-1} \exp(-i\omega t)\, d\omega. \tag{5.2.5}$$

With the identity

$$\frac{1}{2\pi}\int_{-\infty}^{\infty} e^{-i(\omega+\omega')t}\, dt = \delta(\omega+\omega'), \tag{5.2.6}$$

the time-averaged interaction energy obtained from (5.2.2) by substituting (5.2.1), (5.2.4), and (5.2.5) reduces to

$$U(r) = -\frac{1}{32\pi^3\varepsilon_0}\int_{-\infty}^{\infty} \alpha(\omega)\mathbf{p}(\omega)\cdot\nabla\nabla r^{-1}\cdot\nabla\nabla r^{-1}\cdot\mathbf{p}(-\omega)\, d\omega. \tag{5.2.7}$$

Evaluation of $U(r)$ requires knowledge of the magnitude of $\mathbf{p}(\omega)\mathbf{p}(-\omega)$. The fluctuations arise when individual molecules absorb and emit photons, thereby perturbing their electronic orbitals and producing a transient dipole moment. The intensity of the fluctuations at frequency ω, therefore, depends on the ability of the molecules to absorb photons at that frequency,

† This is the one-dimensional version of the transform used in §2.5.

expressed by $\alpha''(\omega)$, and the density of photons with energy $\hbar\omega$ ($2\pi\hbar$ = Planck's constant). The fluctuation–dissipation theorem,

$$\mathbf{p}(\omega)\mathbf{p}(-\omega)=\hbar\varepsilon_0\alpha''(\omega)\coth\frac{\hbar\omega}{2kT}\boldsymbol{\delta}, \tag{5.2.8}$$

defines this relationship. This expression is analogous to (3.2.8), which relates the Brownian forces acting on a particle to the viscosity of the fluid.

Substitution of (5.2.8) into (5.2.7) yields the interaction energy but is difficult to evaluate explicitly. To obtain a more convenient form, note that $\alpha'(-\omega)=\alpha'(\omega)$ and $\alpha''(-\omega)=-\alpha''(\omega)$ according to (5.2.3), so that

$$\begin{aligned} U(r) &= -\frac{3\hbar}{16\pi^3 r^6}\int_{-\infty}^{\infty}\alpha(\omega)\alpha''(\omega)\coth\frac{\hbar\omega}{2kT}\,d\omega \\ &= -\frac{3\hbar i}{32\pi^3 r^6}\int_{-\infty}^{\infty}\alpha^2(\omega)\coth\frac{\hbar\omega}{2kT}\,d\omega. \end{aligned} \tag{5.2.9}$$

Since $\alpha(\omega)\to 0$ as $|\omega|\to\infty$ for both real and imaginary values of ω, the integral in (5.2.9) represents a contour integration around the upper half of the complex plane. Consequently, application of the theory of residues leads to

$$U(r)=-\frac{6kT}{(4\pi)^2 r^6}\sum_{n=0}^{\infty}{}'\alpha^2(i\xi_n), \tag{5.2.10}$$

with $i\xi_n = 2\pi inkT/\hbar$ denoting the poles of $\coth(\hbar\omega/2kT)$ in the upper half plane. The prime over the summation indicates that the $n=0$ term is multiplied by $\frac{1}{2}$. Examination of (5.2.3) shows that evaluation of the complex polarizability at an imaginary frequency yields a real number.

So the correlation between the instantaneous and induced dipoles for like molecules yields a short-range attraction. The energy associated with this attraction is significant (e.g., Israelachvili, 1985, §6.2). For methane with

$$\frac{3}{8\pi^2}kT\sum_{n=0}^{\infty}{}'\alpha^2(i\xi_n)=1.0\times10^{-77}\,\mathrm{J\,m^6},$$

and a molecular diameter of $\sigma=0.4$ nm, (5.2.10) predicts an interaction energy for two molecules in contact of

$$U(\sigma)=-2.4\times10^{-21}\,\mathrm{J}.$$

For a larger molecule such as carbon tetrachloride, with

$$\frac{3}{8\pi^2}kT\sum_{n=0}^{\infty}{}'\alpha^2(i\xi_n)=1.5\times10^{-76}\,\mathrm{J\,m^6},$$

and $\sigma = 0.55\,\text{nm}$, the result is

$$U(\sigma) = -5.4 \times 10^{-21}\,\text{J}.$$

In general, condensed phases form when this attraction becomes comparable to the thermal energy of the molecule. Indeed, at the normal boiling point, $U(\sigma)/kT$ equals -1.55 for methane and -1.12 for carbon tetrachloride.

The treatment above has two restrictions. First, representing the dipoles as points limits the results to separations greater than the molecular size. At smaller separations the electron clouds of atoms within one molecule overlap with those of the second, producing a strong repulsion. Second, the static description of the fields suffices only for separations so small ($r \ll c/\omega$ (c = speed of light)) that the propagation time for radiation between the molecules introduces negligible phase shift. Otherwise the solution to Maxwell's equations, rather than Laplace's, would be required.

The microscopic theory constructs the dispersion potential between two colloidal particles from (5.2.10) by assuming that each molecule in one interacts individually with every molecule in the other without interference by the intervening material. Then the interaction potential for identical particles separated by a vacuum becomes

$$\begin{aligned} \Phi &= \int_{V_1}\int_{V_2} N^2 U(r)\,d\mathbf{x}_1 d\mathbf{x}_2 \\ &= -\frac{A}{\pi^2}\int_{V_1}\int_{V_2} \frac{d\mathbf{x}_1\,d\mathbf{x}_2}{r^6}, \end{aligned} \tag{5.2.11}$$

where

$$r^2 = (\mathbf{x}_1 - \mathbf{x}_2)\cdot(\mathbf{x}_1 - \mathbf{x}_2),$$
N = number density of molecules,
$$A = \tfrac{3}{8} N^2 kT \sum_{n=0}^{\infty}{}' \alpha^2(i\xi_n).$$

Note that the Hamaker constant A accounts for all the material properties independent of the geometrical dependence embodied in the double integral. The result is generalized to unlike materials merely by replacing $N^2\alpha^2$ with $N\alpha\bar{N}\bar{\alpha}$.

The potential for two particles immersed in a fluid with a different polarizability can be constructed rather simply. Two isolated spheres effectively interact individually with spheres of fluid, as depicted in Fig. 5.1(*a*). The interaction then exchanges a fluid sphere and a particle to obtain

Fig. 5.1(b). The interaction potential is the difference between the like and the unlike interactions, with the effect of the intervening fluid cancelling out. Hence, the form of the potential remains as in (5.2.11) but the Hamaker constant depends on the differences of the polarizabilities of the two materials according to

$$A = \tfrac{3}{8}kT \sum_{n=0}^{\infty}{}' \, [N\alpha(i\xi_n) - \bar{N}\bar{\alpha}(i\xi_n)]^2. \tag{5.2.12}$$

Alternatively, the result can be expressed in terms of the dielectric constants through $\varepsilon = 1 + N\alpha$ as

$$A = \tfrac{3}{8}kT \sum_{n=0}^{\infty}{}' \, [\varepsilon(i\xi_n) - \bar{\varepsilon}(i\xi_n)]^2, \tag{5.2.13}$$

although pairwise additivity remains valid only for $\varepsilon - 1 \ll 1$.

Results for a number of geometries are summarized in Table 5.1. Two will be of later use:

(i) Semi-infinite parallel plates at separation h:

$$\Phi_{\text{fp}} = -\frac{A}{12\pi h^2} \qquad \text{(per unit area).} \tag{5.2.14}$$

(ii) Equal spheres of radius a at center-to-center separation r:

$$\begin{aligned} \Phi &= -\tfrac{1}{6}A\left(\frac{2a^2}{r^2 - 4a^2} + \frac{2a^2}{r^2} + \ln\frac{r^2 - 4a^2}{r^2}\right) \\ &\equiv -AH(r/a). \end{aligned} \tag{5.2.15}$$

Fig. 5.1. Decomposition of pair interaction to account for effect of fluid: (a) Isolated particles interacting with fluid spheres. (b) Interacting particles and fluid spheres.

Table 5.1.

Geometry	Potential	
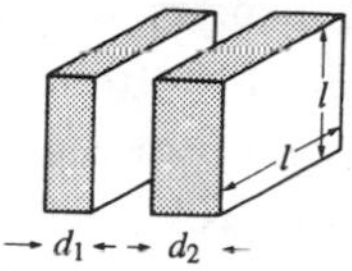 Two slabs (thicknesses d_1, d_2, side l)	$-\frac{l^2 A}{12\pi}\left(\frac{1}{h^2}+\frac{1}{(h+d_1+d_2)^2}-\frac{1}{(h+d_1)^2}-\frac{1}{(h+d_2)^2}\right)$	
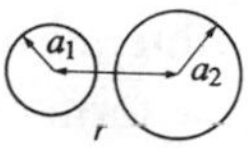 Two spheres (radii a_1, a_2, centre separation r)	$-\frac{A}{6}\left(\frac{2a_1a_2}{r^2-(a_1+a_2)^2}+\frac{2a_1a_2}{r^2-(a_1-a_2)^2}+\ln\frac{r^2-(a_1+a_2)^2}{r^2-(a_1-a_2)^2}\right)$	
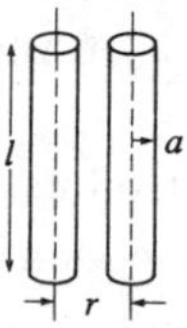 Two parallel cylinders (length l, radius a, separation r)	$-\frac{A}{24}\frac{l}{a}\left(\frac{a}{r-2a}\right)^{3/2}\left(1-\frac{r-2a}{a}+\frac{1}{\sqrt{2\pi}}\ln\frac{r-2a}{a}+\ldots\right)$	$a \gg r-2a$
	$-\frac{3\pi}{8}A\frac{la^4}{r^5}\left(1+\frac{25}{4}\frac{a^2}{r^2}+31.9\frac{a^4}{r^4}+150.7\frac{a^6}{r^6}+\ldots\right)$	$l \gg r \gg a$
	$-A\frac{a^4l^2}{r^6}\left(1-\frac{l^2}{2r^2}+\ldots\right)$	$r \gg l$
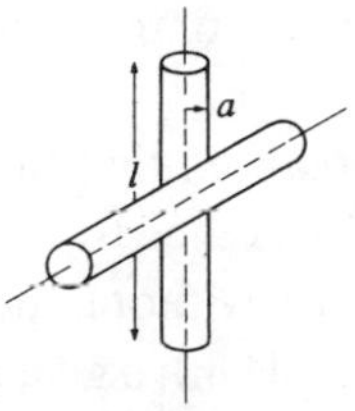 Two crossed cylinders (length l, radius a)	$-\frac{A}{6}\frac{a}{r-2a}\left(1-\frac{3}{2}\frac{r-2a}{a}+\ldots\right)$	$a \gg r-2a$
	$-\frac{\pi}{2}A\left(\frac{a}{r}\right)^4\left(1+\frac{5a^2}{r^2}+21.875\frac{a^4}{r^4}+\ldots\right)$	$a \gg r$

From Mahanty & Ninham 1976.

These demonstrate that the potentials are long-range relative to the intermolecular potential (5.2.9), decaying as $(1/h)^2$ and $a/(r-2a)$ for separations smaller than the particle dimensions. Near contact, both expressions diverge, since the underlying intermolecular potential ignores the short-range repulsion.

The lasting value of the microscopic theory derives from the simple forms of (5.2.14) and (5.2.15), which appear repeatedly in the following chapters. The expression for A generally yields accurate values only for interactions across a vacuum (Israelachvili, 1985). In addition, retardation, introduced by the phase shift between dipoles when $h > c/\omega$, causes A to vary with separation. For a more complete quantitative treatment we turn to the continuum theory.

5.3 Overview of the continuum theory

The microscopic theory establishes the origin of dispersion forces as the fluctuating dipoles induced in non-polar molecules by absorption of photons from the background radiation field. These dipoles generate transient fields, with spectra centered about absorption frequencies, which polarize the surrounding material on wavelengths, $\lambda = 2\pi c/\omega$, of 10–10^4 nm. Hence, the perturbing field arising from a fluctuation at a point transcends the molecular dimensions, producing many-body interactions best described through a continuum theory. From this point of view, objects interact via the fluctuating electromagnetic field that exists in the interior of any absorbing medium and extends beyond its boundaries in the form of travelling waves radiated into the surrounding medium and standing waves that decay exponentially away from the surface. The materials are fully characterized by their dielectric permittivities, the analogues of the molecular polarizabilities.

The original version of the continuum theory (Lifshitz, 1956) introduced into Maxwell's equations a random electric field with Fourier components having magnitudes governed by a fluctuation–dissipation theorem similar to (5.2.8). Solutions were obtained for both travelling and standing waves. The force of attraction then was calculated as the appropriate component of the Maxwell stress tensor.

Subsequently, van Kampen, Nijboer, & Schram (1968) realized that the entire force derives from the standing waves or normal modes excited by the fluctuating fields. Consequently, the theory has been recast without the fluctuations by determining only the normal modes associated with a given geometry for materials with specified dielectric properties. This comprises a boundary value problem with homogeneous equations and boundary conditions for which the spectrum of discrete frequencies $\omega_j (j = 1, 2, \ldots)$ associated with modes of a particular wavevector $\mathbf{q}$ represent the eigenvalues. As demonstrated for flat plates in §5.5, these eigenfrequencies are the zeros of determinants, i.e., $D(\omega_j) = 0$, which depend on the separation between the particles, the dielectric properties of the individual phases, and the wavenumber.

The interaction potential Φ is calculated by considering the particles as coupled harmonic oscillators for which each normal mode contributes $kT \ln(2 \sinh(\hbar\omega_j/2kT))$ to the Helmholtz free energy (Feynman, 1972, §25). Then the change in free energy between interacting and infinitely separated particles is

$$\Phi = kT \sum_j \int \left\{ \ln\left(2 \sinh \frac{\hbar\omega_j}{2kT}\right) - \ln\left(2 \sinh \frac{\hbar\omega_{j\infty}}{2kT}\right) \right\} \mathbf{dq}, \tag{5.3.1}$$

with $\omega_{j\infty}$ denoting the eigenvalues at infinite separation. This expression for the interaction potential is equivalent to that obtained by determining the intensity of the fields through a fluctuation–dissipation theorem and calculating the force from the Maxwell stress.

Explicit evaluation of the ω_j and $\omega_{j\infty}$ would be quite difficult, but can be avoided by using the theory of residues (Mahanty & Ninham, 1976, §2.7) and a sequence of mathematical manipulations to convert (5.3.1) to

$$\Phi = kT\sum_{n=0}^{\infty}{}' \int \ln \frac{D}{D_\infty}(i\xi_n)\,d\mathbf{q}, \tag{5.3.2}$$

with D_∞ representing the form of the determinant at infinite separation, $\xi_n = 2\pi nkT/\hbar$, and the prime indicating that the $n=0$ term is multiplied by $\frac{1}{2}$. Note that the modal frequencies have disappeared, replaced by discrete, equally spaced imaginary frequencies $i\xi_n$. Since $2\pi kT/\hbar = 2.4 \times 10^{14}$ rad/s, this distribution includes a static term ($n=0$), a few terms in the infrared ($n=1$–6), and many in the ultraviolet as illustrated in Fig. 5.2. The fact that the dielectric permittivities embedded in D/D_∞ must be evaluated only at the $i\xi_n$ simplifies profoundly the task of calculating Φ for specific materials.

Our presentation of the continuum theory is organized as follows. The next section (§5.4) discusses the dielectric response of materials, first the general features and then the sources of data and construction of spectra for specific materials. In §§5.5 and 5.6 we derive the interaction potential for flat plates and use numerical results for the interaction of polystyrene across water to illustrate important aspects of the attraction. Section 5.7 contains a simplified approximation to the full theory for flat plates. The effects of shape or curvature are first addressed in §5.8 for macroscopic bodies and used in §5.9 to interpret direct measurements which test the theory. Finally, §5.10 presents an approximate form for the interaction potential between spheres.

Fig. 5.2. Electromagnetic spectra and the frequencies sampled in the calculation of the interaction potential.

Wavelengths/μm
10^{-4} 10^{-3} 10^{-2} 10^{-1} 1 10 10^2 10^3 10^4
X-rays
Ultraviolet
Visible
Near infrared
Far infrared
Microwave
1000 100 10 $1 = n$
Sampling frequencies ξ_n
10^{19} 10^{18} 10^{17} 10^{16} 10^{15} 10^{14} 10^{13} 10^{12} 10^{11} 10^{10}
Frequency/rad s^{-1}

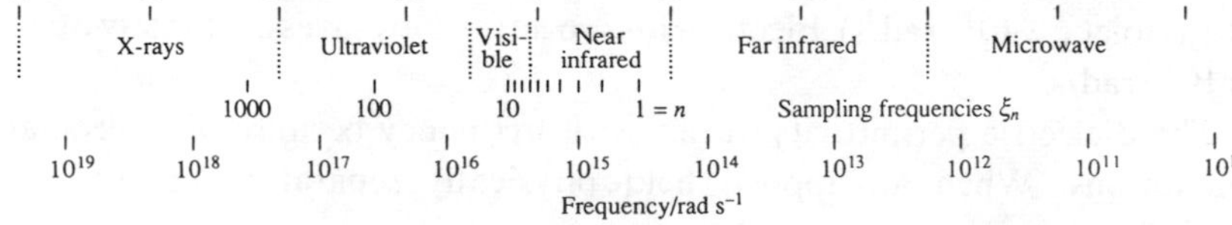

5.4 Dielectric response of materials

The important material properties in the continuum theory are the dielectric permittivities, the analogues of the molecular polarizabilities. For a weak, time varying field $\mathbf{E}(t)$ in an isotropic material the electric displacement

$$\mathbf{D}(t)=\varepsilon_0\mathbf{E}(t)+\varepsilon_0\int_0^\infty f(\tau)\mathbf{E}(t-\tau)\,\mathrm{d}\tau, \tag{5.4.1}$$

includes a history-dependent polarization characterized by the scalar memory function $f(\tau)$. The linearity of (5.4.1) with respect to the electric field enables the decomposition into individual Fourier components with $\mathbf{E}(t)=\mathbf{E}(\omega)\exp(-i\omega t)$ such that $\mathbf{D}(\omega)=\varepsilon_0\varepsilon(\omega)\mathbf{E}(\omega)$. Since

$$\begin{aligned}\varepsilon(\omega)&=\varepsilon'(\omega)+i\varepsilon''(\omega)\\ &=1+\int_0^\infty f(\tau)\exp(i\omega\tau)\,\mathrm{d}\tau,\end{aligned} \tag{5.4.2}$$

the real (capacitive) and imaginary (dissipative) parts are related through

$$\varepsilon'(\omega)=1+\frac{2}{\pi}\int_0^\infty \frac{x\varepsilon''(x)}{x^2-\omega^2}\,\mathrm{d}x \tag{5.4.3}$$

while

$$\varepsilon(i\xi)=1+\frac{2}{\pi}\int_0^\infty \frac{x\varepsilon''(x)}{x^2+\xi^2}\,\mathrm{d}x \tag{5.4.4}$$

(Landau & Lifschitz, 1960, §62).

In the low-frequency limit ($\omega\to0$) for non-conductors, $\varepsilon'\sim\varepsilon(0)$ and $\varepsilon''\sim0$, with $\varepsilon(0)$ the static dielectric constant. At X-ray frequencies or above, $\varepsilon'\sim1$ and $\varepsilon''\sim0$. In regions of the spectrum where absorption is negligible, $\varepsilon'(\omega)=\varepsilon(\omega)=n^2(\omega)$, with n the refractive index.

Dielectric data are available from several sources. Capacitance bridge methods detect ε' and ε'' from zero frequency through the microwave range ($\approx10^{12}$–10^{13} rad/s). Optical reflectance techniques yield the refractive index n and the extinction coefficient from the infrared through the ultraviolet ($\approx10^{17}$ rad/s). Electron loss spectroscopy gives ε'' directly up to $\approx10^{17}$ rad/s.

The dielectric permittivity varies with frequency because of molecular relaxations. When an applied field physically separates positive and negative charges against a short range restoring force, e.g., by shifting either nuclei or electrons within the molecule, then the response

$$\varepsilon(\omega) \approx \frac{f_j}{1 - i(\omega/\omega_j)g_j - (\omega/\omega_j)^2}$$

resembles that of a damped oscillator with strength f_j, resonance frequency ω_j, and bandwidth g_j. The corresponding real and imaginary parts of ε vary near the resonance frequency, as shown in Fig. 5.3. Fortunately, the final form for the interaction energy derived from the continuum theory depends only on $\varepsilon(i\xi)$, which decreases monotonically with frequency from the static dielectric constant at $\omega = 0$ to unity at $\omega = \infty$.

Several slightly different means exist for constructing the complete spectrum from limited data. The data are invariably fit with

$$\varepsilon(i\xi) = 1 + \sum_j \frac{d_j}{1 + \xi\tau_j} + \sum_j \frac{f_j}{1 + g_j(\xi/\omega_j) + (\xi/\omega_j)^2}. \tag{5.4.5}$$

Fig. 5.3. Typical frequency dependence of the dielectric permittivity (Parsegian, 1975).

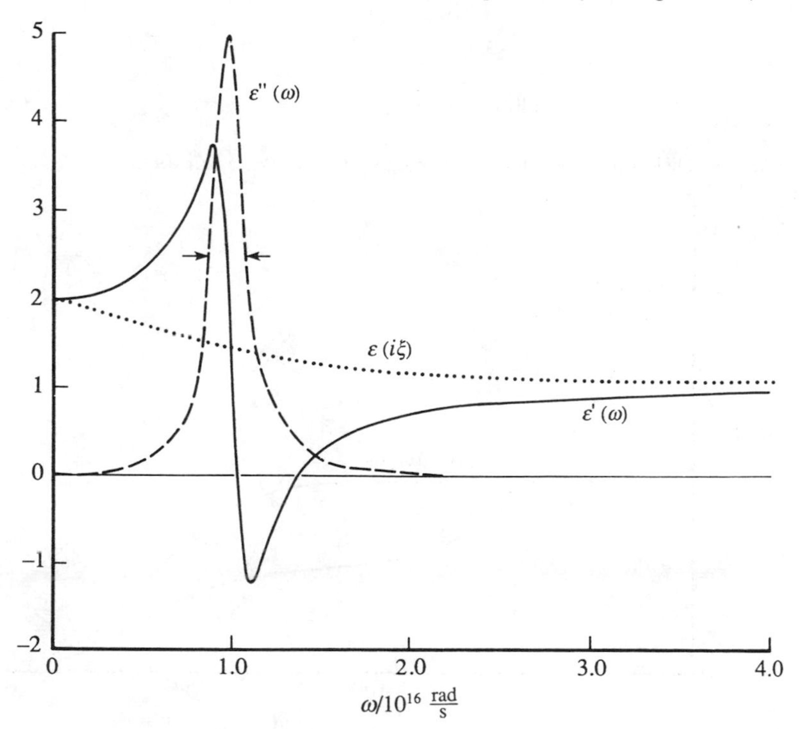

The first sum represents Debye relaxations for polar materials and the second the damped oscillator term. In the low-frequency limit,

$$\varepsilon(0)=1+\Sigma d_j+\Sigma f_j. \qquad (5.4.6)$$

For widely spaced relaxation frequencies, $\varepsilon(i\xi)$ decreases by an amount equal to d_j or f_j as the frequency passes $1/\tau_j$ or ω_j, eventually reaching an asymptote of unity.

For well-studied materials such as water (Gingell & Parsegian, 1972), polystyrene (Parsegian & Weiss, 1981), and quartz (Chan & Richmond, 1977) the spectra have been defined in some detail. Spectral parameters for the first two are tabulated in Table 5.2 and $\varepsilon(i\xi)$ plotted in Fig. 5.4. Only for highly polar fluids such as water does the relaxation in the microwave appear significant. Resolution of the ultraviolet is generally quite important, however, because of the closely spaced ξ_n there (Fig. 5.2).

For other materials with simpler spectra and/or limited data, e.g. $\varepsilon(0)$ plus $n(\omega)$ in the visible, one relaxation in the infrared and one in the ultraviolet may suffice (Hough & White, 1980). In the visible, without absorption,

$$\varepsilon(\omega)=n^2(\omega)=1+\frac{f_{uv}}{1-(\omega/\omega_{uv})^2},$$

so that

$$n^2-1=(n^2-1)(\omega/\omega_{uv})^2+f_{uv}. \qquad (5.4.7)$$

Fig. 5.4. Dielectric spectra for water and polystyrene (Parsegian, 1975).

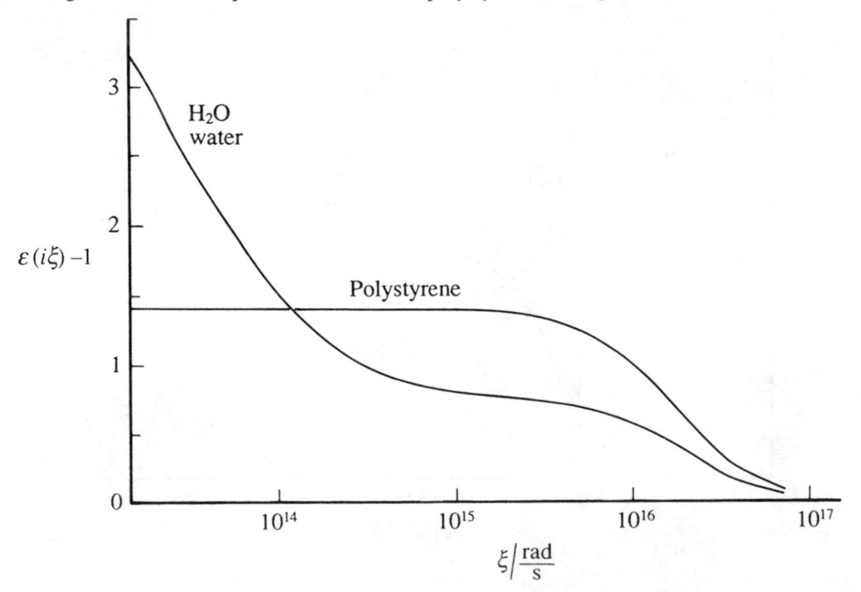

Table 5.2.

(a) Water (Parsegian, 1975)

	d_j	$1/\tau_j$	
microwave	74.8	1.05×10^{11} rad/s	
	ω_j/(rad/s)	f_j	g_j
infrared	3.34×10^{13}	1.46	0.725
	1.11×10^{14}	0.73	0.55
	1.49×10^{14}	0.151	0.30
	3.23×10^{14}	0.142	0.13
	6.78×10^{14}	0.077	0.13
ultraviolet	1.33×10^{16}	0.0393	0.062
	1.62×10^{16}	0.0567	0.088
	1.84×10^{16}	0.0992	0.135
	2.10×10^{16}	0.156	0.158
	2.41×10^{16}	0.152	0.200
	3.00×10^{16}	0.271	0.34

(b) Polystyrene (Parsegian, 1975)

ω_j/(10^{16} rad/s)	f_j	g_j
1.03	0.362	0.102
1.78	0.367	0.318
2.26	0.494	0.357
3.25	0.339	0.572

(c) (Israelachvili, 1985)

	$\varepsilon(0)$	n_0^2	ω_{uv}/(10^{16} rad/s)
water	80	1.777	1.88
n-pentane	1.84	1.820	1.88
n-octane	1.95	1.924	1.88
n-dodecane	2.01	1.991	1.88
n-tetradecane	2.03	2.011	1.82
n-hexadecane	2.05	2.025	1.82
cyclohexane	2.08	2.033	1.82
benzene	2.28	2.253	1.32
carbon tetrachloride	2.24	2.132	1.70
acetone	21	1.847	1.82
ethanol	26	1.852	1.88
polystyrene	2.55	2.424	1.45
poly(vinyl chloride)	3.2	2.331	1.82
poly(tetrafluoroethylene)	2.1	1.847	1.82
fused quartz	3.8	2.097	2.01
mica	7.0	2.560	1.88
CaF_2	7.4	2.036	2.39

Thus, plotting (n^2-1) vs. $(n^2-1)\omega^2$ should yield a straight line with slope ω_{uv}^{-2} and intercept f_{uv}. The strength of the relaxation in the infrared then follows from (5.4.6) as

$$f_{ir}=\varepsilon(0)-1-f_{uv}. \quad (5.4.8)$$

Without data in the infrared, ω_{ir} must be assigned arbitrarily. For weak relaxations, as with hydrocarbons, Hough & White (1980) argue that this introduces little error into the potential.

5.5 Theory for flat plates

Solution of the boundary value problem

To illustrate the derivation of D/D_∞ we now consider interactions between flat plates (Fig. 5.5). The transient electromagnetic fields responsible for dispersion forces must satisfy Maxwell's equations (Stratton, 1941, §1):

$$\nabla\cdot\mathbf{D}=\rho^e \quad (5.5.1)$$

$$\nabla\cdot\mathbf{B}=0 \quad (5.5.2)$$

$$\varepsilon_0 c^2\nabla\times\mathbf{B}=\mathbf{j}+\frac{\partial}{\partial t}\mathbf{D} \quad (5.5.3)$$

$$\nabla\times\mathbf{E}=-\frac{\partial}{\partial t}\mathbf{B}. \quad (5.5.4)$$

The magnetic field **B** is defined through the force,

$$\mathbf{F}=e(\mathbf{E}+\mathbf{u}\times\mathbf{B}), \quad (5.5.5)$$

Fig. 5.5. Geometry for interaction between two identical half-spaces across a second material.

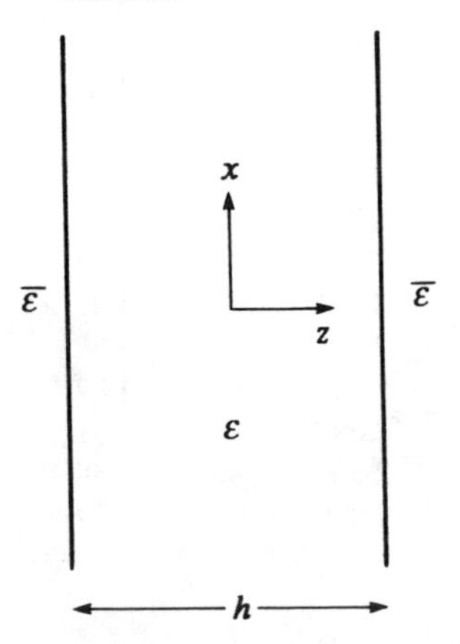

acting on an elementary charge e moving with velocity $\mathbf{u}$. The solenoidal character of $\mathbf{B}$ indicates the absence of magnetic monopoles, the analogues of electric charges. Instead, magnetic fields arise according to (5.5.3) from electric currents $\mathbf{j}$ and time-varying electric fields. This equation represents Ampere's law for non-magnetic materials, known since 1820, plus the $\partial\mathbf{D}/\partial t$ term introduced by Maxwell. The fourth equation, Faraday's law, relates the line integral of the electric field around a loop to the time derivative of the magnetic flux through the loop. Hence, a time-varying magnetic field also induces an electric field.

The presence of the time derivative in (5.5.3) provides the possibility of self-sustaining coupled oscillations between the electric and magnetic fields. Propagating electromagnetic waves with velocity $c = 3.00 \times 10^8$ m/s result. The reader should consult other texts for more complete descriptions of the associated physics (Feynman, Leighton & Sands, 1963) and the mathematical theory (Stratton, 1941). Here we concentrate on their application to dispersion forces.

For isotropic homogeneous materials with no free or fixed charges either in the bulk or at the interfaces, $\rho^e = 0$, $\mathbf{j} = \mathbf{0}$, and $\mathbf{D} = \epsilon_0 \epsilon \mathbf{E}$. We seek periodic solutions for the normal modes localized within the gap between two half-spaces (Fig. 5.5), which can be constructed from the vector potential $\mathbf{Z} = \mathbf{Z}(\omega)\exp(-\omega t)$ (Stratton, 1941, §1.11) as

$$\mathbf{E} = -\frac{\varepsilon}{c^2}\frac{\partial^2 \mathbf{Z}}{\partial t^2}, \qquad \mathbf{B} = \frac{\varepsilon}{c^2}\frac{\partial}{\partial t}\nabla \times \mathbf{Z}, \tag{5.5.6}$$

$$\nabla^2 \mathbf{Z} + K^2 \mathbf{Z} = \mathbf{0}, \tag{5.5.7}$$

$$\nabla \cdot \mathbf{Z} = 0, \tag{5.5.8}$$

with $K^2 = \varepsilon\omega^2/c^2$.

Application of the same conservation laws that lead to Maxwell's equations to a flat interface between two materials establishes the boundary conditions as (Stratton, 1941, §1.13)

$$\mathbf{E}, \mathbf{B} \to \mathbf{0} \qquad \text{as } z \to \pm\infty$$

and

$$\bar{\mathbf{B}} = \mathbf{B}, \quad \bar{\varepsilon}\bar{\mathbf{E}} \cdot \mathbf{n} = \varepsilon \mathbf{E} \cdot \mathbf{n}, \quad \bar{\mathbf{E}} - \bar{\mathbf{E}} \cdot \mathbf{nn} = \mathbf{E} - \mathbf{E} \cdot \mathbf{nn} \tag{5.5.9}$$

at $z=\pm h/2$. The overbar distinguishes the plates from the intervening material, and $\mathbf{n}=(0,0,\pm 1)$ is the unit normal.

Separation of variables, coupled with the decay of the field as $z\to\pm\infty$, produces the solutions

$$\bar{\mathbf{Z}}=\exp(\mathbf{iq}\cdot\mathbf{x})\begin{cases}\bar{\mathbf{a}}\exp(-\bar{s}z) & z\geq h/2\\ \bar{\mathbf{b}}\exp(+\bar{s}z) & z\leq h/2\end{cases} \tag{5.5.10}$$

$$\mathbf{Z}=\exp(i\mathbf{q}\cdot\mathbf{x})[\mathbf{a}\exp(-sz)+\mathbf{b}\exp(sz)] \qquad -h/2\leq z\leq h/2,$$

with

$$\mathbf{q}=(q_1,q_2,0),$$
$$q^2=\mathbf{q}\cdot\mathbf{q},$$
$$s^2=q^2-K^2,$$

and

$$a_3=\frac{i}{s}\mathbf{q}\cdot\mathbf{a},$$

$$b_3=-\frac{i}{s}\mathbf{q}\cdot\mathbf{a},$$

Thus both $\mathbf{Z}$ and the corresponding fields vary periodically in the transverse directions with wavenumbers q_1 and q_2, but decay exponentially normal to the interface with $s^2>0$.

Application of the remaining boundary conditions at $z=\pm h/2$ generates a set of homogeneous linear equations for the unknown constants a_1, a_2, b_1, and b_2 for each material. Non-trivial solutions and, hence, the normal modes of interest exist only if the determinant of the coefficients vanishes, i.e.,

$$\frac{D}{D_\infty}(i\xi)=\left[1-\left(\frac{\bar{s}-s}{\bar{s}+s}\right)^2\exp(-2sh)\right]\cdot\left[1-\frac{\left(\frac{\bar{\varepsilon}}{\bar{s}}-\frac{\varepsilon}{s}\right)^2}{\left(\frac{\bar{\varepsilon}}{\bar{s}}+\frac{\varepsilon}{s}\right)^2}\exp(-2sh)\right]=0. \tag{5.5.11}$$

This equation in effect identifies a spectrum of discrete eigenfrequencies associated with each wave vector $\mathbf{q}$. The frequency-dependent permittivities enter directly through the boundary conditions and indirectly through the $s^2=q^2+\varepsilon\xi^2/c^2$, with $\varepsilon=\varepsilon(i\xi)$.

Interaction potential

Explicit calculation of the eigenfrequencies would be quite tedious. Fortunately, the interaction energy (5.3.2) requires only the dispersion relation itself, yielding the potential per unit area for flat plates as

$$\Phi_{\text{fp}} = \frac{kT}{2\pi} \int_0^\infty q \sum_{n=0}^{\infty}{}' \ln \frac{D}{D_\infty}(i\xi_n)\, dq. \tag{5.5.12}$$

Comparison with (5.2.14) suggests defining an effective Hamaker constant as

$$A_{\text{eff}}(h) = -12\pi h^2 \Phi_{\text{fp}}(h). \tag{5.5.13}$$

As we will see, the values differ from (5.2.10) because of many body effects and retardation.

Several limiting cases illustrate the dependence of Φ_{fp} on separation embedded in the exponentials in D/D_∞. As $h \to 0$, the dominant contributions to the integral correspond to $q^2 \gg \varepsilon\xi^2/c^2$ for which

$$\frac{D}{D_\infty} \approx 1 - \left(\frac{\bar{\varepsilon} - \varepsilon}{\bar{\varepsilon} + \varepsilon}\right)^2 \exp(-2qh). \tag{5.5.14}$$

With the expression

$$\ln(1+u) = \sum_{m=1}^{\infty} \frac{u^m}{m},$$

(5.5.12) can be integrated analytically to obtain the *non-retarded* limit

$$A_{\text{eff}}(0) = \tfrac{3}{2} kT \sum_{n=0}^{\infty}{}' \sum_{m=1}^{\infty} m^{-3} \Delta_n^{2m}, \tag{5.5.15}$$

with $\Delta_n = \dfrac{\bar{\varepsilon} - \varepsilon}{\bar{\varepsilon} + \varepsilon}(i\xi_n)$. Thus $\Phi_{\text{fp}} \propto h^{-2}$, in accord with the corresponding result from the microscopic theory. Since the electric fields become pseudo-static in this limit, the interaction energy could have been derived simply from solutions to Laplace's equation.

If $\varepsilon - 1 \ll 1$ and $\bar{\varepsilon} - 1 \ll 1$, then (5.5.15) reduces to

$$A_{\text{eff}}(0) \sim \tfrac{3}{8} kT \sum_{n=0}^{\infty}{}' [\bar{\varepsilon}(i\xi_n) - \varepsilon(i\xi_n)]^2, \tag{5.5.16}$$

which coincides with Hamaker's result in (5.2.13). The difference between (5.5.15) and (5.5.16) reflects the effect of many-body interactions, i.e., the influence of the medium on the fluctuating fields.

As h increases from zero, the exponentials in D/D_∞ eliminate contributions to the interaction energy from those normal modes with $s \gg h^{-1}$. Consequently, at intermediate separations Φ_{fp} decays faster than h^{-2}. For $\varepsilon\xi_n^2/c^2 \gg h^{-2}$

$$\Phi_{fp} \sim \frac{kT}{4\pi}\int_0^\infty q \ln \frac{D}{D_\infty}(0)\, dq, \tag{5.5.17}$$

or, upon expanding,

$$A_{eff}(\infty) = \tfrac{3}{4}kT \sum_{m=1}^{\infty} m^{-3}\Delta_0^{2m}. \tag{5.5.18}$$

Thus the potential again decays as h^{-2}, but with contribution from only the static ($n=0$) mode. This progressive damping of higher-frequency terms in the summation occurs when the time required for a wave to traverse the gap exceeds its temporal period, i.e., $h\varepsilon^{1/2}/c > 1/\xi_n$. The separation at which this *retardation* significantly affects Φ_{fp} depends on the relative magnitudes of the terms in the summation.

Effect of electrolyte

In aqueous solutions, electrolyte provides a mode of conduction omitted from the preceding analysis. The key question is whether the fluctuating electric fields move the ions sufficiently to generate significant currents or space charges. The magnitude of the current $\mathbf{j} = \Sigma_k e z^k n^k \omega^k \mathbf{E}$ relative to the time-varying displacement field $\partial \mathbf{D}/\partial t$ decreases with increasing frequency. At the lowest sampling frequency, $\xi_1 = 2\pi kT/\hbar$, the ratio,

$$\sum_k \frac{e z^k n^k \omega^k}{\varepsilon\varepsilon_0}\xi_1 \approx 10^{-9}\text{–}10^{-6}, \tag{5.5.19}$$

corresponding to 10^{-3}–1 M, is quite small. Hence, only for $n=0$ can the ions respond to the field and affect the interaction potential.

For $n=0$, the electric field is pseudo-steady, and, consequently, there is no magnetic field, leaving a balance between diffusion and conduction driven by the electric field. The latter is weak, so the perturbations in the ion densities will be small as well. Hence the normal modes at zero frequency are determined as above, except with the space charge ρ^e governed by the Boltzmann distribution; thus Maxwell's equations reduce to the linearized Poisson–Boltzmann equation. The resulting contribution to the interaction potential in the presence of electrolyte is (Mahanty & Ninham, 1976, §7.4)

$$\frac{kT}{4\pi}\int_0^\infty q\ln\left[1-\left(\frac{\dfrac{\bar{\varepsilon}(0)}{s}-\dfrac{\varepsilon(0)}{q}}{\dfrac{\bar{\varepsilon}(0)}{s}+\dfrac{\varepsilon(0)}{q}}\right)^2\exp(-2sh)\right]\mathrm{d}q \qquad (5.5.20)$$

with $s^2=q^2+\kappa^2$. The limiting form for $\kappa h \gg 1$,

$$-\frac{kT}{8\pi h}\kappa\exp(-2\kappa h),$$

clearly illustrates the screening of the $n=0$ term due to ionic conduction. High ionic strengths, e.g., 1 M NaCl with $\kappa^{-1}=0.3$ nm, effectively suppress the $n=0$ term, decreasing the attraction. For $\kappa h \ll 1$, on the other hand, the ions have no effect.

5.6 Calculations for specific materials

The detailed calculations of Parsegian (1975) for polystyrene plates interacting across a vacuum, pure water, and a 0.1 M salt solution illustrate several characteristic features of the potential (Fig. 5.6). The

Fig. 5.6. Calculated values for the effective Hamaker constant (5.5.13) for interaction between polystyrene half-spaces across pure and salt water (Parsegian, 1975).

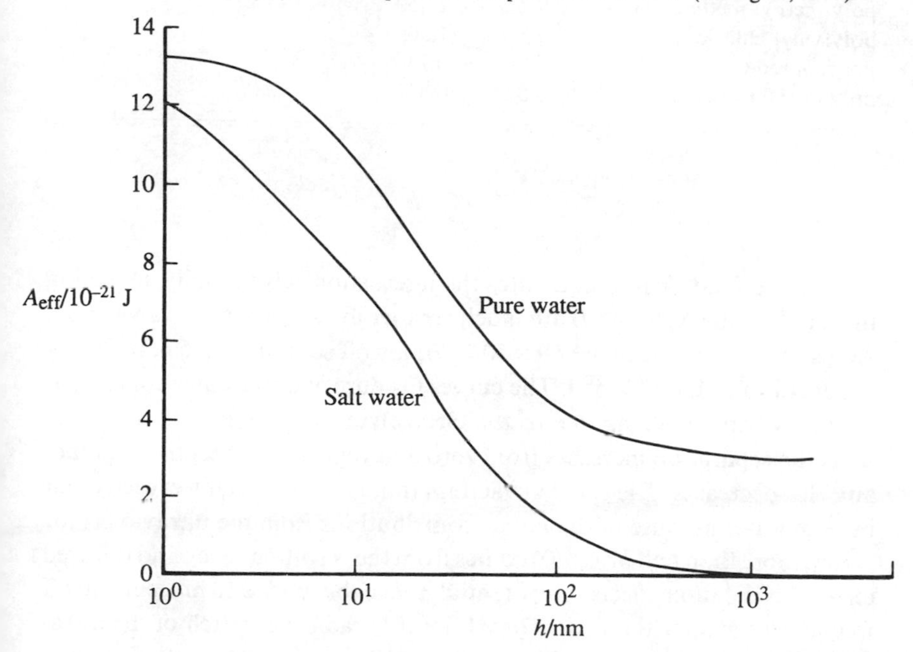

Table 5.3.

	$A_{eff}(0)/10^{-20}$ J		
	Vacuum	Water	
polystyrene	7.9	1.3	
hexadecane	5.4	—	Parsegian & Weiss (1981)
gold	40	30	
silver	50	40	
copper	40	30	
water	4.0	—	
pentane	3.8	0.34	
decane	4.8	0.46	
hexadecane	5.2	0.54	
water	3.7	—	Hough & White (1980)
quartz			
fused	6.5	0.83	
crystalline	8.8	1.70	
fused silica	6.6	0.85	
calcite	10.1	2.23	
calcium fluoride	7.2	1.04	
sapphire	15.6	5.32	
poly(methyl methacrylate)	7.1	1.05	
poly(vinyl chloride)	7.8	1.30	
polyisoprene	6.0	0.74	
poly(tetrafluoroethylene)	3.8	0.33	

intervening fluid clearly moderates the interaction substantially, reflecting the fact that the Δ_n for $n>0$ are much smaller for water than for a vacuum. Across a vacuum, $A_{eff}(0)=7.9\times 10^{-20}$ J, far off scale in Fig. 5.6, while for water $A_{eff}(0)=1.3\times 10^{-20}$ J. The curves for pure and salt water coincide at contact where screening due to the electrolyte vanishes.

As the separation increases from zero and retardation becomes significant, A_{eff} decreases. The effect in vacuum (not shown) actually exceeds that in pure water because of the larger contributions from the ultraviolet; for water, more than half of $A_{eff}(0)$ comes from the zero-frequency and infrared terms. Retardation affects the potential noticeably by $h\approx 10$ nm, cutting off frequencies greater than $\xi\approx c/2h\approx 1.5\times 10^{16}$ rad/s, i.e., much of the ultraviolet. At separations $h\approx c/2\xi_1\approx 500$ nm, all the higher-frequency contributions become at least partially retarded. At larger separations, retardation completely suppresses the contributions from $n \geq 1$, leaving only the non-retarded $n=0$ term and $A_{eff}(\infty) \ll A_{eff}(0)$ as expected from (5.5.18).

Because of the significant low-frequency contributions with water, ionic screening in salt water substantially reduces the potential, as illustrated by the difference between the two curves in Fig. 5.6. At infinite ionic strength, i.e., without the $n=0$ term, $A_{eff}(0) \approx 1.0 \times 10^{-20}$ J. At 0.1 M, the $n=0$ term disappears for $h \geq 2/\kappa = 2.0$ nm. Consequently, retardation eventually reduces A_{eff} to zero at large separations.

Although presented here for specific materials, retardation and free electrolyte would affect other systems similarly. Results for $A_{eff}(0)$ are available for a variety of other systems (Gingell & Parsegian, 1972; Hough & White, 1980; Parsegian & Weiss, 1981). Table 5.3 lists the results for a number of organic and inorganic solids interacting across either a vacuum or water. The presence of water typically reduces the attraction by a factor of 3 to 10. Hamaker constants in water range from $\leq 1\,kT$ for some hydrocarbons to $\geq 10^2\,kT$ for metals. Calculations with different approximations for $\varepsilon(i\xi)$ based on the same data suggest ± 5–10 per cent uncertainty in the values quoted.

5.7 Geometrical effects: the Derjaguin approximation

In discussing electrostatic forces in Chapter 4 we noted that for interactions between spheres with thin double layers the fields within the gap become one-dimensional. The interaction potential then follows from integrating the potential for flat plates, evaluated at the local surface-to-surface separation, over the entire gap. This method, originally proposed by Derjaguin (1934), applies in general when the minimum separation and the range of the interaction are both small relative to the radii of curvature of the surfaces.

Consider two spheres with radii a_1 and a_2 at a separation $h = r - a_1 - a_2 \ll a_1, a_2$ (cf. Horn & Israelachvili, 1981). In cylindrical polar coordinates (r, θ, z), the surface-to-surface separation within the gap is

$$z(r) = h + \tfrac{1}{2} r^2 \left(\frac{1}{a_1} + \frac{1}{a_2} \right) + O\left(\frac{r^4}{a_1^3}, \frac{r^4}{a_2^3} \right), \qquad (5.7.1)$$

so that

$$\mathrm{d}z \approx \left(\frac{1}{a_1} + \frac{1}{a_2} \right) r\,\mathrm{d}r.$$

This permits the interaction energy to be expressed as

$$\Phi \approx 2\pi \int_0^\infty \Phi_{fp}(z) r \, dr$$
$$\approx 2\pi \frac{a_1 a_2}{a_1 + a_2} \int_h^\infty \Phi_{fp}(z) \, dz \tag{5.7.2}$$

and the force as

$$F = -\frac{\partial \Phi}{\partial h} \approx 2\pi \frac{a_1 a_2}{a_1 + a_2} \Phi_{fp}(h). \tag{5.7.3}$$

Note that taking the upper limit of the integration as infinity assumes the range of the interaction to be small relative to the radii.

For dispersion forces between equal spheres, (5.7.2) also can be written as

$$\Phi = -\frac{a}{12} \int_h^\infty A_{eff}(z) \frac{dz}{z^2}$$
$$= -\frac{a}{12} \left\{ \frac{A_{eff}(h)}{h} + \int_h^\infty \frac{dA_{eff}}{dz} \frac{dz}{z} \right\}. \tag{5.7.4}$$

As $h \to 0$ or $h \to \infty$ (without electrolyte), $dA_{eff}/dh \to 0$; hence $\Phi \propto h^{-1}$ in both the non-retarded and the fully retarded limits. At intermediate separations, however, the second term is non-zero, since the extent of retardation varies with position in the gap.

The accuracy of the Derjaguin approximation is difficult to assess without calculating the first correction due to curvature. The approximation clearly should fail for micron-sized particles, since the wavelengths characterizing the range of the interaction are comparable to the size. But it should be entirely adequate for the interpretation of force measurements between macroscopic objects with radii of millimeters or larger as described in the next section.

5.8 Direct measurements

The preceding sections describe a rigorous quantitative theory for the dispersion force. Only since the late 1960s have the means been available for testing these results by direct measurement. The efforts began years earlier (Derjaguin *et al.*, 1954; Overbeek & Sparnaay, 1954; Kitchener & Prosser, 1957; Derjaguin *et al.*, 1964) but were limited to retarded interactions at large surface-to-surface separations because of the roughness of the fused quartz and polished glass surfaces employed. Finally Tabor & Winterton (1969) recognized that cleaved muscovite mica provides a molecularly smooth surface ideal for such measurements. Subsequent measurements have confirmed the Lifshitz theory for the dispersion forces both in vacuum (Tabor & Winterton, 1969; Israelachvili & Tabor, 1973) and in water (Israelachvili & Adams, 1978).

Before discussing the results, we briefly describe the instrument (Israelachvili & Adams, 1978). The force is actually measured between macroscopic surfaces, in the configuration of crossed cylinders, but at nanometer separations. The surfaces consist of ≈ 1-μm-thick mica sheets silvered on the back and glued to quartz pieces with a radius of curvature of ≈ 10 mm (Fig. 5.7). The separation is measured by constructing a multiple-beam

Fig. 5.7. Schematic of instrument for measuring forces between mica surfaces (Israelachvili & Adams, 1978).

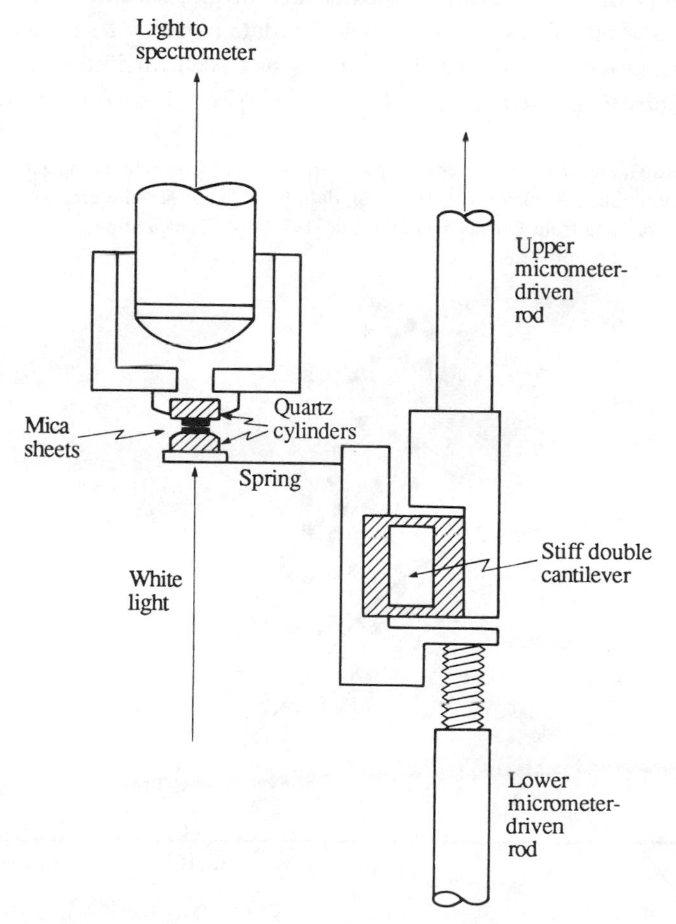

interferometer with the silvered mica sheets. White light entering the gap from the bottom experiences multiple reflections between the surfaces, emerging only at discrete wavelengths (Israelachvili, 1973). Resolving these with a spectrometer determines the separation within 0.1 nm routinely.

The lower surface is connected through a leaf spring to a clamp positioned within ± 0.1 nm via a double cantilever spring and precision potentiometer. As the surfaces are brought together from a large separation, the leaf spring must deflect toward contact in order to balance the attractive force. Knowledge of the spring constant plus these two positions then determines the force with a sensitivity of $\approx 10^{-7}$ N. The result is a conceptually simple device capable of quite precise measurements at separations approaching the molecular level.

The measurements of dispersion forces in vacuum summarized in Fig. 5.8 agree reasonably well, at separations greater than 20 nm, with theoretical calculations based on the best available dielectric data (Chan & Richmond, 1977). The discrepancy at smaller separations has been attributed to a monolayer of adsorbed water along with a change in the radius of curvature

Fig. 5.8. Effective Hamaker constant versus separation for mica plates interacting across a vacuum (Chan & Richmond, 1977): ●, data from Tabor & Winterton (1969); ■, data from Israelachvili & Tabor (1973); ——, calculation.

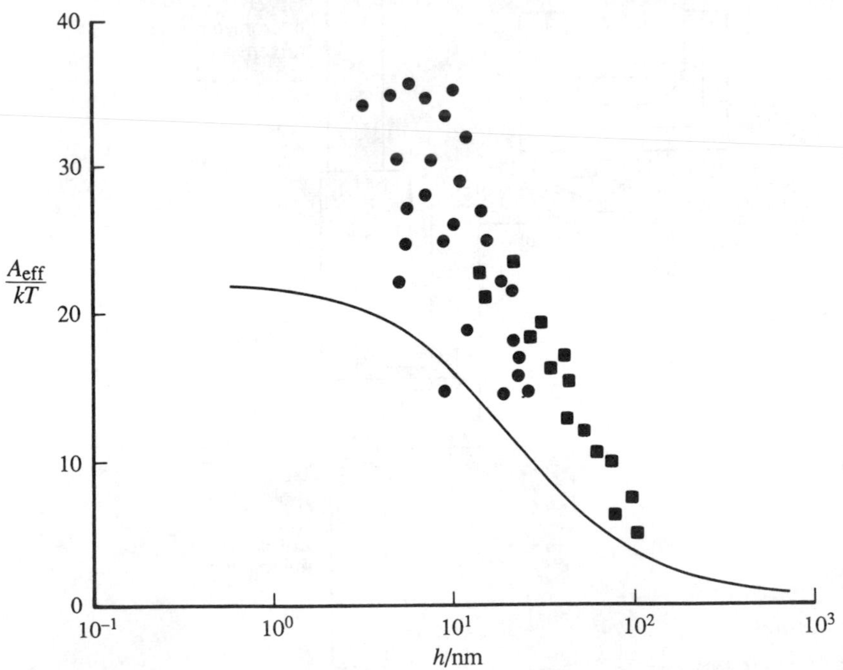

of the mica sheets (≈ 30 per cent) due to deformation of the glue by the strong attractive forces. Allowing for these brings theory and experiment within 10 per cent of one another over the full range of separations (White, Israelachvili, & Ninham, 1976).

In water the attraction is much weaker, as expected from the theory, so deformation is not a problem. Figure 5.9 shows data for a range of ionic strengths, demonstrating agreement with the theory to within 30 per cent. The effective Hamaker constant at contact, $\approx 2.2 \times 10^{-20}$ J, is within 25 per cent of the predicted value and almost an order of magnitude smaller than in vacuum. The two sets of data illustrate the accuracy of the continuum theory for dispersion forces, at least for separations greater than a few nanometers.

5.9 A simplified approximation for flat plates

The contrast between the simple one-parameter representation of dispersion forces embodied in the non-retarded Hamaker form for the interaction potential (5.2.11) and the exact result from the continuum theory (5.5.12) is marked. The latter requires some numerical effort to evaluate and is specific to the particular materials. In this section we present

Fig. 5.9. Force measured between mica plates in aqueous electrolyte solutions (Israelachvili, 1985), compared with predictions; R denotes the radius of curvature of the mica plates.

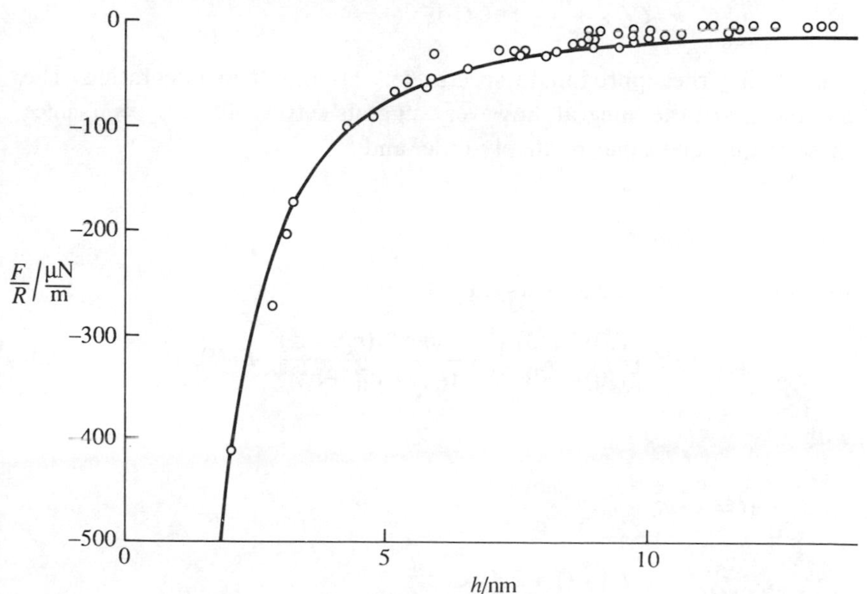

a simplified approximation that sacrifices some accuracy but eliminates the numerical effort and reduces considerably the number of parameters characterizing the individual materials (e.g., Prieve and Russel, 1988).

The development is motivated by observations (e.g. Hough & White, 1980; Israelachvili, 1985, §§11.3 and 11.4) that relaxations in the ultraviolet dominate the portion of the dielectric spectra most important for the dispersion interactions. Thus we retain only a single u.v. relaxation such that for $n>0$

$$\varepsilon(i\xi_n)=1+\frac{n_0^2-1}{1+(\xi_n/\omega_{uv})^2}, \tag{5.9.1}$$

with $n_0^2=1+f_{uv}$, the low-frequency limit of the refractive index in the visible, and ω_{uv} the frequency for the dominant relaxation in the ultraviolet.

The evaluation of (5.5.12) is then simplified by noting that the quantity $2sh$ appearing in the exponentials varies from $r_n h \equiv 2\xi_n h\varepsilon^{1/2}/c$ at the lower limit of integration to $2qh$ at the upper limit. In the non-retarded limit, the former goes to zero, suggesting an expansion about $\xi_n/cq=0$. Substitution and truncation at first order makes analytical integration possible, leaving

$$A_{\text{eff}}=\tfrac{3}{2}kT\sum_{n=0}^{\infty}{}'\left(\frac{\bar{\varepsilon}-\varepsilon}{\bar{\varepsilon}+\varepsilon}\right)^2(r_n+1)\exp(-r_n). \tag{5.9.2}$$

The final form is obtained by converting the summation to an integral as

$$\sum_{n=0}^{\infty}{}' C_n=\tfrac{1}{2}C_0+\frac{\hbar}{2\pi kT}\int_0^\infty C\,\mathrm{d}\xi,$$

substituting the approximate spectra (5.9.1), and then integrating. The complexity of the integral, however, suggests setting $\bar{\omega}_{uv}\approx\omega_{uv}\approx\omega$, since these frequencies differ relatively little, and

$$\left(\frac{n_0^2+(\xi/\omega)^2}{1+(\xi/\omega)^2}\right)^{1/2}\approx n_0,$$

to simplify the exponent. This leads to

$$A_{\text{eff}}=\tfrac{3}{4}kT\left(\frac{\bar{\varepsilon}(0)-\varepsilon(0)}{\bar{\varepsilon}(0)+\varepsilon(0)}\right)^2+\frac{3\hbar\omega}{16\sqrt{2}}\frac{(\bar{n}_0^2-n_0^2)^2}{(\bar{n}_0^2+n_0^2)^{3/2}}F(H), \tag{5.9.3}$$

with

$$H=n_0(\bar{n}_0^2+n_0^2)^{1/2}\frac{h\omega}{c}$$

$$F(H)=\frac{4\sqrt{2}}{\pi}\int_0^\infty\frac{(1+2Hx)e^{-2Hx}}{(1+2x^2)^2}\mathrm{d}x.$$

The limiting forms of F

$$H \to 0, \qquad F \sim 1,$$

$$H \to \infty, \qquad F \sim \frac{4\sqrt{2}}{\pi H},$$

provide the non-retarded,

$$A_{\text{eff}}(0) = \tfrac{3}{4}kT\left(\frac{\bar{\varepsilon}(0) - \varepsilon(0)}{\bar{\varepsilon}(0) + \varepsilon(0)}\right)^2 + \frac{3}{16\sqrt{2}}\hbar\omega \frac{(\bar{n}_0^2 - n_0^2)^2}{(\bar{n}_0^2 + n_0^2)^{3/2}}, \tag{5.9.4}$$

and fully retarded,

$$A_{\text{eff}}(h) = \tfrac{3}{4}kT\left(\frac{\bar{\varepsilon}(0) - \varepsilon(0)}{\bar{\varepsilon}(0) + \varepsilon(0)}\right)^2 + \frac{3}{4\pi}\frac{\hbar c}{n_0 h}\left(\frac{\bar{n}_0^2 - n_0^2}{\bar{n}_0^2 + n_0^2}\right)^2, \tag{5.9.5}$$

limits. At intermediate values the interpolating formula $F(H) \approx (1 + (\pi H/4\sqrt{2})^{3/2}]^{-2/3}$ accurately represents F.

Comparison of the predictions from (5.9.3) with those from the complete theory for polystyrene half-spaces interacting across water provides a strenuous test. The $n=0$ term must be deleted for salt water, since electrolyte screens the interaction. Figure 5.10 reveals that the approximation captures quantitatively the magnitude of A_{eff} and the separation

Fig. 5.10. Comparison of predictions, ---, from the approximation (5.9.3) with the exact results, +++, from Parsegian (1975) for polystyrene half-spaces interacting across pure water and salt water.

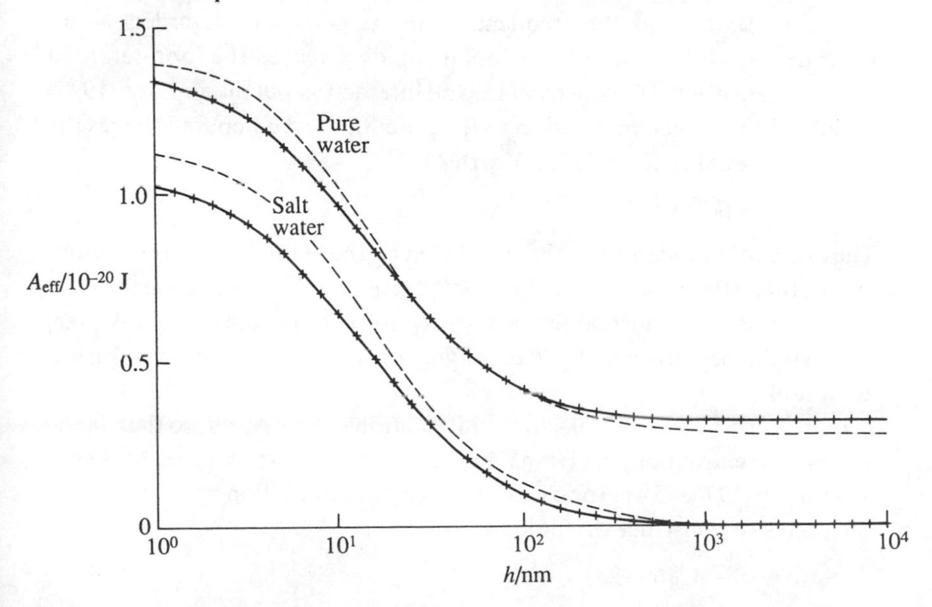

dependence caused by retardation. Note, however, that the complex spectra of water and the similarity of the two spectra in the ultraviolet make the result sensitive to the values of the spectral parameters. Errors introduced by the simplified approximation (5.9.1) are magnified by the difference in (5.9.2). For materials with simpler spectra or larger values for A_{eff}, the error is smaller. The savings are considerable, since $\varepsilon(0)$, n_0^2, and ω_{uv} fully characterize the materials and the effect of retardation enters only through the dimensionless separation H. Israelachvili (1985, §11.3) tabulates these spectral parameters for a variety of material (Table 5.2c).

5.10 Interactions between spheres

The Derjaguin approximation does not suffice for the sub-micron particles of central interest in colloid science. For interacting spheres, bispherical coordinates present a natural coordinate system for solving Maxwell's equations. Indeed, for the pseudo-steady limit, solutions to Laplace's equation by separation of variables yield the non-retarded dispersion potential (Love, 1977). The unsteady fields characterizing retarded interactions, however, necessitate solutions to the Helmholtz equation (5.5.7), which is not separable in bispherical coordinates. Multipole expansions exist (Langbein, 1974), but convergence is excruciatingly slow. Fortunately, recent work (Love, 1977; Kiefer, Parsegian, & Weiss, 1978; Pailthorpe & Russel, 1982) identifies reliable approximations that yield the full potential for interacting spheres with little more effort than is required for flat plates.

The $n=0$ term in the frequency sum remains non-retarded at all separations, with $\Delta_0 \approx \pm 1$ for polar fluids. Hence, the non-retarded solutions are needed to construct the full interaction potential. Love (1977) evaluated the exact contribution to the potential and compared the results with the Hamaker form (5.2.12) with

$$A = \tfrac{3}{4} kT \Delta_0^2.$$

The exact result differs from that predicted by the Hamaker approximation by less than 50 per cent for $(r-2a)/a < 0.2$, even for $\Delta_0 \approx \pm 1$. Since the $n=0$ term represents at most 30 per cent of $A_{eff}(0)$ (e.g., for polystyrene–water) and would be screened by electrolyte when $h \geq \kappa^{-1}$, the error seems acceptable.

For the $n \geq 1$ contributions to the interaction energy $\Delta_n \ll 1$, so that, in the absence of retardation, the Hamaker form (5.2.12, 5.5.15) suffices. Mahanty & Ninham (1976, §5.4) proposed correcting for retardation by using A_{eff} for flat plates to construct the potential as

$$\Phi = -A_{eff}(r-2a)H(r/a). \tag{5.10.1}$$

To test this approximation Pailthorpe & Russel (1982) evaluated the exact multipole expansion, including retardation effects (Langbein, 1974, §6.1) for polystyrene spheres in salt water. They used the earlier, less accurate polystyrene spectra of Gingell & Parsegian (1972) and assumed the $n=0$ term to be completely screened by salt.

In Fig. 5.11 we plot the exact solution, expressed as $-\Phi/H(r/a)$, together

Fig. 5.11. Effect of retardation for the interaction between polystyrene spheres in salt water for $a=100$ nm and 250 nm showing exact results, ——, (Pailthorpe & Russel, 1982), and approximation from (5.10.1), - - -.

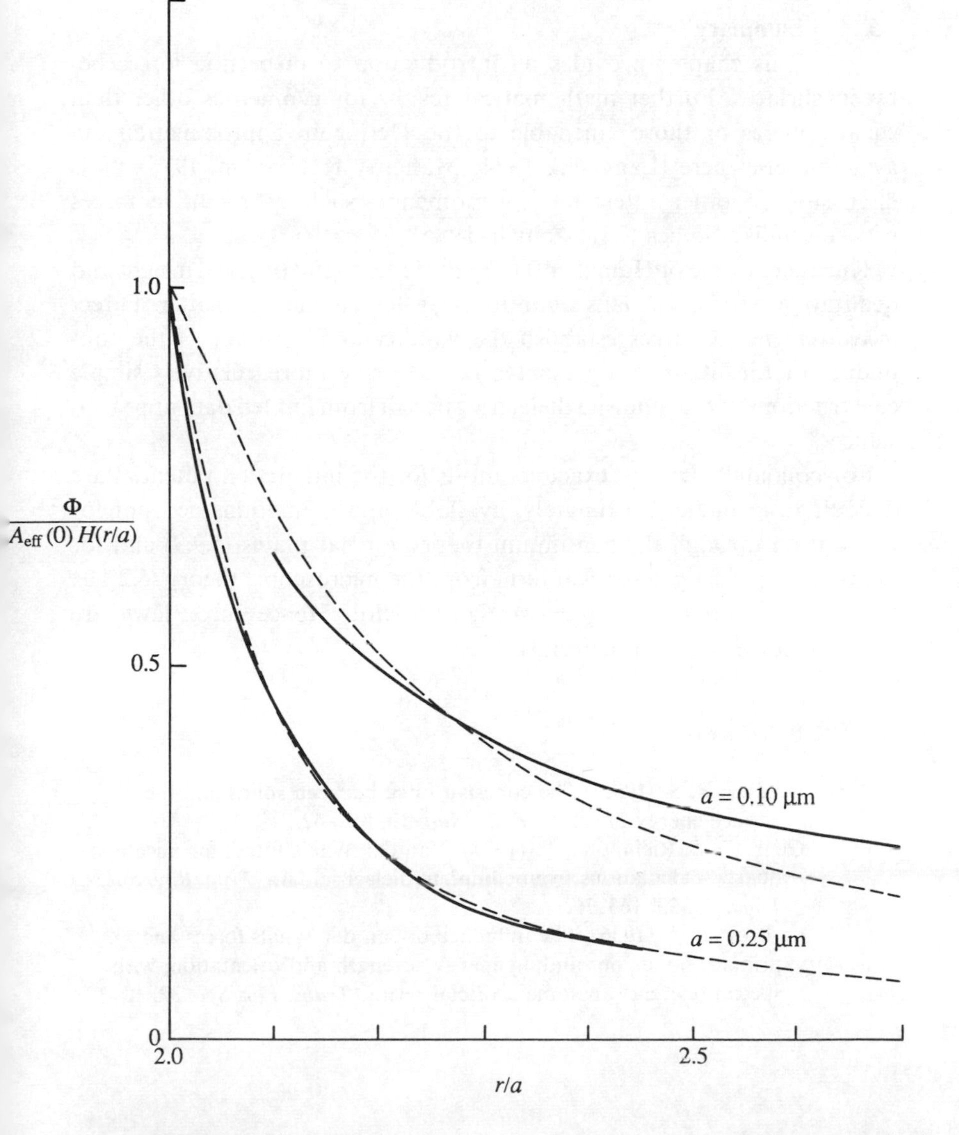

with the approximation (5.9.3) for A_{eff}. Both are normalized on values at contact to eliminate the effect of different dielectric data and highlight the effect of retardation. The comparison suggests that this simple approximation may be sufficiently accurate for many purposes.

Thus, screening of the zero-frequency contribution and retardation of the higher frequency terms cause the potential for spheres to deviate from the Hamaker form (5.2.15) at separations greater than 5–10 nm. However, the former effect is relatively small and the approximation (5.10.1) provides a computationally simple means for accounting for the latter and constructing the complete potential.

5.11 Summary

This chapter provides an introduction to dispersion forces between surfaces. Further mathematical results for geometries other than equal spheres or those amenable to the Derjaguin approximation are available elsewhere (Langbein, 1974; Mahanty & Ninham, 1976), as is discussion of other interesting phenomenon such as repulsive forces between unlike bodies (Israelachvili, 1985).

The microscopic or Hamaker theory provides useful physical insight and qualitative trends, but fails quantitatively for condensed matter. Direct measurements of forces establish the validity and accuracy of the continuum or Lifshitz theory presented here. Furthermore, relatively simple constructions of the requisite dielectric spectra from limited data appear to suffice.

For colloidal particles, exact solutions for the interaction potential are difficult to evaluate. Fortunately, available approximations account for retardation through the continuum theory for flat plates (5.9.3) and for shape through the geometrical term from the microscopic theory (5.2.11). Numerical evaluation is then straightforward. Hence force laws are available for a range of materials.

References

Bradley, R. S. (1932). The cohesive force between solids and the surface energy of solids. *Phil. Mag.* **13**, 853–62.

Chan, D. & Richmond, P. (1977). Van der Waals forces for mica and quartz: calculations from complete dielectric data, *Proc. Roy. Soc. Lond. A* **353**, 163–76.

de Boer, J. H. (1936). The influence of van der Waals forces and primary bonds on binding energy, strength and orientation, with special reference to some artificial resins. *Trans. Far. Soc.* **32**, 10–38.

Derjaguin, B. V. (1934). Friction and adhesion. IV. The theory of adhesion of small particles. *Kolloid-Z.* **69**, 155–64.

Derjaguin, B. V., Titijevskaia, A. S., Abrikossova, I. I. & Malkina, A. D. (1954). Investigations of the forces of interaction of surfaces in different media and their application to the problem of colloid stability. *Disc. Far. Soc.* **18**, 24–41.

Derjaguin, B. V., Voropayeva, T. N., Kabonov, B. N. & Titijevskaia, A. S. (1964). Surface forces and the stability of colloids and disperse systems. *J. Colloid Interface Sci.* **19**, 113–35.

Dzaloshinskii, I. E., Lifshitz, E. M. & Pitaevskii, L. P. (1961). The general theory of van der Waals forces. *Adv. Phys.* **10**, 165–208.

Feynman, R. P., Leighton, R. B. & Sands, M. (1963). *Lectures on Physics.* Addison-Wesley.

Feynman, R. P. (1972). *Statistical Mechanics.* W. A. Benjamin.

Gingell, D. & Parsegian, V. A. (1972). Computation of van der Waals interactions in aqueous systems using reflectivity data. *J. Theor. Biol.* **36**, 41–52.

Hamaker, H. C. (1937). London–van der Waals attraction between spherical particles. *Physica* **4**, 1058–72.

Horn, R. G. & Israelachvili, J. N. (1981). Direct measurement of structural forces between two surfaces in a nonpolar liquid. *J. Chem. Phys.* **75**, 1400–11.

Hough, D. B. & White, L. R. (1980). The calculation of Hamaker constants from Lifshitz theory with applications to wetting phenomena, *Adv. Colloid Interface Sci.* **14**, 3–41.

Israelachvili, J. N. (1973). Thin film studies using multiple beam interferometry. *J. Colloid Interface Sci.* **44**, 259–72.

Israelachvili, J. N. (1985). *Intermolecular and Surface Forces.* Academic Press.

Israelachvili, J. N. & Adams, G. E. (1978). Measurement of forces between two mica surfaces in aqueous electrolyte solutions in the range 1–100 nm. *J. Chem. Soc. Far. Trans. I* **74**, 975–1001.

Israelachvili, J. N. & Tabor, D. (1973). van der Waals forces. Theory and experiment. *Prog. Surf. Membr. Sci.* **7**, 1–55.

Kallman, H. & Willstäter, M. (1932). The theory of the structure of colloidal systems. *Naturwiss.* **20**, 952–3.

Kiefer, J. E., Parsegian, V. A. & Weiss, G. H. (1978). Some convenient bounds and approximations for the many body van der Waals attraction between two spheres. *J. Colloid Interface Sci.* **67**, 140–53.

Kitchener, J. A. & Prosser, A. P. (1957). Direct measurement of long range van der Waals forces. *Proc. Roy. Soc. Lond. A* **242**, 403–9.

Landau, L. D. & Lifshitz, E. M. (1960). *Electrodynamics of Continuous Media.* 4th edn. Pergamon.

Langbein, D. (1974). *Theory of van der Waals Attraction.* 'Springer Tracts in Modern Physics'. Springer.

Lifshitz, E. M. (1956). The theory of molecular attractive forces between solids. *Soviet Physics JETP* **2**, 73–83.

Love, J. D. (1977). On the van der Waals force between two spheres or a sphere and a wall. *J. Chem. Soc. Far. Trans. II* **73**, 669–88.
Mahanty, J. & Ninham, B. W. (1976). *Dispersion Forces*. Academic Press.
Ninham, B. W. & Parsegian, V. A. (1970). van der Waals forces. Special characteristics in lipid–water systems and a general method of calculation based on the Lifshitz theory. *Biophys. J.* **10**, 646–63.
Overbeek, J. Th. G. & Sparnaay, M. J. (1954). Coagulation and flocculation. II. Classical coagulation. London–van der Waals attraction between macroscopic objects. *Disc. Far. Soc.* **18**, 12–24.
Pailthorpe, B. A. & Russel, W. B. (1982). The retarded van der Waals interactions between spheres. *J. Colloid Interface Sci.* **89**, 563–6.
Parsegian, V. A. (1975). Long range van der Waals forces, in *Physical Chemistry: Enriching Topics in Colloid and Surface Science* (eds H. van Olphen and K. J. Mysels). Theorex, pp. 27–72.
Parsegian, V. A. & Ninham, B. W. (1969). Application of the Lifshitz theory to the calculation of van der Waals forces across thin lipid films. *Nature* (*Lond.*) **224**, 1197–8.
Parsegian, V. A. & Weiss, G. H. (1981). Spectroscopic parameters for computation of van der Waals forces. *J. Colloid Interface Sci.* **81**, 285–9.
Prieve, D. C. & Russel, W. B. (1988). Simplified predictions of Hamaker constants from Lifshitz theory. *J. Colloid Interface Sci.* **125**, 1–13.
Stratton, J. A. (1941). *Electromagnetic Theory*. McGraw-Hill.
Tabor, D. & Winterton, R. H. S. (1969). Direct measurements of normal and retarded van der Waals forces. *Proc. Roy. Soc. Lond. A* **312**, 435–50.
van Kampen, N. G., Nijboer, B. R. A. & Schram, K. (1968). On the macroscopic theory of van der Waals forces. *Phys. Lett.* **26A**, 307–8.
White, L. R., Israelachvili, J. N. & Ninham, B. W. (1976). Dispersion interaction of crossed mica cylinders: a reanalysis of the Israelachvili-Tabor experiments. *J. Chem. Soc. Far. Trans. I* **72**, 2526–36.

Problems

1 The interaction potential from the pairwise additive or microscopic theory can be derived from (5.2.11) for any geometry. Formulate and evaluate the integral for a thin rod of length l and radius $a \ll l$ interacting with a semi-infinite flat plate in both the parallel and perpendicular orientations at separations large relative to the radius. Note which orientation is more favorable.

2 The magnitude of the dispersion attraction and the effect of retardation vary with the dielectric properties of the fluid and particle.

Select from Table 5.2(c) the combinations with (i) the largest and smallest values of $A_{eff}(0)$ and (ii) the largest and smallest differences between $A_{eff}(0)$ and $A_{eff}(\infty)$. Plot the dispersion potentials predicted for the latter from (5.9.3) as functions of separation.

3 In many cases colloidal particles are stabilized in water simply by adsorbing layers of non-ionic surfactants on their surfaces. With surfactants of chain length l in a close-packed layer, two particles cannot approach closer than $2l$. Use the simplified form for the potential between spheres, (5.10.1), to estimate the surfactant chain length required to reduce the attraction to $-2kT$ for poly(tetrafluoroethylene) spheres of radii (a) 25 nm and (b) 250 nm for both very low and very high ionic strengths.

4 Silica spheres bearing a dense layer of grafted octadecyl chains and dispersed in cyclohexane provide a popular non-aqueous model system. Relatively small particles, $a \leq 40$ nm, appear to behave as hard spheres. Assess the particle size above which the dispersion attractions might become significant, e.g. $>0.25kT$.

5 Titanium dioxide, the most common pigment, frequently must be dispersed in water or any of a variety of aqueous solvents. Complete dielectric data is unavailable but the rutile form is known to have a primary absorption frequency of 1.1×10^{16} rad/s and dielectric constants and refractive indices are found in the *Handbook of Chemistry and Physics* (Chemical Rubber Publishing Co.). With these parameters estimate $A_{eff}(0)$ for TiO_2 in water and n-pentane.

6 One can gain some insight into the effects of attractive potentials on the dynamics of dispersions by calculating the trajectories of interacting particles. Incorporate the non-retarded dispersion potential into the trajectory equations for spheres accounting for far-field hydrodynamic interactions (Problem 2.5) and determine the maximum upstream position R_2 leading to doublet formation for $R_3=0$ and values of $A_{eff}(0)/6\pi\mu a^3\gamma$ of (i) 0.1 and (ii) 10.

6

FORCES DUE TO SOLUBLE POLYMER

6.1 Introduction

The preceding chapters address interactions between colloidal particles dispersed in pure liquid or electrolyte solution. The hydrodynamic and dispersion forces depend only on the bulk properties of the individual phases, i.e. the viscosity and the dielectric permittivities. Electrostatic forces arising from the surface charges, however, are accompanied by free electrolyte. The associated electric fields distribute these additional species non-uniformly in the surrounding fluid, thereby producing a spatially varying osmotic pressure. Electrostatic interactions between particles alter these ion distributions, affecting the electric and pressure fields and generating an interparticle force.

We now consider another component commonly present in colloidal systems, soluble polymer. In many ways, the phenomena and the theoretical treatment resemble those for electrostatics. The interactions between polymer and particle generate non-uniform distributions of polymer throughout the solution. Particle–particle interactions alter this equilibrium distribution, producing a force whose sign and magnitude depend on the nature of the particle–polymer interaction. The major difference from the ionic solutions lies in the internal degrees of freedom of the polymer, which necessitate detailed consideration of the solution thermodynamics.

The reasons for adding soluble polymer to colloidal dispersions are several. The earliest known role, as stabilizer, aids or preserves the dispersion through adsorption of the macromolecule onto the surfaces of the particles to produce a strongly repulsive interaction. Homopolymers achieve this by adsorbing to particles non-specifically at multiple points along their backbone, while block or graft copolymers adsorb irreversibly at one end with the other remaining in solution (Napper, 1983). However, under some conditions, adsorption of polymer flocculates otherwise stable dispersions, owing to bridging of polymer between particles (Vincent, 1974). In addition, even in the absence of adsorption, polymer can cause flocculation as demonstrated as early as 1939 by Bondy (Napper, 1983). Later Asakura & Oosawa (1954, 1958) identified the mechanism as attraction induced by the exclusion of polymer from the gap between two particles.

As these observations demonstrate, the nature of the interparticle potential depends critically on the polymer–particle interaction. This chapter deals with three situations:

(i) polymer bound irreversibly to the surface by one end only,
(ii) dissolved non-adsorbing polymer, and
(iii) polymer that adsorbs weakly at random points along the backbone.

We can illustrate qualitatively the nature of the resulting interparticle forces through consideration of the local osmotic pressure of the polymer in solution. One need know only that the osmotic pressure generally increases with polymer concentration and that integration of this normal force over the surface yields the net force. For terminally anchored polymer, interaction between particles at separations less than the chain dimension requires interpenetration or compression of the chains. The associated increase in the local polymer concentration in the gap produces an excess osmotic pressure and a repulsive force. Non-adsorbing polymer, on the other hand, is driven from the gap, leaving a deficit in the osmotic pressure and generating an attractive force. With randomly adsorbing polymer, an additional factor enters, since segments of a chain originally adsorbed to one surface can also attach to the other. Then the nature of the potential depends on the magnitude of the resulting attraction relative to the increased osmotic pressure.

Theories to predict these effects quantitatively must address the internal degrees of freedom of the macromolecules, interactions between segments, and the geometry of the interparticle interaction. The treatment presented here accommodates these through a differential equation describing the chain configurations, a mean-field approximation for the segment–segment interactions, and the Derjaquin approximation to convert results for flat plates to geometries with curvature. We begin with an introduction to the thermodynamics of polymer solutions (6.2), then discuss the application to interactions with surfaces. The results for terminally anchored chains (6.3), non-adsorbing polymer (6.4), and reversibly adsorbing polymers (6.5) illustrate the important features of the phenomena and provide some quantitative predictions. Selected experimental data complement and test the theory. More extensive treatments of these and related subjects are available elsewhere (deGennes, 1979; Takahashi & Kawaguchi, 1982; Napper, 1983).

6.2 Polymers in solution

General features

The more detailed treatments of polymer solutions strive to predict equilibrium molecular conformations and thermodynamic properties from knowledge of the chemical structure and the intermolecular forces. Fortunately, most of the features relevant to interactions between colloidal particles in the presence of soluble polymer can be understood and, to some extent, predicted by less detailed theories. These only require the concentration of polymer in solution plus measures of the contour length of the molecule, its ability to bend, and the interactions between different parts of the molecule (Yamakawa, 1971; deGennes, 1979).

The polymer is modelled as a freely jointed chain of N segments of length l, which reflect the intrinsic stiffness of the backbone. For flexible linear polymers, such as shown in Fig. 6.1, a segment includes five to ten bonds, i.e., several monomers. Hence, molecular weights ranging from 10^3 to 10^7 correspond to values of N of 10 to 10^5.

If the segments were aligned then the chain would be fully extended with a mean square end-to-end distance of $\langle r^2 \rangle = (Nl)^2$. However, Brownian

Fig. 6.1. Chemical structure of several linear polymers.

$[-O-CH_2-CH_2-]_n$

Poly(oxyethylene)

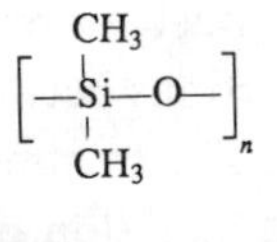

$\left[-\overset{CH_3}{\underset{CH_3}{\overset{|}{\underset{|}{Si}}}} -O- \right]_n$

Poly(dimethyl siloxane)

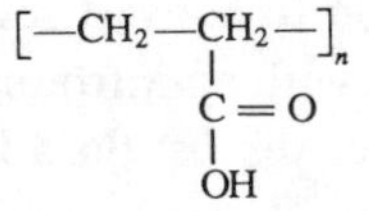

$[-CH_2-\underset{\underset{OH}{|}}{\underset{C=O}{\underset{|}{CH_2}}}-]_n$

Poly(acrylamide) (hydrolyzed)

motion disorients the individual segments, causing the chain to contract to a random-walk configuration with $\langle r^2 \rangle = Nl^2$ when segments do not interact. In real solutions, segments on distant portions of the chain do interact because of their physical volume l^3 and short-range attractions such as van der Waals forces. The net effect is to exclude each segment immersed in a solution with segment density n from a fraction $\frac{1}{2}nv + \frac{1}{3}n^2 w + \ldots$ of space. The excluded-volume parameter v accounts for pair interactions and is generally less than or equal to l^3, owing to the attractions. The triplet interactions represented by w are dominated by the physical volume of the segments such that $w \approx l^6$. To motivate the quantitative treatment in subsequent sections, we now outline, in a qualitative manner similar to Daoud & Janninck (1976) and Schaefer (1984), the consequences of these segment–segment interactions.

At the theta state, defined by $v = 0$, attractions exactly cancel the effect of physical volume for pair interactions. Within an isolated molecule $n \approx N(Nl^2)^{-3/2}$, so that $n^2 w \approx N^{-1} \ll 1$ and $\langle r^2 \rangle \sim Nl^2$, as expected from the random walk. In good solvents for which $v > 0$, the excluded volume reduces the configurations available to the segments. This decreases the entropy and increases the free energy by an amount related to the thermal energy of the segments times the fraction of the volume excluded such that

$$NkT(\tfrac{1}{2}vn + \tfrac{1}{6}wn^2) \sim \tfrac{1}{2}kT\frac{N^2 v}{\langle r^2 \rangle^{3/2}}. \tag{6.2.1}$$

In order to reduce this contribution to the free energy, the coil expands, but the chain responds as an elastic spring to this extension and the free energy increases by $\frac{3}{2}kT(\langle r^2 \rangle/(Nl^2) - 1)$. The equilibrium configuration, obtained by minimizing the sum of these two contributions to the free energy, yields the mean-square end-to-end distance of the swollen coil in a good solvent as

$$\frac{\langle r^2 \rangle}{Nl^2} \approx \left(\frac{v}{l^3}\right)^{2/5} N^{1/5}. \tag{6.2.2}$$

In poor solvents with $v < 0$, the coils tend to collapse until the terms on the left hand side of (6.2.1), corresponding to the physical volume and the attraction, balance such that

$$\frac{\langle r^2 \rangle}{Nl^2} \approx N^{-1/3}\left(\frac{w}{vl^3}\right)^{2/3}. \tag{6.2.3}$$

Note that $Nl^3/\langle r^2 \rangle^{3/2} \approx -vl^3/w$ suggesting that the coil collapses completely, i.e. to bulk density, only when $-v/l^3 \approx w/l^6 = O(1)$.

More careful treatments of coil expansion and collapse are available elsewhere (Yamakawa, 1971; Moore, 1977; deGennes, 1979), but (6.2.2) and (6.2.3) serve to suggest that at infinite dilution the random-coil state with $\langle r^2\rangle \sim Nl^2$ persists only for $-N^{-1/2} < v/l^3 < N^{-1/2}$. Outside this range, excluded volume affects the coil dimension. At sufficiently dilute concentrations, though, interactions between segments in different chains remain negligible; then the osmotic pressure of the solution derives solely from the thermal energy of the chain and follows van't Hoff's law,

$$P = \frac{n}{N}kT. \tag{6.2.4}$$

Fig. 6.2. Phase diagram for a polymer solution with segment density n and excluded-volume parameter $v/w^{1/2}$ showing the following concentration regimes (Daoud & Janninck, 1976): I′, ideal and dilute with $-2/N^{1/2} < v/w^{1/2} < 1/N^{1/2}$ and $nw^{1/2} < N^{-1/2}$; I, non-ideal and dilute with $nw^{1/2} < (v/w^{1/2})^{-3/5}N^{-4/5}$; II, semi-dilute with $(v/w^{1/2})^{-3/5}N^{-4/5} < nw^{1/2} < v/w^{1/2}$; III, concentrated with $\pm v/w^{1/2} < nw^{1/2}$; IV, phase separated.

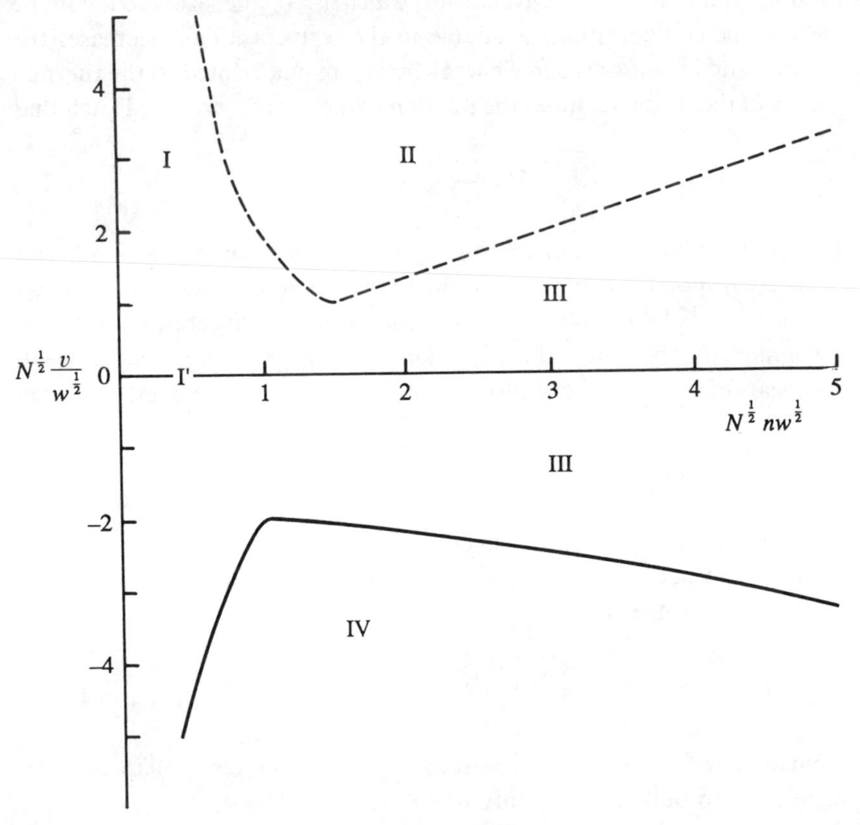

Intermolecular interactions become important when the bulk concentration becomes comparable to that within an individual coil, i.e., $n \approx N/\langle r^2 \rangle^{3/2}$, so that chains begin to overlap. For good solvents, this defines the transition between the *dilute* and *semi-dilute* regimes as $nl^3 \approx (v/l^3)^{-3/5} N^{-4/5}$. At semi-dilute concentrations, interactions with segments on other chains begin to screen those within the individual coil, allowing the chains to relax toward their ideal state. These interactions also cause the osmotic pressure to increase non-linearly with concentration.

At sufficiently high concentrations for which $wn^2 > vn$ or $nl^3 > vl^3/w$, the three-body effects dominate the pair interactions, eliminating correlations among segments within an individual chain. In this *concentrated* regime, the chains then assume their ideal state, but excluded volume still affects the osmotic pressure. Near the theta state, i.e. as $v/l^3 \to 0$, the semidilute regime disappears and overlap produces a transition directly from dilute to concentrated for $nl^3 \approx N^{-1/2}$. Figure 6.2 summarizes these results for the various concentration regimes.

The values of l, N, v, and w for specific polymers are derived from knowledge of the molecular weight M, the length l_0 and mass m_0 per bond, the specific volume $\tilde{v}$, the excluded volume per bond $(\tilde{v}m_0/N_A)(1-2\chi)$, with χ the Flory parameter (Yamakawa, 1971), and the mean-square end-to-end distance in ideal solution, expressed as $\langle r^2 \rangle = c_\infty l_0^2 M/m_0$. Requiring that the effective chain have the same contour length, mean-square end-to-end distance, physical volume, and total excluded volume as the real chain leads to

$$l = c_\infty l_0,$$

$$N = \frac{M}{c_\infty m_0},$$

$$w^{1/2} = c_\infty \frac{\tilde{v} m_0}{N_A},$$

$$v = w^{1/2}(1-2\chi).$$

Examples of available data and the segment lengths and physical volumes are tabulated in Table 6.1.

A theory constructed by Edwards (1965, 1966) for the thermodynamics of bulk solutions incorporates excluded volume via a mean-field, but self-consistent, approximation which ignores correlations among the positions of the segments but accommodates spatial variations in the segment density on the scale of the polymer coil. The approach is valid when the correlations

Table 6.1. *Properties of polymer in solution*

	l_0/nm	$\frac{10^3 m_0}{\text{kg/mol}}$	$\frac{10^3 \tilde{v}}{\text{m}^3/\text{kg}}$	$r_0^2/N_0 l_0^2$	l/nm	$\frac{10^3 m}{\text{kg/mol}}$	$w^{1/2}/l^3$
poly(oxy-ethylene)	0.148	14.7	0.79	4.1	0.60	60	0.39
poly(12-hydroxy-stearic acid)	0.150	20	—	6.1	0.91	122	—
poly(dimethyl siloxane)	0.162	37	0.98	5.2	0.84	192	0.55
polystyrene	0.154	52	0.95	9.5	1.46	494	0.25

From Brandrup & Immergut, 1975.

are either weak, as at dilute concentrations, or short-range, as in the concentrated regime. It has been extended to the semi-dilute regime both qualitatively and quantitatively (deGennes, 1979; Muthukumar & Edwards, 1982) and also provides meaningful results for poor solvents if the three-body interactions are included (Moore, 1977).

Development of the self-consistent field theory for polymers at interfaces has proceeded along several different paths (e.g., deGennes, 1969, 1979, §IX.2; Helfand & Tagami, 1971; Dolan & Edwards, 1974; Helfand, 1975; Scheutjens & Fleer, 1979, 1985). The following synthesizes the basic formulation of Helfand (1975) with approximate solutions of the form suggested by deGennes (1969).

Some justification for proceeding with this mean-field treatment of excluded-volume effects can be found in the conditions of interest for polymer-particle interactions. For terminally anchored chains, the densities within the layers are high and often fall in the concentrated regime. For non-adsorbing polymer, the ideal and concentrated systems demonstrate the essentials, though semidilute solutions hold considerable interest. For reversibly adsorbing polymer, all concentration regimes are important and the phenomena quite subtle (cf. deGennes, 1987), but the mean-field theory illustrates the basic features. Hence the simplest form of the theory serves our purposes. In most cases, analogous treatments of the semi-dilute regime are available elsewhere.

Thermodynamic functions

Now the task of relating thermodynamic properties to the parameters characterizing the dissolved polymer begins. After a general development of the basic thermodynamic functions, which follows

McQuarrie (1976, §§1.4, 2.4) and serves for analyses of equilibrium phase behavior (Chapter 10) as well, we proceed to incorporate the effects of internal degrees of freedom critical to predicting the behavior of polymer solutions.

This approach treats the polymer solution as a pseudo-one-component system. For chains of N segments with length l, the dimensionless segment density nl^3 and the excluded-volume parameter v/l^3 comprise the two independent variables. The latter plays the role of a dimensionless inverse temperature, as does the depth of a potential well in a molecular system. Thus nl^3 and v/l^3 determine the pressure P for a single phase system. In a two phase region, v/l^3 sets both P and the segment densities of the coexisting phases.

The first law of thermodynamics establishes the internal energy E as a state function, such that for a reversible process

$$dE = dq - P\,dV, \tag{6.2.5}$$

with dq the heat absorbed by the system, P the pressure, and dV the volume change. The second law specifies the existence of the entropy S and the temperature T such that

$$dS = \frac{dq}{T} \tag{6.2.6}$$

along any reversible path. Consequently,

$$dE = T\,dS - P\,dV$$

and

$$\left(\frac{\partial E}{\partial S}\right)_V = T, \qquad \left(\frac{\partial E}{\partial V}\right)_S = -P. \tag{6.2.7}$$

Although S and V rarely prove convenient independent variables, two related state functions are useful:

(i) The Helmholtz free energy, $A(T, V)$, defined by

$$A = E - TS$$

so that

$$dA = -P\,dV - S\,dT, \tag{6.2.8}$$

and

$$\left(\frac{\partial A}{\partial V}\right)_T = -P, \qquad \left(\frac{\partial A}{\partial T}\right)_V = -S.$$

(ii) The Gibbs free energy, $G(T, P)$, defined by

$$G = A + PV$$

so that

$$\mathrm{d}G = -S\,\mathrm{d}T + V\,\mathrm{d}P, \tag{6.2.9}$$

and

$$\left(\frac{\partial G}{\partial T}\right)_P = -S, \quad \left(\frac{\partial G}{\partial P}\right)_T = V.$$

Clearly, knowledge of the Helmholtz free energy $A(T, V)$ alone determines both the pressure and the Gibbs free energy.

Evaluation of $A(T, V)$ from knowledge of molecular interactions requires two additional statements, though. First, the probability of finding a volume V containing $M = nV/N$ polymer chains in a state with internal energy E_j is specified by the Boltzmann distribution

$$P_j = \frac{1}{Q}\exp\left(-\frac{E_j}{kT}\right),$$

with

$$Q = \sum_j \exp\left(-\frac{E_j}{kT}\right) \tag{6.2.10}$$

defining the configuration partition function. Second, E equals the ensemble average over the individual states:

$$\begin{aligned} E &= \sum_j E_j P_j \\ &= kT^2\left(\frac{\partial \ln Q}{\partial T}\right)_{M,V}. \end{aligned} \tag{6.2.11}$$

For indistinguishable chains,

$$Q = \frac{Z_M}{M!}, \tag{6.2.12}$$

with Z_M the configuration integral.

The entropy follows from (6.2.7) and (6.2.11) as

$$\begin{aligned} S &= \int \frac{1}{T}\,\mathrm{d}\left[kT^2\left(\frac{\partial \ln Q}{\partial T}\right)_{M,V}\right] \\ &= kT\left(\frac{\partial \ln Q}{\partial T}\right)_{M,V} + k\ln Q, \end{aligned} \tag{6.2.13}$$

so that substitution into (6.2.8) yields

$$A = -kT \ln Q, \tag{6.2.14}$$

representing a general starting point for the calculation of thermodynamic properties.

For polymer chains, the configuration integral can be divided into translational and internal parts as

$$Z_M = Z_{tr} Z_{int}. \tag{6.2.15}$$

Neglecting any effect of interactions on the translational degrees of freedom on the basis of diluteness, $nl^3 \ll 1$, gives

$$Z_{tr} = V^M \tag{6.2.16}$$

corresponding to ideal gas behavior (McQuarrie, 1976, §5-1), while

$$Z_{int} = W^M \tag{6.2.17}$$

defines W, the partition function for an individual chain.

With Stirling's approximation

$$\ln M! = M \ln M - M, \tag{6.2.18}$$

(6.2.12), and (6.2.15) to (6.2.17), the Helmholtz free energy can be written as

$$\frac{A}{MkT} = \ln \frac{n}{N} - 1 - \ln W. \tag{6.2.19}$$

The osmotic pressure and chemical potential are

$$\begin{aligned} \frac{P}{nkT} &= \frac{n}{N} \frac{\partial}{\partial n}\left(\frac{A}{MkT}\right) \\ &= \frac{1}{N}\left(1 - n\frac{\partial \ln W}{\partial n}\right) \\ \frac{\mu}{kT} &= \frac{\partial}{\partial n}\left(\frac{n}{N} \frac{A}{MkT}\right) \\ &= \frac{1}{N}\left(\ln n - \frac{\partial}{\partial n}(n \ln W)\right). \end{aligned} \tag{6.2.20}$$

Since the solution is in fact a two component system, the free energy per unit volume is also related to the chemical potentials of the polymer segments, μ_p, and solvent, μ_s, by

$$\frac{A}{V} = \mu_p n + \mu_s (w^{-1/2} - n). \tag{6.2.21}$$

The Gibbs–Duhem relation, $n\,d\mu_p + (w^{-1/2} - n)\,d\mu_s = 0$, then determines

$$\frac{A}{V} = \mu n - P,$$

where (6.2.22)

$$\mu = \mu_p - \mu_s$$
$$P = -\mu_s / w^{1/2}.$$

The exchange potential, μ, characterizes the cost in Gibbs free energy of introducing an additional polymer segment and removing a solvent molecule. The last equation relates the osmotic pressure to the chemical potential of the solvent.

For polymer solutions at either dilute or concentrated conditions, i.e., region I or III in Fig. 6.2, the Helmholtz free energy can be expressed in a virial series as (deGennes, 1979, §III.1)

$$\frac{A}{MkT} = \ln\frac{n}{N} - 1 + \tfrac{1}{2}Nvn + \tfrac{1}{6}Nwn^2 \ldots \tag{6.2.23}$$

with v and w the binary and ternary cluster integrals for segment–segment interactions as noted above. The osmotic pressure and chemical potential follow from (6.2.20), as

$$\frac{P}{nkT} = \frac{1}{N} + \frac{nv}{2} + \frac{n^2 w}{3} + \ldots, \tag{6.2.24}$$

$$\frac{\mu}{kT} = \frac{1}{N}\ln\frac{n}{N} + nv + \frac{n^2 w}{2} + \ldots. \tag{6.2.25}$$

In each, the first term corresponds to ideal behavior. Note that the non-ideal terms in the osmotic pressure comprise the excluded volume per segment discussed earlier.

Self-consistent field theory

In this theory, a polymer chain in solution is characterized by the density $G(\mathbf{r}, \mathbf{r}', s)$ of configurations available to a subchain of $s \le N$ segments, beginning at $\mathbf{r}'$ and ending at $\mathbf{r}$ (Fig. 6.3(a)). Once G is known, the mean-square end-to-end distance for a chain follows directly as

$$\langle r^2 \rangle = \frac{\int r^2 G(\mathbf{r}, 0, N)\,d\mathbf{r}}{\int G(\mathbf{r}, 0, N)\,d\mathbf{r}}. \tag{6.2.26}$$

Likewise, the total number of configurations available to a chain beginning at $\mathbf{r}'$ is

$$\int G(\mathbf{r}, \mathbf{r}', N)\,d\mathbf{r},$$

and the average number of configurations available to a chain in the system, i.e. the partition function for an individual chain, takes the form

$$W = \iint G(\mathbf{r}, \mathbf{r}', N)\,d\mathbf{r}\,d\mathbf{r}'. \tag{6.2.27}$$

Hence G also determines the Helmholtz free energy and other thermodynamic properties of the solution.

Fig. 6.3. (*a*) Coordinates of chain beginning at $\mathbf{r}'$ with the sth segment at $\mathbf{r}$. (*b*) Definition of vector $\mathbf{n}$ from s to $s+1$ segment.

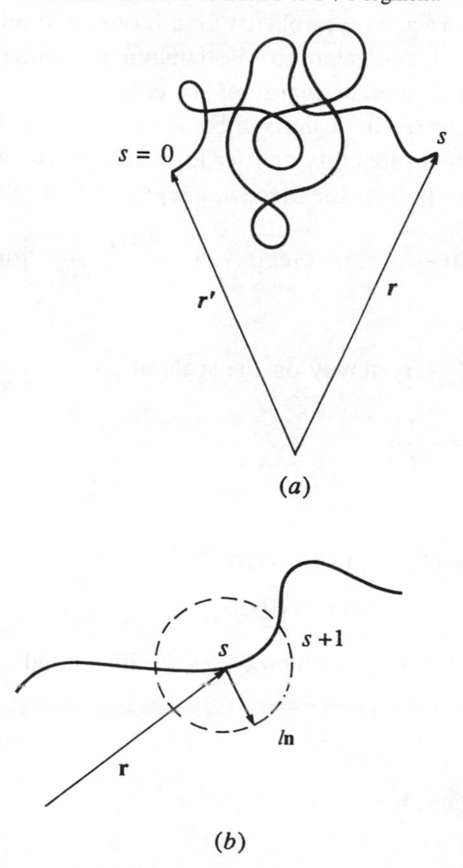

An individual chain of length N can be divided into subchains of lengths s and $N-s$. Then the probability of finding the sth segment at $\mathbf{r}$ for a chain that begins at $\mathbf{r}'$ and ends at $\mathbf{r}''$ is proportional to $G(\mathbf{r},\mathbf{r}',s)G(\mathbf{r}'',\mathbf{r},N-s)$. Integration over all positions for the endpoints and all positions along the chain with proper normalization yields the local segment density as

$$n(\mathbf{r})=\frac{n}{NW}\int\int_0^N G(\mathbf{r},\mathbf{r}',s)\int G(\mathbf{r}'',\mathbf{r},N-s)\,\mathrm{d}\mathbf{r}''\,\mathrm{d}s\,\mathrm{d}\mathbf{r}', \tag{6.2.28}$$

with $n=MN/V$ the average segment density.

The determination of G begins with the recognition that at equilibrium the configurations for a chain are specified by

$$G(\mathbf{r},\mathbf{r}',s)=\exp\left(-\int_0^s\frac{U(\mathbf{r}(s'))}{kT}\,\mathrm{d}s'\right). \tag{6.2.29}$$

The potential $U(\mathbf{r})$ per segment accounts for interactions with all other segments and the expression accounts implicitly for all configurations of the intermediate segments. This is equivalent to a Boltzmann distribution, with the integral representing the internal energy of the chain.

A differential equation for G can be derived by assuming it and U to be smooth functions. First, the probability density for a subsequent segment along the chain is related to that for the preceding segment as (Fig. 6.3(b)):

$$G(\mathbf{r},\mathbf{r}',s)=\frac{1}{4\pi}\int G(\mathbf{r}+l\mathbf{n},\mathbf{r}',s-1)\exp\left(-\int_{s-1}^{s}\frac{U(\mathbf{r}(s'))}{kT}\partial s'\right)\mathrm{d}\mathbf{n}. \tag{6.2.30}$$

Then, since both G and U vary slowly on the scale of l,

$$\int_{s-1}^{s}\frac{U(\mathbf{r}(s'))}{kT}\,\mathrm{d}s'=\frac{U(\mathbf{r})}{kT}+\ldots \tag{6.2.31}$$

and

$$G(\mathbf{r}+l\mathbf{n},\mathbf{r}',s-1)=G(\mathbf{r},\mathbf{r}',s)+l\mathbf{n}\cdot\nabla G(\mathbf{r},\mathbf{r}',s)$$
$$+\tfrac{1}{2}l^2\mathbf{nn}:\nabla\nabla G(\mathbf{r},\mathbf{r}',s)+\ldots-\frac{\partial G}{\partial s}(\mathbf{r},\mathbf{r}',s)\ldots.$$

Substitution into (6.2.30) and integration produces the differential equation

$$\frac{\partial G}{\partial s}=\frac{l^2}{6}\nabla^2 G+\left[1-\exp\left(\frac{U}{kT}\right)\right]G. \tag{6.2.32}$$

The condition that the chain begins at $\mathbf{r}'$ requires

$$G(\mathbf{r},\mathbf{r}',0)=\delta(\mathbf{r}-\mathbf{r}').$$

We can interpret (6.2.32) by noting that the first term accounts for the connectivity of the chain, the second for the tendency of entropic effects to disperse the segments, and the third for the action of the potential U. For $U/kT=0$, the balance between connectivity and entropy determines the density of configurations. Having $U/kT>0$ locally reduces the density of configurations, effectively excluding the chain from that region of space; $U/kT<0$ has the opposite effect.

The choice of U/kT to account properly for interactions with the other segments is clearly critical. Formal derivations (e.g., Helfand, 1975; Joanny, Leibler & deGennes, 1979; Scheutjens & Fleer, 1979) lead to

$$\exp\left(\frac{U}{kT}\right)-1=\frac{1}{kT}\frac{\partial f}{\partial n},$$

with (6.2.33)

$$f=\mu n-P-\frac{n}{N}kT\left(\ln\frac{n}{N}-1\right)$$

representing the portion of the free energy density, A/V, arising from segment–segment interactions. From (6.2.24),

$$\frac{f}{kT}=\tfrac{1}{2}vn^2+\tfrac{1}{6}wn^3+\ldots,$$

so that (6.2.34)

$$e^{U/kT}-1=vn+\tfrac{1}{2}wn^2+\ldots.$$

Along with this comes a modified form for the free energy,

$$\begin{aligned}\frac{A}{MkT}&=\ln\frac{n}{N}-1-\ln W+\frac{1}{M}\int_V\left(\frac{f}{kT}-\frac{n}{kT}\frac{\partial f}{\partial n}\right)\mathrm{d}V\\&=\ln\frac{n}{N}-1-\ln W-\frac{1}{M}\int_V(\tfrac{1}{2}vn^2+\tfrac{1}{3}wn^3)\,\mathrm{d}V,\end{aligned}\qquad(6.2.35)$$

from (6.2.34). The integral term corrects for the fact that the mean-field theory overcounts the segment–segment interactions.

The differential equation (6.2.32), together with the potential (6.2.33), the expression (6.2.28) for the segment density, and suitable boundary conditions, completes the formulation of the theory. The neglect of correlations among segments makes this a mean-field approximation. Self-consistency is insured through the relationship between the local segment density and the probability density G.

Application to bulk solutions

For ideal chains in Region I′ of Fig. 6.2, $v=0$ and $wn^2 \ll 1$, so the governing equation reduces to

$$\frac{\partial G}{\partial s}=\frac{l^2}{6}\frac{1}{r^2}\frac{\partial}{\partial r}r^2\frac{\partial G}{\partial r}, \tag{6.2.36}$$

with

$$G(r,0,0)=\delta(r)$$
$$G(\infty,0,s)=0.$$

The solution, corresponding to that for transient diffusion (3.3.8)

$$G\equiv G_0=\left(\frac{3}{2\pi l^2 s}\right)^{3/2}\exp\left(-\frac{3r^2}{2l^2 s}\right), \tag{6.2.37}$$

indicates a Gaussian chain, characteristic of a random walk, with mean-square end-to-end distance $\langle r^2\rangle=Nl^2$ from (6.2.26).

For solutions at sufficiently high concentrations in Region III of Fig. 6.2, $n(\mathbf{r})=n$, independent of position, and the governing equation (6.2.32) again becomes linear with solution $G=G_0\exp[-(vn+\frac{1}{2}wn^2)s]$. Substitution into (6.2.26) shows that the configuration of the chain remains Gaussian, with $\langle r^2\rangle=Nl^2$ due to screening of the excluded volume effects. The partition function (6.2.27) is altered, however, resulting in (6.2.24) and (6.2.25) for the osmotic pressure and chemical potential.

In poor solvents, phase separation can occur as indicated by the solid curve in Fig. 6.2. The phase boundary corresponds to polymer solutions at two concentrations connected by a horizontal tie line, e.g., n and n_*, with the same chemical potentials and osmotic pressures. Solution of the two equations resulting from (6.2.24) and (6.2.25) yields the curve shown in Fig. 6.2 with a critical point at $n=n_*=1/(N^{1/2}w^{1/2})$ and $-v/w^{1/2}=2/N^{1/2}$.

So the mean-field theory predicts that chains remain Gaussian in the dilute and concentrated regimes for bulk solutions, although the thermodynamic properties are non-ideal except at infinite dilution. The following sections discuss the application to polymers interacting with interfaces which generate non-Gaussian conformations and spatially non-uniform segment densities over this same range of concentrations.

6.3 Terminally anchored polymers

Structure of isolated layers

When otherwise soluble polymers are attached by one end to a surface, their conformation depends on the surface density of chains n_p as well as N and the excluded volume v/l^3 (deGennes, 1980). At low densities, $n_p\langle r^2\rangle < 1$, the isolated chains extend $\approx \langle r^2\rangle^{1/2}$ into the solution, creating a layer with the density profile shown in Fig. 6.4(a) and a thickness of $L \approx N^{1/2} l$ for ideal chains and from (6.2.2) $L \approx N^{3/5} l$ in good solvents. When $n_p\langle r^2\rangle > 1$, the coils overlap. At both theta conditions and in good solvents the interactions between neighboring chains will reduce the volume available to each. Consequently, the chains expand away from the surface into the bulk. The configurations of the individual molecules and the density profile within the layer differ markedly from the dilute situation (Fig. 6.4(b)). When $n_p l^2 \approx 1$ the molecules become fully stretched.

Here we begin with the ideal case and calculate the configurations directly by applying the self-consistent field approach to terminally anchored chains in the spirit of Dolan & Edwards (1974). For high surface densities, i.e., $n_p\langle r^2\rangle > 1$, interpenetration of the polymer chains produces segment densities that are independent of lateral position. Hence G varies only with distance from the interface and position along the chain and is governed by the one-dimensional version of (6.2.32)

Fig. 6.4. Schematic of the conformations and corresponding segment densities for chains in good solvents terminally anchored to a wall at (a) dilute and (b) semi-dilute concentrations (deGennes, 1980; Milner, Witten, & Cates, 1989).

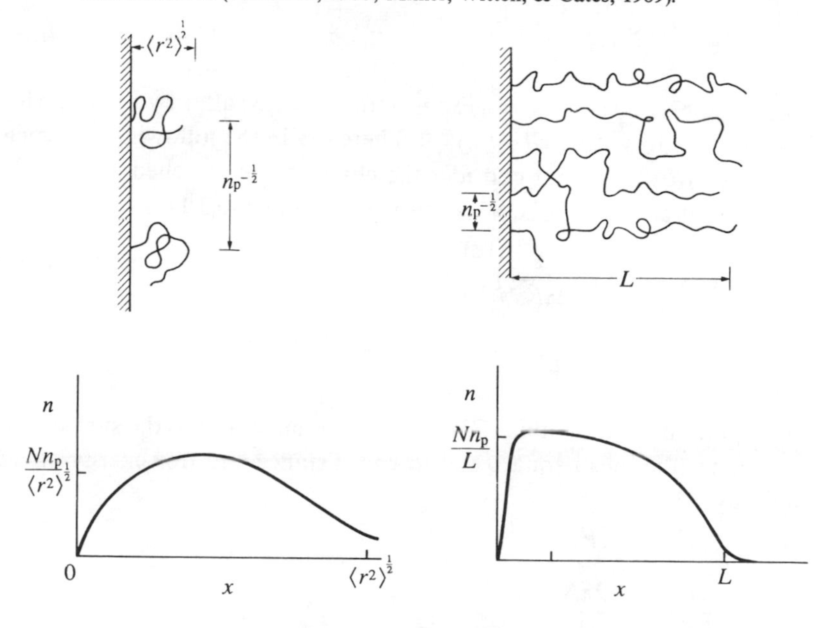

$$\frac{\partial G}{\partial s}=\frac{l^2}{6}\frac{\partial^2 G}{\partial x^2}+\left[1-\exp\left(\frac{U}{kT}\right)\right]G, \tag{6.3.1}$$

with

$$G(x,x',0)=\delta(x-x')$$

and

$$n(x)=n_p\frac{\int_0^N G(x,0,s)\int_0^\infty G(x,x'',N-s)\,dx''\,ds}{\int_0^\infty G(x'',0,N)\,dx''}. \tag{6.3.2}$$

The free energy for a terminally anchored chain is given by (6.2.35) without the $\ln(n/N)-1$ term from the translational degrees of freedom.

The interface at $x=0$ restricts the segments to $x>0$. The appropriate boundary condition can be derived from (6.2.29) in the same manner as the differential equation by integrating over only the hemisphere within the fluid. The result is

$$\frac{\partial G}{\partial s}=\frac{l}{2}\frac{\partial G}{\partial x}-\exp\left(\frac{U}{kT}\right)G \tag{6.3.3}$$

at $x=0$. However, $\partial \ln G/\partial s \ll 1$ and $U/kT \ll 1$ near the interface, reducing this to

$$0=\frac{l}{2}\frac{\partial G}{\partial x}-G. \tag{6.3.4}$$

This resembles a Taylor series expansion, truncated after two terms, which is equivalent to setting $G(-l/2)=0$. Therefore in the following we replace (6.3.4) with $G(0)=0$, but consider the chains to be attached at $x'=l/2$.

For ideal conditions, i.e., $U=0$, the solution to (6.3.1) is

$$G(x,x',s)=\left(\frac{3}{2\pi l^2 s}\right)^{1/2}\left[\exp\left(-\frac{3(x-x')^2}{2l^2 s}\right)-\exp\left(-\frac{3(x+x')^2}{2l^2 s}\right)\right], \tag{6.3.5}$$

corresponding to individual Gaussian chains anchored to the surface. This determines the mean-square end-to-end distance and free energy of each chain as

$$\langle x^2\rangle=\tfrac{2}{3}Nl^2$$

$$\frac{A}{kT}=\tfrac{1}{2}\ln\frac{2\pi N}{3}. \tag{6.3.6}$$

Since $\langle x^2\rangle = \langle r^2\rangle/3 = Nl^2/3$ and $A/kT = 0$ for ideal chains in solution, the presence of the wall increases both $\langle x^2\rangle$ and A/kT.

The effect of solvent quality on the thickness of the layer L and free energy of the individual chains can be estimated through a simple mean-field theory along the lines discussed by deGennes (1975) and Alexander (1977). Consider a layer having uniform segment density $n = Nn_p/L$, with each chain beginning at $x = l/2$ and ending at $x = L - l/2$. Then the segments within the layer will experience the constant potential

$$\exp\left(\frac{U}{kT}\right) - 1 = \frac{Nn_p v}{L} + \frac{w}{2}\left(\frac{Nn_p}{L}\right)^2 + \ldots, \tag{6.3.7}$$

and the solutions to (6.3.1) will have the same form, $G = G_0 \exp(-(vn + \frac{1}{2}wn^2)s)$, as for the concentrated bulk solutions. In this case, though, G_0 pertains to chains confined within $0 \leq x \leq L$ and is given by (6.3.5) for $L^2 \gg Nl^2$ and

$$G_0(x, x', s) = \frac{2}{L}\sin\frac{\pi x}{L}\sin\frac{\pi x'}{L}\exp\left[-\left(\frac{\pi l}{L}\right)^2\frac{s}{6}\right] \tag{6.3.8}$$

for $L^2 \ll Nl^2$ (Dolan & Edwards, 1974). The corresponding asymptotes for the free energy per chain are

$$\frac{A}{kT} = \tfrac{1}{2}Nvn + \tfrac{1}{6}Nwn^2 + \begin{cases} \dfrac{\pi^2 Nl^2}{6L^2}, & \dfrac{L^2}{Nl^2} \ll 1 \\[2ex] \dfrac{3L^2}{2Nl^2}, & \dfrac{L^2}{Nl^2} \gg 1. \end{cases} \tag{6.3.9}$$

These results suggest approximating the free energy over the full range of L^2/Nl^2 by

$$\frac{A}{kT} = \frac{3}{2}\left(\frac{L^2}{Nl^2} + \frac{Nl^2}{L^2} - 2\right) + \frac{Nvn}{2} + \frac{Nwn^2}{6}. \tag{6.3.10}$$

The equilibrium layer thickness then corresponds to the free energy for which the chemical potential of the solvent in the layer equals that in the bulk, or from (6.2.20, 22).

$$\frac{\partial A}{\partial n} = -\frac{L^2}{\mathrm{N}n_p}\frac{\partial A}{\partial L} = 0, \tag{6.3.11}$$

subject to the constraint that $nL = Nn_p$. Expressing the equilibrium condition in terms of the dimensionless layer thickness $\alpha_0 = L/N^{1/2}l$ produces

$$\alpha_0^3 - (1 + \tfrac{1}{9}\phi_p^2)\alpha_0^{-1} = \tfrac{1}{6}z. \qquad (6.3.12)$$

The dimensionless surface density, $\phi_p = Nn_p w^{1/2}/l$, represents the volume fraction of segments that would result from collapsing the chains into a layer of thickness l. The second parameter, $z = N^{3/2}n_p v/l$, is the ratio of the excluded volume per chain ($N^2 v$) to the volume occupied by the chain at ideal conditions ($N^{1/2}l/n_p$).

Figure 6.5 illustrates the transition from stretched to collapsed layers predicted as a function of the excluded volume and the surface density. At theta conditions and low surface densities $\alpha_0 = 1$, but, at greater than monolayer density, the physical volume of the segments expands the layer until $\alpha_0 \sim 0.58\phi_p^{1/2}$ or $L \sim 0.58Nl(n_p w^{1/2}/l)^{1/2}$ for $\phi_p^2 \gg 1$. In good solvents with $z \gg 1$, the layer thickness also increases linearly with chain length since $\alpha_0 \sim 0.55z^{1/3}$ or $L \sim 0.55Nl(n_p vl)^{1/3}$. In poor solvents layers contract until

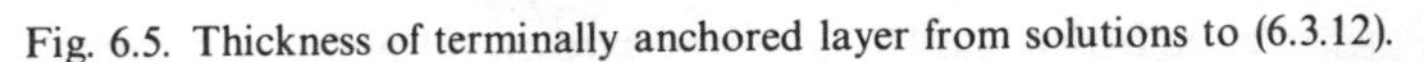
Fig. 6.5. Thickness of terminally anchored layer from solutions to (6.3.12).

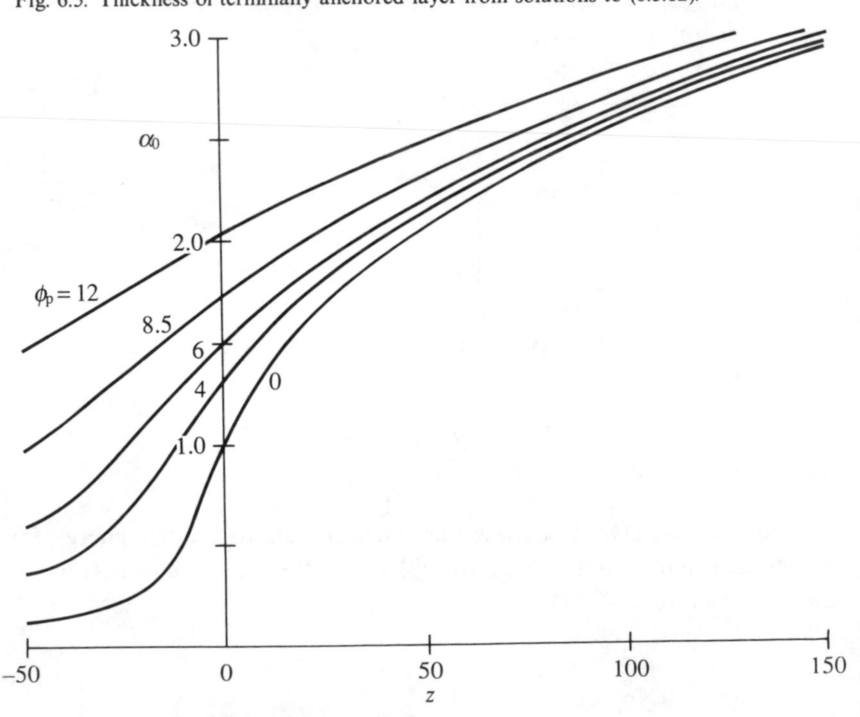

either the elastic compression or the physical volume of the segments balances the negative excluded volume for $\alpha_0 \sim -6(1+\frac{1}{9}\phi_p^2)/z$.

Interactions between layers: ideal solutions

When two surfaces approach, the attached polymer layers first interact at separations of the order of twice the layer thickness. The molecules anchored on one surface interact directly with the opposing surface, however, only at separations less than $N^{1/2}l$. Both the polymer–polymer and polymer–surface interactions alter the free energy by changing the volume available to an individual chain. For $v>0$, interpenetration of the two layers reduces the available volume, and, hence, the number of configurations. The associated increase in free energy produces a repulsive force. However, with negative excluded volume, the interaction increases the volume available per chain, thereby decreasing the free energy and causing an attraction. In both cases, reducing the separation to less than one layer thickness constrains the chains sufficiently to produce a strong repulsion.

The interaction potential equals the change in free energy per unit area that follows from (6.2.35) as

$$\Phi = -2n_p kT \ln W(h) - kT\int_0^h (\tfrac{1}{2}vn^2 + \tfrac{1}{3}wn^3)\,\mathrm{d}x - A_\infty, \tag{6.3.13}$$

with

$$W(h) = \int_0^h G(x, l/2, N)\,\mathrm{d}x$$

and

$$A_\infty = n_p kT \ln\left(2\pi\frac{N}{3}\right) + 2n_p A_0(z, \phi_p).$$

A_∞ is the free energy of the isolated layers with the first term comprising the ideal result and $A_0(z, \phi_p)$ accounting for the effects of segment–segment interactions.

The equation (6.3.1) is a non-linear integro-differential equation, because of the dependence of U on G through (6.2.28) and (6.2.31). For ideal solutions, though, the potential term vanishes, leaving a linear equation with an analytical solution that can be expressed either as the eigenfunction expansion (Dolan & Edwards, 1974)

$$G(x, x', s) = \frac{2}{h}\sum_{m=1}^{\infty} \sin\frac{m\pi x}{h} \sin\frac{m\pi x'}{h} \exp\left[-\left(\frac{\pi m l}{h}\right)^2 \frac{s}{6}\right], \tag{6.3.14}$$

or as the inverse of a Laplace transform that reduces to (6.3.5) for large separations.

In the small separation limit, the first term of (6.3.14) suffices so that from (6.2.28)

$$n(x)=\frac{4Nn_p}{h}\sin^2\frac{\pi x}{h},$$

and from (6.3.13),

$$\Phi=n_pkT\left[\frac{\pi^2}{3H^2}+\ln\left(\frac{3H^2}{8\pi}\right)\right], \tag{6.3.15}$$

with $H=h/N^{1/2}l$. Thus, at small separations, $\Phi/n_pkT \sim \pi^2/3H^2$. Comparison with expression (6.3.10) for the free energy of an isolated layer associates this strong repulsion with the compression of chains between the surfaces.

Figure 6.6 compares (6.3.15) with the exact free energy of interaction (Dolan & Edwards, 1974). Note that the interaction potential per adsorbed molecule, Φ/n_pkT, becomes $O(1)$ only when $H<1$, i.e., at separations less than $N^{1/2}l$, and then (6.3.15) provides an accurate approximation.

Fig. 6.6. Interaction potential between flat plates bearing n_p terminally anchored polymer chains at ideal conditions: ——, exact analytical solution (Dolan & Edwards, 1974); - - -, approximation (6.3.15) for small separations.

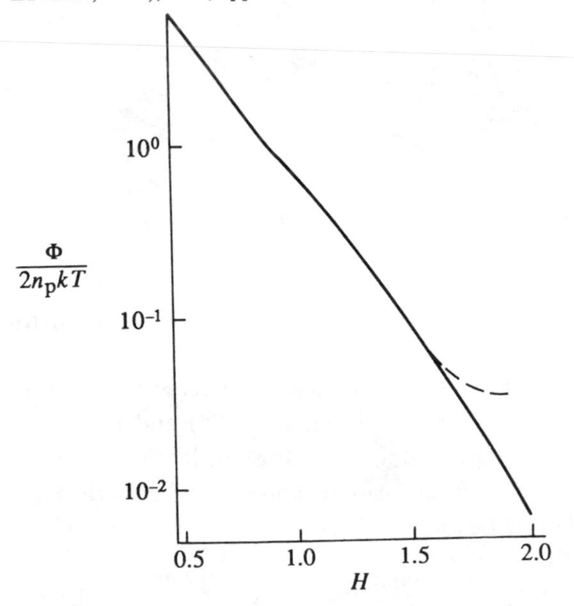

Interactions between layers: good and poor solvents

The mean-field theory described above for isolated layers can also be applied to interactions by assuming a piecewise constant segment density within the gap. The approach easily accommodates the higher-order interactions between segments that are quite important in theta and poor solvents, but approximates rather crudely the density profile within the layer.

The equilibrium thickness for the individual layers still corresponds to $\partial A/\partial L = 0$. For uniform densities within each layer, the total concentration of segments has the form

$$n = \frac{Nn_{\mathrm{p}}}{L}\begin{cases} 1 & 0 \le x \le h-L \quad \text{or} \quad L \le x \le h \\ 2 & h-L \le x \le L \end{cases}$$

provided $h/2 \le L \le h$. If $L < h/2$, the layers do not interact. This density profile determines the free energy per chain in the layers as

$$\frac{A}{kT} = \tfrac{3}{2}(\alpha^2 + \alpha^{-2} - 2) + \frac{3}{2}\frac{z}{\alpha}\left(1 - \frac{H}{3\alpha}\right) + \frac{7}{6}\frac{\phi_{\mathrm{p}}^2}{\alpha^2}\left(1 - \frac{3}{7}\frac{H}{\alpha}\right), \tag{6.3.16}$$

with $\alpha = L/N^{1/2}l$ the dimensionless layer thickness. The equilibrium condition follows as

$$\alpha^3 - \left[1 + \frac{7\phi_{\mathrm{p}}^2}{9}\left(1 - \frac{9}{14}\frac{H}{\alpha}\right)\right]\alpha^{-1} = \tfrac{1}{2}z\left[1 - \frac{2}{3}\frac{H}{\alpha}\right]. \tag{6.3.17}$$

Numerical solutions for α indicate that the mode of interaction changes qualitatively with increasing surface density of chains (Fig. 6.7). At low coverages, e.g., $\phi_{\mathrm{p}} < 1$, the layers interpenetrate at $H \approx 2\alpha_0$ for $z = 0$. In poor solvents, interpenetration reduces the free energy, so the chains actually extend slightly such that $\alpha > \alpha_0$. With decreasing separation, the layer thickness decreases in poor solvents but remains constant in theta solvents until $\alpha = H$. Thereafter, the surfaces compress the chains. At higher coverages, e.g., $1 \le \phi_{\mathrm{p}} \le 3$, the physical volume of the segments opposes interpenetration, favoring compression of the individual layers with $\alpha = H/2$. Eventually, interpenetration becomes favorable and the chains gradually extend until they reach the opposite surface. Increasing the surface density further causes the transition from compressed to extended chains to occur over a narrow range of separations, as illustrated for $\phi_{\mathrm{p}} = 5$.

The interaction potential

$$\Phi = 2n_{\mathrm{p}}[A(H, z, \phi_{\mathrm{p}}) - A_0(z, \phi_{\mathrm{p}})] \tag{6.3.18}$$

Fig. 6.7. Mean-field predictions for layer thickness as a function of separation: (*a*) $z = 0$, (*b*) $z = -2.5$. The horizontal lines indicate non-interacting layers with $\alpha = \alpha_0$, the portions with $\alpha = H/2$ imply interacting but non-interpenetrating, and those with $\alpha = H$ correspond to fully interpenetrating layers compressed between the surfaces.

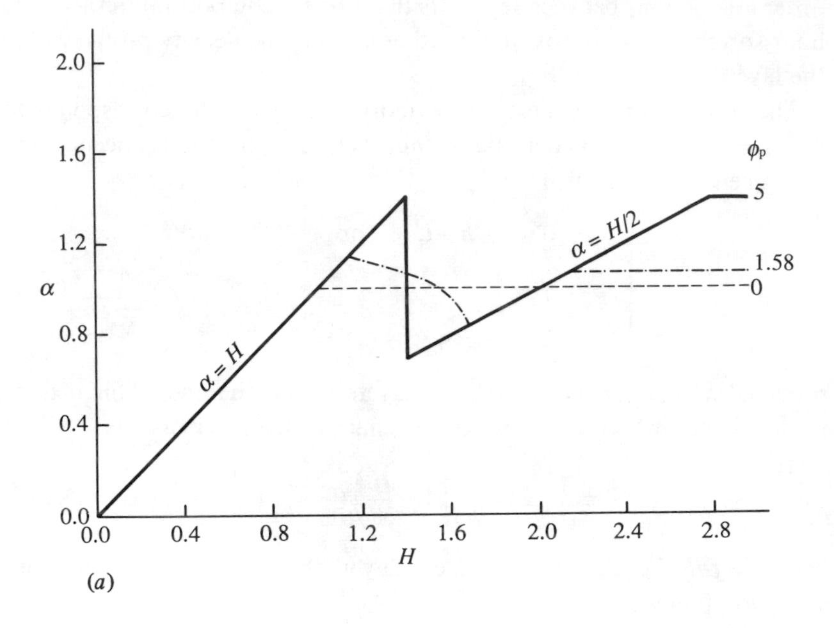

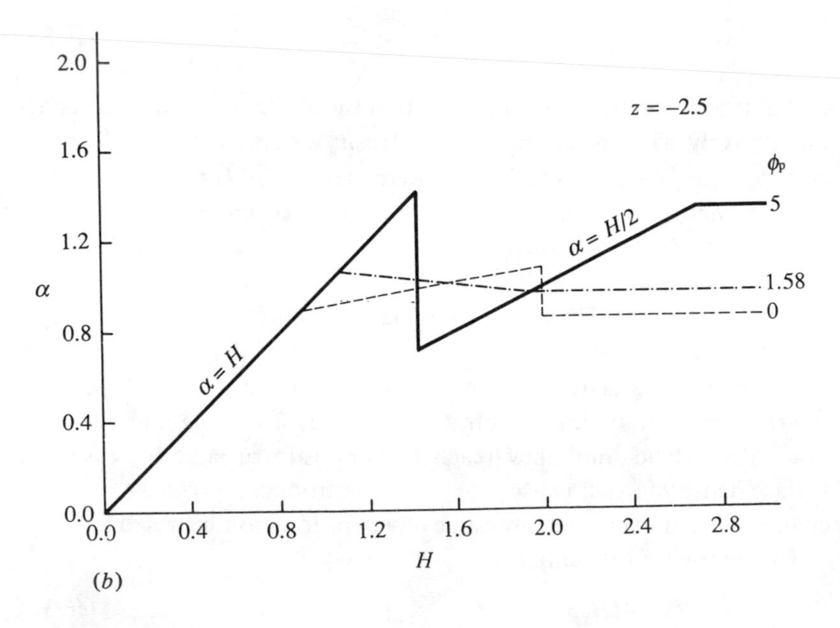

can be evaluated with $A(H, z, \phi_p)$ from (6.3.16) and (6.3.17) and $A_0(z, \phi_p)$ from (6.3.10) and (6.3.12). Note that for small separations, when $\alpha = H \ll 1$,

$$\Phi = 3n_p kT(1 + \tfrac{4}{9}\phi_p^2)H^{-2}, \tag{6.3.19}$$

demonstrating that the interaction is independent of z and repulsive, owing to the elasticity of the chains and the physical volume of the segments.

The plot of the potential as a function of separation for good solvents (Fig. 6.8) demonstrates the strong repulsions for $H < 2\alpha_0$. Since the chains are stretched by the excluded-volume interactions, the range of the repulsion increases with increasing z, exceeding significantly that for the ideal case.

For interactions between dense layers with $z \gg 1$, interpenetration occurs only for $H < 2^{1/2}$ (Fig. 6.7). For $2^{1/2} < H < 2\alpha_0$, the interaction potential assumes the analytical form

$$\frac{\Phi}{n_p kT} = 3\left(\frac{H^2}{4} - \alpha_0^2 + \frac{4}{H^2} - \frac{1}{\alpha_0^2}\right) + z\left(\frac{2}{H} - \frac{1}{\alpha_0}\right) + \frac{\phi_p^2}{3}\left(\frac{4}{H^2} - \frac{1}{\alpha_0^2}\right). \tag{6.3.20}$$

Fig. 6.8. Interaction potential from mean-field theory ((6.3.16) to (6.3.18)) for good solvents with $\phi_p = O(1)$.

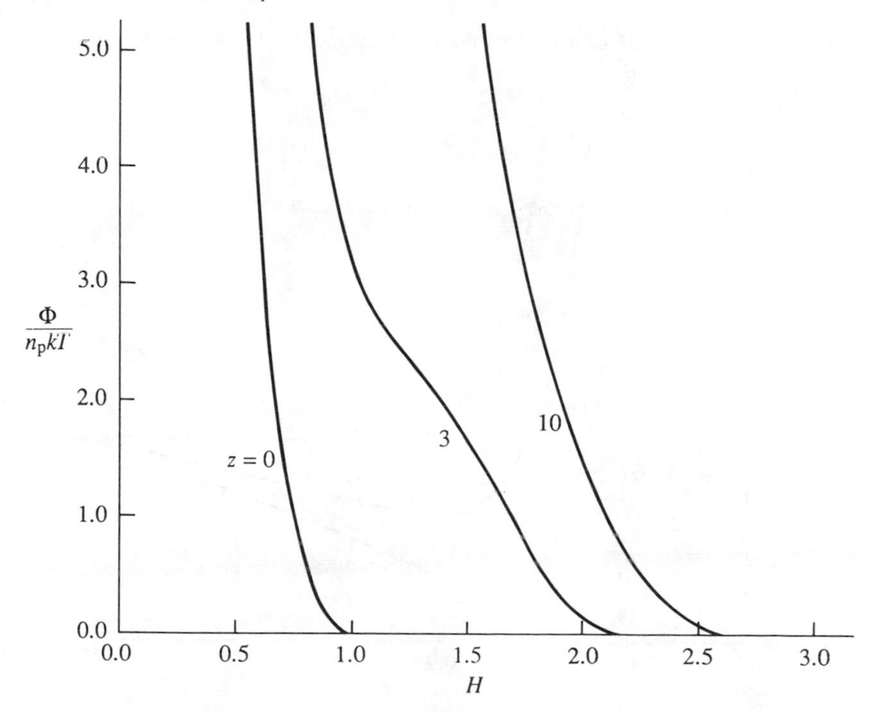

For $1-H/2\alpha_0 \ll 1$ and $\alpha_0 \sim (z/6)^{1/3}$, this reduces to

$$\frac{\Phi}{n_p kT} \sim \frac{3^{4/3}}{2^{2/3}} N \left(\frac{n_p v}{l}\right)^{2/3} \left[1 - \left(\frac{3}{4}\right)^{1/3} \frac{h}{Nl} \left(\frac{l}{n_p v}\right)^{1/3}\right]^2. \tag{6.3.21}$$

Thus the magnitude of the potential scales as $Nn_p{}^{5/3}$ and the separation as $Nn_p{}^{1/3}$.

For poor solvents, the potentials in Fig. 6.9 have an attractive minimum only for $-z > \phi_p^2$ or $-N^{1/2}v/w^{1/2} > \phi_p$. Then the attraction overrides the repulsion due to the physical volume of the segments and causes the layers to extend and interpenetrate for $H \approx 2\alpha_0$. For fixed $v/l^3 < 0$, the strength of the attraction decreases with increasing ϕ_p.

Experimental results

Hadziioannou *et al.* (1968) have measured forces between block poly(vinyl-2-pyridine)/polystyrene copolymer layers adsorbed on mica surfaces with the device described in §5.9. The solvents chosen, toluene and cyclohexane, insured that the polystyrene blocks themselves would not adsorb and that the poly(vinyl-2-pyridine) would not desorb during the

Fig. 6.9. Predictions of the mean-field theory ((6.3.16) to (6.3.18)) for the interaction potential in poor solvents with $z = -2.5$.

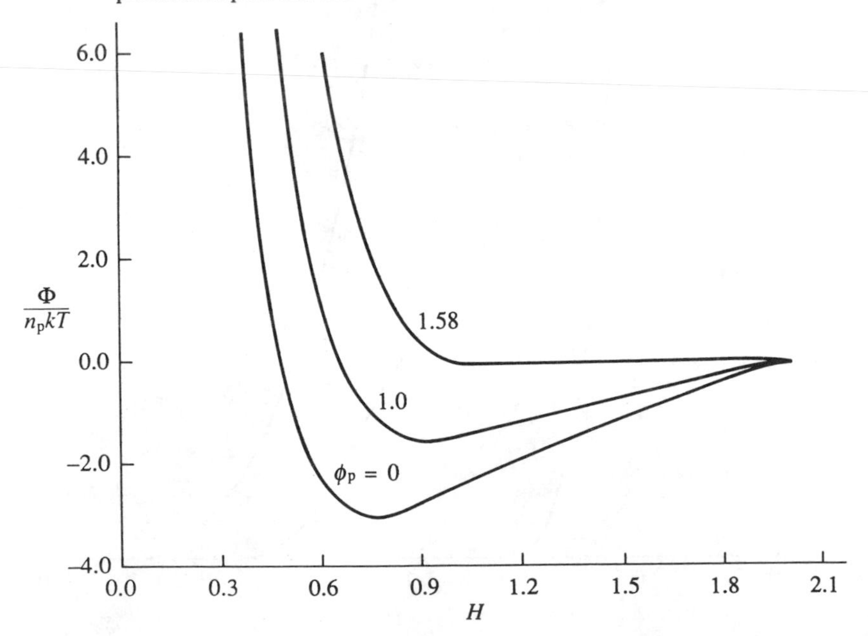

experiment. Hence the situation should approximate that assumed for the theory.

The polymer solution was contacted with the mica surfaces until the adsorption reached saturation; then the solution was replaced by pure solvent. The reproducibility and time independence of the forces indicates irreversible adsorption and rapid equilibration of the segments within the polystyrene blocks.

Figures 6.10 and 6.11 display the forces measured for equal monodisperse blocks of molecular weight 60 kg/mol in cyclohexane at 38 °C, close to theta conditions, and toluene at 32 °C, a good solvent. In each case, the interaction becomes strongly repulsive at separations less than 20 nm. Near

Fig. 6.10. Interaction potential for poly(vinyl-2-pyridine)/polystyrene block copolymer layers in cyclohexane at 311 K: ●, data of Hadziioannou *et al.* (1986); – – –, predictions from (6.3.16) to (6.3.18) with $z=0$, $N=121$, $l=1.46$ nm, $\phi_p=4.0$ and 5.0.

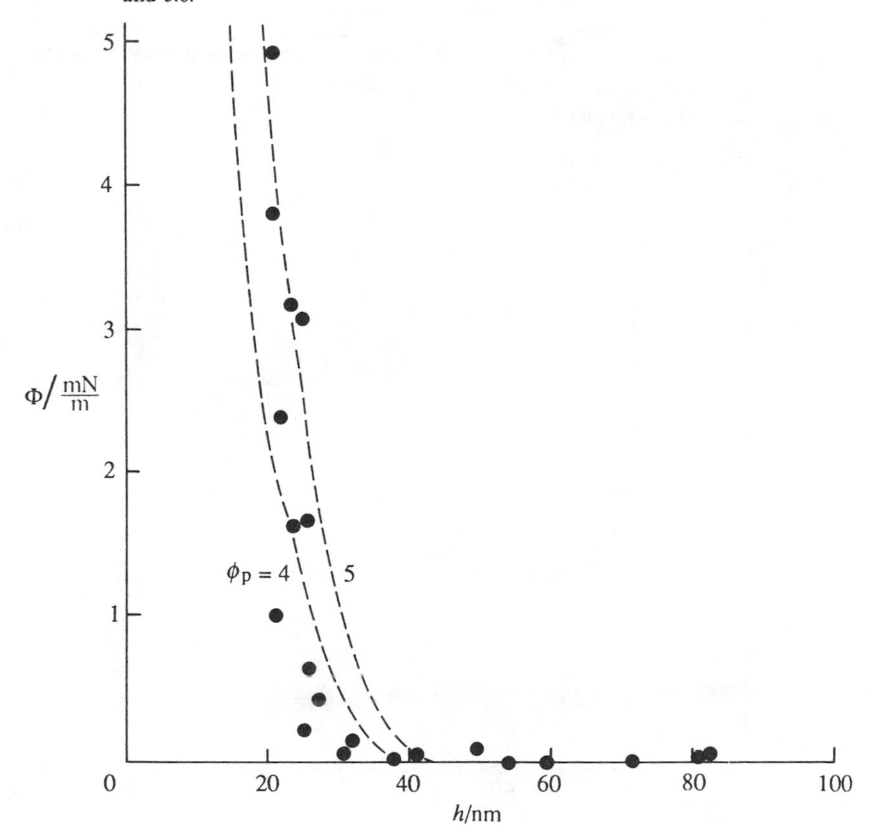

theta conditions, the potential drops rapidly to zero beyond 20 nm. At good solvent conditions, the repulsion persists to 40–50 nm, as expected for stretched chains. Each of these features agrees qualitatively with the expectations from the theory, as explained below.

According to the Derjaguin approximation, $F/\pi R$ equals the interaction potential per unit area between flat plates. Direct comparison with the theory, though, requires values for n_p, N, l, and v. Here we obtain N and l from the molecular weight and data in the literature (Brandrup & Immergut, 1975) as 121 and 1.46 nm, respectively, n_p from the forces at theta conditions, and estimate v from the temperature.

At the high adsorption densities of the experiments the third virial term in the free energy is significant. Fitting the theory for $v=0$ (curves in Fig. 6.11) to the measurements in cyclohexane at 38 °C suggests $\phi_p \approx 4$. For toluene at 27°C, data (Brandrup & Immergut, 1975, §14) indicates $v/w^{1/2} \approx 0.12$; with $\phi_p \approx 4$ and $N \approx 121$ this corresponds to $N^{1/2}\, v/w^{1/2} \approx 1.3$ and $z \approx 5$.

Fig. 6.11. Interaction potential for block copolymer layers in toluene at 305 K: ○, data of Hadziioannou *et al.* (1986); - - -, predictions from (6.3.16) to (6.3.18) with $\phi_p = 4.0$, $z = 1.0$ and 5.0.

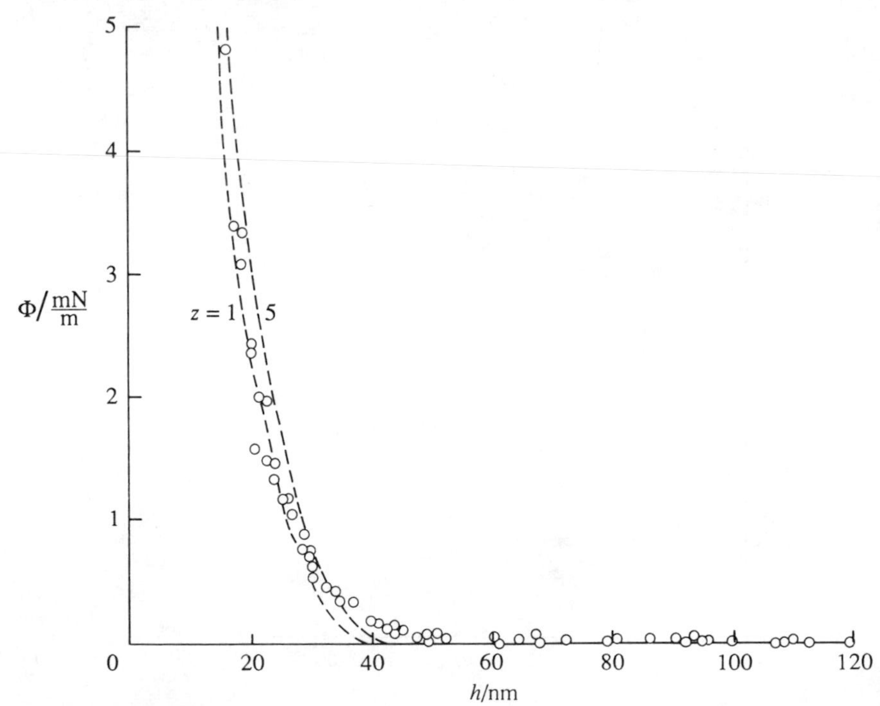

The theoretical curve for good solvents, scaled as suggested by (6.3.21), in Fig. 6.12 follows the data reasonably well. The discrepancy for separations greater than twice the nominal thickness of the isolated layers arises in part from the assumption of constant segment densities within the layers. More refined theories (Milner, Witten, and Cates, 1988) predict a quadratic variation in segment density with distance from the wall and a softer repulsion in better accord with the data.

The experiments confirm the existence of strong repulsions in both theta and good solvents for layers consisting of terminally anchored chains. The nature of attractive interactions in poor solvents remains to be defined.

6.4 Non-adsorbing polymer

Dissolved, non-adsorbing polymer molecules must alter their configuration if their center of mass is to approach a surface closer than about one coil radius. The resulting increase in the configurational free energy renders this process unfavorable in dilute solutions, so a layer of thickness $L \approx \langle r^2 \rangle^{1/2}$ adjacent to any surface should be depleted of segments. Experiments confirm this expectation (Ausserc, Hervet, & Rondelez, 1982).

The consequences of these phenomena for suspended particles can be understood from either a mechanical or a thermodynamic standpoint. A

Fig. 6.12. Interaction potential for block copolymer layers in toluene at 304 K: ○, data of Hadziioannou *et al.* (1986); - - -, predictions from (6.3.16) to (6.3.18) with $\phi_p = 4.0$ and $z = 5.0$.

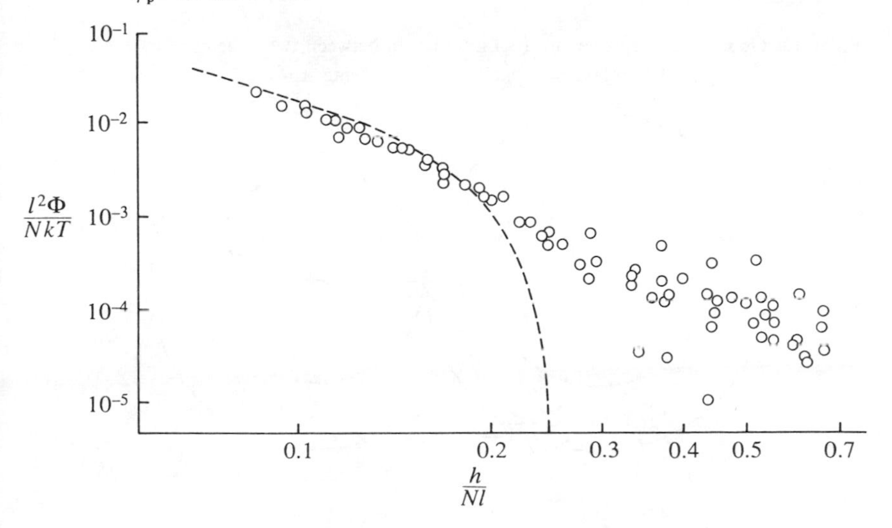

particle immersed in a polymer solution experiences an osmotic pressure acting normal to its surface. For an isolated particle, the integral of the pressure over the entire surface nets zero force. But when the depletion layers of two particles overlap, polymer will be excluded from a portion of the gap (Fig. 6.13). Consequently, the pressure due to the polymer solution becomes unbalanced, resulting in an attraction. The same conclusion follows from consideration of the Helmholtz free energy. Overlap of the depletion layers reduces the total volume depleted of polymer, thereby diluting the bulk solution and decreasing the free energy.

The original geometrical analysis proposed by Asakura & Oosawa (1954, 1958) and generalized by Vrij (1976) and others neglects the internal degrees of freedom of the polymer molecules to obtain simple, useful expressions for the interaction potential. Here we determine the configurational probability through the self-consistent field theory, along the lines of Joanny, Leibler, & deGennes (1979), and thereby demonstrate the validity of simpler approaches. The free energy of interaction then follows from the chemical potential of the solvent, or equivalently the osmotic pressure of the bulk solution.

The configuration density function G, governed as before by (6.3.1) with $G=0$ at $x=0, h$, is now related to the segment density through (6.2.28) with n the average density of segments within the gap such that

$$\int_0^h n(x)\,\mathrm{d}x = hn. \tag{6.4.1}$$

Fig. 6.13. Geometrical representation of interaction between two spheres of radius a having depletion layers of thickness L and center-to-center separation r.

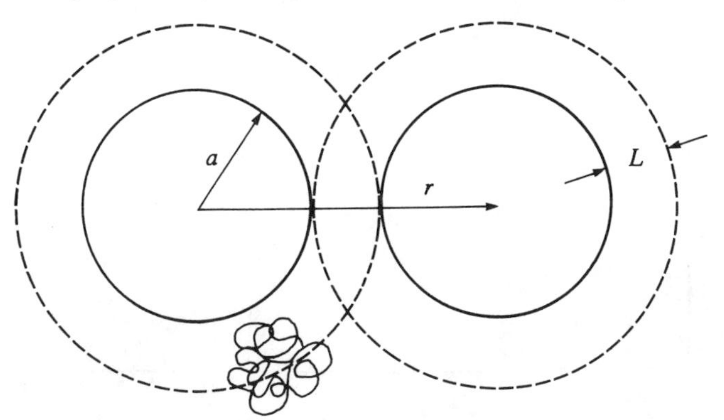

Equilibrium with a bulk solution at segment density n_b requires that the chemical potential μ from (6.2.20) for chains in the gap equal that corresponding to the bulk. With (6.2.25), this becomes

$$n = n_b W \exp\left(N v n_b + \frac{\partial \ln W}{\partial \ln n} + N \frac{\partial}{\partial n} \int \tfrac{1}{2} v n^2 \frac{\mathrm{d}V}{V} \right). \tag{6.4.2}$$

The solution for G is obtained by assuming a separable solution to the non-linear equation (6.3.1) of the form (Gerber & Moore, 1977)

$$G(x, x', s) = \exp\left(\frac{s}{N_0}\right) g(x) g(x'), \tag{6.4.3}$$

with

$$\int_0^h g^2(x)\,\mathrm{d}x = 1.$$

The evolution equation then becomes

$$\frac{\mathrm{d}^2 g}{\mathrm{d}x^2} - \frac{6g}{N_0 l^2} - \frac{12 v n g^3}{l} = 0, \tag{6.4.4}$$

with $g = 0$ at $x = 0, h$. The solution, constructed by multiplying through by $\mathrm{d}g/\mathrm{d}x$ and integrating twice, cannot be evaluated analytically. However, it can be expressed as the Jacobian elliptic function (Abramowitz & Stegun, 1964, §16)

$$g(x) = g_0 \sin \phi,$$

with ϕ related to x through

$$ax = \int_0^{\phi} \frac{\mathrm{d}\theta}{(1 - m \sin^2 \theta)^{1/2}}. \tag{6.4.5}$$

This particular function is periodic with $\sin \phi = 0$ for ax equal to integral multiples of $2K(m)$, where $K(m)$ is the complete elliptic integral of the first kind. Thus the boundary conditions are satisfied if $ah = 2K(m)$. The normalization condition on g determines

$$h g_0^2 = \frac{m K(m)}{K(m) - E(m)} \tag{6.4.6}$$

with $E(m)$ the complete elliptic integral of the second kind. Differentiation twice and substitution into (6.4.4) indicates that m and N_0 must satisfy

$$K(m)[K(m)-E(m)]=\tfrac{3}{2}N^{1/2}nvH$$

and (6.4.7)

$$\frac{N}{N_0}=\tfrac{2}{3}H^{-2}(1+m)K^2(m).$$

For separations sufficiently small that $m \to 0, n(x) \sim 2n\sin^2\frac{\pi x}{h}$ and

$$W \sim \exp\left(-\frac{\pi^2 N l^2}{6h^2}\right). \tag{6.4.8}$$

Application of the equilibrium condition (6.4.2) then determines

$$n \sim n_b \exp\left(Nvn_b - \frac{\pi^2 N l^2}{6h^2}\right). \tag{6.4.9}$$

Thus the segment density within the gap effectively vanishes when $h<\pi N^{1/2}l/[6(1+Nvn_b)]^{1/2}$. For theta conditions this corresponds to $h<\pi\langle r^2\rangle^{1/2}/6^{1/2}$ as suggested above. For good solvents with $Nvn_b>1$ the interactions in the bulk favor configurational changes near the interface, and (6.4.9) indicates that chains remain in the gap until $h<\pi l/(6vn_b)^{1/2}$. For good solvents in the semi-dilute regime, the phenomenon is similar but the concentration dependence of the separation at which chains leave the gap differs as demonstrated by Joanny, Leibler, & deGennes (1979).

In both theta and good solvents chains are effectively excluded from the gap when $h<2L$, with L the depletion layer thickness, allowing the interaction potential to be written with (6.2.22) as

$$\begin{aligned}\Phi &= A(h)-A(2L)\\ &= -(2L-h)P\end{aligned}$$

so that (6.4.10)

$$\begin{aligned}F &= -\mathrm{d}\Phi/\mathrm{d}h\\ &= -P.\end{aligned}$$

These simple results conform to those originally obtained by Asakura & Oosawa (1954, 1958).

Combining the mean-field expression for the osmotic pressure of the bulk solution (6.2.24) with (6.4.10) evaluated at $h=0$ yields the minimum in the potential energy at contact as

$$\frac{\Phi_{\min}}{kT}=-\frac{n_b}{N}\left(1+\tfrac{1}{2}Nvn_b\right)\frac{\pi l}{\left[6\left(\frac{1}{N}+vn_b\right)\right]^{1/2}} \tag{6.4.11}$$

$$\sim -\pi\frac{ln_b}{(6N)^{1/2}} \qquad \text{dilute or ideal solutions}$$

$$\sim -\pi\frac{ln_b}{2}\left(\frac{vn_b}{6}\right)^{1/2} \qquad \text{concentrated solutions.}$$

Comparison of these potentials with those for the terminally anchored chains shows the interaction to be relatively weak. For example, experiments with polystyrene in cyclohexane, which does not adsorb on mica, yielded no detectable forces between mica surfaces owing to the polymer (Luckham & Klein, 1985). Indeed, estimates of the potential from (6.4.11) at

Fig. 6.14. Interaction potential between spheres of radius a in solution of non-adsorbing polymer with depletion layer of thickness L and osmotic pressure P from (6.4.13).

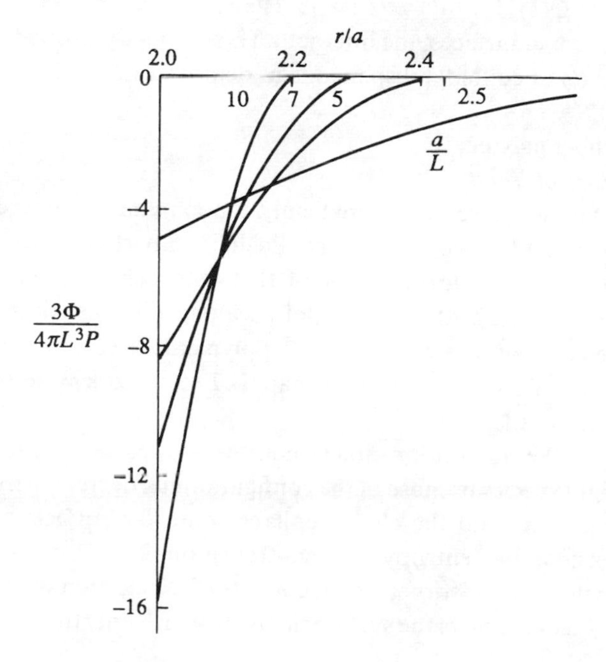

the experimental conditions fall several orders of magnitude below the detection limit for the instrument.

Since the interaction potential takes the simple form

$$\Phi = -PV_{\text{excl}}, \tag{6.4.12}$$

with V_{excl} the volume of overlap between the two depletion layers, the generalization to other geometries is straightforward (Vrij, 1976). For equal spheres of radius a at center-to-center separation r,

$$\Phi(r) = -\frac{4\pi}{3}(a+L)^3\left(1 - \frac{3r}{4(a+L)} + \frac{r^3}{16(a+L)^3}\right)P, \tag{6.4.13}$$

for $2a < r < 2(a+L)$. This potential decreases monotonically from zero at $r = 2(a+L)$ to a minimum at contact (Fig. 6.14),

$$\frac{\Phi_{\min}}{kT} = -\frac{4\pi L^3}{3}\left(1 + \frac{3a}{2L}\right)\frac{P}{kT}, \tag{6.4.14}$$

which depends on the size ratio a/L and the polymer concentration and molecular weight through L^3P. In ideal solutions with $a/L \gg 1$,

$$\frac{\Phi_{\min}}{kT} \sim \frac{\pi^3}{12} al^2 n_b \tag{6.4.15}$$

which becomes $O(1)$ for $nl^3 \approx l/a \ll 1$. Hence, while not measurable between macroscopic surfaces, the interaction is sufficiently strong to affect the phase behavior of colloidal dispersions as demonstrated in Chapter 10.

6.5 Adsorbing polymer

Structure of isolated layers

Thus far, we have considered only the extremes of irreversibly anchored and non-adsorbing polymers. Each is important and can be achieved through careful formulation of the system. More commonly, though, polymers adsorb at random points along their backbone to a degree that depends on the nature of the polymer, the solvent, and the surface. An interaction energy of less than $1kT$ per segment suffices to generate significant adsorption.

The nature of the adsorption differs considerably from that for low-molecular-weight species because of the configurational degrees of freedom of the polymer. Rarely do the chains collapse onto the surface, thereby sacrificing considerable entropy in the transition from three to two dimensions. Instead the adsorbed chain consists of a collection of *trains*, in which each segment contacts the surface; *loops*, in which only the initial and final segments contact the surface; and *tails*, which begin at the surface but terminate in the solution (Fig. 6.15). Such configurations minimize the entropy loss while maintaining enough segments on the surface to achieve an adsorption energy large relative to kT for the polymer molecule as a whole. Since each segment is not tightly bound, the equilibrium is a dynamic one with continuous exchange of adsorbed and free segments. However, the probability of all adsorbed segments of a chain leaving the surface simultaneously is small, so desorption may be quite slow, e.g. upon washing with pure solvent. Exchange does occur, though, because segments from molecules in the bulk can occupy sites vacated by those of an adsorbed molecule, thereby displacing the adsorbed molecule in a relatively short time.

Predicting the structure of the adsorbed layer is difficult because the segment densities can vary from highly concentrated at the surface to infinitely dilute in the bulk solution. The complete theory, which does not yet exist, would be robust in dealing with $nl^3 \approx 1$, but would also accommodate the subtleties of semidilute solutions. Indeed, this spatial variation in densities also confounds measurements, since techniques sensitive to low concentrations, such as hydrodynamic measurements, indicate much greater layer thicknesses than those that detect a lower moment of thc density, such as ellipsometry (cf. Cohen Stuart, Cosgrove, & Vincent, 1986).

Theoretical approaches developed over the past three decades (cf. Takahashi & Kawaguchi, 1982; deGennes, 1987) approximate to differing degrees the combined effects of adsorption and excluded volume on the chain configuration. The mean field form of the self-consistent field theory has been developed extensively by Scheutjens and Fleer (1979, 1985, 1986).

Fig. 6.15. Schematic of adsorbed polymer molecule identifying trains, loops, and tails.

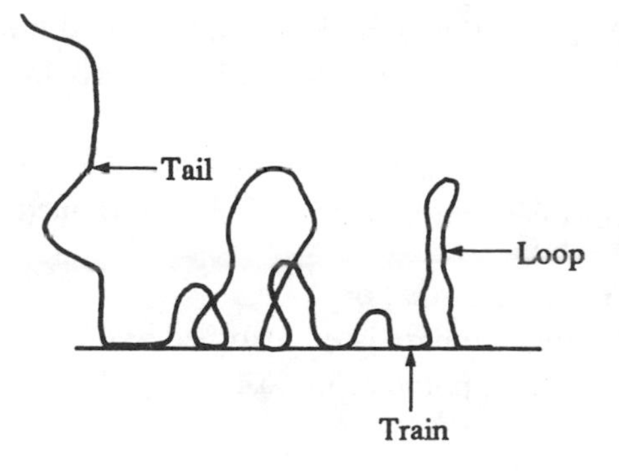

Here we present a simplified version (Ploehn and Russel, 1989) to illustrate the features of isolated adsorbed layers and experimental results to identify the characteristics of interactions between such layers.

The situation differs from those treated in the previous sections in the reversible adsorption of segments at the surface, with energy χ_s expressed in units of kT per segment. Assuming the range of the attraction to be short relative to the segment length leads to

$$\exp(-U/kT)=\exp(-U_c/kT)+l\delta(x)\exp(\chi_s-U_s/kT), \tag{6.5.1}$$

with $\delta(x)$ the Dirac delta function. U_c and U_s account for segment–segment interactions, off and on the surface, respectively.

The attraction at the surface causes G to be discontinuous such that

$$G(x,x',s)=G_c(x,x',s)+G_s(x',s)\delta(x). \tag{6.5.2}$$

Expansion of the continuous part of G in a Taylor series, substitution into (6.2.30) and integration as before still determines the governing differential equation (6.2.32) for G_c for $x>0$ and for $x=0$ yields the surface boundary condition. Recall that $U(x)$ depends on the local segment density, so the value $U(0)\equiv U_s$ at the surface includes the adsorbed segments while the value $U(0+)=U_c(0)$ just off the surface includes only the nonadsorbed segments. Equating individually the singular and continuous parts of the form for $x=0$ determines G_s,

$$G_s(x',s)=lKG_c(0,x',s) \tag{6.5.3}$$

with $K=\exp[(U_c(0)-U_s)/kT+\chi_s]$, and the boundary condition on G_c,

$$\frac{\partial G_c}{\partial s}=\frac{l}{2(1+K)}\frac{\partial G_c}{\partial x}+\left[1-\frac{2}{1+K}\exp\left(\frac{U_c(0)}{kT}\right)\right]G_c. \tag{6.5.4}$$

Thus the boundary value problem to be solved for G_c consists of the differential equation (6.2.32) together with the boundary conditions (6.5.4) at $x=0$ and $G_c\to G_b$ as $x\to\infty$, with G_b describing chains in the bulk solution.

The segment density follows from G according to (6.3.2) with n_p, the number of adsorbed chains per unit area of surface, determined by either

(i) requiring that $n\to n_b$ as $x\to\infty$, or
(ii) equating the chemical potentials of adsorbed and free chains, as done for non-adsorbing polymer in §6.4.

From (6.2.20) and (6.2.35) the latter leads to

$$n_{\mathrm{p}}=\frac{ln_{\mathrm{b}}}{N}W\exp\left(\frac{\partial \ln W}{\partial \ln n_{\mathrm{p}}}+\frac{\partial}{\partial n_{\mathrm{p}}}\int_0^h(\tfrac{1}{2}vn^2+\tfrac{1}{3}wn^3)\,\mathrm{d}x\right) \tag{6.5.5}$$

for $l^3n_{\mathrm{b}}\ll 1$.

An appropriate analytical solution to (6.2.32) can be obtained by constructing a matched asymptotic expansion consisting of an inner region with $x=O(l)$ and $n\gg n_{\mathrm{b}}$ and an outer region with $x=O(N^{1/2}l)$ and $l^3(n-n_{\mathrm{b}})\ll 1$.

In the inner region the approximation (6.4.3) used for non-adsorbing chains is appropriate. In this case, though, segment densities become quite high adjacent to the surface, so the $O(n^2)$ term in the self-consistent potential must be retained. Then with the potential expanded in a virial series and truncated as in (6.2.34), the equation for g takes the form

$$\frac{\mathrm{d}^2g}{\mathrm{d}x^2}-\frac{6}{N_0l^2}g-\frac{6vn_{\mathrm{b}}}{l}g^3-3wn_{\mathrm{b}}^2g^5=0, \tag{6.5.6}$$

with $g\to 0$ as $x\to\infty$. Integrating twice to obtain g yields

$$g^2(x)=\frac{4g_0^2E(x)}{N_0(lg_0^2-\tfrac{1}{2}vn_{\mathrm{b}}E(x))-\tfrac{2}{3}wn_{\mathrm{b}}^2E^2(x)}$$

with (6.5.7)

$$E(x)=\exp\left[-\left(\frac{6N}{N_0}\right)^{1/2}\frac{2x}{N^{1/2}l}\right].$$

Note that $g(x)\to 0$ as $x\to\infty$, indicating that the inner region consists of bound chains.

In the outer region, the position along the chain and the distance from the surface must be rescaled as $t=s/N$ and $y=x/N^{1/2}l$. Then, for $Nv(n-n_{\mathrm{b}})\ll 1$, equation (6.2.32) reduces to

$$\frac{\partial G_0}{\partial t}=\frac{1}{6}\frac{\partial^2 G_0}{\partial y^2}, \tag{6.5.8}$$

with $G_0(0,y',t)=0$ and $G_0(y,y',0)=\delta(y-y')$. Thus the chains assume Gaussian configurations perturbed by the presence of the surface as described by (6.3.5). These comprise the tails associated with the bound chains in the inner region, plus the free chains in solution.

Since the inner solution decays to zero in the outer region, as $x \to \infty$, and the outer solution goes to zero in the inner region, as $y \to 0$, the sum of the two provides a uniformly valid approximation for all x and s. Thus the segment density calculated from

$$G(x,x',s) = g(x)g(x')e^{s/N_0}\{1+K[\delta(x)+\delta(x')]\} + G_0(x/N^{1/2}l, x'/N^{1/2}l, s/N) \quad (6.5.9)$$

satisfies $n(x) \to n_b$ as $x \to \infty$.

The three constants in the inner solution – g_0^2, from the integration; N_0, the eigenvalue; and n_p, the adsorbed amount – are determined from the normalization

$$\int_0^\infty (n(x) - n_b)\mathrm{d}x = Nn_p, \quad (6.5.10)$$

the boundary condition (6.5.4), and the equilibrium condition (6.5.5). Simultaneous solution of these yields a complete description of the adsorbed layer, i.e. the segment density profile as well as the distributions of trains, loops, and tails.

The resulting density profile in Fig. 6.16 corresponds to polystyrene of

Fig. 6.16. Predicted profiles for total density and segment in tails, loops and non-absorbed chains for $N=10^4$, $v/l^3=0$, $\chi_s=1$, $w^{1/2}/l^3=1.0$ (Ploehn & Russel, 1989).

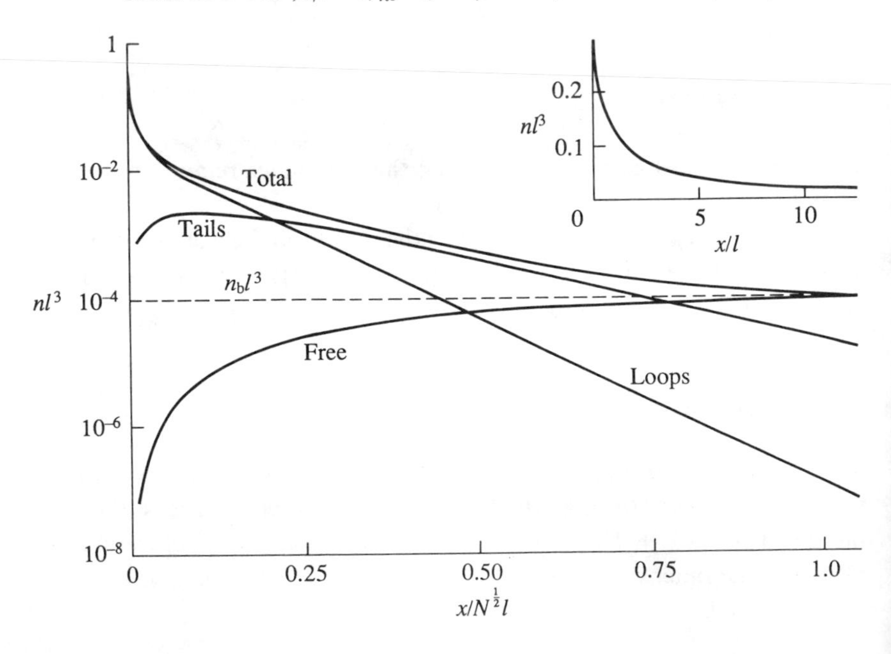

roughly 5×10^3 kg/mol adsorbed with energy $1.5\,kT$ per segment from cyclohexane at theta conditions with a solution concentration of 10^{-4} kg/dm^3. The layer appears relatively compact with the concentration falling to 10^{-2} kg/dm^3 at $x = 10l$–$15l$ or $y \ll 1$. This comprises the inner region where loops dominate. The segment density exceeds the bulk value, however, much farther from the surface due to the tails associated with the outer solution.

The theoretical predictions in Figs 6.17 and 6.18 (Ploehn & Russel, 1989) demonstrate the dependence of $\phi_{ads} = Nn_p l^2$ on concentration, solvent quality, molecular weight, and adsorption energy. For $\chi_s > 0.1$, the

Fig. 6.17. Predictions for adsorbed amount $Nn_p l^2$ as a function of bulk concentration, molecular weight, and solvent quality with $\chi_s = 1$ and $w^{1/2}/l^3 = 1.0$ (Ploehn & Russel, 1989):

	N	v/l^3
A	10^4	0
B	10^4	0.04
C	10^3	0
D	10^3	0.04

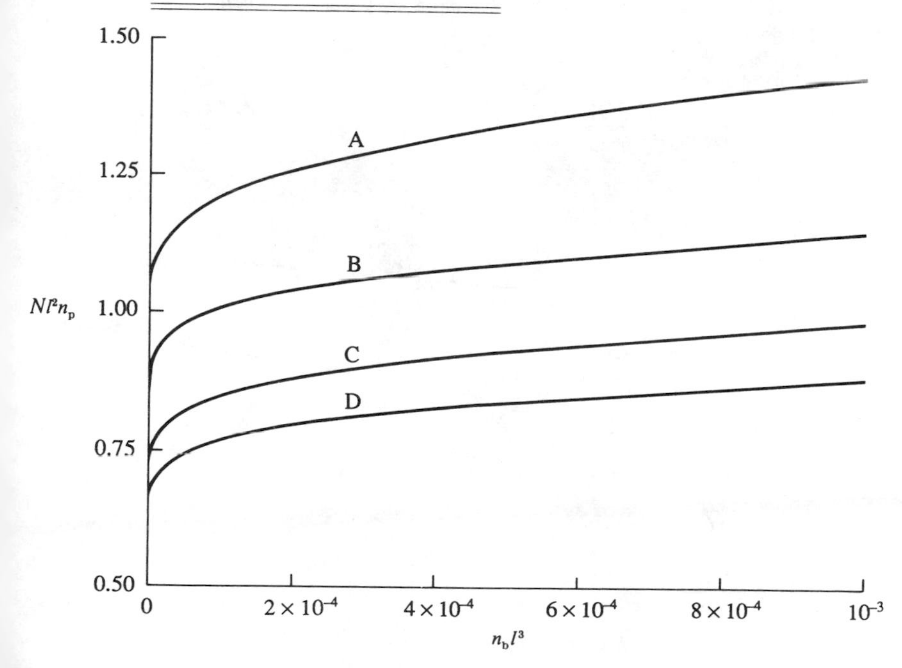

adsorption reaches 'full coverage' at very low bulk concentrations and then increases only slightly with further increments in $\phi_b = n_b l^3$. The amount of adsorption at full coverage increases with molecular weight but decreases with improving solvent quality. Note that the total adsorption energy required per chain must compensate for the loss of configurational degrees of freedom and the additional segment–segment interactions within the layer. Clearly, polymers of higher molecular weight sacrifice fewer configurations in the process and adsorb more easily. Improving the solvent quality, however, increases the free energy of adsorbed chains, requiring that more segments of each bind to the surface and thereby reducing the number of adsorbed chains.

These predictions can be compared with ellipsometric measurements of adsorption on macroscopic surfaces. For example, for polystyrene adsorbing onto chrome from cyclohexane at theta conditions, Takahashi *et al.* (1980) found similar forms for the isotherms with the adsorbed amount generally increasing with molecular weight (Fig. 6.18). Both the ellipsomet-

Fig. 6.18. Adsorbed amount as a function of molecular weight: Predictions (Ploehn & Russel, 1989) for $v/l^3 = 0$, $w^{1/2}/l^3 = 1$, $\tilde{v}n_b = 1.9 \times 10^{-4}$, and $\chi_s = 1$(A) and 15(B) Data for polystyrene adsorbed from cyclohexane at $T = \theta \approx 35\,°C$ onto: ◇, chrome (Takahashi, *et al.*, 1980); ▽, silica (vander Linden & van Leemput, 1978); ×, silica (Kawaguchi, Hayakawa, & Takahashi, 1980).

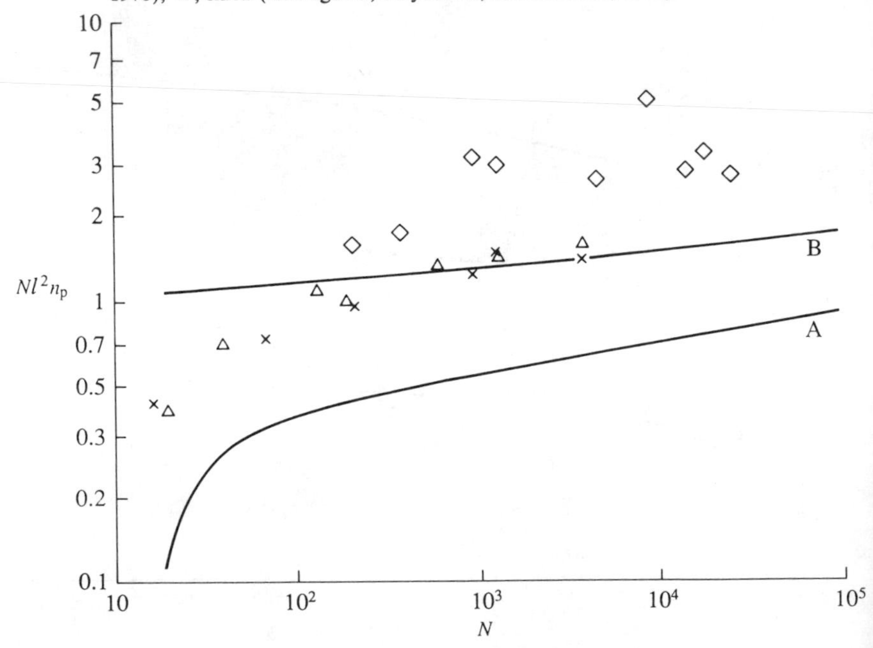

ric thickness $\langle x \rangle$, comparable to the first moment of the density profile, and the hydrodynamic thickness t_{hyd} increase as the square root of the molecular weight (Fig. 6.19). The latter exceeds the first moment considerably, since fluid flow senses even the very low segment densities in the tails that extend beyond the bulk of the layer. Scheutjens and Fleer (1979) compare numerical solutions of the full equations with additional data, demonstrating the capabilities of the approach.

In concluding, we note the primary features of adsorbed polymer layers:

(i) For modest values of the adsorption energy per segment, the isotherms indicate essentially full coverage at quite low solution concentrations.
(ii) The layers appear compact relative to the corresponding free polymer coils but have tails that extend much farther into solution.
(iii) The layer thickness increases with molecular weight.
(iv) Desorption requires a long time relative to the equilibration of segments within the adsorbed layer.

Interactions between adsorbed layers

The interaction between surfaces is more complicated with adsorbed polymer layers than with terminally anchored chains for two reasons:

chains originally attached to one surface can adsorb to the other as well, termed 'bridging', and
given sufficient time, chains can desorb from the surface and migrate out of the gap.

Both phenomena profoundly affect the interaction potential as illustrated by direct measurements described below.

With *strong adsorption* at *full coverage* for polymers in a *good solvent*, e.g. poly(oxyethylene) in toluene or aqueous solutions (Klein & Luckham, 1984, 1985), interaction produces a purely repulsive potential (Fig. 6.20). Furthermore, the layers interact at separations somewhat greater than twice the end-to-end distance of the free chains, consistent with the existence of tails extending well beyond the bulk of the adsorbed layer. In addition, the forces measured were time-dependent, requiring approximately fifteen minutes after an initial compression to relax to the equilibrium curve shown. The purely repulsive potential and the nature of the relaxation phenomena indicate that the interaction primarily compressed the layers, increasing somewhat the fraction of segments bound to the surface but not producing significant bridging.

Fig. 6.19. (*a*) Ellipsometric layer thickness as a function of chain length: □, polystyrene on chrome from cyclohexane at 35 °C (Takahashi *et al.*, 1980); ——, predictions for $v/l^3=0$, $w^{1/2}/l^3=1.0$, and $\tilde{v}n_b=2.8\times10^{-3}$ (Ploehn & Russel, 1989). (*b*) Hydrodynamic layer thickness as a function of chain length for poly(ethylene oxide) adsorbed on polystyrene latices at 25 °C: □, Cohen Stuart, Cosgrove, & Vincent (1984); +, Kato *et al.* (1981); ——, prediction for $v/l^3=0.026$, $\chi_s=1$, $\tilde{v}n_b=2.4\times10^{-3}$ (Ploehn & Russel, 1989).

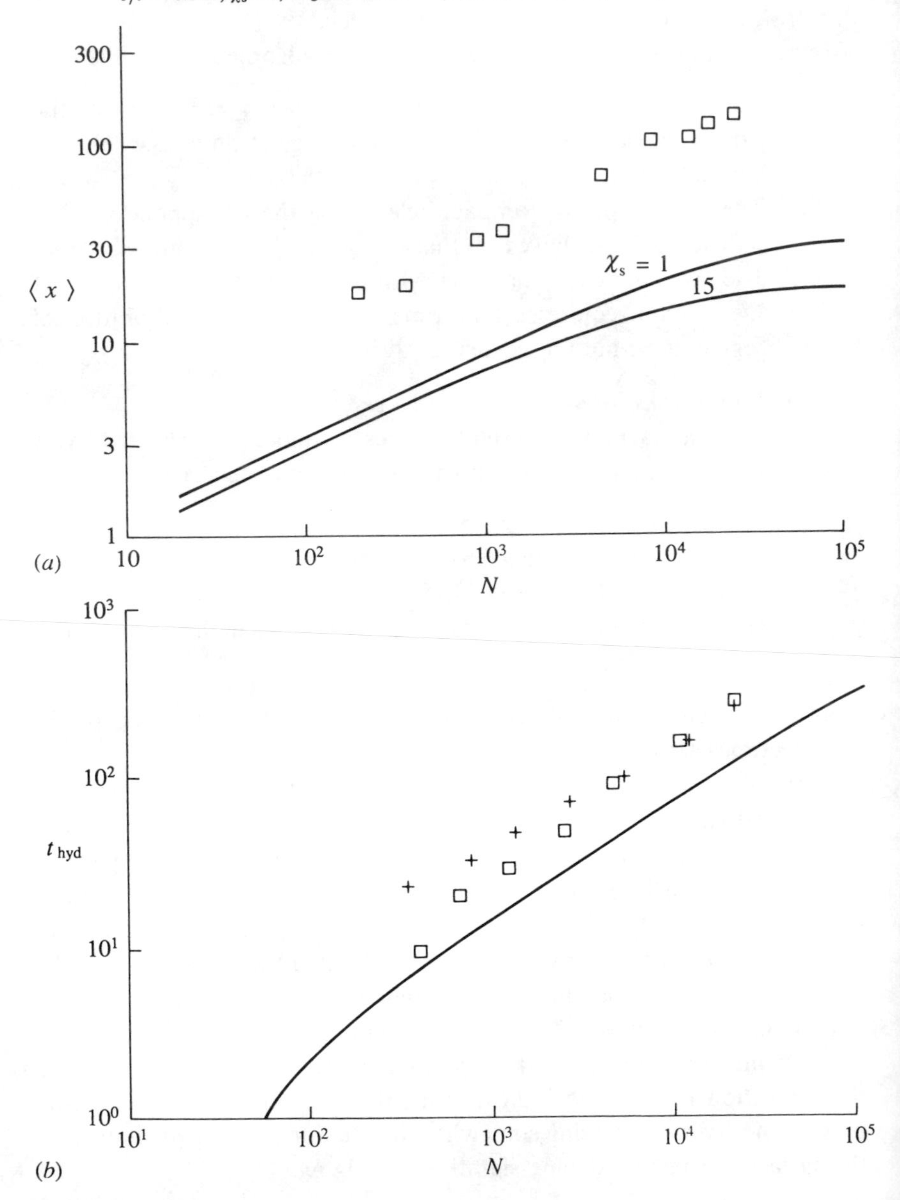

With *strong adsorption* but only *partial coverage*, the situation differs considerably. Then the layers interpenetrate more easily and excess surface is available for adsorption of tails from the opposing layer. Almog & Klein (1985) moderated the adsorption of polystyrene from cyclopentane onto the mica surfaces and measured the interaction potentials illustrated in Fig. 6.21. In all cases, the potential remained repulsive at small separations, demonstrating that the strongly adsorbed polystyrene does not leave the gap. At low coverages, an attractive minimum appeared but diminished and eventually disappeared with increasing coverage. Experiments without polymer established that dispersion forces were insignificant, leaving only bridging to account for the attraction. The absence of any time dependence suggests weaker adsorption than with poly(oxyethylene).

With still *weaker adsorption* and only slightly better than *theta solvents*, as for polystyrene in cyclohexane, even interaction at full coverage produces attraction (Israelachvili *et al.*, 1984). The magnitude (Fig. 6.22) is somewhat less than with strong adsorption and partially covered surfaces, but still quite significant. The layers equilibrate rapidly relative to the case of strong adsorption, and the repulsion at small separations establishes that complete desorption does not occur. The attraction reflects bridging, facilitated by greater interpenetration of the layer near theta conditions and

Fig. 6.20. Interaction potential between poly(oxyethylene) ($M = 31$ kg/mol, $\langle r^2 \rangle^{1/2} = 44$ nm) layers adsorbed onto mica plates at full coverage from toluene at 23 °C. The inset shows results on an expanded scale for two different experiments (Klein & Luckham, 1985).

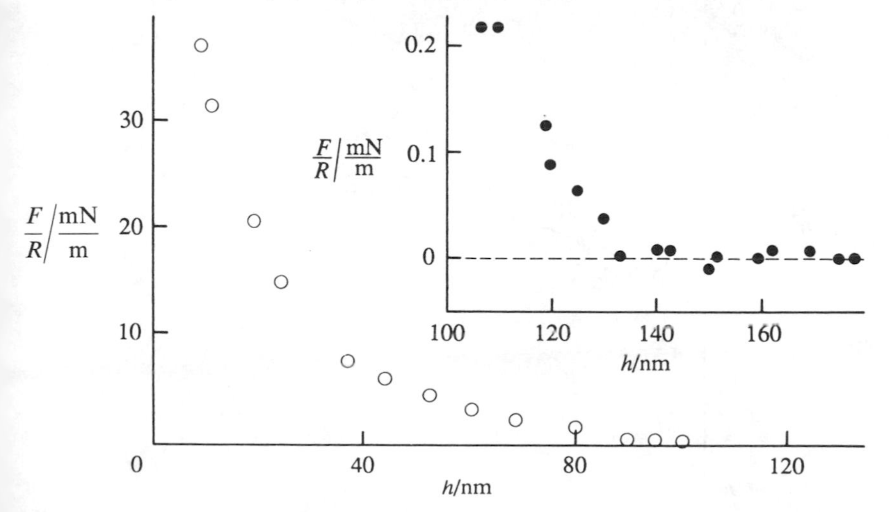

Fig. 6.21. Interaction potential between polystyrene ($M = 5000$ kg/mol, $\langle r^2 \rangle^{1/2} = 94$ nm) layers adsorbed onto mica plates from cyclopentane at 23 °C at coverages that vary with the exposure (Almog & Klein, 1985): ●, 12 h at 30 μm separation; ○, 12 h at 100 μm separation; ▽, an additional 2.5 h at 500 μm separation; □, an additional 16 h at 250 μm separation.

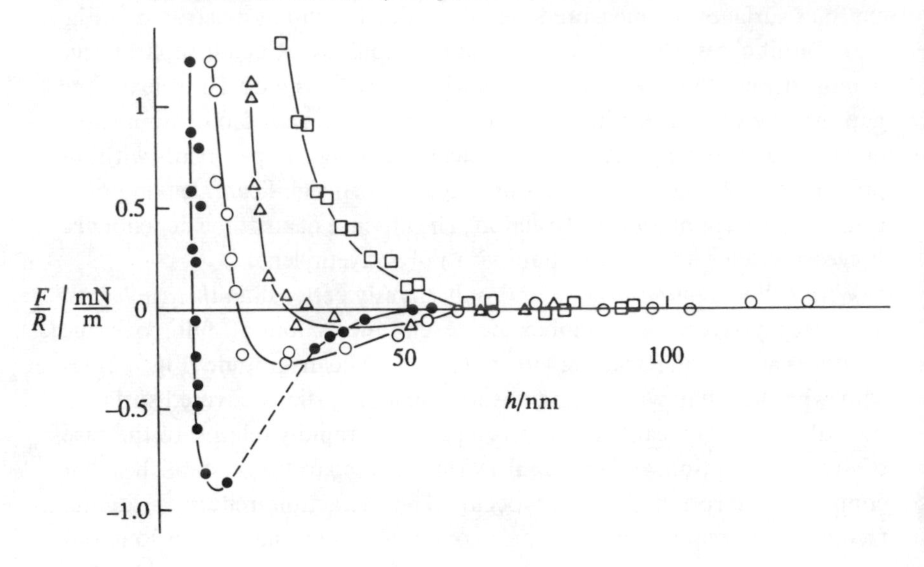

Fig. 6.22. Interaction potential between polystyrene (○, $M_w = 600$ kg/mol, $\langle r^2 \rangle^{1/2} = 51$ nm; ●, $M_w = 900$ kg/mol, $\langle r^2 \rangle^{1/2} = 63$ nm) layers adsorbed onto mica plates from cyclohexane at 37 °C (Israelachvili *et al.*, 1984).

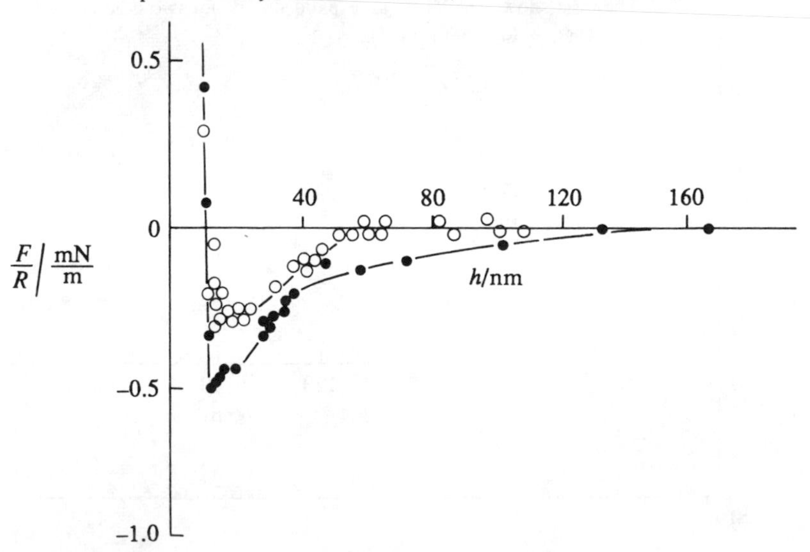

faster exchange between bound and free segments owing to the weak adsorption.

These experiments demonstrate that adsorbed polymer layers produce significant interactions between surfaces at separations comparable to twice the end-to-end distance of the macromolecule in solution. The long-range interaction is repulsive with strong adsorption and full coverage in theta or good solvents, but becomes attractive with weak adsorption or incomplete coverage. At small separations, the potential remains strongly repulsive unless the polymer leaves the gap. The attraction arises from adsorption of molecules to both surfaces simultaneously. The repulsion reflects the same mechanisms as for terminally anchored layers, i.e. the excluded-volume interactions among segments and the compression of chains. Of course, strong attraction also occurs in poor solvents as for the terminally anchored chains.

Theoretical explanations of these interaction potentials exist in part. DeGennes (1982) distinguished between purely attractive interactions at full equilibrium, for which polymer desorbs and leaves the gap as the separation is decreased, and repulsive interactions at restricted equilibrium, for which chains must remain in the gap but can rearrange locally. Subsequent scaling theories for the latter situation have predicted strong repulsions qualitatively similar to those observed, and numerical solutions of the self-consistent field model by Scheutjens & Fleer (1985, 1986) mimic the trends described above.

6.6 Summary

This chapter addresses three modes of polymer–surface interaction in order to illustrate the nature of the interaction potential expected between colloidal particles. The results for the potentials can be classified as in Table 6.2.

For the terminally anchored and adsorbing polymer, the forces are sufficiently large to be measurable between macroscopic surfaces, but the attractions with non-adsorbing polymer are relatively weak. For colloidal particles, though, even the latter produce interaction potentials comparable to or larger than the thermal energy kT. Hence the forces can affect quite dramatically the behavior of dispersions, as will be described in subsequent sections.

The self-consistent field theory represents a general approach to the analysis of polymer–surface interactions, though presented here in a restricted sense. The basic limitation is the mean-field nature of the formulation, which neglects correlations among segments and variations in segment density parallel to the surface. If solved numerically without

Table 6.2.

	Repulsion	Attraction
terminally anchored polymers	good solvents poor solvent with high coverage	poor solvents with low to moderate coverage
non-adsorbing polymer	under no conditions	good solvents at all concentrations
adsorbing polymer	strong adsorption at full coverage in good or theta solvents	poor solvents, partial coverage, or weak adsorption

linearizing $\exp(U/kT)$ or assuming the variables to be separable, it is equivalent to the lattice theory of Scheutjens & Fleer (1985). With a form for U appropriate for semidilute solutions, it provides the basis for the scaling theories of DeGennes and co-workers. Calculations for geometries other than flat plates, and for more complex situations such as interactions between terminally anchored and free polymer (Gast & Leibler, 1986) are possible. So the approach appears capable of describing in detail a broad range of phenomena.

References

Abramowitz, M. & Stegun, I. A. (1970). *Handbook of Mathematical Functions.* US Department of Commerce.

Alexander, S. (1977). Adsorption of chain molecules with a polar head: A scaling description. *J. Phys. (Paris)* **38**, 983–7.

Almog, Y. & Klein, J. (1985). Interactions between mica surfaces in a polystyrene–cyclopentane solution near the θ-condition. *J. Colloid Interface Sci.* **106**, 33–44.

Asakura, S. & Oosawa, F. (1954). On interaction between two bodies immersed in a solution of macromolecules. *J. Chem. Phys.* **22**, 1255–6.

Asakura, S. & Oosawa, F. (1958). Interactions between particles suspended in solutions of macromolecules. *J. Polym. Sci.* **33**, 183–92.

Aussere, D., Hervet, H. & Rondelez, F. (1982). Concentration dependence of the interfacial depletion layer thickness for polymer solutions in contact with nonadsorbing walls. *Macromolecules* **149**, 85–8.

Brandrup, J. & Immergut, E. H. (eds) (1975). *Polymer Handbook*, 2nd ed. Wiley.

Cohen Stuart, M. A., Cosgrove, T. & Vincent, B. (1986). Experimental aspects of polymer adsorption on solid/solution interfaces. *Adv. Colloid Interface Sci.* **24**, 143–240.

Daoud, M. & Janninck, G. (1976). Temperature-concentration diagram of polymer solutions. *J. Phys. (Paris)* **37**, 973–9.

deGennes, P.-G. (1969). Some conformational problems for long macromolecules. *Rep. Prog. Phys.* **32**, 187–205.

deGennes, P.-G. (1975). Collapse of a polymer chain in poor solvents. *J. Phys. Lett. (Paris)* **36**, L55–7.

deGennes, P.-G. (1979). *Scaling Concepts in Polymer Physics*. Cornell University Press.

deGennes, P.-G. (1980). Conformations of polymers attached to an interface. *Macromolecules* **13**, 1069–75.

deGennes, P.-G. (1982). Polymers at an interface. 2. Interaction between two plates carrying adsorbed polymer layers. *Macromolecules* **15**, 492–500.

deGennes, P.-G. (1987). Polymer at an interface; a simplified view. *Adv. Colloid Interface Sci.* **27**, 189–209.

Dolan, A. K. & Edwards, S. F. (1974). Theory of stabilization of colloids by adsorbed polymer. *Proc. Roy. Soc. Lond. A* **337**, 509–16.

Edwards, S. F. (1965). The statistical mechanics of polymer with excluded volume. *Proc. Phys. Soc.* **85**, 613–24.

Edwards, S. F. (1966). The theory of polymer solutions at intermediate concentrations. *Proc. Phys. Soc.* **88**, 265–80.

Fleer, G. J. & Scheutjens, J. M. H. M. (1986). Interactions between adsorbed layers of macromolecules. *J. Colloid Interface Sci.* **111**, 504–15.

Gast, A. P. & Leibler, L. (1986). Interactions of sterically stabilized particles suspended in a polymer solution. *Macromolecules* **19**, 686–91.

Gerber, P. R. & Moore, M. A. (1977). Comments on the theory of steric stabilization. *Macromolecules* **10**, 476–81.

Hadziioannou, G., Patel, S., Granick, S. & Tirrell, M. (1986). Forces between surfaces of block copolymers adsorbed on mica. *J. Am. Chem. Soc.* **108**, 2869–76.

Helfand, E. (1975). Theory of inhomogeneous polymers: Fundamentals of Gaussian random-walk model. *J. Chem. Phys.* **62**, 999–1005.

Helfand, E. & Tagami, Y. (1971). Theory of the interface between immiscible polymers. *J. Poly. Sci.: Poly. Lett.* **9**, 741–6.

Israelachvili, J. N., Tirrell, M., Klein, J. & Almog, Y. (1984). Forces between two layers of adsorbed polystyrene immersed in cyclohexane below and above the θ-temperature. *Macromolecules* **17**, 204–9.

Joanny, J. F., Leibler, L. & deGennes, P.-G. (1979). Effects of polymer solutions on colloid stability. *J. Poly. Sci.: Poly. Phys. Ed.* **17**, 1073–84.

Kato, T., Nakamura, K., Kawaguchi, M. & Takahashi, A. (1981). Quasielastic light scattering measurements of polystyrene latices and conformation of poly(ethylene oxide) adsorbed on the latices. *Polymer J.* **13**, 1037–43.

Kawaguchi, M., Hayakawa, K. & Takahashi, A. (1980). Adsorption of polystyrene onto silica at the theta temperature. *Polymer J.* **12**, 265–70.

Klein, J. & Luckham, P. F. (1984). Forces between two adsorbed poly(ethylene oxide) layers in a good aqueous solvent in the range 0–150 nm. *Macromolecules* **17**, 1041–8.

Luckham, P. F. & Klein, J. (1985). Interactions between smooth solid surfaces in solutions of adsorbing and non-adsorbing polymers in good solvent conditions. *Macromolecules* **18**, 721–8.

McQuarrie, D. A. (1976). *Statistical Mechanics.* Harper & Row.

Milner, S. T., Witten, T. A. & Cates, M. E. (1988). Theory of the grafted polymer brush. *Macromolecules* **21**, 2610–19.

Moore, M. A. (1977). Theory of the polymer coil–globule transition. *J. Phys. A.: Math. Gen.* **10**, 305–14.

Muthukumar, M. & Edwards, S. F. (1982). Extrapolation formulas for polymer solution properties. *J. Chem. Phys.* **76**, 2720–30.

Napper, D. H. (1983). *Polymeric Stabilization of Colloidal Dispersions.* Academic Press.

Ploehn, H. J. & Russel, W. B. (1989). Self-consistent field model of polymer adsorption: Matched asymptotic expansion describing tails. *Macromolecules* **22**, 266–75.

Schaefer, D. W. (1984). A unified model for the structure of polymers in semi-dilute solution. *Polymer* **25**, 387–94.

Scheutjens, J. M. H. M. & Fleer, G. J. (1979). Statistical theory of the adsorption of interacting chain molecules. 1. Partition function, segment density distribution, and adsorption isotherms. *J. Phys. Chem.* **83**, 1619–35.

Scheutjens, J. M. H. M. & Fleer, G. J. (1985). Interactions between two adsorbed polymer layers. *Macromolecules* **18**, 1882–1900.

Takahashi, A. & Kawaguchi, M. (1982). The structure of macromolecules adsorbed on interfaces. *Adv. Poly. Sci.* **46**, 1–66.

Takahashi, A., Kawaguchi, M., Hirota, H. & Kato, T. (1980). Adsorption of polystyrene at the θ temperature. *Macromolecules* **13**, 884–9.

Vander Linden, C. & van Leemput, R. (1978). Adsorption studies of polystyrene on silica. I. Monodisperse adsorbate. *J. Colloid Interface Sci.* **67**, 48–62.

Vincent, B. (1974). Adsorbed polymers and dispersion stability. *Adv. Colloid Interface Sci.* **4**, 193–277.

Vrij, A. (1976). Polymers at interfaces and the interactions in colloidal dispersions. *Pure and Appl. Chem.* **48**, 471–83.

Yamakawa, H. (1971). *Modern Theory of Polymer Solutions*. Harper & Row.

Problems

1 Show how to derive the phase boundary in Fig. 6.2 from the expressions (6.2.24) and (6.2.25) for the pressure and chemical potential of homogeneous polymer solutions. Obtain analytical expressions for the coexistence between very dilute and very concentrated solutions for $N^{1/2}v/w^{1/2} \ll -2$.

2 Extract from (6.3.10) and (6.3.12) the asymptotic forms for the layer thickness L and the free energy A of terminally anchored layers for (i) $z \gg 1$ and (ii) $z \ll -1$. From each identify the explicit dependence of L on N.

3 When a block copolymer adsorbs, the insoluble block forms a thin layer on the surface and the remainder of the molecule behaves much like a terminally anchored chain. Several hypotheses have been proposed to relate the surface density of chains n_p to the molecular weight N^* of the insoluble block. Develop such an expression for close-packed two-dimensional Gaussian coils, i.e. $n_p r^2 \approx 1$, and predict the dependence of the layer thickness L on N^* for fixed values of N and the other parameters.

4 Poly(12-hydroxy stearic acid) is often grafted onto the backbone of a linear polymer to form a copolymer with the configuration of a comb. Adsorbing the polymer backbone onto a colloidal particle then provides terminally anchored chains to act as stabilizers. For stabilizer chains with molecular weight 6 kg/mole and $n_p = 0.75 \times 10^{17}$ m^{-2} on a flat surface calculate (i) the layer thickness at theta conditions (assuming $\tilde{v} = 10^{-3}$ m^3/kg) and (ii) the value of v/l^3 required to induce an attractive minimum in the interaction potential.

5 For non-adsorbing polymer at theta conditions interacting with a sphere of radius a, the configurations are governed by (6.2.32) with $U=0$ and the Laplacian expressed in spherical coordinates. Since chains have equal probability of beginning anywhere in the fluid, the function

$$G(r,s) = \int G(r, r', s)\, d\mathbf{r}'$$

has the auxiliary conditions

$$\begin{aligned} &G(r,0)=1 \qquad a \le r < \infty \\ &G(a,s)=0 \\ &G(\infty,s)=1. \end{aligned}$$

Construct solutions either asymptotically or by transform techniques to determine the number density $n(r)$ and then calculate the depletion layer thickness as

$$L=\int_a^\infty (r-a)[n_b-n(r)]\,\mathrm{d}r\Big/\int_a^\infty [n_b-n(r)]\,\mathrm{d}r.$$

Plot $L/N^{1/2}l$ as a function of $N^{1/2}l/a$ to illustrate the effect of curvature on the depletion layer.

6 The bulk concentration of non-adsorbing polymer in a good solvent affects the interaction potential between colloidal particles through the depletion layer thickness L and the osmotic pressure P.

(i) From the results of §6.4, develop expressions for the magnitude, $-\Phi_{min}/kT$, and the range, $2L$, of the potential as functions of the polymer concentration, l^3n_b, and the ratio of particle to polymer sizes, $a/N^{1/2}l$.

(ii) In aqueous solutions the particles also interact electrostatically. Sketch qualitatively the variation of the attractive minimum in the total interaction potential, i.e. the superposition of the electrostatic and polymeric components, with l^3n_b for a situation in which $\kappa l \approx 0.1$ and $v/l^3 \approx 0.1$.

7

ELECTROKINETIC PHENOMENA

7.1 Introduction

Electrokinetic processes derive from interactions between macroscopic motion and diffuse electric charge. Here the emphasis is on phenomena used to characterize the electrical properties of particles in aqueous suspensions, and to introduce the subject some simple models will be described. The detailed discussions begin with electrophoresis, the motion of individual particles due to an external field. This technique is widely used to measure particle charge. The electrical conductivity of a suspension also reflects electrokinetic processes and is addressed next. The theory of conductivity is based on the same basic model as electrophoresis, extended to cover contributions from many particles. Measurements of mobility and conductivity are complementary, since one reflects events around an individual particle directly while the other averages over the particle population. The final topic concerns the response of colloidal dispersions to unsteady electric fields. Double layers are polarized by external fields and an unsteady field engenders relaxation processes at relatively low frequencies. Thus, studies of dielectric relaxation provide additional insight into electrokinetic behavior.

Examples of electrokinetic phenomena

Suppose that charged colloidal particles suspended in an ionic solution are somehow confined to a channel connecting two reservoirs. Flow of the solution sweeps part of the diffuse charge from one reservoir to the other, producing a streaming current. When there is no external connection between the reservoirs, charge accumulation produces a difference in electrical potential known as the streaming potential. At steady state, this potential difference drives a return flow of ions that balances that due to convection so the net current vanishes. The analogue of the streaming potential is the sedimentation potential, which occurs in a suspension of settling particles.

Another set of electrokinetic phenomena is associated with the extra work required to deform diffuse charge clouds. Since the fluid appears to be more viscous, these are called electroviscous effects. Electroviscous effects derived from inter- and intra-particle phenomena alter the rheology of suspensions.

Next, imagine that an external electric field is applied. It causes relative motion between the suspended particles and bulk fluid (electrophoresis) and between the particles and their diffuse charge clouds (electro-osmosis). Since the charge on a particle and that in the diffuse layer have different signs, the charge cloud and particle tend to move in opposite directions. This dipole slows particle migration. Particle motion can often be followed optically, so electrophoresis is used to measure the mobility of charged particles to deduce their electrical charge. External electric fields also drive currents, and phenomena that change the local ion concentrations or migration rates affect the local conductivity. Therefore, contributions from suspended particles change the average conductivity relative to the solution.

In measuring the conductivity of a suspension, a steady field is imposed and an ohmic conductivity serves to characterize matters. More complicated behavior ensues when the imposed electric field varies sinusoidally in time. Now, the polarization of the diffuse cloud around each particle reverses periodically but there is a phase lag due to the finite speed of the transport process. The overall response of the suspended particles is akin to a suspension of dipoles, and a complex admittance or impedance can be used to represent the system.

In these examples, electrokinetic phenomena manifest themselves in different ways. The theory of electrokinetics provides a general model, tailored to particular situations by specifying initial and boundary conditions.

A model problem

Some insight can be gained from studying the situation depicted in Fig. 7.1, the flow of an ionic solution past a charged surface. The total current due to convection is

$$L\int_0^{\infty} \rho^{(\mathrm{f})} u(y)\,\mathrm{d}y.$$

Results from §4.7 relate the charge density to the potential at the shear surface[†] (the ζ-potential) as

$$\rho^{(f)} = -\varepsilon\varepsilon_0\kappa^2\zeta e^{-\kappa y}. \tag{7.1.1}$$

Thus the streaming current through a rectangle $L \times d$ ($d \gg \kappa^{-1}$) is

$$-L\varepsilon\varepsilon_0\zeta\gamma.$$

To calculate the streaming potential, the streaming current is balanced

[†] The shear surface is defined as the envelope where shear appears in the fluid adjacent to a rigid body when fluid and solid are in relative motion. The potential at this surface, the ζ-potential, may differ from that on the surface of the rigid body due to layers of adsorbed ions and molecules (the Stern layer). We omit consideration of the Stern layer henceforth and identify the ζ-potential with the surface potential, ψ_s (cf. §4.5).

Fig. 7.1. Definition sketches for: (*a*) Shear flow past a plane. (*b*) Electric field parallel to a plane.

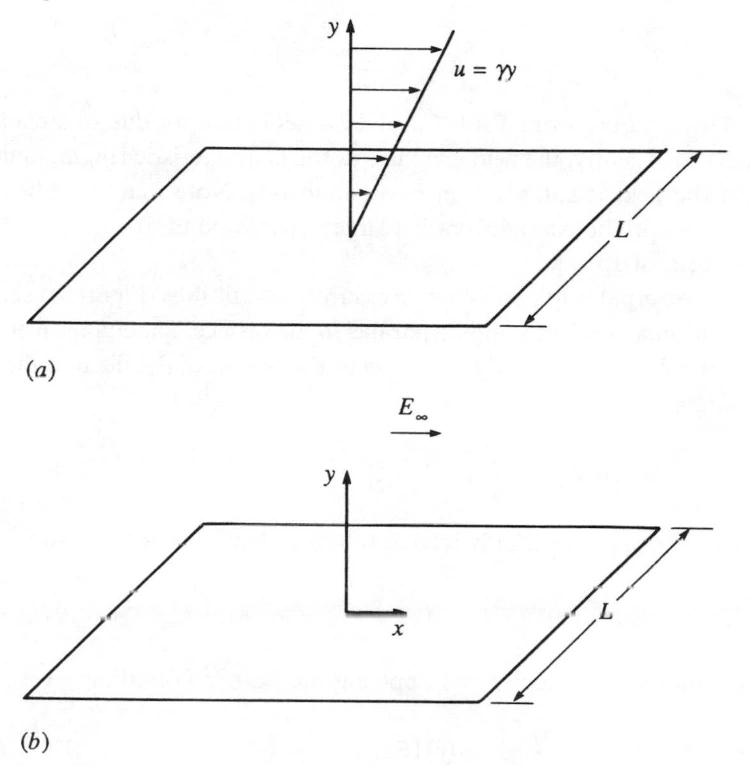

Table 7.1. *Representative values for electrokinetic phenomena*

Shear rate, γ	$1\,s^{-1}$
Length scale, d	10^{-3} m
ζ-potential, ζ	100 mV
Ion diffusivity, ωkT	$10^{-9}\,m^2/s$
Debye length, κ^{-1}	0.1 μm

against that due to electromigration. The total charge carried by the various ions in a uniform field E is

$$Ld\sum_{1}^{N} e^2\omega^k(z^k)^2 n_b^k E;$$

ω^k is the mobility (velocity per unit force) of the kth ion and n_b^k is its concentration. For identical mobilities, i.e., $\omega^k = \omega$,

$$E = \frac{\gamma\zeta}{d\kappa^2\omega kT}. \tag{7.1.2}$$

Using values from Table 7.1 gives a field strength due to streaming of 1 mV/m. Clearly, the field increases as the characteristic length diminishes, but the field is still small in most situations. Note that the diffuse layer thickness in the example is rather large; it corresponds (roughly) to an ionic strength of 10^{-5} M.

An external field also drives an electro-osmotic flow. Figure 7.1 shows the situation with a field imposed parallel to the surface. The equation of motion for the fluid, with a body force due to the action of the field on the diffuse charge, is

$$0 = \rho^{(f)}E_\infty + \mu\frac{\partial^2 u}{\partial y^2}. \tag{7.1.3}$$

Since the space charge is related to the potential according to

$$\varepsilon\varepsilon_0\frac{\partial^2\psi}{\partial y^2} = -\rho^{(f)}, \tag{7.1.4}$$

solving for the velocity and applying the no-slip condition gives

$$u(y) = -\frac{\varepsilon\varepsilon_0}{\mu}[\zeta - \psi(y)]E_\infty. \tag{7.1.5}$$

With the linearized potential we find

$$u(y) = -\frac{\varepsilon\varepsilon_0}{\mu}(1 - e^{-\kappa y})\zeta E_\infty, \tag{7.1.6}$$

indicating that the free stream velocity is approached exponentially. Here fluid appears to slip past the surface when viewed on length scales larger than the Debye thickness. A field of 100 V/m parallel to a surface with a ζ-potential of 100 mV in water gives an electro-osmotic velocity of 7 μm/s.

It turns out that (7.1.6) also represents the relation between the applied field and velocity for a particle moving through a viscous fluid (electrophoresis) under conditions where the diffuse layer is thin compared to the particle dimension (Smoluchowski, 1903). The arguments will be given in detail shortly, but it should be obvious that measurements of electrophoretic mobility when $\kappa a \gg 1$ yield a ζ-potential free of complications due to polarization of the diffuse layer.

These examples illustrate why one works on microscopic scales to deduce surface characteristics from electrokinetic measurements. Nevertheless, although the magnitudes are small, it is a mistake to conclude that electrokinetic phenomena are negligible in situations characterized by large length scales. For example, charge separation due to pumping poorly conducting liquids can have catastrophic consequences (Klinkenberg & van der Minne, 1958).

7.2 **Electrophoresis**

Electrophoresis, or, in some of the older literature, cataphoresis, is the steady translation of a particle under the influence of an external electric field. There is a vast literature on the subject since measurements of electrophoretic mobility are widely used to measure the charge on small particles. According to both Abramson, Moyer & Gorin (1942) and Dukhin & Derjaguin (1974), electrophoresis was first described by F. Reuss in a paper published in 1809. Reuss observed electrophoresis of clay particles as well as electro-osmosis of water through a bed of quartz sand. The early investigators found that practically every surface became charged when placed in contact with an aqueous solution; however, quantitative theories were slow in developing. Both Helmholtz (1879) and Smoluchowski (1903) made important contributions to the theory of electrokinetics by deriving relations between the velocity, the electrostatic potential at the surface, and the applied field for thin double-layers. We will formulate the problem in general terms from which their results emerge as special cases.

Scale analysis

To visualize the situation imagine a smooth sphere immersed in an ionic solution (see Fig. 7.2). The sphere has a surface charge due to one of the mechanisms noted in §4.4 and an associated diffuse charge. A uniform electric field imposed on the system causes electrophoresis. The balance laws express the conservation of mass, momentum, and ionic species, along with a relation between charge and potential, the Poisson equation:

$$\textit{Continuity}: \nabla \cdot \mathbf{u} = 0; \tag{7.2.1}$$

$$\textit{Momentum}: \rho \mathbf{u} \cdot \nabla \mathbf{u} = -\nabla p + \rho^{(\mathrm{f})}\mathbf{E} + \mu \nabla^2 \mathbf{u}; \tag{7.2.2}$$

$$\textit{Conservation of ions}: \nabla \cdot \mathbf{j}^k = 0; \tag{7.2.3}$$

$$\textit{Electrostatics}: \varepsilon\varepsilon_0 \nabla^2 \psi = -\rho^{(\mathrm{f})}. \tag{7.2.4}$$

Here, $\mathbf{u}$ represents the local fluid velocity, $\mathbf{j}^k$ is the flux of the kth ionic species and

$$\rho^{(\mathrm{f})} = \sum_1^N e z^k n^k. \tag{7.2.5}$$

The transport of the ions by the combined effects of flow, electromigration, and diffusion can be expressed by adding convective transport to the Nernst–Planck relation. That relation gives the ionic flux as the product of the ion mobility and the thermodynamic force, the gradient of the electrochemical potential (cf. 4.6.2). Thus

$$\mathbf{j}^k = n^k \mathbf{u} - e z^k n^k \omega^k \nabla \psi - kT\omega^k \nabla n^k. \tag{7.2.6}$$

Fig. 7.2. Definition sketch for electrophoresis – a sphere in a uniform field.

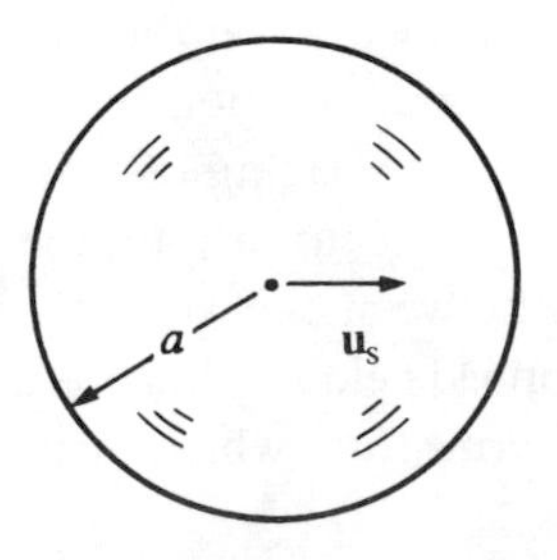

The boundary conditions follow from stipulating that the sphere moves as a rigid object with no ions penetrating the surface and retains a uniform charge independent of the imposed field. In addition, the solution far from the sphere remains undisturbed. Thus, on the surface,

$$\mathbf{u}=\mathbf{u}_s, \qquad (\mathbf{j}^k-n^k\mathbf{u}_s)\cdot\mathbf{n}=0, \qquad -\varepsilon\varepsilon_0\nabla\psi\cdot\mathbf{n}=\frac{Q}{4\pi a^2}. \tag{7.2.7}$$

Far from the sphere,

$$\mathbf{u}\to 0, \qquad n^k\to n_b^k, \qquad -\nabla\psi\to E_\infty\mathbf{e}_3. \tag{7.2.8}$$

Further progress requires consistent simplification of the mathematical formulation through a scale analysis (Saville, 1977). The basic premise is that the imposed field is weak so that the equilibrium double-layer structure is only slightly perturbed. If the scale for potential is chosen as kT/e and the length scale is a, then a stress scale based on electrical parameters is $\varepsilon\varepsilon_0(kT/ae)^2$. Balancing the electric stress against the viscous stress and the pressure makes the velocity scale $\varepsilon\varepsilon_0(kT/e)^2/\mu a$. In addition, ion mobilities are scaled using one of them, ω^0, and ion concentrations scaled on $\sum_k(z^k)^2n_b^k\equiv n^0$. This produces a set of dimensionless equations and identifies several dimensionless groups. Accordingly,

$$\nabla\cdot\mathbf{u}=0, \tag{7.2.9}$$

$$Re\,\mathbf{u}\cdot\nabla\mathbf{u}=-\nabla p+\nabla^2\psi\nabla\psi+\nabla^2\mathbf{u}, \tag{7.2.10}$$

$$Pe\,\mathbf{u}\cdot\nabla n^k=\omega^k\nabla\cdot(z^kn^k\nabla\psi+\nabla n^k), \tag{7.2.11}$$

$$\nabla^2\psi=-(a\kappa)^2\sum_1^N z^kn^k. \tag{7.2.12}$$

The boundary conditions on the surface are

$$\begin{aligned}&\mathbf{u}=\mathbf{u}_s,\\ &(z^kn^k\nabla\psi+\nabla n^k)\cdot\mathbf{n}=0,\\ &-\nabla\psi\cdot\mathbf{n}=q.\end{aligned} \tag{7.2.13}$$

Far from the sphere we have

$$\begin{aligned}&\mathbf{u}\to\mathbf{0}\\ &n^k\to n_\infty^n\\ &-\nabla\psi\to\beta\mathbf{e}_3.\end{aligned} \tag{7.2.14}$$

Table 7.2.

Physical parameters	
particle radius, a	0.1 μm
particle charge density, $Q/4\pi a^2$	8×10^{-2} C/m^2
characteristic ion density, n_b^k	10^{-5} mol/(dm)^3
characteristic ion mobility, ω^0	10^{-9} m^2/s
dielectric constant, ε	80
viscosity, μ	10^{-3} N s/m^2
density, ρ	10^3 kg/m^3
field strength, E_∞	10^3 V/m
Dimensionless groups	
Reynolds number, Re	4.4×10^{-4}
Peclet number, Pe	0.4
charge density, q	4.5×10^2
double layer, κa	0.96
field strength, β	4.0×10^{-3}

Here $n_\infty^k = n_b^k/n^0$. The dimensionless groups are as follows:

Reynolds number	$Re \equiv \varepsilon\varepsilon_0(kT)^2\rho/(\mu e)^2$
Peclet number	$Pe \equiv \varepsilon\varepsilon_0 kT/e^2\mu\omega^0$
charge density	$q \equiv eQ/4\pi a\varepsilon\varepsilon_0 kT$
field strength	$\beta \equiv aeE_\infty/kT$
double-layer thickness	$(\kappa a)^2 \equiv e^2\sum_1^N (z^k)^2 n_b^k a^2/\varepsilon\varepsilon_0 kT$

Some typical magnitudes (Table 7.2) illustrate why inertial effects can safely be ignored. Convective effects, however, influence the ion distributions since the Peclet number is $O(1)$. Furthermore, whenever the surface charge density is large, the equilibrium situation must be calculated from the non-linear Poisson–Boltzmann equation.

The model can be simplified by introducing perturbation expansions of the form

$$\begin{aligned} \mathbf{u} &= \mathbf{0} + \beta\mathbf{u}_\beta + \ldots, \\ p &= p_0 + \beta p_\beta + \ldots, \\ \psi &= \psi_0 + \beta\psi_\beta + \ldots, \\ n^k &= n_0^k + \beta n_\beta^k + \ldots. \end{aligned} \qquad (7.2.15)$$

The first terms represent the equilibrium double-layer structure, while the $O(\beta)$-alterations arise from the imposed field. From the differential equations and boundary conditions we derive two sets of equations, using a coordinate system centered on the sphere.

First, the $O(1)$-problem for the equilibrium double-layer is

$$\begin{aligned} \mathbf{0} &= -\nabla p_0 + \nabla^2\psi_0\nabla\psi_0, \\ \mathbf{0} &= \nabla\cdot[z^k n_0^k \nabla\psi_0 + \nabla n_0^k], \\ \nabla^2\psi_0 &= -(\kappa a)^2 \sum_1^N z^k n_0^k, \end{aligned} \tag{7.2.16}$$

with boundary conditions

$$(z^k n_0^k \nabla\psi_0 + \nabla n_0^k)\cdot\mathbf{n} = 0, \qquad -\nabla\psi_0\cdot\mathbf{n} = q \qquad \text{on} \qquad r = 1 \tag{7.2.17}$$

and

$$n_0^k \to n_\infty^k, \qquad \psi_0 \to 0, \qquad \text{as } r\to\infty. \tag{7.2.18}$$

The $O(\beta)$-problem for the electrokinetic processes is

$$\begin{aligned} &\nabla\cdot\mathbf{u}_\beta = 0, \\ &\mathbf{0} = -\nabla p_\beta + \nabla^2\psi_0\nabla\psi_\beta + \nabla^2\psi_\beta\nabla\psi_0 + \nabla^2\mathbf{u}_\beta, \\ &Pe\,\mathbf{u}_\beta\cdot\nabla n_0^k = \omega^k\nabla\cdot[z^k n_0^k\nabla\psi_\beta + z^k n_\beta^k\nabla\psi_0 + \nabla n_\beta^k], \\ &\nabla^2\psi_\beta = -(\kappa a)^2\sum_1^N z^k n_\beta^k. \end{aligned} \tag{7.2.19}$$

Here the boundary conditions are

$$\mathbf{u}_\beta = \mathbf{0}, \qquad (z^k n_0^k\nabla\psi_\beta + z^k n_\beta^k\nabla\psi_0 + \nabla n_\beta^k)\cdot\mathbf{n} = 0, \qquad -\nabla\psi_\beta\cdot\mathbf{n} = 0 \qquad \text{at } r = 1; \tag{7.2.20}$$

$$\mathbf{u}_\beta \to -\mathbf{u}_s, \qquad n_\beta^k \to 0, \qquad -\nabla\psi_\beta \to \mathbf{e}_3 \qquad \text{as } r\to\infty. \tag{7.2.21}$$

The first of these two problems has already been treated in §4.8. A variety of techniques have been employed for the second problem, beginning with those introduced by Helmholtz (1879) and Smoluchowski (1903) in studies of particles with thin double layers. We will invert the chronological ordering and begin with situations where the diffuse layer is thick.

A thick diffuse layer

In the limit $\kappa a \to 0$ the Poisson–Boltzmann equation simplifies to Laplaces' equation and the electrical body force disappears from the equation of motion. Accordingly, the electric field reduces to that derived for a charged sphere, (4.4.6), and the electric force is given by (4.4.9). This force is balanced against viscous drag (Stokes' law) and so, as Hückel (1924) found,

$$6\pi\mu a[\varepsilon\varepsilon_0(kT)^2/e^2\mu a]\mathbf{u}_s = QE_\infty\mathbf{e}_3. \tag{7.2.22}$$

Thus we obtain

$$\mathbf{u}_s = \tfrac{2}{3}\beta q\mathbf{e}_3, \tag{7.2.23}$$

in dimensionless form with the velocity scaled on $\varepsilon\varepsilon_0(kT)^2/e^2\mu a$. Here electrokinetic effects are absent, owing to the negligible influence of the diffuse layer. Conversely, when the diffuse layer is thin one expects a strong effect since electro-osmotic drag will predominate.

A thin diffuse layer

To analyze this configuration a different methodology is used. First, two situations are envisioned. In one, the particle is held fixed in the presence of a uniform field that drives an electro-osmotic flow through its action on the diffuse charge cloud and produces a drag force on the particle. In the other situation, the particle is held fixed in the presence of a uniform flow without any external field. Because the effect of convection on the charge cloud is negligible, the drag is given by Stokes' law. Finally, balancing the two drag forces yields an expression for the electrophoretic velocity.

First we study the situation where the particle is held fixed in a uniform electric field. As $\kappa a \to \infty$, the charge density is exponentially small (cf. §§4.7 and 4.8), so there is an outer region where the electrical body force vanishes. Symmetric solutions of the Stokes equations that describe the $O(\beta)$-velocity field yield the velocity components in the radial and angular directions as

$$u_\beta = 2\cos\theta(Ar^{-1} + Br^{-3}), \qquad v_\beta = -\sin\theta(Ar^{-1} - Br^{-3}). \tag{7.2.24}$$

Here A and B are constants to be determined by matching with velocities in the inner region close to the particle where electrical effects are important. Once this has been done, we can calculate the drag force by integrating the stress over a surface enclosing the particle.

The potential in the outer region is found from the situation where the net free charge vanishes (cf. §4.4). Here perturbations to the equilibrium ion densities vanish and the potential is

$$\psi_\beta = -(r + Cr^{-2})\cos\theta. \tag{7.2.25}$$

The constant C must be determined from the potential in the inner region.

The analysis for the inner region requires rescaling the equations to take advantage of the fact that the diffuse layer is thin and retains the equilibrium double-layer structure, cf. §4.8. Radial distances are scaled as

$$r = 1 + y/\kappa a. \tag{7.2.26}$$

Then, within the thin charge cloud, derivatives normal to the particle surface are large compared with the tangential ones and charge convection has a negligible effect on the equilibrium structure of the diffuse layer. Matching the potential given by (7.2.25) with the potential in the diffuse layer shows that C is $\frac{1}{2}$.

The next task is to calculate the velocity field generated by the electrical body force distribution. The dominant terms in the equation for the tangential velocity, once the pressure term has been removed by differentiation, yield

$$\frac{\partial^3 v_\beta}{\partial y^3} = -\frac{\partial^3 \psi_0}{\partial y^3}\frac{\partial \psi_\beta}{\partial \theta}. \tag{7.2.27}$$

Recall that ψ_0 denotes the equilibrium potential. From this expression, the velocity field follows as

$$v_\beta = \tfrac{3}{2}\sin\theta[\psi_0(0) - \psi_0(y)]. \tag{7.2.28}$$

Matching this with the outer velocity field discloses that $A = -B = -3\psi_0(0)/4$; $\psi_0(0) = e\zeta/kT$.

The force on the particle can be calculated from the formula

$$\mathbf{F} = \int_S \boldsymbol{\sigma}\cdot\mathbf{n}\,dS, \tag{7.2.29}$$

where $\boldsymbol{\sigma}$ represents the stresses due to pressure, viscous effects, and electrical forces. Since the stresses are in equilibrium, the divergence of the stress tensor vanishes, permitting integration over a large surface as in §4.4. Thus

$$\mathbf{F} = \lim_{r\to\infty}\int_{S(r)} r^2(\boldsymbol{\sigma}\cdot\mathbf{n})\,d\Omega. \tag{7.2.30}$$

After deriving an expression for the pressure and evaluating the integral, we have

$$\mathbf{F} = 8\pi\beta A\mathbf{e}_3. \tag{7.2.31}$$

This force and the Stokes drag from the second problem balance, yielding

$$6\pi\mathbf{u}_s - 6\pi\beta\psi_0(0)\mathbf{e}_3 = 0. \tag{7.2.32}$$

When re-expressed in dimensional form, the formula shows that the speed of translation is independent of particle size (Smoluchowski, 1903). In dimensional form, we have

$$u_s = \frac{\varepsilon\varepsilon_0\zeta}{\mu} E_\infty. \tag{7.2.33}$$

This expression also appears to suggest that the velocity is independent of ionic strength when the charge cloud is thin compared with the particle radius. It must be remembered, however, that the diffuse-layer thickness affects the potential at the inner edge of the layer, so an accurate picture requires some knowledge of the chemistry of the interfacial region. Another way of emphasizing this point is to convert the expression to one involving the charge instead of the potential. Using the Debye–Hückel approximation, we obtain the dimensional velocity as

$$u_s = \frac{eQE_\infty}{4\pi a^2 \mu} \kappa^{-1}. \tag{7.2.34}$$

Thus, keeping the surface charge constant yields a vanishingly small velocity as the thickness of the diffuse charge cloud is diminished since $\kappa^{-1} \to 0$.

Electrophoresis with an equilibrium diffuse layer

To investigate situations intermediate to those just studied, we restrict ourselves to small potentials where convective effects on the charge cloud can be omitted. The first complete solution for the electrophoresis of a sphere was given by Henry (1931), who also gave results for a circular cylinder. Our approach differs from Henry's in several respects, not the least of which is that we invoke the Debye–Hückel linearization at the outset.

The technique is based on the fact that as long as the diffuse charge cloud is undeformed, the electric force on the sphere undergoing electrophoresis is the same as that on a charged sphere in a non-conducting medium. This force balances the viscous drag arising from uniform translation plus the electro-osmotic flow. Calculation of the flow field and drag due to electro-osmosis requires some results from Chapter 2.

First, the drag on an isolated sphere placed in an arbitrary flow, $\mathbf{u}(\mathbf{x})$, can be calculated through Faxen's law as

$$\mathbf{F} = 6\pi\mu a\left(1 + \frac{a^2}{6}\nabla^2\right)\mathbf{u}(\mathbf{x}). \tag{7.2.35}$$

The flow represented by $\mathbf{u}(\mathbf{x})$ is generated by point forces representing the action of the incident field on the equilibrium charge cloud. This force distribution is

$$\mathbf{f} = -\rho_0^{(f)} \nabla \psi_\beta. \tag{7.2.36}$$

The velocity due to a superposition of point forces is

$$\mathbf{u}(\mathbf{x}) = \frac{1}{8\pi\mu} \int_V \mathbf{I}(\mathbf{x}-\mathbf{x}') \cdot \mathbf{f} \, dV, \tag{7.2.37}$$

with

$$\mathbf{I}(\mathbf{x}-\mathbf{x}') = \frac{1}{r}\left(\boldsymbol{\delta} + \frac{\mathbf{rr}}{r^2}\right), \qquad \mathbf{r} = \mathbf{x} - \mathbf{x}'. \tag{7.2.38}$$

Substitution into (7.2.35) and integration yields the force due to drag from the electro-osmotic flow as

$$-4\pi\mu a q \frac{(\kappa a)^2}{1+\kappa a} e^{\kappa a} \int_1^\infty (1 - \tfrac{1}{4} r^{-3} + \tfrac{1}{4} r^5) e^{-\kappa a r} \, dr \, \beta \mathbf{e}_3.$$

Combining this force with those due to a steady translation and the electric field yields the dimensionless electrophoretic velocity, viz.

$$\begin{aligned} u_s = \frac{2}{3} \frac{\beta q}{(1+\kappa a)} \{ & 1 + \tfrac{1}{16}(\kappa a)^2 - \tfrac{5}{48}(\kappa a)^3 - \tfrac{1}{96}(\kappa a)^4 \\ & + \tfrac{1}{96}(\kappa a)^5 + [\tfrac{1}{8}(\kappa a)^4 - \tfrac{1}{96}(\kappa a)^6] e^{\kappa a} E_1(\kappa a) \} \end{aligned} \tag{7.2.39}$$

Here $E_1(\cdot)$ denotes the exponential integral.

Henry's formula, (7.2.39), expresses the electrophoretic mobility for arbitrary Debye thicknesses in terms of the particle charge, or, alternatively, the ζ-potential. Figure 7.3 shows how changes in the thickness of the diffuse layer affect the mobility when either the charge or the potential is held constant. In the latter case, the mobility stays non-zero as the thickness diminishes because the particle charge increases. This increase balances the increased electro-osmotic drag, which tends to drive the mobility to zero.

Effects due to deformation of the diffuse layer

Convection and electromigration of ions tend to polarize the particle and its charge cloud. This confers a dipole character to the double-layer system. Diffusion acts to relax the asymmetry. These processes become important at higher potentials and require a much more detailed analysis, owing to the non-linear nature of the problem. The first successful attempts, by Overbeek (1943) and Booth (1950), used perturbation methods to expand the variables as power series in the particle potential or charge. Owing to the analytical complexity of the problem, only a few terms were calculated. Consequently, although qualitative trends are depicted faithfully, the range of applicability is limited. The advent of digital computation methods has supplanted these perturbation techniques and eliminated many simplifications made for mathematical convenience.

Numerical solutions by Wiersema, Loeb, & Overbeek (1966), over a range of particle charges and diffuse layer thicknesses, show that the perturbation solutions overestimated effects of polarization and relaxation. More recently, O'Brien & White (1978) developed a computational scheme to cover an even wider range of conditions. Figures 7.4 to 7.6 depict results dealing with the effects of the particle potential, diffuse layer thickness, and electrolyte properties.

Figures 7.4 and 7.5 show how the potential at the shear surface affects the mobility when the thickness of the diffuse layer is held constant. For thick charge clouds the increase is monotonic but not linear owing to the influence of electro-osmosis. As the potential increases, the charge on the particle and in the diffuse layer increase. Increases in mobility due to a larger charge are, therefore, offset by the increased electro-osmotic drag.

Fig. 7.3. Electrophoretic mobility as a function of the diffuse layer thickness. (*a*) Henry's solution for constant charge. (*b*) Henry's solution for constant potential.

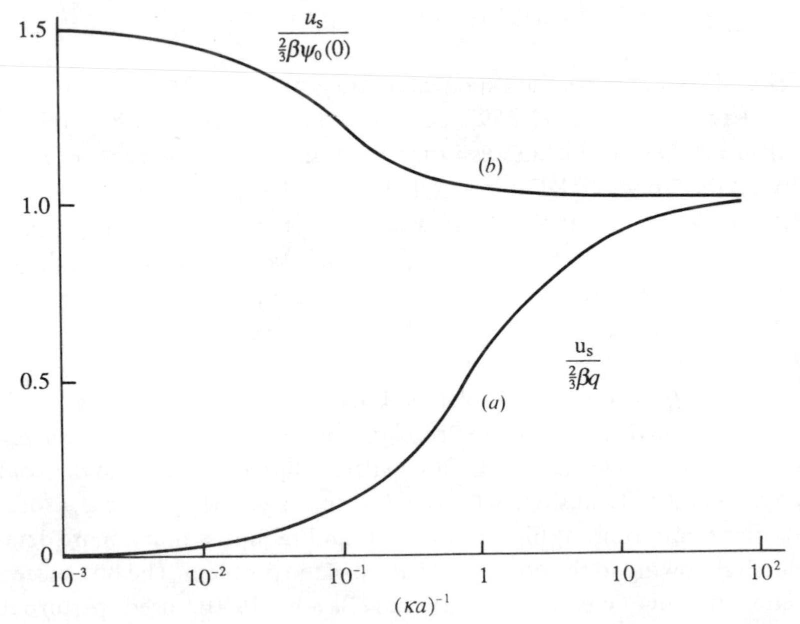

With thinner diffuse layers, the mobility is no longer monotonic and a maximum appears. As the diffuse-layer thickness diminishes, the maximum occurs at higher and higher potentials. Figure 7.6 shows effects due to the valence character of the electrolyte. Note here that the maximum shifts to lower mobilities and potentials as the valence of the counterion increases.

Although vestiges of the maxima are apparent in the work of Overbeek (1943), Booth (1950), and Wiersema, Loeb, & Overbeek (1966), uncertainty remained as to whether maxima were intrinsic features or arose

Fig. 7.4. Electrophoretic mobility of a spherical particle in KCl ($\kappa a < 3$), (O'Brien & White, 1978).

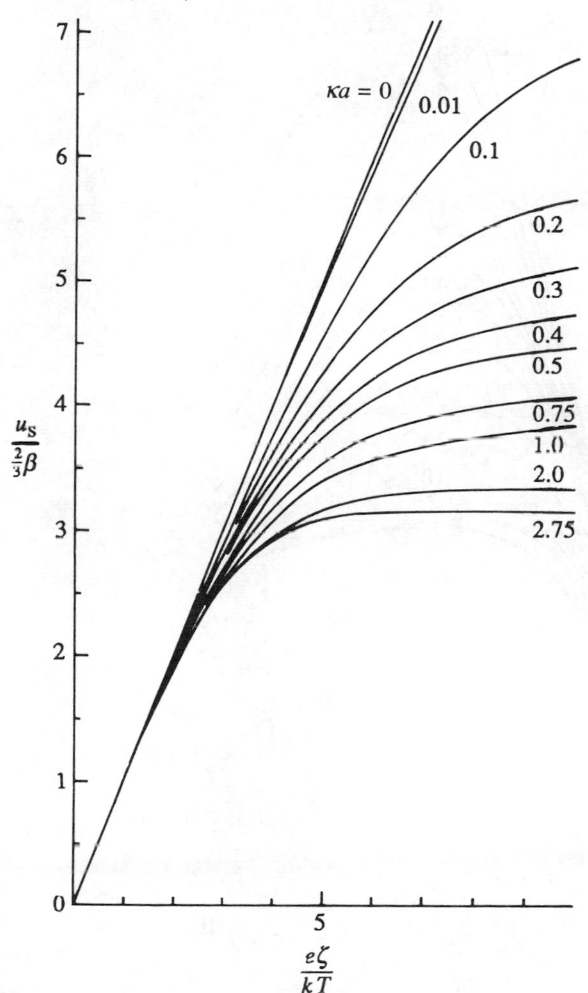

from inaccurate approximations. Using a perturbation technique, Dukhin & Semenikhin (1970) showed that the maxima arise from polarization of charge in a thin diffuse layer and electro-osmotic drag (see also O'Brien, 1983).

Although treated as independent parameters in the theory, particle charge or potential and diffuse-layer thickness are closely related in an experiment. For a given particle, changes in ionic strength alter the particle

Fig. 7.5. Electrophoretic mobility of a spherical particle in KCl ($\kappa a > 3$) (O'Brien & White, 1978).

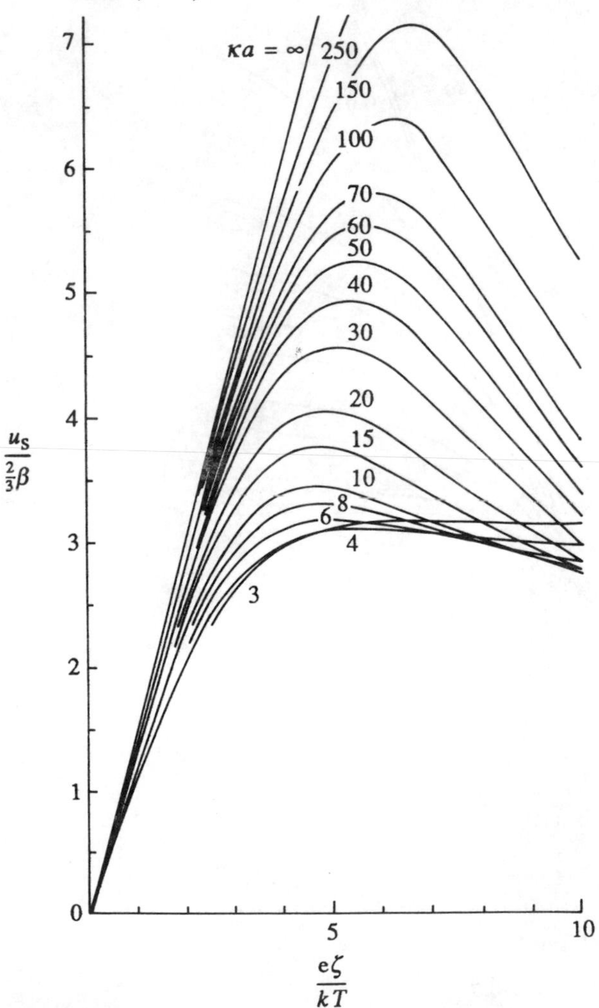

charge as well as the diffuse-layer thickness. Thus it is difficult, to say the least, to compare the theoretical predictions with experimental behavior without a model for the chemistry of the surface layer.

Measurements of electrophoretic mobilities

Our treatment of the experimental techniques is brief; those seeking extensive presentations should consult Hunter's book on the ζ-potential (Hunter, 1981). Biological applications are reviewed by Mehrishi (1972) and Seaman (1975). The technique discussed here, microelectrophoresis, requires particles large enough to be microscopically visible, but not so large that they sediment rapidly. The essential feature is a transparent chamber of cylindrical or rectangular cross-section equipped with electrodes. After the chamber is filled with a very dilute mixture containing particles suspended in an electrolyte, a field is applied. Particle motion is followed by eye using a microscope or with an automated tracking system. Measurements of the field strength and particle migration rate yield the mobility. Figure 7.7 is a schematic diagram of an apparatus.

Two sorts of complications arise: electro-osmosis carries particles at speeds other than their intrinsic electrophoretic velocities and buoyancy causes motion of particles and fluid. Buoyancy due to Joule heating is

Fig. 7.6. Effect of counterion valence on the electrophoretic mobility of a spherical particle at $\kappa a = 5$ (O'Brien & White, 1978).

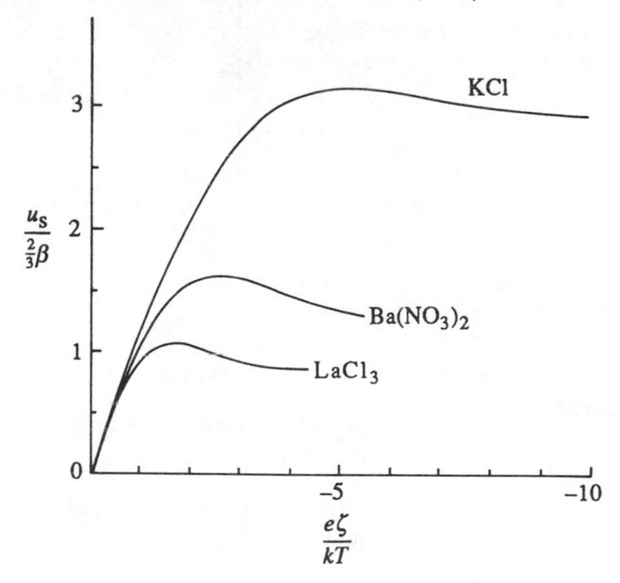

minimized simply by using small capillaries and low field-strengths (<10 V/cm), but a major complication arises from electro-osmosis. The force exerted by the electric field on charge in the fluid adjacent to the inner surface of the glass tube causes bulk flow. Because the tube is closed, the net flux must be zero so there is a return down the center. By solving the Navier–Stokes equations, the velocity profile outside the diffuse layer in a rectangular channel is found to be

$$u = \frac{\varepsilon\varepsilon_0 \zeta E_\infty}{2\mu}(1 - 3y^2/h^2). \tag{7.2.40}$$

Here ζ represents the equilibrium electrostatic potential at the surface of the channel and $2h$ is its thickness; E_∞ is the field strength tangential to the surface. Since particles are carried along by this flow, the particle velocities observed in a microelectrophoresis device are the sum of the electrophoretic motion and the fluid velocity.

Two techniques are used to separate the intrinsic electrophoretic migration from the composite motion. The most accurate method involves measuring the particle velocity at several points across the channel to give the velocity as a function of position. The area under this curve divided by

Fig. 7.7. Schematic diagram of an electrophoresis chamber (Shaw, 1969). (Note that circular and rectangular cross sections can be used.)

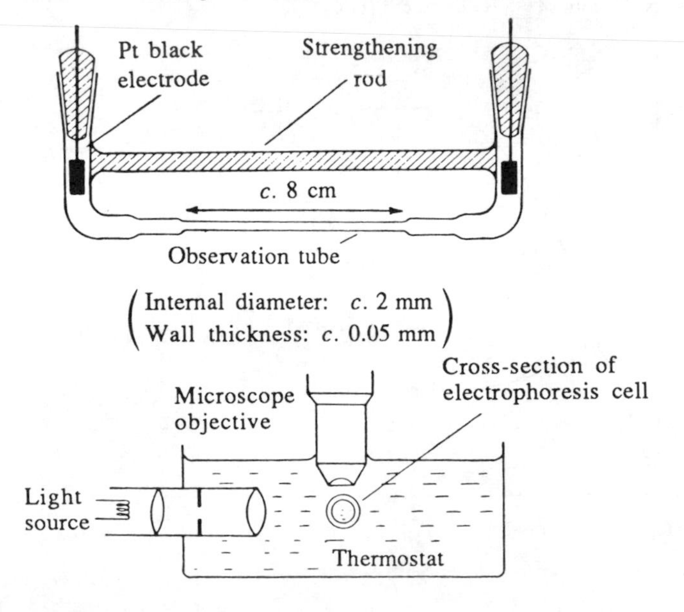

the cross-sectional area is equal to the electrophoretic velocity since the net flow of liquid is zero. A second, somewhat less accurate, technique is to measure at the position where the fluid velocity is zero, the stationary level ($y = h\sqrt{3}$ in a rectangular channel). Figure 7.8 shows measurements of the velocity of red blood cells in a rectangular channel; note the parabolic shape. Analysis of the data gives a mobility of 1.25 (μm/s)/(V/cm).

Comparisons between theory and experiment

Although the study of electrokinetics is over a century old, development continues. Much recent work centers on the character of the charged surface, since, as noted earlier, the Gouy–Chapman model of the diffuse layer appears generally valid. Few surfaces behave as though either the charge density or the potential is constant as the ionic strength of the

Fig. 7.8. Electrophoretic velocities of red blood cells in a rectangular channel at a field strength of 5 V/cm (Zukoski & Saville, 1987).

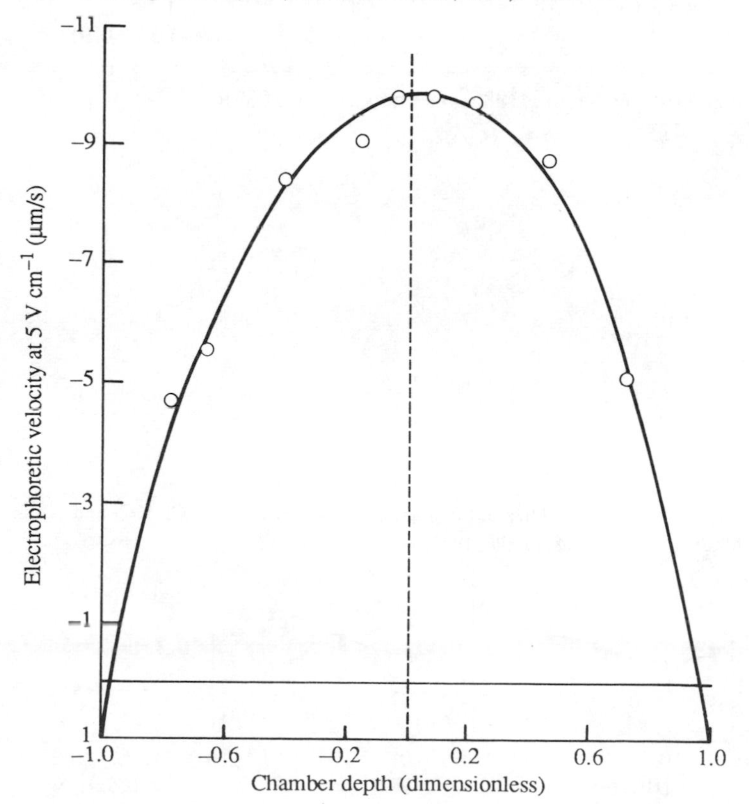

Table 7.3. ζ-potentials calculated from mobility measurements

Salt	Salt concentration/M	κa	$-e\zeta/kT$
HCl	1×10^{-2}	26.9	2.37
	5×10^{-3}	19.1	2.33
	1×10^{-3}	8.5	2.44
	5×10^{-4}	6.0	2.17
	1×10^{-4}	1.9	2.40
	5×10^{-5}	1.2	2.65
			Ave. 2.37 ± 0.11
KCl	1×10^{-2}	26.9	3.06
	1×10^{-3}	8.5	2.96
	1×10^{-4}	2.7	2.68
			Ave. 2.90 ± 0.15
$MgSO_4$	1×10^{-3}	17.0	1.25
	5×10^{-4}	12.0	1.32
	1×10^{-4}	5.4	1.42
	5×10^{-5}	3.8	1.54
			Ave. 1.38 ± 0.10

Data from Zukoski & Saville (1985) on Latex A, $a=0.083\,\mu$m.

Fig. 7.9. Electrophoretic mobilities of polystyrene latex particles in HCl, KCl and $MgSO_4$. (Zukoski & Saville, 1985).

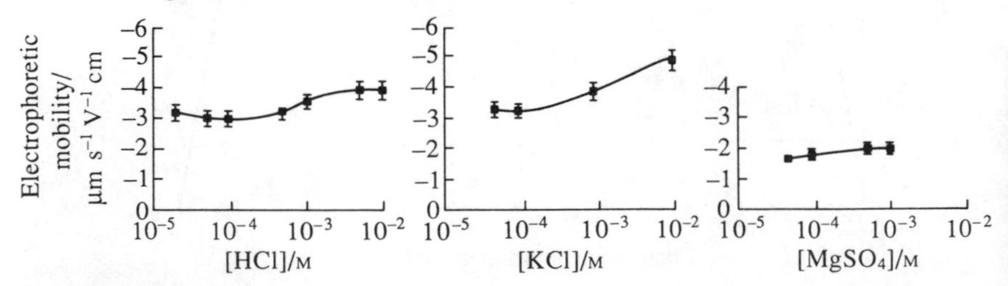

surrounding electrolyte changes. Instead, the charge appears governed by complicated chemical processes involving adsorption as well as the dissociation of ionogenic groups within the Stern layer (cf. Healy & White, 1978). Lacking model particles with well-defined charging characteristics, the surface charge or potential must be treated as a parameter defined by the electrokinetic measurement itself.

Polystyrene particles are often used as model particles, since they can be prepared in relatively monodisperse form and their surfaces are hard and smooth on a macroscopic scale. Extensive studies by, for example, Ottewill & Shaw (1972) and Zukoski & Saville (1985) show that the charge measured by direct titration often exceeds that from electrophoretic mobility measurements. Figure 7.9 shows how the mobility varies with the ionic strength with latex particles. Table 7.3 indicates that these particles behave as though the ζ-potential is almost constant in a given electrolyte. However, measurements of electrical conductivity (to be discussed shortly) give a conflicting picture.

7.3 Electrical conductivity of dilute suspensions

The theory set out in §7.2 also provides a model for the collective behavior of dilute suspensions where, instead of focusing on the behavior of a single particle, electrokinetic effects are averaged over a multitude of particles. The presence of particles and their charge clouds affects passage of current through a suspension in several ways. Non-conducting particles themselves offer obstacles to the ion flux and polarize the incident field, altering the bulk conductivity according to the theory worked out by Maxwell (1892). When the particles are surrounded by a diffuse charge cloud, additional effects occur. The basic theoretical task is to average the fields described in §7.2 over the suspension volume. To illustrate the method we derive the Maxwell equation for a suspension of uncharged, non-conducting particles.

Maxwell's theory

For a solution of uniform concentration with conductivity σ the electric potential is derived from solutions of Laplace's equation. Outside a spherical, non-conducting particle the potential is

$$\psi = -E_\infty(1 + Ar^{-3})\mathbf{x}\cdot\mathbf{e}_3, \tag{7.3.1}$$

where $\mathbf{e}_3$ denotes the orientation of the incident field and $A = a^3/2$. To calculate the average current due to free charge transport, we evaluate the integral of the local current $\mathbf{j}$, i.e.,

$$\mathbf{J}=\frac{1}{V}\int_V \mathbf{j}\,\mathrm{d}V. \tag{7.3.2}$$

Then the effective conductivity, σ_*, is defined in terms of the average, or macroscopic, field strength, as

$$\mathbf{J}=\sigma_*\langle\mathbf{E}\rangle, \tag{7.3.3}$$

where

$$\langle\mathbf{E}\rangle=\frac{1}{V}\int_V(-\nabla\psi)\,\mathrm{d}V. \tag{7.3.4}$$

Since the integral in (7.3.4) is over a volume containing particles and fluid, the average field inside the particles will be needed, even though the current there is zero. The potential inside an uncharged particle in a dilute suspension is

$$\psi=-\tfrac{3}{2}E_\infty\mathbf{x}\cdot\mathbf{e}_3 \tag{7.3.5}$$

so the field there is uniform, i.e.,

$$-\nabla\psi=\tfrac{3}{2}E_\infty\mathbf{e}_3. \tag{7.3.6}$$

The integral in (7.3.2) is converted to one over the particles by adding and subtracting the current that would exist were particles absent, $\sigma\langle\mathbf{E}\rangle$. Thus

$$\mathbf{J}=\sigma\langle\mathbf{E}\rangle+\sigma\int_{V_\mathrm{p}}\nabla\psi\,\mathrm{d}V, \tag{7.3.7}$$

where the last integral is taken over by the volume occupied by the particles, V_p. Next we suppress particle–particle interactions by dealing with dilute systems and use (7.3.6) for the field strength to obtain

$$\mathbf{J}=\sigma_*E_\infty\mathbf{e}_3, \qquad \sigma_*=\sigma(1-\tfrac{3}{2}\phi). \tag{7.3.8}$$

Here ϕ denotes the volume fraction of particles. According to (7.3.8), the effective conductivity is diminished by more than would be expected if it were simply proportional to the relative amounts of the two phases. The extra factor of $\frac{1}{2}\phi$ comes from polarization of the field by individual particles.

Diffuse layer effects

Calculation of the average conductivity of a suspension of charged particles in an ionic solution follows along the same lines, although complications arise from motion of the particles, variations in ion densities within the diffuse layer, and polarization of the ion cloud. Procedures to deal with these matters were worked out by Saville (1979, 1983) and O'Brien (1981).

As before, the local current is averaged over a representative volume of the suspension. Inside the particles the current still vanishes, but outside the particles the (dimensional) current is

$$\mathbf{j}=\sum_{1}^{N}[ez^k n^k \mathbf{u} - e^2(z^k)^2\omega^k n^k \nabla\psi - ez^k\omega^k kT\nabla n^k]. \tag{7.3.9}$$

Solutions to the equations set out in §7.2 furnish expressions for the velocity, potential, and concentration fields. These are then used to calculate the average current and an effective conductivity. For the average current in a dilute suspension we obtain (in dimensionless form)

$$\mathbf{J}=\underbrace{\beta\sum_{1}^{N}(z^k)^2\omega^k n_1^k \mathbf{e}_3}_{\text{(i)}}+\underbrace{\frac{1}{V}\beta\int_{V_\mathrm{p}}\sum_{1}^{N}\{(z^k)^2\omega^k n_1^k\nabla\psi_\beta+z^k\omega^k\nabla n_\beta^k\}\,\mathrm{d}V}_{\text{(ii)}}$$

$$\underbrace{-\frac{1}{V}\beta\int_{V_\mathrm{F}}\sum_{1}^{N}\{(z^k)^2\omega^k(n_0^k-n_1^k)\nabla\psi_\beta+(z^k)^2\omega^k n_\beta^k\nabla\psi_0-Pe\,z^k n_0^k\mathbf{u}_\beta\}\,\mathrm{d}V.}_{\text{(iii)}} \tag{7.3.10}$$

Here n_1^k denotes the uniform concentration of ions beyond the diffuse layer, and V_F denotes the fluid volume; the other symbols were defined in connection with the material on electrophoresis. The terms in (7.3.10) account for the following processes:

(i) conductivity of the solution outside the diffuse layer,
(ii) polarization of the ion densities and the incident electric field,
(iii) current within the diffuse layer due to electromigration and electro-osmosis.

It is important to recognize that in a space-filling suspension the conductivity of the solution outside the diffuse layer surrounding each particle differs from that around an isolated particle owing to two effects: counterions added as part of particle-charging processes and the attraction of counterions towards and repulsion of co-ions away from the charged surfaces (non specific adsorption). These produce $O(\phi)$-changes as large as those arising from all the other processes (Saville, 1983).

Using (7.3.10), an effective (dimensionless) conductivity can be expressed as

$$\sigma_*=\sigma_\infty+3\left[\sum_{1}^{N}(z^k)^2\omega^k n_\infty^k H(z^k,\kappa a,q)-qz^c\omega^c(\kappa a)^{-2}+\Delta\right]\phi. \tag{7.3.11}$$

Table 7.4. *Components of the conductivity increment for negatively charged particles in HCl solutions* $(e\zeta/kT) = -2.0$

	κa				
	1	2.5	5	10	25
$3\sum_{1}^{N}(z^k)^2\omega^k n_\infty^k H(z^k, \kappa a, q)/\sigma_\infty$	−10.67	−3.35	−1.52	−0.72	−0.28
$-3qz^c\varpi^c/(a\kappa)^2\sigma_\infty$	19.67	5.51	2.36	1.08	0.41
$3\Delta/\sigma_\infty$	9.08	2.01	0.23	−0.57	−1.10
$(\sigma_* - \sigma_\infty)/\sigma_\infty\phi$	18.08	4.17	1.07	−0.21	−0.97

The first term represents the conductivity of the electrolyte solution in the absence of suspended particles, the second accounts for the effects of non-specific adsorption, the third arises from the added counterions with valence z^c and mobility ω^c, and the fourth, sometimes called the dipole term, embodies all the other effects due to the particles. The dipole term can be calculated from the polarization of the potential and concentration fields, i.e.

$$-\mathbf{e}_3\,\Delta = \lim_{r\to\infty}\frac{1}{4\pi}\int_{S(r)}[\Omega_\beta\mathbf{n} - \nabla\Omega\cdot\mathbf{xn}]\,\mathrm{d}S, \tag{7.3.12}$$

with

$$\Omega_\beta = \sum_{1}^{N}[(z^k)^2\omega^k n_0^k\psi_\beta + z^k\omega^k n_\beta^k]. \tag{7.3.13}$$

See O'Brien (1981) or Saville (1983) for details. Table 7.4 lists contributions to the conductivity increment $(\sigma_* - \sigma_\infty)/\sigma_\infty\phi$ from non-specific adsorption, added counterions, and electrokinetic effects embodied in the dipole term for representative situations. The non-specific adsorption term is negative here because the hydronium ions attracted towards the surface have a higher mobility than the chloride ions that replace them in the bulk of the solution. Note that the magnitudes of the non-specific adsorption and added counterion terms diminish as the diffuse layer is thinned and the dipole term approaches $-\frac{3}{2}$, the Maxwell contribution for uncharged particles.

Table 7.5. *Comparison of theoretical and experimental results for the conductivity increment*

System	$-e\zeta/kT$	κa	$[\sigma_* - \sigma_\infty)/\sigma_\infty \phi]$		
			Theory		Experiment
1	2.41	1.13	18.8[a]	18.1[b]	24[c]
2	2.38	1.61	17.8	18.8	17
11	3.86	0.91	42.9	41.8	64

[a]Figures in this column were calculated from a perturbation expansion correct to the second order in the ζ-potential. Note that these figures differ slightly from those reported by Saville (1983), owing to use of a more accurate expression for the non-specific adsorption contribution here.
[b]Figures in this column were calculated from numerical solutions of the equations.
[c]Figures in this column are abstracted from results reported by Watillon & Stone-Masui (1972).

Comparisons between theory and experiment

Although it is fairly easy to measure the electrical conductivity of a solution or dilute suspension, few studies exist where the suspensions were carefully cleaned and the particles fully characterized. Results from one such study by Watillion & Stone-Masui (1972) are shown in Table 7.5. This compares limiting slopes (the conductivity increment) calculated from the experiments with the dilute suspension theory applied to arbitrary potentials, (7.3.11), and an expansion for low potentials. The agreement for systems 1 and 2 is satisfactory, but the conductivity predicted for system 11 falls more than 35 per cent below the measured value. Note also that the theoretical results from the perturbation expansion are fairly close to those from the exact, numerical solution.

A recent set of experiments (Zukoski & Saville, 1985) was designed to test the theory more extensively by measuring the conductivity and mobility. Care was taken in cleaning the suspensions to insure reproducible results, since conductivity measurements are strongly influenced by contaminants. Figure 7.10 illustrates both the reproducibility and linearity of those results. Measurements were made on 23 suspensions involving 3 different salts and particles synthesized by 2 different techniques. Figure 7.11 shows the conductivity increment for different salts demonstrating the expected

Fig. 7.10. The electrical conductivity of suspensions of latex particles in 5×10^{-4} M HCl. Different symbols represent experiments with different samples (Zukoski & Saville, 1985).

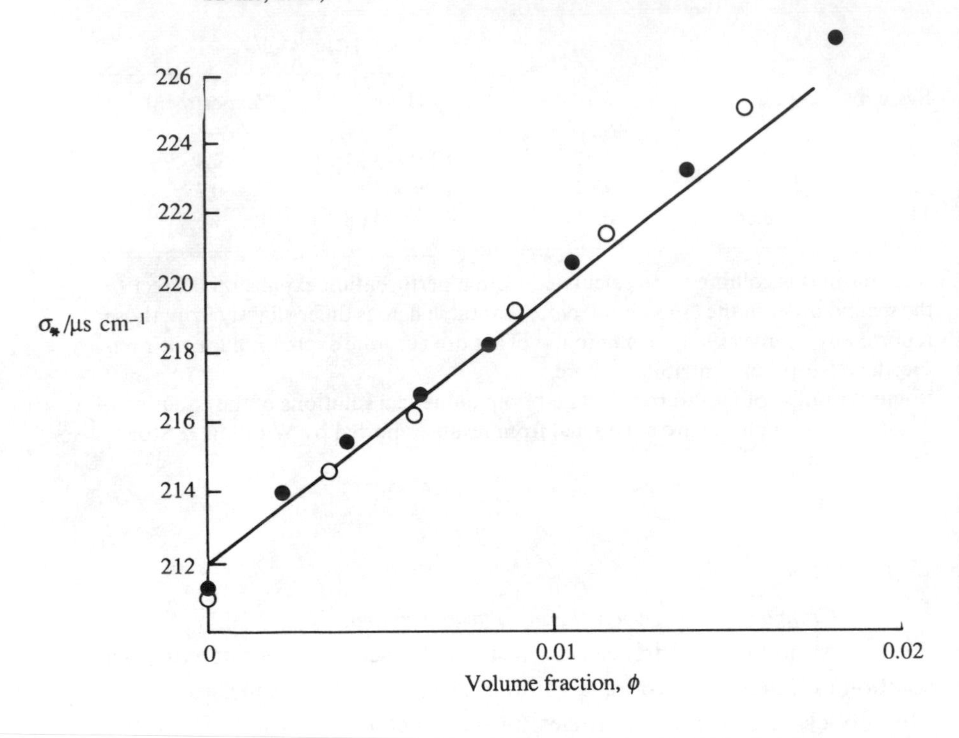

Fig. 7.11. The conductivity increment in various electrolytes with latex A (Zukoski & Saville, 1985).

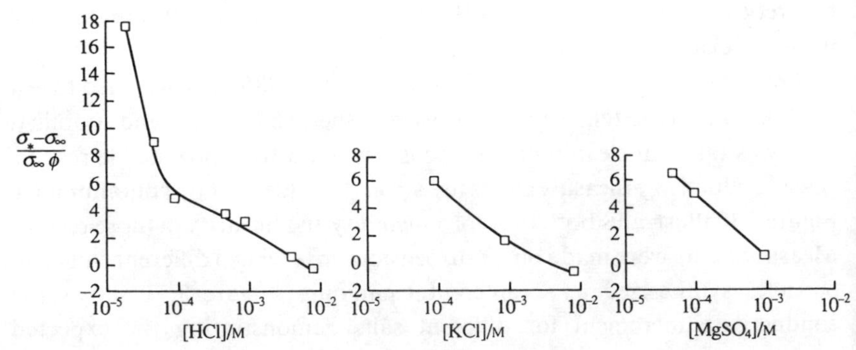

Fig. 7.12. ζ-potentials calculated for latex A in various salts: □, calculated from mobility measurements, ○, calculated from conductivity measurements (Zukoski & Saville, 1985).

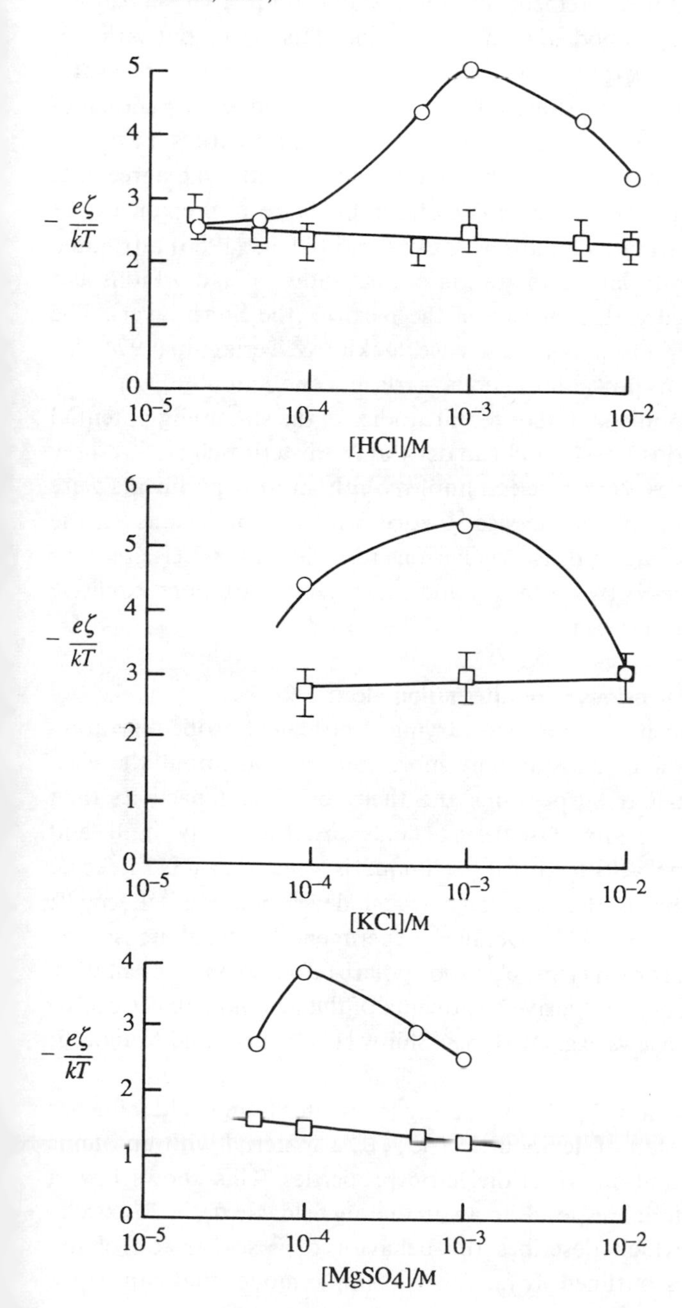

decrease with increasing salt concentration as the diffuse layer is made thinner.

The independent measurements of mobility and conductivity constitute a test of the premises embodied in the electrokinetic theory set out earlier. If the theory is correct, ζ-potentials calculated from the mobility measurements ought to agree with those inferred from the conductivity measurements. Figure 7.12, which shows the dimensionless ζ-potentials from each measurement, demonstrates that this is not the case. Although agreement improves at either low or high ionic strengths, large discrepancies are observed at intermediate conditions. Zukoski & Saville (1986) attributed these discrepancies to lateral migration of ions within a layer of ions and molecules adsorbed at the surface of the particle (the Stern layer). The concept is not a new one in colloid science. Dukhin & Derjaguin (1974), for example, suggested how 'anomalous surface conduction' might alter electrokinetic phenomena. Other recent studies of the streaming potential by O'Brien & Perrins (1984), and van der Put & Bijsterbosch (1983) show similar discrepancies. Nevertheless, until recently most experiments were interpreted in terms of the classical theory, with immobile ions on the surface. It is premature to draw conclusions regarding the precise reasons for the discrepancies between theory and experiment until more extensive experimental work has been done.

7.4 Dilute suspensions with alternating electric fields

Measurements with a time-varying electric field probe relaxation processes and produce substantially more information about the electrokinetic properties of suspensions and their constituent particles than static-field measurements. Oscillating fields are frequently used and simplest to analyze. Although the technique is venerable (cf. Fricke & Curtis, 1937), rigorous theories are a recent development (DeLacey & White, 1981; O'Brien, 1982). Detailed experimental results are sparse, owing to difficulties arising from electrode polarization, so our presentation focuses on the theory. Extensive treatments of the general subject can be found in various reviews, e.g. Dukhin & Shilov (1974, 1980) and Mandel & Odjik (1984).

First we analyze how dipoles oscillating in an alternating field influence current flow in a slab of 'leaky dielectric', i.e., a material with an ohmic conductivity in addition to its dielectric properties. This shows how a homogeneous material responds to an oscillating field. Next, the Maxwell–Wagner theory, which describes the behavior of a suspension of uncharged spheres, is outlined. It furnishes a simple model that can be set out in closed form to illustrate the methodology for colloidal particles. The theory for colloidal suspensions then follows along the lines discussed in connection with the static conductivity, with additional effects due to the alternating field. Extant comparisons between theory and experiment support the qualitative features of the theory, but are not discussed extensively, owing to the absence of experiments with fully characterized systems.

Dipoles are induced in matter by the action of an external field. The polarizability of an individual atom or molecule, $\alpha(\omega)$, is a complex function of the frequency, ω, of the forcing field, and so is the dielectric constant, i.e.,

$$\varepsilon(\omega)=\varepsilon'(\omega)+i\varepsilon''(\omega). \tag{7.4.1}$$

Electronic polarization derives from deformation of the electron cloud by the incident field whereas in materials with permanent dipoles the field changes the orientation distribution. Orientation polarization is a balance between the orientating effect of the external field and randomization by thermal motion. In either case, the dielectric constant is frequency-dependent, owing to inertial effects on atomic or molecular scales, and the material is said to be dispersive. It takes a certain amount of time for a molecule to align itself, owing to its moment of inertia, and at high frequencies the molecule cannot follow, so the dielectric constant decreases. The smaller moment of inertia of the electron makes the electronic polarizability insensitive to frequency changes up to optical frequencies.

The leaky dielectric

Although the dielectric constants of low-molecular-weight solutions are relatively insensitive to the frequency of the electromagnetic field below 10^9 Hz, an alternating field still produces a current owing to oscillation of the individual dipoles. The polarization current is (Feynman, Leighton, & Sands, 1964)

$$\mathbf{j}^{(p)}=\frac{\partial \mathbf{P}}{\partial t}. \tag{7.4.2}$$

Analysis of a parallel plate apparatus (Fig. 7.13) illustrates how this current manifests itself. The device is filled with a linear, leaky dielectric, where σ is the conductivity and the current density due to the migration of free charge is

$$\mathbf{j}^{(f)}=\sigma\mathbf{E}. \tag{7.4.3}$$

An alternating voltage is impressed on one of the plates and the other is grounded (Fig. 7.13), i.e.,

$$\psi_0 = \Delta\psi e^{-i\omega t}, \psi_h = 0. \tag{7.4.4}$$

To derive a relation between the current, $I(t)$, the potential, $\psi(t)$, and the properties of the dielectric we proceed as follows.

According to Maxwell's theory of electromagnetism (cf. Feynman, Leighton, & Sands, 1964, and §4.2), the total charge is conserved, so

$$\frac{\partial}{\partial t}\rho^{(e)} = -\nabla \cdot (\mathbf{j}^{(f)} + \mathbf{j}^{(p)}). \tag{7.4.5}$$

Substituting relations for the current, (7.4.2) and (7.4.3), and using (4.2.1) and (4.2.14), shows that

$$\nabla \cdot (\sigma\mathbf{E} - i\omega\varepsilon_0\varepsilon\mathbf{E}) = 0. \tag{7.4.6}$$

The electric field is irrotational as long as magnetic effects are negligible so a potential can be defined and, if the material is homogeneous, it follows from (7.4.6) that solutions of Laplace's equation define the potential. Thus

$$\psi = \Delta\psi e^{-i\omega t}(1 - y/h). \tag{7.4.7}$$

To relate the current and potential to the properties of the leaky dielectric we balance the current into and out of the system diagrammed in Fig. 7.13 against the charge accumulation rate. Because the material is homogeneous, charge is found only at the plate-dielectric interface. Thus, since the field in the plate is small compared with that in the dielectric, the total charge is, cf. (4.3.1),

$$Q = A\varepsilon_0\mathbf{E}\cdot\mathbf{n}; \tag{7.4.8}$$

Fig. 7.13. Schematic diagram for an admittance measurement.

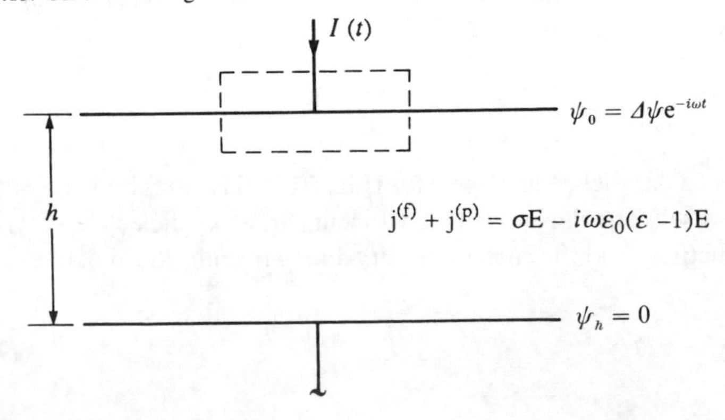

A represents the area of one plate. The complex amplitude of the current follows from the charge balance as

$$I_0 = (\sigma - i\omega\varepsilon_0\varepsilon)\frac{\Delta\psi}{h}A. \tag{7.4.9}$$

The complex number $I_0/\Delta\psi$, the admittance,† can be calculated from measurements of the magnitude and phase lag of the current and related to the conductivity and dielectric constant as

$$\frac{I_0 h}{\Delta\psi A} = \sigma - i\omega\varepsilon_0\varepsilon = \sigma + \omega\varepsilon_0\varepsilon''(\omega) - i\omega\varepsilon_0\varepsilon'(\omega). \tag{7.4.10}$$

This shows that the real part of the admittance is proportional to the conductivity for the current, owing to polarization and free charge. The imaginary part gives the real part of the dielectric constant. As noted above, the dielectric constant is relatively insensitive to frequency changes below 10^9 Hz for single-phase materials, so changes in the admittance will be proportional to the frequency.

Dispersive behavior will be found at frequencies below 10^9 Hz with a suspension between the plates due to polarization of the diffuse charge clouds. To analyze this situation we first note that the analysis leading to (7.4.10) is applicable to a homogeneous suspension if the conductivity and dielectric constant are treated as 'average' or 'effective' properties of the suspension. The task is to relate the average properties to those of the constituents of the suspension.

Maxwell–Wagner theory

To illustrate how suspended particles alter the dielectric response, consider a suspension of spheres exposed to an alternating field. The spheres are non-conducting, but the suspending medium is an ionic conductor. Frequencies are assumed to be below the level where molecular or electronic relaxation occur, so the dielectric constants of both phases are real numbers. Nevertheless, the response of the suspension turns out to be characterized by a frequency-dependent quantity, the complex conductivity. This can be then interpreted in terms of a frequency-dependent conductivity and dielectric constant by analogy with (7.4.10).

Most of the necessary formulas are obtained by reinterpreting results for static fields. If both phases are homogeneous, then the potential follows from solutions of Laplace's equation and retains forms given by (7.3.1) and

† Some authors prefer to work in terms of the complex impedance, the reciprocal of the admittance.

(7.3.5). However, the integration constants are different. Two conditions are needed to define the constants of integration. One is the continuity of the electric potential. The other follows from examining the total current. According to (7.4.6), the total current density in each phase,

$$\mathbf{j}^{(t)} = (\sigma - i\omega\varepsilon_0\varepsilon)\mathbf{E} \equiv K\mathbf{E}, \tag{7.4.11}$$

is solenoidal. Since there are no sources of charge, this current must be continuous across the interface. These two conditions provide expressions from which the integration constants can be evaluated.

Further insight comes from calculating the charge induced on a sphere owing to the field. The free charge which builds up from ionic conduction normal to the interface is, cf. (4.3.3),

$$\frac{Q^{(f)}}{A} = -3\varepsilon_0 \frac{\bar{\varepsilon}}{2 - i\omega\varepsilon_0(\bar{\varepsilon} + 2\varepsilon)/\sigma} \mathbf{E}_\infty \cdot \mathbf{n}. \tag{7.4.12}$$

The polarization charge is

$$\frac{Q^{(p)}}{A} = 3\varepsilon_0 \frac{(\bar{\varepsilon} - 1) - i\omega\varepsilon_0(\bar{\varepsilon} - \varepsilon)/\sigma}{2 - i\omega\varepsilon_0(\bar{\varepsilon} + 2\varepsilon)/\sigma} E_\infty \cdot \mathbf{n}. \tag{7.4.13}$$

The overbars denote properties of the sphere and A stands for its area. These formulas show that interface polarization depends on the frequency of the imposed field, and two characteristic times exist:

$$\varepsilon_0(\bar{\varepsilon} - \varepsilon)/\sigma \qquad \text{and} \qquad \varepsilon_0(\bar{\varepsilon} + 2\varepsilon)/\sigma.$$

At low frequencies, $\omega \ll \sigma/\varepsilon_0$, the interface charge is that for a static interface. At high frequencies, conduction processes are swamped and the free charge disappears, leaving a residual polarization charge. Hence the suspension acts as a dispersive medium exhibiting relaxation at frequencies near the characteristic values. For a 10^{-3} M KCl solution, the conductivity† is 15 mS/m and so $\sigma/\varepsilon_0\varepsilon$ is about 2×10^7 Hz.

An expression for the effective conductivity follows from averaging (7.4.11) in the fashion used to obtain (7.3.8). For the average current density we have

$$\mathbf{J}^{(t)} = K\{1 - 3[(K - \bar{K})/(2K + \bar{K})]\phi\}\mathbf{E}_\infty, \tag{7.4.14}$$

which is (7.3.8) with the new variables

$$K = \sigma - i\omega\varepsilon_0\varepsilon \qquad \text{and} \qquad \bar{K} = -i\omega\varepsilon_0\bar{\varepsilon} \tag{7.4.15}$$

† The conductivity is measured in millisiemens per meter; one siemens per meter is equal to 1 $(\text{ohm m})^{-1}$.

that obtain for non-conducting particles in a conducting medium. Note that the effective 'conductivity' is a complex number whose frequency dependent real and imaginary parts can be related to the complex admittance, (7.4.10).

This example illustrates how the properties of the suspension may be frequency-dependent, even though the properties of the constituent parts are frequency-independent. Attempts to use the Maxwell–Wagner theory (with the addition of a 'surface conductivity') to describe the behavior of colloidal suspensions have not been very successful (c.f. O'Konski, 1960, Schwann *et al.*, 1962, and Schwartz, 1962). A theory is needed to account for double-layer polarization.

Behavior of suspensions of colloidal particles

To describe the time-dependent behavior of the charge cloud around a colloidal particle requires a generalization of the electrokinetic equations. Once this is done, expressions for the average properties of the suspension can be derived by combining the techniques employed for the conductivity problem with those used to develop the Maxwell–Wagner theory. First, we note that temporal accelerations involving either fluid or particle motion remain negligible as long as

$$\frac{\omega^2 a}{\nu} \ll 1 \qquad \text{and} \qquad \rho_p \frac{\omega^2 a}{\mu} \ll 1,$$

(see Chapter 2). For example, for 0.2 μm particles these accelerations are negligible at frequencies below 10^7 Hz.

To assess the importance of temporal changes in the structure of the diffuse layer, we reason as follows. In a steady field, the diffuse layer has a dipole character with a charge separation distance, l, of $O(a+\kappa^{-1})$. If the sense of the field is reversed then the dipole must invert by diffusion and electro-migration which requires a time of $O(l^2/D)$, D being a characteristic ion diffusivity. Thus temporal forcing of the charge cloud at a frequency ω will be on the same time scale as ionic transport when

$$\frac{2\pi D\kappa^2}{(1+a\kappa)^2\omega} \approx 1.$$

Table 7.6 shows frequencies at which this is the case for different diffuse-layer thicknesses.

Table 7.6. *Characteristic frequencies*[a]

κa	$\dfrac{\omega}{2\pi}$ Hz
1	2×10^3
10	8×10^3
100	1×10^4

[a]Calculated for $0.2\,\mu$m particles with $D=10^{-9}\,\text{m}^2/\text{s}$.

For frequencies much smaller than these, temporal terms in the ion balances are small because transport rates are 'fast' relative to the external forcing. Here, concentration fields are quasi-static and follow the imposed field exactly. Conversely, when the dimensionless group is 0(1) or larger, temporal terms are needed to reflect the competition between external forcing and transport. Relaxation ought to be evident near the frequencies indicated.

One additional alteration is needed to account for the polarization current arising from the alternating field and to produce a solenoidal expression for the total current (O'Brien, 1982). Using (7.3.9) for the current carried by the free charge and (7.4.2) for the polarization current yields

$$\mathbf{j}^{(t)}=\sum_{1}^{N}[ez^k n^k\mathbf{u}-e^2(z^k)^2\omega^k n^k\nabla\psi-ez^k\omega^k kT\nabla n^k]+i\omega\varepsilon_0\varepsilon\nabla\psi \quad (7.4.16)$$

in systems with diffuse layers; cf. (7.4.6). This formula can be manipulated with the divergence theorem to produce an expression for the average conductivity of the suspension (O'Brien, 1982).

The mathematical model for transient behavior consists of the equilibrium double-layer equations, (7.2.16) to (7.2.18), and the electrokinetic equations, (7.2.19) to (7.2.21), with temporal terms added to the ion balances. To compute the average current and bulk conductivity, (7.4.16) is averaged over a representative volume. Analytical results are possible for situations where the particle charge is small (O'Brien, 1982) or the diffuse layer thin (see, for example, Hinch *et al.*, 1984), but they are rather complicated. Here we focus on numerical solutions for some representative situations (DeLacey & White, 1981).

Two different methodologies are used to report results. The reason for this equivocation stems from the fact that the complex conductivity can be expressed as

$$\sigma(\omega,\phi)-i\omega\varepsilon_0\varepsilon'(\omega,\phi)$$

or

$$\sigma(0,\phi)+\omega\varepsilon_0\varepsilon''(\omega,\phi)-i\omega\varepsilon_0\varepsilon'(\omega,\phi).$$

In the first instance, the frequency dependence of the real part has been assigned to the conductivity and dielectric response is taken to be 'loss free'. Conversely, all the frequency dependence has been assigned to the imaginary part of the dielectric constant in the second formula. Both methods of presentation are used (see, for example, Schwann *et al.*, 1962). Changes due to the presence of the particles are of interest here, so we look at the following incremental changes for dilute systems:

$$\Delta\sigma\equiv\frac{\sigma(\omega,\phi)-\sigma(\omega,0)}{\sigma(\omega,0)\phi},\qquad \Delta\varepsilon'\equiv\frac{\varepsilon'(\omega,\phi)-\varepsilon'(\omega,0)}{\phi},$$

$$\Delta\varepsilon''\equiv\frac{\sigma(\omega,\phi)-\sigma(0,\phi)}{\omega\varepsilon_0\phi}. \tag{7.4.17}$$

Figures 7.14 to 7.18 depict results calculated for 0.2-μm spheres in KCl solutions by DeLacey & White (1981). The particles are non-conducting with a dielectric constant of 2; surface charge is assumed to be immobile. Figure 7.14 shows how the real part of the dielectric constant varies with surface potential and frequency. Note especially the large magnitude at $\omega=0$, the sensitivity to the ζ-potential, and the relaxation near 10^5 Hz. All this is interpreted as follows. At low surface potentials the polarization is negative, in agreement with Maxwell–Wagner theory, since the dielectric constant of the sphere material is less than that of the suspending fluid. As the surface charge increases, the polarizability of the charge cloud begins to be a factor. Because the ions are mobile, the diffuse layer is easily polarized, producing a large charge separation compared with the microscopic distances associated with polarization of individual atoms and molecules. Accordingly, the dielectric constant is large. Moreover, because the space charge is displaced in the direction of the applied field, polarization is always positive. It increases with the surface potential because the larger the surface potential, the larger the space charge around the particle. Finally, the decay at higher frequencies (near 10^5 Hz) arises when the charge cloud can no longer respond to rapid changes in the direction of the field (cf. Table 7.6).

Figure 7.15 shows an increase in the dielectric constant and the relaxation frequency as the diffuse layer becomes thinner. The first reflects an increase in the concentration of ions in the diffuse layer, proportional to κ^2 (cf. Chapter 4). The second is to be expected if the characteristic frequency is given by

$$\frac{2\pi D\kappa^2}{(1+\kappa a)^2\omega} = \text{constant.} \tag{7.4.17}$$

Fig. 7.14. Real dielectric increment for spherical particles in KCl solutions at different ζ-potentials (DeLacey & White, 1981).

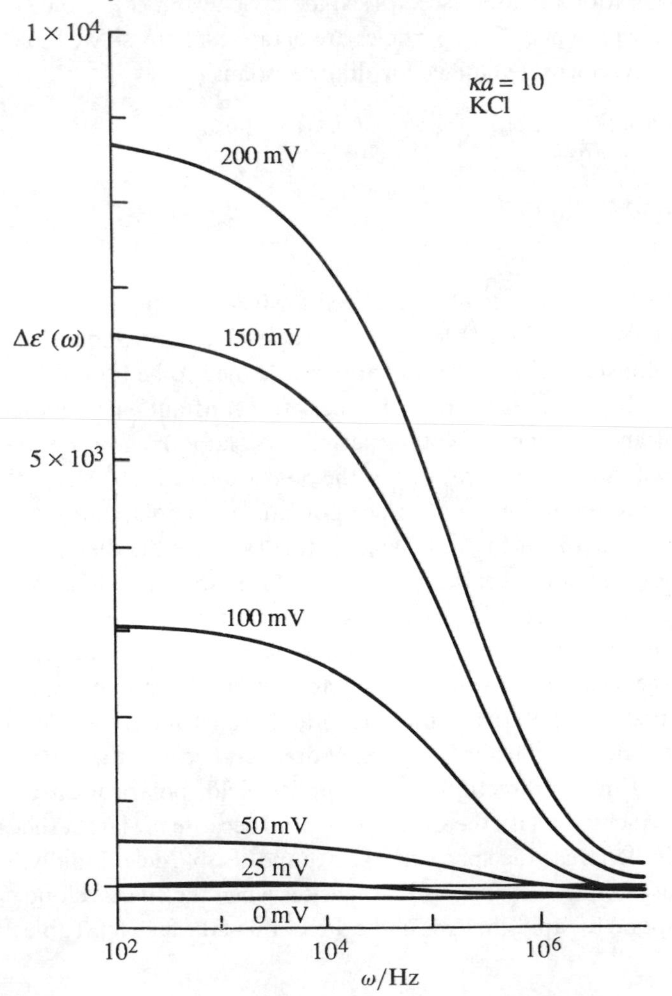

The numerical results also show that the dielectric constant increases with the valence of the counterions, since the amount of mobile charge in the diffuse layer increases.

The increment in the conductivity is much less sensitive to changes in frequency, Fig. 7.16,† than the dielectric constant increment. Adding uncharged particles replaces electrolyte with non-conducting regions and decreases the conductivity, as indicated by the negative conductivity increment. As particle charge increases, effects from the diffuse charge cloud begin to compensate. At higher frequencies, ion trajectories are shortened

† Note that here the effects of added counterions and non-specific adsorption are omitted from the results. With dilute suspensions these effects contribute to the conductivity of the solution beyond the double layer and do not alter the qualitative features shown here.

Fig. 7.15. Real dielectric increment for spherical particles in KCl solutions at different diffuse layer thicknesses (DeLacey & White, 1981).

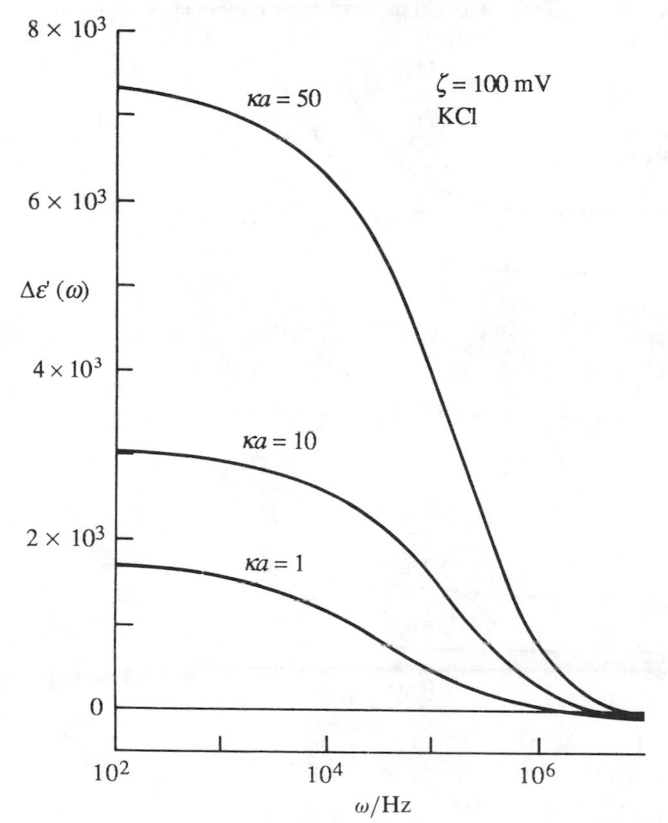

and the blocking effect of the particle diminishes, leading to a higher conductivity.

Effects of the double-layer thickness are depicted on Fig. 7.17. Here we see that the conductivity increment decreases as the diffuse-layer thickness diminishes, in part because the increment is normalized against the conductivity of the bulk solution, which increases with the bulk ionic strength. Another factor is the larger blocking effect of the particle relative to the thin diffuse layer.

Figure 7.18 shows the response in terms of the imaginary part of the dielectric increment, $\Delta\varepsilon''$. This represents the out-of-phase part of the dielectric response produced by ion diffusion. At low frequencies, ions displaced by the field have ample time to set up a new configuration, and since diffusion is in phase with the imposed field, $\Delta\varepsilon''$ is small. As the time scale of the imposed field becomes comparable with the time scale for ion

Fig. 7.16. Conductivity increment for spherical particles in KCl solutions at different ζ-potentials (DeLacey & White, 1981).

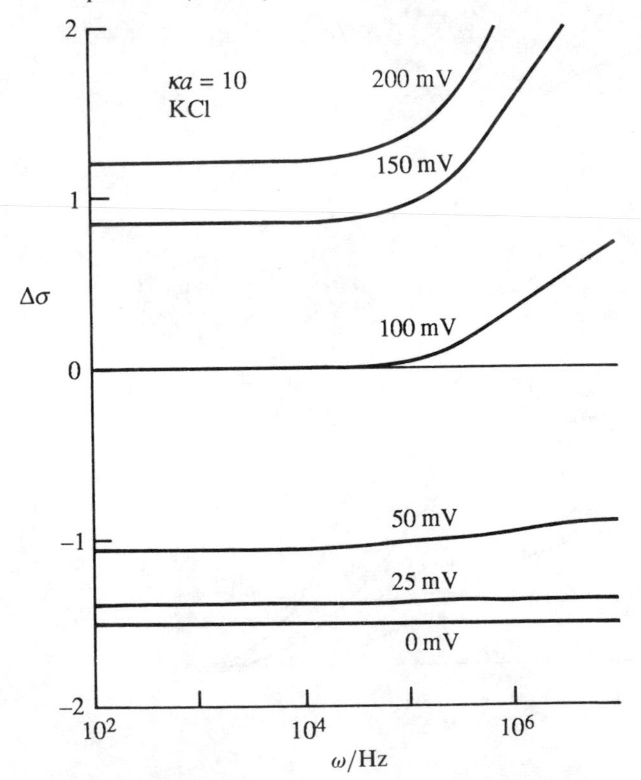

diffusion, diffusion cannot keep up with driving field, leading to an increased out-of-phase response. At high frequencies, displaced ions are returned to their original positions by the field before diffusion occurs and the out-of-phase response diminishes. Other results (not shown) indicate that $\Delta\varepsilon''$ increases with κ and the ζ-potential, as expected.

Experimental data on the frequency-dependent behavior are sparse. Much of the older data was obtained with concentrated systems, outside the pervue of rigorous theories. Schwann *et al.* (1962), for example, reported data on suspensions containing 17–30 per cent (vol.) particles. The qualitative features of their results, Fig. 7.19, agree with those predicted by DeLacey & White (1981) for dilute systems. Data on ε' for dilute suspensions of polymer latices reported by Springer, Korteweg, & Lyklema

Fig. 7.17. Conductivity increment of spherical particles in KCl solutions at different diffuse layer thicknesses (DeLacey & White, 1981).

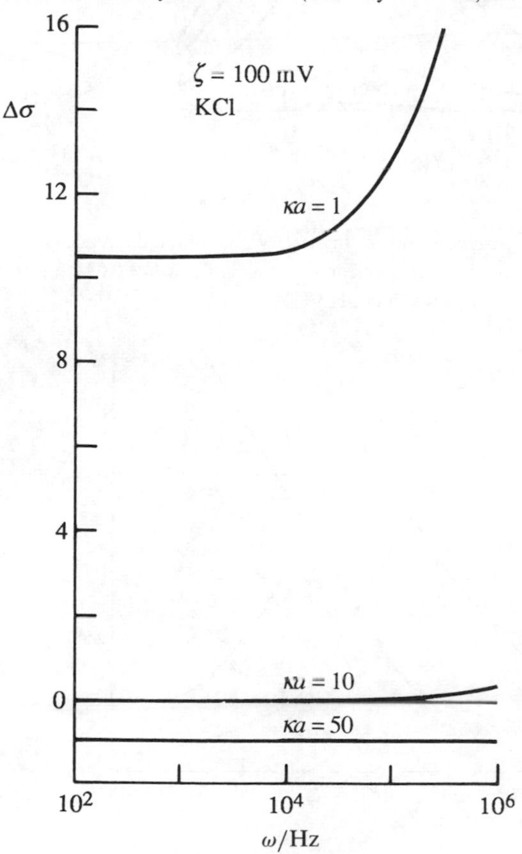

Fig. 7.18. Imaginary dielectric increment for spherical particles in KCl solutions at different ζ-potentials (DeLacey & White, 1981).

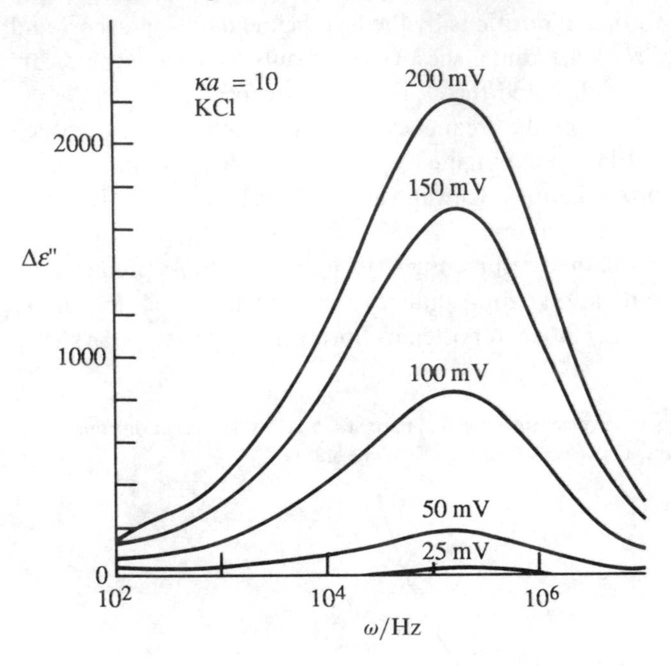

Fig. 7.19. Real part of the dielectric constant, ε', imaginary part of the dielectric constant, ε'', and conductivity, σ_*, for a suspension of latex particles (a=0.094 μm) in a concentrated suspension (Schwann *et al.*, 1962).

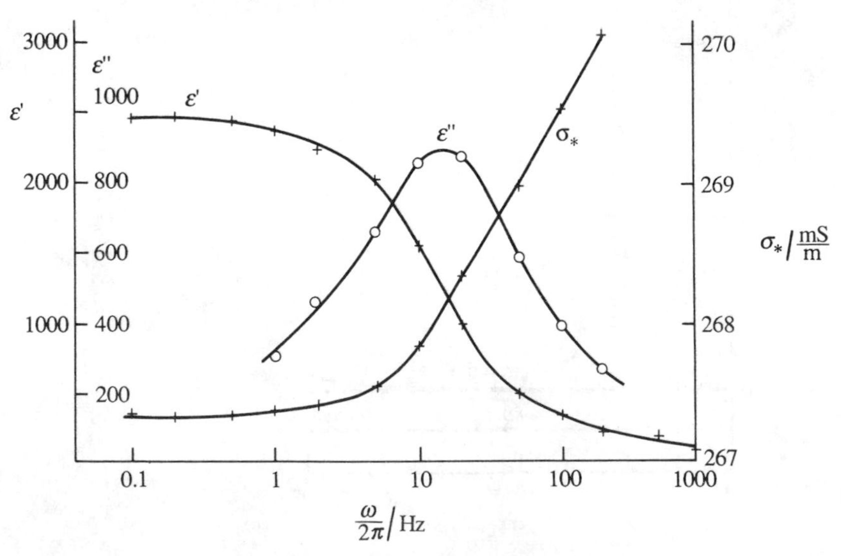

Fig. 7.20. Dielectric behavior of latex particles ($a = 0.095\,\mu\text{m}$) in 10^{-4} M HCl solutions at different latex concentrations $\kappa a = 3.1$, $\zeta = 81$ mV: (*a*) real part of the dielectric constant; (*b*) real dielectric increment (Myers & Saville, 1989).

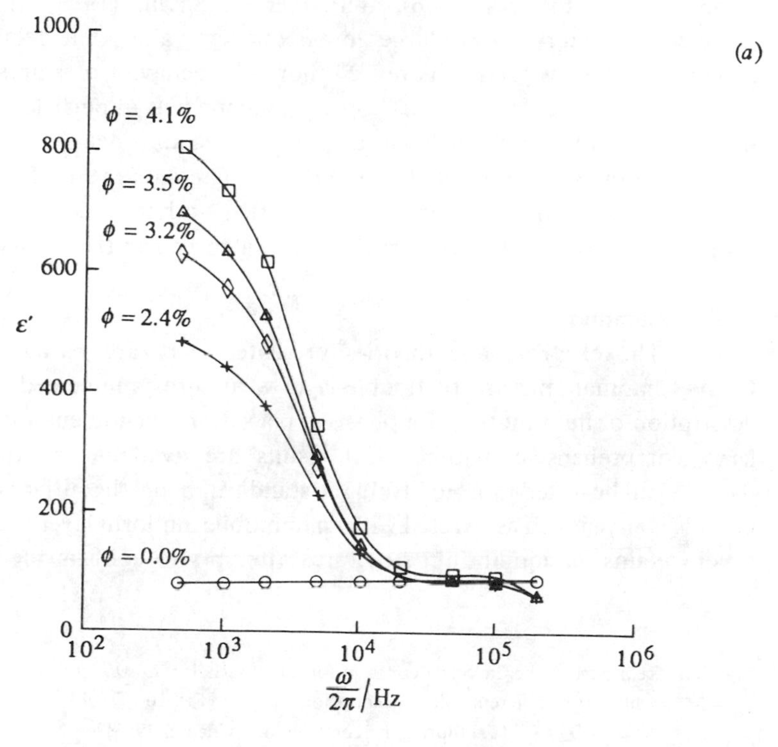

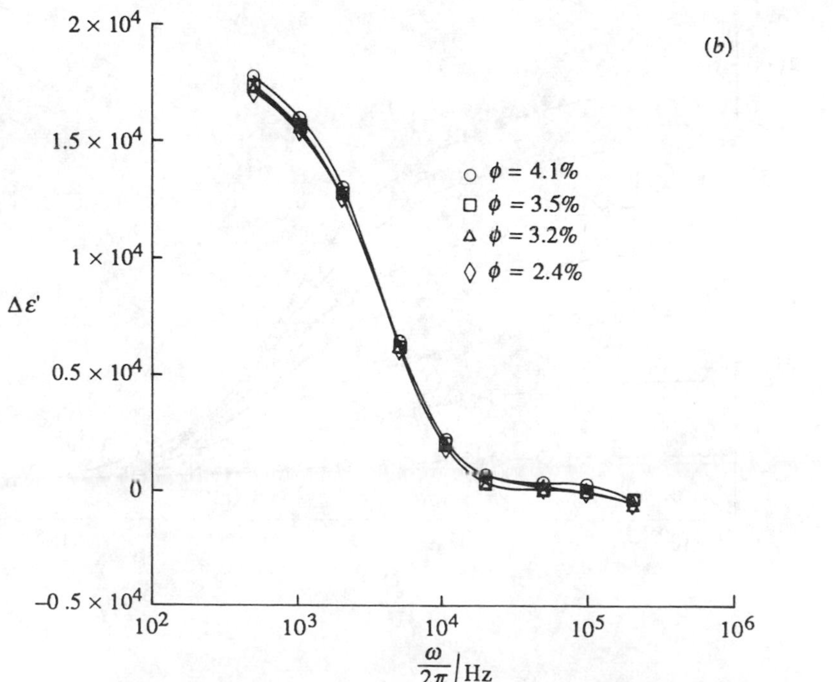

(1983), Lim & Franses (1986), and Myers & Saville (1989) show the expected behavior, viz. a large low-frequency value that relaxes at frequencies of a few KHz. Figure 7.20 shows the behavior of ε' and $\Delta\varepsilon'$ at four-particle concentrations. $\Delta\varepsilon'$ appears independent of particle concentration, indicating that ε' is linear in ϕ, but $\Delta\varepsilon'$ is an order of magnitude larger than the value predicted by the theory. The data in Fig. 7.21 show that ε' increases with κa, in qualitative agreement with the theory, but more data with fully characterized systems are needed to test the theory.

7.5 Summary

The electrokinetic theories presented here are based on the Gouy–Chapman picture of double-layer structure, augmented with a description of fluid motion. Emphasis is placed on phenomena for which fairly comprehensive mathematical results are available, so that the theories can be tested thoroughly. In the standard model the surface of each particle is envisioned as covered with an immobile, uniform layer of charge, which retains its equilibrium structure. Attempts to confirm the theory

Fig. 7.21. Real part of the dielectric constant for latex particles (a=0.22 μm) in KCl solutions at different salt concentrations: κa=0 (○), 10 (□), 14 (*), 18 (∇), 21 (◇), 25 (△), 27 (●) (Springer, Korteweg & Lyklema, 1983).

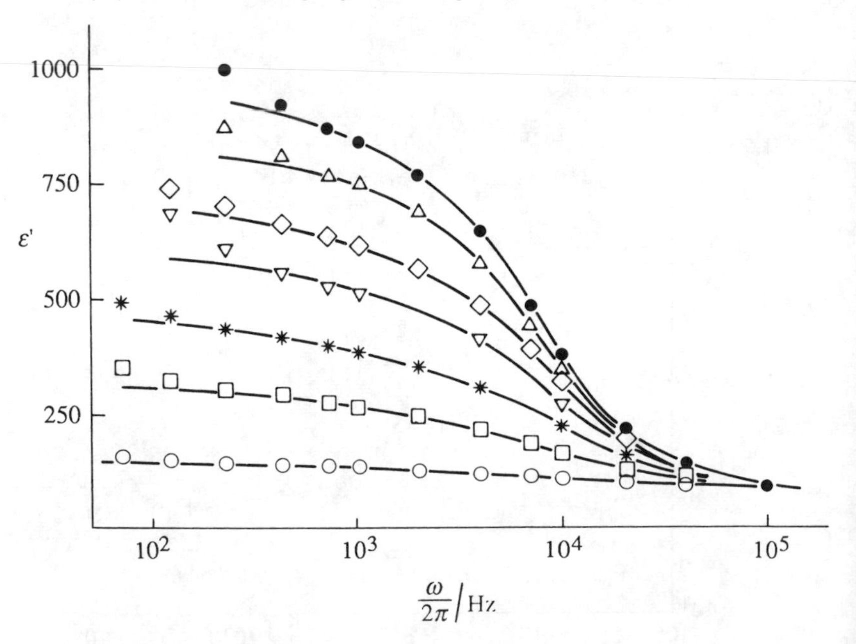

through experiments on latex particles lead to ambiguous results, in that the charge measured by electrophoresis appears to differ from that inferred from conductivity measurements, and is quite different from that measured by titration. The discrepancy appears to stem from ion migration behind the shear envelope. Whether or not similar behavior will appear with other particles is an open question. Experiments with some biological particles appear to yield satisfactory agreement between the charge determined from electrophoretic mobility measurements and titration. Clearly, much more experimental work will be needed to resolve the issue.

Dielectric relaxation seems to offer a powerful tool to probe electrokinetic behavior, but the experimental difficulties are significant. Data sufficient to test the theory quantitatively are not available. Nevertheless, the fact that one can examine the response of the system over a range of frequencies while keeping the chemical environment constant offers intriguing possibilities.

We have steered clear of concentrated systems for an obvious reason – current theories are neither sufficiently general nor rigorous enough to offer unambiguous results. Most of the problem stems from many-body interactions. This is not to imply that such systems are not important – they surely are – but that our understanding of their electrokinetic behavior is fragmentary.

References

Abramson, H. A., Moyer, L. S. & Gorin, M. H. (1942). *Electrophoresis of Proteins.* Reinhold.

Booth, F. (1950). The cataphoresis of spherical, solid non-conducting particles in a symmetrical electrolyte. *Proc. Roy. Soc. Lond. A* **203**, 533–51.

DeLacey, E. H. B. & White, L. R. (1981). The dielectric response and conductivity of a suspension of colloidal particles. *J. Chem. Soc. Faraday Trans. II* **77**, 2001–39.

Dukhin, S. S. & Derjaguin, B. V. (1974). *Electrokinetic Phenomena,* vol. 7 of *Surface & Colloid Science* (ed. E. Matijevic). Wiley.

Dukhin, S. S. & Semenikhin, N. M. (1970). Theory of double-layer polarization and its effect on the electrokinetic and electroptical phenomena and the dielectric constants of dispersed systems. *Kolloid Zh.* **32**, 360–8.

Dukhin, S. S. & Shilov, V. N. (1974). *Dielectric Phenomena and the Double Layer in Disperse Systems and Polyelectrolytes.* Wiley.

Dukhin, S. S. & Shilov, V. N. (1980). Kinetic aspects of electrochemistry of disperse systems II. Induced dipole moment and the non-equilibrium double-layer of a colloid particle. *Adv. in Colloid & Interface Sci.* **13**, 153–95.

Feynman, R. P., Leighton, R. B. & Sands, M. (1964). *The Feynman Lectures on Physics*, vol. II. Addison-Wesley.

Fricke, H. & Curtis, H. J. (1937). Dielectric properties of water–dielectric interphases. *J. Phys. Chem.* **41**, 729–45.

Healy, T. W. & White, L. R. (1978). Ionizable surface group models of aqueous interfaces. *Adv. Colloid Interface Sci.* **9**, 303–45.

Helmholtz, H. von (1879). Studien über electrische grenzschichten. *Ann. Phys.* **7**, 337–82.

Henry, D. C. (1931). The cataphoresis of suspended particles. I. The equation of cataphoresis. *Proc. Roy. Soc. Lond. A* **133**, 106–29.

Hinch, E. J., Sherwood, J. D., Chen, W. C. & Sen, P. N. (1984). Dielectric response of a dilute suspension of spheres with thin double layers in an asymmetric electrolyte. *J. Chem. Soc. Faraday Trans. II* **80**, 535–55.

Huckel, E. (1924). Die kataphorese der Kugel. *Phys. Z.* **25**, 204–10.

Hunter, R. J. (1981). *Zeta Potential in Colloid Science*. Academic Press.

Klinkenberg, A. & van der Minne, J. L. (1958). *Electrostatics in the Petroleum Industry*. Elsevier.

Lim, K.-H. & Franses, E. I. (1986). Electrical properties of aqueous dispersion of polymer microspheres. *J. Colloid Interface Sci.* **110**, 201–13.

Mandel, M. & Odjik, T. (1984). Dielectric properties of polyelectrolyte solutions. *Ann. Rev. Phys. Chem.* **35**, 75–108.

Maxwell, J. C. (1892). *A Treatise on Electricity and Magnetism*, vol. I, 3rd edn. Clarendon Press.

Mehrishi, J. (1972). Molecular aspects of the mammalian cell surface. *Progr. Biophys. Mol. Biol.* **25**, 1–70.

Myers, D. F. & Saville, D. A. (1989). Dielectric spectroscopy of colloidal dispersions. *J. Colloid Interface Sci* **131**, 448–60; 461–70.

O'Brien, R. W. (1981). The electrical conductivity of a dilute suspension of charged particles. *J. Colloid Interface Sci.* **81**, 234–48.

O'Brien, R. W. (1982). The response of a colloidal suspension to an alternating electric field. *Adv. in Colloid & Interface Sci.* **16**, 281–320.

O'Brien, R. W. (1983). The solution of electrokinetic equations for colloidal particles with thin double-layers. *J. Colloid Interface Sci.* **92**, 204–16.

O'Brien, R. W. & Perrins, W. T. (1984). The electrical conductivity of a porous plug. *J. Colloid Interface Sci.* **99**, 20–31.

O'Brien, R. W. & White, L. R. (1978). Electrophoretic mobility of a spherical colloidal particle. *J. Chem. Soc. Faraday Trans. II* **74**, 1607–26.

O'Konski, C. T. (1960). Electric properties of macromolecules. V. Theory of ionic polarization in polyelectrolytes. *J. Phys. Chem.* **64**, 605–19.

Ottewill, R. H. & Shaw, J. N. (1972). Electrophoretic studies on polystyrene latices. *J. Electroanal. Chem.* **37**, 133–42.

Overbeek, J. Th. G. (1943). Theorie der elektrophorese. *Kolloid-Beih.* **54**, 287–364.

Saville, D. A. (1977). Electrokinetic effects with small particles. *Ann. Rev. Fluid Mech.* **9**, 321–37.

Saville, D. A. (1979). Electrical conductivity of charged spheres in ionic solutions. *J. Colloid Interface Sci.* **71**, 477–90.

Saville, D. A. (1983). The electrical conductivity of suspensions of charged particles in ionic solutions: the roles of added counterions and non-specific adsorption. *J. Colloid Interface Sci.* **91**, 34–50.

Schwann, H. P., Schwartz, G., Maczuk, J. & Pauly, H. (1962). On the low-frequency dispersion of colloidal particles in electrolyte solutions. *J. Phys. Chem.* **66**, 2626–35.

Schwartz, G. (1962). A theory of the low-frequency dielectric dispersion of colloidal particles in electrolyte solution. *J. Phys. Chem.* **66**, 2636–42.

Seaman, G. V. F. (1975). Electrokinetic behavior of red blood cells in *The Red Blood Cell*, vol. II (ed. D. Surgernor), pp. 1135–229. Academic Press.

Shaw. D. J. (1969). *Electrophoresis.* Academic Press.

Smoluchowski, M. von (1903). Contribution à la théorie de l'endosmose électrique et de quelques phenomènes corrélatifs. *Bulletin International de l'Académie des Sciences de Cracovie.* **8**, 182–200.

Springer, M. M. A., Korteweg, A. & Lyklema, J. (1983). The relaxation of the double-layer around colloid particles and the low frequency dielectric dispersion. II. Experiments. *J. Electroanal. Chem.* **135**, 55–66.

van der Put, A. G. & Bijsterbosch, B. H. (1983). Electrical conductivity of dilute and concentrated aqueous dispersions of monodisperse polystyrene particles. *J. Colloid Interface Sci.* **75**, 512–24.

Watillon, A. & Stone-Masui, J. (1972). Surface conductance in dispersions of spherical particles. *J. Electroanal. Chem.* **37**, 143–60.

Wiersema, P. H., Loeb, A. L. & Overbeek, J. Th. G. (1966). Calculation of the electrophoretic mobility of a spherical colloid particle. *J. Colloid Interface Sci.* **22**, 70–99.

Zukoski, C. F. & Saville, D. A. (1985). An experimental test of electrokinetic theory. *J. Colloid Interface Sci.* **107**, 322–33.

Zukoski, C. F. & Saville, D. A. (1986). The interpretation of electrokinetic measurements using a dynamic model of the Stern layer. I. The dynamic model. *J. Colloid Interface Sci.* **114**, 32–44. II. Comparisons between theory and experiment. *Ibid.*, 45–53.

Zukoski, C. F. & Saville, D. A. (1987). Electrokinetic properties of particles in concentrated suspensions. *J. Colloid Interface Sci.* **115**, 422–36.

Problems

1 Smoluchowski's formula for the mobility of a sphere with a thin double layer can be derived in several ways. Use the method employed in connection with (7.2.39) to derive the formula.

2 When a particle settles through a viscous fluid, the velocity field far from the particle decays as r^{-1} due to the influence of viscosity (see §2.5 or 2.6). A different decay rate is found when a particle with a thin diffuse layer undergoes electrophoresis where the velocity field decays as r^{-3}.

Show that the flow outside the diffuse layer is irrotational when a particle undergoes electrophoresis and hence the velocity field is that given by 2.3.10.

3 Derive the formula for the velocity field for electro-osmosis in a closed, long, thin channel, i.e. (7.2.40).

The velocity of a particle is the sum of the electrophoretic velocity and the local velocity of the fluid. Use the formula just derived to calculate the electro-osmotic velocity and the electrophoretic velocity from the following data by plotting the particle velocity against $1-(y/h)^2$.

Data on particle velocities (μm/s) measured at 5.92 V/cm in a chamber with $2h = 1.07$ mm; $2a = 4.5$ μm; the salt strength is 0.045 M.

y/h:	−0.915	−0.770	−0.531	−0.407	−0.159	−0.034	0.090	0.214	0.338	0.586	0.834	0.958
u_s:	−15.3	−7.96	7.01	10.83	17.76	18.53	17.91	16.20	13.90	4.6	−10.80	−19.66

Check your result for the particle mobility by showing that the average particle velocity is equal to the electrophoretic velocity.

Calculate the ζ-potentials of the particles and the chamber walls.

4 To gain insight into the way the diffuse charge cloud alters the behavior of a suspension, a 'surface conductivity' model can be used. Here the diffuse layer is envisioned to be a thin, homogeneous layer, with a conductivity different from the bulk fluid. This changes the boundary condition at the surface of a sphere so that the normal component of the current is balanced by conduction along the surface. Derive an expression for the boundary condition, calculate the electric potential inside and outside a sphere, and derive the counterpart of (7.3.8) for situations where surface conduction is present.

5 The Maxwell–Wagner theory can also be modified to take account of surface conduction. Derive expressions for the free charge and polarization charge densities at the surface of a sphere when there is surface conduction, cf. (7.4.12) and (7.4.13). Plot the charge densities as functions of frequency for several values of $\sigma_s/\sigma a$; σ_s denotes the surface conductivity and σ stands for the conductivity of the bulk solution.

Derive an expression for the complex conductivity, (7.4.14), when surface conduction takes place. Graph the conductivity increment, $\Delta\sigma$, and the dielectric constant increment, $\Delta\varepsilon$, as functions of frequency for several values of $\sigma_s/\sigma a$. Compare the magnitudes with those measured for suspensions of polystyrene latex particles.

8

ELECTROSTATIC STABILIZATION

8.1 Introduction

As noted in Chapter 5, dispersion forces acting between similar particles suspended in a chemically different liquid are inevitably attractive, providing a driving force toward macroscopic phase separation. Hence, maintenance of a dispersed state requires an opposing interparticle repulsion, most commonly achieved through electrostatic forces in aqueous dispersions or the adsorption of soluble polymer in either aqueous or non-aqueous environments. Since all characteristics of colloidal systems change markedly in the transition from the dispersed to the aggregated state, the question of stability occupies a central position in colloid science (e.g. Verwey & Overbeek, 1948; Napper, 1983; Hunter, 1987).

Even among unstable or aggregated systems, the nature or degree of aggregation varies. Following the convention of La Mer (1964), many authors have attempted to distinguish between flocculation, referring to loose aggregation, with highly porous flocs and/or particles held relatively far apart, and coagulation, with more closely packed flocs of particles in contact. Unfortunately, floc structure has been quantified only recently, leaving the classification ambiguous in many cases.

In the following, we distinguish instead on the basis of the strength of the attractive potential responsible for aggregation. Then the criterion becomes whether the system attains equilibrium in the period of interest. For attractions strong relative to the thermal energy kT, Brownian motion eventually eliminates all individual particles, producing a non-equilibrium phase whose structure is governed by the range of the attractive potential and the mode of aggregation. This we refer to as flocculation. In the other limit, attractions with magnitudes comparable to kT lead to an equilibrium phase separation in which particles coexist with equal chemical potentials in two distinct phases. In the dilute phase, individual particles have sufficient entropy to compensate for the lower internal energy of those in the condensed phase. The structure of the latter ranges from disordered, i.e. a fluid, to ordered, as for a solid, depending on the range and magnitude of the attraction.

This distinction is also imprecise, since all systems evolve toward equilibrium over time. However, we can convey the current level of understanding by classifying systems according to their sensitivity to changes in the solution chemistry. Those transformed from a dispersed to a flocculated state by a relatively small change in conditions are identified by the mode of stabilization, electrostatic or polymeric. This irreversible change in state accompanies the appearance of an accessible attractive minimum in the pair potential with $-\Phi_{min}/kT \geq 5$–10. Particles reaching this minimum require very long times, $\geq a^2/D_0 \exp(-\Phi_{min}/kT)$, to rearrange into an equilibrium condensed phase. Equilibrium phase transitions occur on reasonable time scales only for interaction potentials with $-\Phi_{min}/kT \approx 1$–3.

Hence, this chapter addresses electrostatic stabilization of aqueous dispersions and the flocculation due to dispersion forces when electrolyte suppresses the repulsion. Chapter 9 presents a similar but more limited treatment of stabilization by adsorbed or terminally anchored polymer. In that case, solvent quality is the analogue of electrolyte concentration. Chapter 10 then treats equilibrium phase transitions that arise from either dispersion or polymer-induced forces in systems that allow a gradual variation in the strength of the attraction.

Early experiments (Schulze, 1882, 1883; Hardy, 1900; Freundlich, 1910) demonstrated the ability of electrolyte to flocculate a variety of aqueous dispersions, suggesting that stability depends primarily on the magnitude or range of the electrostatic repulsion. Shortly thereafter, Smoluchowski (1917) derived expressions for the rates of doublet formation in unstable dispersions due to Brownian and shear-induced collisions, respectively. At that time, however, neither the origin of the attraction nor the functional form of the repulsion were known. Only after the interparticle potentials described in §4.3 and §5.2 became available could Derjaguin & Landau (1941) and Verwey & Overbeek (1948) synthesize their theory of colloid stability to rationalize the observations.

We first assess the stabilization of quiescent dispersions on the basis of the magnitude of the electrostatic repulsion relative to the dispersion attraction. This provides some useful rules of thumb and illustrates the inherent thermodynamic instability of such systems. A repulsive barrier large relative to the thermal energy kT produces only kinetic stability, i.e. a metastable state with a small but finite rate of flocculation. Subsequent sections then analyze the rate processes leading to the formation of doublets in order to predict the degree of stability. For Brownian flocculation, comparison of the predictions with experimental results illustrates the general success of the theory and identifies some lingering questions. Extension of the approach to the growth of larger flocs is outlined along with the requisite assumptions. Comparison with several sets of recent data, including measurements of floc structure, suggests a predictive capability. For shear induced flocculation the theory is more difficult and not yet complete, but establishes some guidelines with respect to both the enhancement of rates in unstable dispersions and mechanical stability. The aggregation of unlike particles due to both Brownian motion (Hogg, Healy & Fuerstenau, 1966) and shear (Adler, 1981), termed heteroflocculation, can be treated similarly.

8.2 Interparticle potential and criteria for stability

In aqueous systems without dissolved polymer, dispersion and electrostatic interactions generally determine the interparticle potential. As described earlier, both arise from electromagnetic phenomena, the former from correlations in fluctuating fields induced by dielectric differences between the particles and the fluid and the latter from fields due to surface charges acquired by most particles in polar fluids such as water. Superposition of the two provides a dimensionless potential, Φ/kT, which depends on A_{eff}/kT, the effective Hamaker constant; $\varepsilon\varepsilon_0\psi_s^2 a/kT$ and $a\kappa$, characterizing the magnitude and range of the repulsion; and r/a, the separation.

The interaction potentials in Fig. 8.1 reflect the results of superimposing the linearized Derjaguin approximation with constant potential boundary conditions (Table 4.3) for the electrostatic repulsion and the approximations described in §§5.2 and 5.10 for the attraction. A finite potential at contact can be obtained by specifying a minimum separation, $\delta=(r-2a)_{min}=0.1$–0.2 nm, to account for molecular dimensions not recognized by the continuum theory (Hough & White, 1980). The curves illustrate the effects of both ionic strength and the Hamaker constant with the latter chosen to model metals such as gold with $A_{eff}(0)/kT=25$ and polystyrene with $A_{eff}(0)/kT=2.5$.

The key features of the interaction potentials are the primary and secondary minima, Φ_{min} and Φ_{sec}, and the primary maximum, Φ_{max}. In the absence of other short-range forces, the deep primary minimum,

$$-\frac{\Phi_{min}}{kT}=\frac{A_{eff}(0)}{12kT}\frac{a}{\delta}-2\pi\ln 2\frac{\varepsilon\varepsilon_0\psi_s^2 a}{kT}\gg 1, \tag{8.2.1}$$

comprises the thermodynamically stable state. Pairs residing there would be bound essentially irreversibly, with very slow escape via thermal fluctuations alone. However, at moderate to low ionic strengths the repulsive barrier is large, $\Phi_{max}/kT\gg 1$, making diffusion of initially dispersed particles into the primary minimum very slow as well. Thus the possibility exists for stabilization in a kinetic sense, i.e., preservation of a dispersed non-equilibrium state. At high ionic strengths the repulsive barrier disappears permitting rapid and irreversible flocculation into the primary minimum.

The transition from kinetic stability to rapid flocculation occurs over a narrow range of ionic strengths, including the critical flocculation concentration n_{crit} at which $\Phi_{max}/kT=0$. Typically (e.g. Fig. 8.1), the corresponding Debye length satisfies $a\kappa\gg 1$ and Φ_{max} appears at $\kappa(r-2a)\geq O(1)$. Thus n_{crit} can be calculated more generally from the non-linear superposition

Fig. 8.1. Interaction potentials for spheres with $a=0.1$ μm and $e\psi_s/kT=1.0$ in water at a range of ionic strengths: (*a*) Sols with $A_{eff}(0)/kT=25$ (Verwey & Overbeek, 1948). (*b*) Polystyrene latices with $A_{eff}(0)/kT=2.5$ (Parsegian, 1975).

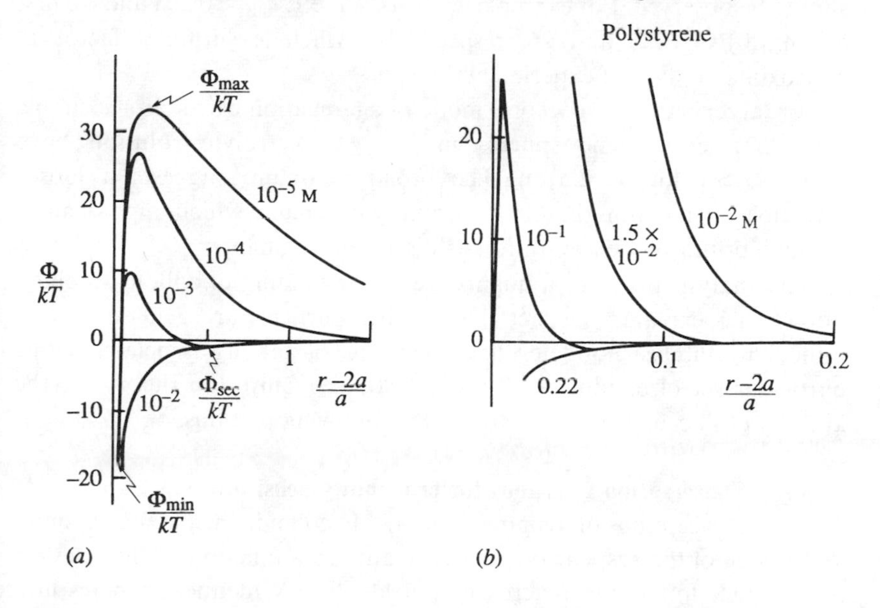

approximation for the electrostatic potential (Table 4.3) and the non-retarded limit of the dispersion potential (5.10.1) (without the $n=0$ term). For a symmetric electrolyte,

$$\frac{\Phi}{kT}=\frac{8a}{z^2 l_b}\tanh^2\frac{ez\psi_s}{4kT}\exp[-\kappa(r-2a)]-\frac{aA_{eff}(0)}{12kT(r-2a)}, \tag{8.2.2}$$

with $l_b=e^2/4\pi\varepsilon\varepsilon_0 kT$ and κ^{-1}, the Debye length, related to the electrolyte concentration by (4.7.2). Setting $\Phi=-d\Phi/dr=0$ determines $\kappa(r-2a)=1$ and

$$n_{crit}=\frac{49.6}{z^6 l_b^3}\left[\frac{kT}{A_{eff}(0)}\right]^2\tanh^4\frac{ez\psi_s}{4kT}. \tag{8.2.3}$$

This prediction of the critical flocculation concentration has several interesting features. For $ez\psi_s/4kT>1$, $n_{crit}\propto z^{-6}$, explaining the Schulze–Hardy rule suggested by early experiments (Verwey & Overbeek, 1948, Ch. VII, §2). But when $ez\psi_s/4kT\ll 1$, $n_{crit}\propto\psi_s^2 z^{-2}$. Also, this simple formula accounts only implicitly for the effect of chemistry in controlling the variation of ψ_s with ionic strength and ion type. In fact, multivalent ionic species can form complexes in solution or adsorb onto surfaces to change the magnitude and even the sign of the surface charge. Thus the z^{-6} dependence is far from universal. With univalent electrolytes, (8.2.3) indicates semi-quantitatively the concentrations at which flocculation should be observed. For example, with $e\psi_s/kT\approx 1$, $n_{crit}\approx 0.2$ M and 1.3 mM for $A_{eff}(0)/kT=2.5$ and 25, respectively, which is within a factor of approximately two of experimental results.

For larger particles a second mode of aggregation exists. For example (Fig. 8.2), polystyrene spheres in 10^{-3} M electrolyte solution have $-\Phi_{sec}/kT>1$ for $a>2.5\,\mu$m. The broad minimum suggests a looser reversible aggregation than in the primary minimum, which might lead to an equilibrium phase separation as discussed in Chapter 10.

This analysis provides a qualitative understanding of colloid stability, but neither explains the fact that smaller particles are generally more difficult to stabilize, nor determines the degree of stability associated with a particular set of conditions. These questions require the theory for the kinetics of flocculation described in the following sections.

8.3 Conservation equations for probability densities

Descriptions of dynamic processes in colloidal dispersions require knowledge of the suspension microstructure as a function of the relevant interparticle forces and macroscopic fields. For N identical spheres in a volume V (Fig. 8.3), the function $P_N(\mathbf{x}_1,\ldots,\mathbf{x}_N,t)$ specifies the probability

Fig. 8.2. Interaction potentials between polystyrene spheres with $\psi_s = 25\,\mathrm{mV}$ as a function of radius at $I = 10^{-3}\,\mathrm{M}$, constructed from the linearized Derjaguin approximation for the electrostatic repulsion and A_{eff} for flat plates combined with the Hamaker geometrical factor for the dispersion attraction.

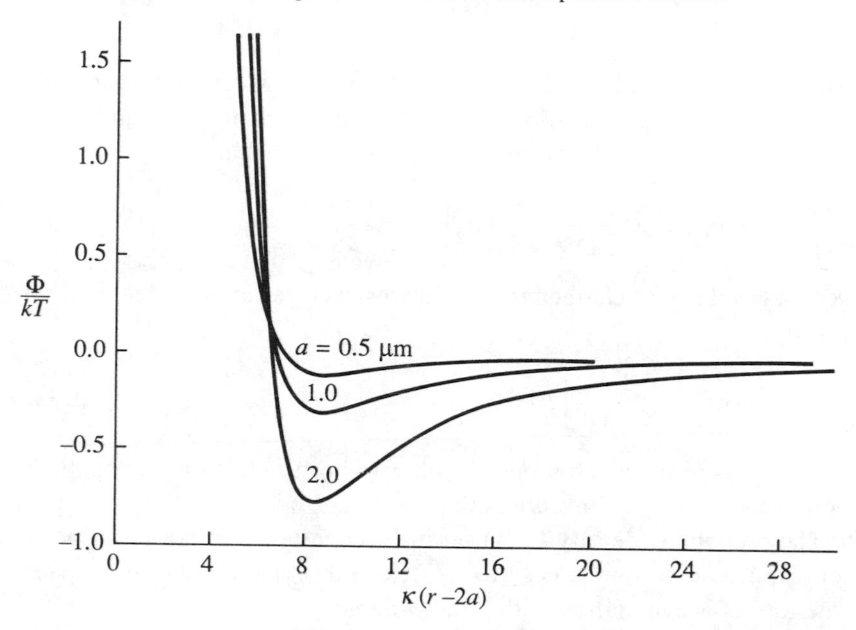

Fig. 8.3. Coordinate system for N identical spheres in volume V.

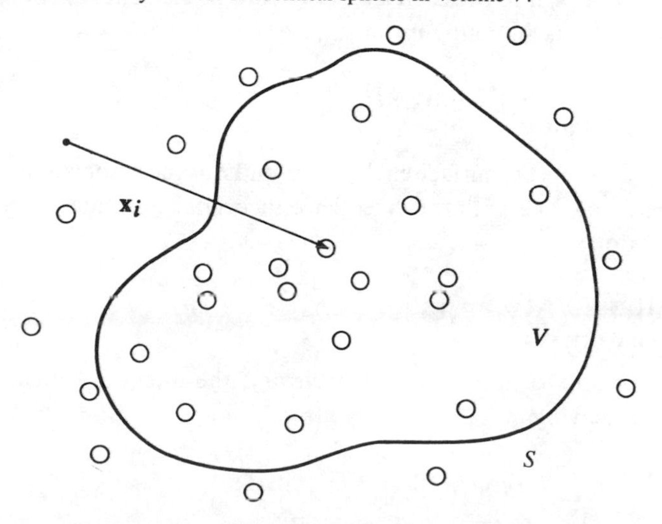

of finding spheres centered simultaneously at $(\mathbf{x}_1, \ldots, \mathbf{x}_N)$. For indistinguishable spheres, the appropriate normalization is

$$\int P_N \, \mathrm{d}\mathbf{C}_N = N!,$$

with $\mathrm{d}\mathbf{C}_N = \mathrm{d}\mathbf{x}_1 \ldots \mathrm{d}\mathbf{x}_N$ (McQuarrie, 1976, §13–2). Most analyses require considerably less information, though, as suggested by the equilibrium theories in Chapter 10. For example, the pair probability

$$P_2(\mathbf{x}_1, \mathbf{x}_2, t) = \frac{1}{(N-2)!} \int P_N \, \mathrm{d}\mathbf{C}_{N-2} \tag{8.3.1}$$

generally suffices to characterize the microstructure, and the local number density,

$$n(\mathbf{x}_1) = \frac{1}{(N-1)!} \int P_N \, \mathrm{d}\mathbf{C}_{N-1}, \tag{8.3.2}$$

corresponds to the macroscopically observed concentration. For spatially homogeneous dispersions, $n = N/V$.

The probability density P_N must satisfy a conservation law constructed for a differential volume $(\mathbf{x}_1 \pm \mathrm{d}\mathbf{x}_1, \ldots, \mathbf{x}_N \pm \mathrm{d}\mathbf{x}_N)$ in configuration space. The rate of accumulation within this volume

$$\int \frac{\partial P_N}{\partial t} \mathrm{d}\mathbf{C}_N$$

is non-zero if there exist net fluxes $P_N \mathbf{U}_i$, with $\mathbf{U}_i$ the translational velocity of the particle at $\mathbf{x}_i$ into the corresponding volume elements $\mathrm{d}\mathbf{x}_i$ in physical space. The net flux in configuration space,

$$-\sum_{i=1}^{N} \int_{A_i} \int P_N \mathbf{U}_i \cdot \mathbf{n}_i \, \mathrm{d}\mathbf{x}_i \, \mathrm{d}\mathbf{C}_{N-1},$$

where A_i represents the surface enclosing $\mathbf{x}_i$ and $\mathbf{n}_i$ is the associated outward unit normal, together with the divergence theorem determine the conservation equation

$$\int \left(\frac{\partial P_N}{\partial t} + \sum_{i=1}^{N} \nabla_i \cdot P_N \mathbf{U}_i \right) \mathrm{d}\mathbf{C}_N = 0. \tag{8.3.3}$$

Since the volume of integration is arbitrary, the integrand must vanish everywhere, leaving

$$\frac{\partial P_N}{\partial t} + \sum_{i=1}^{N} \nabla_i \cdot P_N \mathbf{U}_i = 0. \tag{8.3.4}$$

The pair conservation equation governing P_2 is derived by averaging (8.3.4) over the positions of the other $N-2$ spheres. In general, couplings exist between the fluxes of any pair and the positions of the other $N-2$ spheres. However, at dilute volume fractions, $\phi = 4\pi a^3 n/3 \ll 1$,

$$\frac{1}{(N-2)!}\int \nabla_i \cdot P_N \mathbf{U}_i \, d\mathbf{C}_{N-2} = \nabla_i \cdot P_2 \mathbf{U}_i \{1 + O(\phi)\}$$

for $i = 1, 2$, with $\mathbf{U}_i$ depending only on $\mathbf{x}_1$ and $\mathbf{x}_2$. The corresponding terms for $i \neq 1, 2$ represent three-body couplings which are also $O(\phi)$ smaller than the direct pair interactions. Thus, with (8.3.1) the pair conservation equation takes the form

$$\frac{\partial P_2}{\partial t} + \nabla_1 \cdot P_2 \mathbf{U}_1 + \nabla_2 \cdot P_2 \mathbf{U}_2 = 0. \tag{8.3.5}$$

Transforming into center-of-mass coordinates, with $\mathbf{x} = (\mathbf{x}_1 + \mathbf{x}_2)/2$ and $\mathbf{r} = \mathbf{x}_2 - \mathbf{x}_1$, yields

$$\frac{\partial P_2}{\partial t} + \nabla_{\mathbf{x}} \cdot P_2 \mathbf{U}_{\mathbf{x}} + \nabla_{\mathbf{r}} \cdot P_2 \mathbf{U}_r = 0 \tag{8.3.6}$$

with $\mathbf{U}_{\mathbf{x}} = (\mathbf{U}_1 + \mathbf{U}_2)/2$ and $\mathbf{U}_{\mathbf{r}} = \mathbf{U}_2 - \mathbf{U}_1$.

The velocities, arising from convection by an imposed velocity gradient $\nabla \mathbf{u}_\infty$, any external or interparticle forces $\mathbf{F}_i$, and Brownian motion, can be expressed as

$$\begin{aligned}\mathbf{U}_i = \nabla \mathbf{u}_\infty \cdot \mathbf{x}_i - \mathbf{C}_{ij}(\mathbf{r}_{ij}) \cdot \mathbf{E} \cdot \mathbf{r}_{ij} + \boldsymbol{\omega}_{11}(\mathbf{r}_{ij}) \cdot (\mathbf{F}_i - kT\nabla_i \ln P_2) \\ + \boldsymbol{\omega}_{12}(\mathbf{r}_{ij}) \cdot (\mathbf{F}_j - kT\nabla_j \ln P_2)\end{aligned} \tag{8.3.7}$$

with $i, j = 1, 2$ but $i \neq j$ and $\mathbf{E} = (\nabla \mathbf{u}_\infty + \nabla \mathbf{u}_\infty^T)/2$. The hydrodynamic functions accounting for interactions between spheres in a shear flow, $\mathbf{C}_{ij}$, and relating the translational velocities to the applied forces, $\boldsymbol{\omega}_{11}$ and $\boldsymbol{\omega}_{12}$, are defined in §2.8 and 2.9. The interparticle forces $\mathbf{F}_i$ correspond to those discussed in Chapters 4 to 6. The effective Brownian force, $-kT\nabla_i \ln P_2$, derives from the analysis in §3.5, which identified the diffusion coefficient as kT times the hydrodynamic mobility. Equating the flux expressed as, for example, $-kT\boldsymbol{\omega}_{ji} \cdot \nabla_i P_2$ to $P_2 \boldsymbol{\omega}_{ji} \cdot \mathbf{F}_i$ leads to this expression for the force.

In the absence of spatial gradients in the pair density or an external force

$$\nabla_{\mathbf{x}} \cdot (P_2 \mathbf{U}_{\mathbf{x}}) = 0,$$

leaving the pair conservation equation as

$$\frac{\partial P_2}{\partial t} + \nabla_{\mathbf{r}} \cdot P_2(\mathbf{U} - \boldsymbol{\omega} \cdot \nabla_r \Phi) = \nabla_{\mathbf{r}} \cdot \boldsymbol{\omega} kT \cdot \nabla_r P_2, \tag{8.3.8}$$

with

$$\mathbf{U} = \nabla\mathbf{u}_\infty \cdot \mathbf{r} - \mathbf{C} \cdot \mathbf{E} \cdot \mathbf{r},$$
$$\boldsymbol{\omega} = 2(\boldsymbol{\omega}_{11} - \boldsymbol{\omega}_{12})$$
$$\equiv \frac{1}{3\pi\mu a}\left\{G(r)\frac{\mathbf{rr}}{r^2} + H(r)\left(\boldsymbol{\delta} - \frac{\mathbf{rr}}{r^2}\right)\right\}.$$

This provides the basis for addressing a variety of dynamic processes in dilute suspensions, from the kinetics of flocculation in this chapter to the rheological response of dispersions in Chapter 14.

To set the stage for these analyses, we first identify dimensionless groups that gauge the relative importance of the individual driving forces. For a process with time scale t_p, both formal scaling of the equation and dimensional analysis lead to five groups for systems in which only electrostatic and dispersion forces act between the particles:

$$\frac{t_d}{t_p} = \frac{3\pi\mu a^3}{kTt_p},$$

$$Pe = \frac{3\pi\mu a^3}{kT}\gamma,$$

$$N_f = \frac{6\pi\mu a^3}{A_{eff}(0)}\gamma \quad \text{or} \quad \frac{A_{eff}(0)}{kT},$$

$$N_r = \frac{\varepsilon\varepsilon_0\psi_s^2 a}{A_{eff}(0)} \quad \text{or} \quad \frac{\varepsilon\varepsilon_0\psi_s^2 a}{kT},$$

$$a\kappa$$

with $\gamma = (\mathbf{E}:\mathbf{E})^{1/2}$. The ratio of the diffusion time t_d to the process time indicates whether the explicit time dependence of P_2 will be significant. Pe and N_f assess the importance of convection relative to diffusion and the dispersion force, respectively. N_r and $1/a\kappa$ compare, respectively, the magnitudes and the ranges of the electrostatic and dispersion forces.

To illustrate typical magnitudes, we consider spheres with $a = 0.1$–$1.0\,\mu\text{m}$, $A_{eff}(0)/kT = 2.5$–25, and $e\psi_s/kT = 1$ in water. The corresponding diffusion times, $t_d = 0.0025$–2.5 s, indicate that a pseudo-steady theory suffices for process times exceeding a few seconds. For flowing systems, Brownian motion dominates for most conditions with the smaller particles (e.g. $Pe < 1$ for $\gamma < 400\,\text{s}^{-1}$), but for the larger particles shear is normally more important since $Pe > 1$ for $\gamma > 0.4\,\text{s}^{-1}$. The relative importance of the interparticle forces varies considerably over this range of sizes and Hamaker constants, with $N_r = 0.4$–40 and $N_f = (0.08$–$0.8)Pe$. Thus each of the forces can be important, with the dominant ones depending on the specific conditions. In addition, these order-of-magnitude estimates, while useful, overlook the quite different dependence of the individual forces on separation, as illustrated in §8.2 by the variation of Φ_{max}/kT with $a\kappa$ as well as N_r.

Subsequent sections address flocculation under conditions corresponding to limiting values of these dimensionless groups. For example, purely Brownian flocculation occurs for $Pe=N_f=0$, i.e. no flow. Then rapid or diffusion-limited aggregation refers to $N_r=0$ or $a\kappa > a\kappa_{crit}$, while $N_r \gg 1$ and $a\kappa < a\kappa_{crit}$ leads to a slow or reaction-limited process (§8.4). At small but finite Pe, shear enhances the rate somewhat (§8.8). For $Pe \gg 1$, however, collisions due to shear dominate Brownian encounters. Without repulsion ($N_r=0$), the rate depends primarily on N_f, with Brownian motion either enhancing or retarding the process slightly (§8.8). With no Brownian motion ($Pe=\infty$), repulsion can prevent collisions, making the mechanical stability of a dispersion a complex function of N_f, N_r, and $a\kappa$ (§8.9).

8.4 Initial stage of Brownian flocculation

Flocculation is a transient, non-equilibrium process that eventually converts a dispersion of individual particles into a disordered solid. At intermediate stages, aggregates range in size from a few to many thousands of particles and in structure from dense flocs to tenuous networks. Here we examine doublet formation, the initial step of the process. This rigorous theory comprises the basis for the quantitative understanding of colloid stability. Section 8.7 then treats the generalization, in a more approximate fashion, to the growth of larger aggregates.

The theory for doublet formation has evolved in three stages. First, Smoluchowski (1917) calculated the rate of rapid flocculation J_0 (doublets/volume–time) from the free diffusion of spheres. Fuchs (1934) then extended the theory formally, allowing the flux J to depend on an arbitrary interaction potential and expressing the results in terms of the stability ratio $W=J_0/J$. Derjaguin & Landau (1941) and Verwey & Overbeek (1948) subsequently incorporated the dispersion attraction and electrostatic repulsion to calculate W explicitly. Since $J \ll J_0$ when $\Phi_{max}/kT \gg 1$, their calculations provided a quantitative measure of stability. Finally, at the suggestion of Derjaguin (Derjaguin, 1966; Derjaguin & Muller, 1967), Spielman (1970) and Honig, Roebersen & Wiersema (1971) included hydrodynamic interactions in calculating the rates of rapid or diffusion-limited ($J \approx J_0$ and slow or reaction-limited ($J < J_0$) flocculation. We present this final version of the theory below.

The objective of the theory is to predict the rate at which individual spheres disappear, by colliding to form doublets. In a homogeneous suspension, the conservation equation for the singlet density $n(t)$ is derived by multiplying (8.3.4) by $1/(N-1)!$ and integrating over the positions of $N-1$ spheres. The derivation requires noting that $P_N=0$ for those volumes in configuration space in which particles overlap, i.e. $r_{ij}=|\mathbf{x}_i-\mathbf{x}_j|<2a$ for $i\neq j$. The integrals over the remaining volume can then be converted to integrals over the surfaces $r_{ij}=2a$ which move with velocity $\mathbf{U}_j$. For example, application of the transport theorem together with the dilute approximation converts the integral of the transient term to the following:

$$\begin{aligned}
&\frac{1}{(N-1)!}\int\frac{\partial P_N}{\partial t}\,\mathrm{d}\mathbf{C}_{N-1}\\
&=\frac{\mathrm{d}n}{\mathrm{d}t}+\sum_{i\neq 1}\sum_{j\neq i}\frac{1}{(N-1)!}\int_{r_{ij}=2a}P_N\mathbf{U}_j\cdot\mathbf{n}\,\mathrm{d}\mathbf{x}_i\mathrm{d}\mathbf{C}_{N-2}\\
&=\frac{\mathrm{d}n}{\mathrm{d}t}+\int_{r_{12}=2a}P_2\mathbf{U}_1\cdot\mathbf{n}\,\mathrm{d}\mathbf{x}_2\{1+O(\phi)\}.
\end{aligned}\tag{8.4.1}$$

Then with the divergence theorem the remaining term takes the form

$$\begin{aligned}
&\sum_i\frac{1}{(N-1)!}\int\nabla_i\cdot P_N\mathbf{U}_i\,\mathrm{d}\mathbf{C}_{N-1}\\
&=\nabla_1\cdot\int P_N\mathbf{U}_1\,\mathrm{d}\mathbf{C}_{N-1}+\sum_{i\neq 1}\frac{1}{(N-1)!}\int_{r_{i1}=2a}\int P_N\mathbf{U}_i\cdot\mathbf{n}\,\mathrm{d}\mathbf{x}_i\,\mathrm{d}\mathbf{C}_{N-2}\\
&=-\int_{r_{21}=2a}P_2\mathbf{U}_2\cdot\mathbf{n}\,\mathrm{d}\mathbf{x}_2\{1+O(\phi)\}
\end{aligned}\tag{8.4.2}$$

for a dilute, spatially homogeneous dispersion with $\mathbf{n}=(\mathbf{r}_i-\mathbf{r}_j)/2a$.

Substituting (8.4.1) and (8.4.2) into the integrated form of (8.3.4) leads to

$$\frac{\mathrm{d}n}{\mathrm{d}t}=-J,\tag{8.4.3}$$

with

$$J=-\int_{r=2a}P_2\{\mathbf{U}-\boldsymbol{\omega}\cdot\nabla_r(\Phi+kT\ln P_2)\}\cdot\mathbf{n}\,\mathrm{d}\mathbf{r}$$

and $n=n_0$ at $t=0$. Lubrication stresses between two spheres approaching one another cause both $\mathbf{U}\cdot\mathbf{n}$ and $\boldsymbol{\omega}\cdot\mathbf{n}$ to vanish as $r-2a\to 0$; thus, at least in a mathematical sense, flocculation occurs only if $\mathrm{d}\Phi/\mathrm{d}r \to \infty$ at contact as for dispersion forces. The form of (8.4.3) defines the time scale for the depletion of singlets as $t_p = 2n_0/J$, indicating that the pair density will achieve a pseudo-steady state if $3\pi\mu a^3 J/2n_0 kT \ll 1$.

Calculation of the rate J requires solution of (8.3.8) for P_2 subject to the boundary conditions

$$P_2 = \begin{cases} n^2 & r \to \infty \\ 0 & r = 2a. \end{cases} \tag{8.4.4}$$

The former assumes a random, spatially homogeneous microstructure and the latter irreversible flocculation. With the pseudo-steady approximation and the spherical symmetry of the boundary conditions and the interparticle forces for Brownian flocculation ($\mathbf{U} = \mathbf{0}$), P_2 should depend only on r, reducing (8.3.8) to

$$\frac{1}{r^2}\frac{\mathrm{d}}{\mathrm{d}r} r^2 G\left(P_2 \frac{\mathrm{d}}{\mathrm{d}r}\frac{\Phi}{kT} + \frac{\mathrm{d}P_2}{\mathrm{d}r}\right) = 0. \tag{8.4.5}$$

Integration and application of the boundary conditions determines

$$P_2 = n^2 \frac{\exp(-\Phi/kT)\displaystyle\int_{2a}^{r} \frac{\exp(\Phi/kT)}{r^2 G(r)}\,\mathrm{d}r}{\displaystyle\int_{2a}^{\infty} \frac{\exp(\Phi/kT)}{r^2 G(r)}\,\mathrm{d}r}. \tag{8.4.6}$$

The flux follows from (8.4.3) as

$$J = \frac{4kT}{3\mu a} \frac{n^2}{\displaystyle\int_{2a}^{\infty} \frac{\exp(\Phi/kT)}{r^2 G(r)}\,\mathrm{d}r}. \tag{8.4.7}$$

A useful estimate for the maximum rate of flocculation derives from setting $\Phi = 0$ and $G = 1$. Then

$$P_2 = n^2\left(1 - \frac{2a}{r}\right),$$

and

$$J_0 = \frac{8kTn^2}{3\mu} = n\frac{2\phi kT}{\pi\mu a^3}. \tag{8.4.8}$$

Table 8.1. *Time scales for doublet formation in water with* $W=1$

t_p	$\phi=10^{-5}$	$\phi=10^{-1}$
$a=0.1\,\mu m$	70 s	7 ms
$1.0\,\mu m$	20 h	7 s

The combination of (8.4.6) and (8.4.7) then determines the stability ratio as

$$W=\frac{J_0}{J}=2a\int_{2a}^{\infty}\frac{\exp(\Phi/kT)}{r^2G(r)}\,dr. \tag{8.4.9}$$

The characteristic time for doublet formation, which now takes the form

$$t_p=\frac{\pi\mu a^3 W}{\phi kT},$$

varies with particle size and volume fraction. The values in Table 8.1 demonstrate that one can obtain convenient time scales in the laboratory by operating at very low volume fractions, but at moderate concentrations flocculation is almost instantaneous in the absence of repulsion. The process remains pseudo-steady, however, provided $t_d/t_p=3\phi/W\ll 1$.

Explicit prediction of J requires, of course, an expression for the interaction potential Φ. Flocculation primarily occurs for thin double layers, so the Derjaguin approximation is appropriate for the repulsion. A correspondingly simple form for the dispersion potential is available in the non-retarded limit, a valid approximation if the attraction becomes insignificant at separations greater than 5–10 nm (e.g. Fig. 8.1). This implies $aA_{eff}(0)/kT<120$–240 nm or $a<50$–100 nm for polystyrene and $a<5$–10 nm for metals such as gold. For larger particles, retardation should reduce the rate of rapid flocculation below that predicted with the non-retarded potential. With repulsion, the attraction becomes significant only when $r-2a\approx 1/\kappa$; hence, the non-retarded form should suffice for slow flocculation when $I>10^{-3}$ M.

8.5 Predictions of the stability ratio

At high ionic strengths the electrostatic repulsion becomes insignificant and the potential reduces to that due to dispersion effects alone. In this regime, $W \equiv W_\infty$ still differs from unity because of the finite range of the attraction and the hydrodynamic interactions omitted from Smoluchowski's calculation of J_0. For non-retarded interactions, evaluation of (8.4.9) yields the stability ratio as a function of A_{eff}/kT, as shown in Fig. 8.4. For $A_{eff}/kT < 75$, which includes virtually all systems of interest, $W_\infty > 1$, indicating that the lubrication stresses in the gap between the spheres retard flocculation more than the attraction enhances it.

Decreasing the electrolyte concentration below n_{crit} introduces a significant repulsive barrier and increases the stability ratio. Then W also varies with $a\kappa$, $\varepsilon\varepsilon_0\psi_s^2 a/kT$, and the manner in which the surface potential or charge regulates during the interaction. Figure 8.5 (Prieve & Ruckenstein, 1980) illustrates the parametric dependence for constant-charge boundary conditions and univalent electrolytes. The increase in the stability ratio with increasing particle size or charge and decreasing Hamaker constant simply reflects the effects of these parameters on Φ_{max}/kT.

Fig. 8.4. Rate of rapid Brownian flocculation J, normalized on the Smoluchowski result $J_0 = 8kTn^2/3\mu$, for a non-retarded dispersion potential of the Hamaker form (Spielman, 1970).

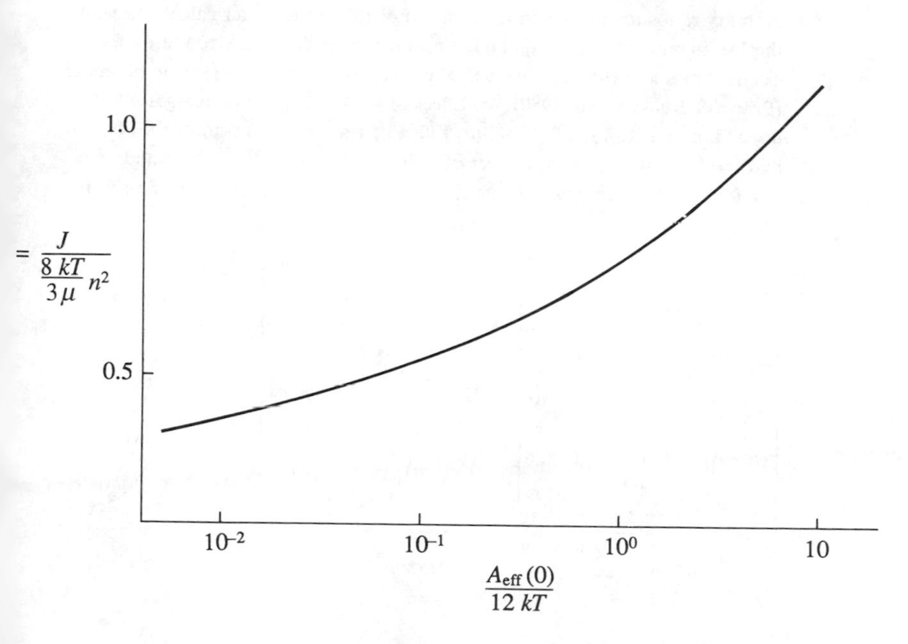

In general, log–log plots of the stability ratio against the salt concentration display two nearly linear regions. The horizontal asymptote at high salt concentrations corresponds to rapid flocculation and W_∞. The slow flocculation regime, with $\mathrm{d}\ln W/\mathrm{d}\ln a\kappa$ approximately constant, appears at low ionic strengths. This slope varies somewhat with both the charge density and the Hamaker constant but increases in direct proportion to the particle size. The critical flocculation concentration lies in the transition region between these two asymptotes.

As expected from (8.2.3), n_{crit} increases with increasing charge density or decreasing Hamaker constant, both of which enhance repulsion with respect to attraction, and varies weakly with particle size. Combining the linearized form of (8.2.3) for a univalent electrolyte,

$$n_{\mathrm{crit}}=0.194 l_{\mathrm{b}}^{-3}\left(\frac{kT}{A_{\mathrm{eff}}}\right)^{2}\left(\frac{e\psi_{\mathrm{s}}}{kT}\right)^{4}, \tag{8.5.1}$$

and the corresponding relation between surface potential and charge density q,

$$\frac{e\psi_{\mathrm{s}}}{kT}=\frac{eq}{\varepsilon\varepsilon_0\kappa kT},$$

Fig. 8.5. Stability ratios for Brownian flocculation of spheres in water calculated with the Derjaguin approximation and constant-charge boundary condition for the electrostatics and the non-retarded Hamaker form for the dispersion potential (Prieve & Ruckenstein, 1980): (*a*) Effect of surface density of charges q for $a=0.1\,\mu\mathrm{m}$ and $A_{\mathrm{eff}}(0)/kT=0.48$. (*b*) Effect of Hamaker constant for $a=0.1\,\mu\mathrm{m}$ and $q=4.8\times10^{-3}\,\mathrm{C/m^2}$. (*c*) Effect of radius for $q=4.8\times10^{-3}\,\mathrm{C/m^2}$ and $A_{\mathrm{eff}}(0)/kT=0.48$. Arrows indicate the critical flocculation concentration from (8.5.2).

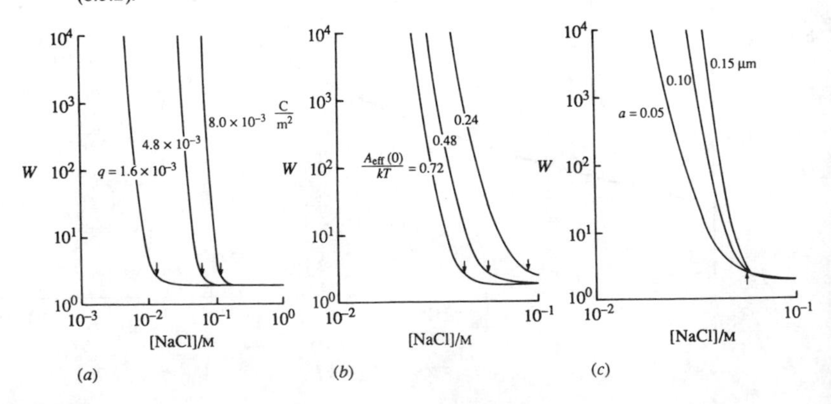

with

$$\kappa^2 = \frac{2e^2 n_{crit}}{\varepsilon\varepsilon_0 kT}$$

yields

$$n_{crit} = 9.0\left(\frac{kT}{A_{eff}}\right)^{2/3}\left(\frac{ql_b^2}{e}\right)^{4/3} \quad \text{mol/l.} \tag{8.5.2}$$

Thus the critical flocculation concentration depends more strongly on the charge density than on the Hamaker constant. The resulting predictions (shown by the arrows in Fig. 8.5) provide reasonable estimates of the onset of rapid flocculation.

Comparison of the predictions for W with the potentials in Fig. 8.1 confirms that the stability ratio becomes large when $\Phi_{max}/kT \gg 1$. The major contribution to the integral in (8.4.9) then comes from the vicinity of the maximum, motivating the expansion (Reerinck & Overbeek, 1954)

$$\exp\left(\frac{\Phi}{kT}\right) \sim \exp\left(\frac{\Phi_{max}}{kT}\right)\exp\left[\tfrac{1}{2}(r - r_{max})^2 \frac{d^2}{dr^2}\frac{\Phi_{max}}{kT}\right],$$

$$G(r) \sim \frac{r_{max} - 2a}{a},$$

so that

$$W \sim \left(-\frac{2}{\pi}\frac{d^2}{dr^2}\frac{\Phi_{max}}{kT}\right)^{-1/2} \frac{\exp\left(\dfrac{\Phi_{max}}{kT}\right)}{r_{max} - 2a}. \tag{8.5.3}$$

This suggests correlating the stability ratios calculated from numerical evaluation of (8.4.9) with the height of the energy barrier. For a limited range of conditions, Prieve & Ruckenstein (1980) found

$$W = W_\infty + 0.25\left[\exp\left(\frac{\Phi_{max}}{kT}\right) - 1\right] \tag{8.5.4}$$

to accurately represent their numerical results (Fig. 8.6). Hence, the stability ratio can be estimated from the value for rapid flocculation and the height of the energy barrier at the conditions of interest.

From these results, one can estimate the conditions required for a certain degree of stability. For example, preserving a suspension of 0.1 μm radius polystyrene spheres at $\phi = 0.1$ for one month requires $W \geq 4 \times 10^8$, or $\Phi_{max}/kT \geq 21$. Figure 8.1 indicates that an ionic strength somewhat less than

10^{-1} M should suffice. Thus the theory explains the feasibility of sustained electrostatic stabilization of aqueous dispersions.

Calculation of the stability ratio for other conditions or other systems involves a straightforward numerical integration of (8.4.9) with appropriate forms of the interaction potentials and the hydrodynamic functions. With, for example, the non-retarded Hamaker form for the dispersion interactions and the linearized Derjaguin approximation for the electrostatics, only A_{eff}/kT, $a\kappa$, $\varepsilon\varepsilon_0\psi_s^2 a/kT$, and the choice between constant-charge and constant-potential boundary conditions are required as input. One should exercise discretion in use of the linearized electrostatic potential though, since the range of validity is limited (§4.10).

8.6 Measurements of doublet formation rates

Measurement of flocculation rates to characterize the stability of colloidal dispersions dates to the nineteenth-century work cited earlier. These and later studies typically involved the measurement of turbidity, sediment weight or volume, viscosity, or other properties that primarily reflect the formation of large aggregates. While indicating the relative degree of stability of specific dispersions, such data are clearly unsuitable for testing quantitative theories for the rate of doublet formation.

Fig. 8.6. Correlation of predictions of the stability ratio for Brownian flocculation of spheres in water with the height of the repulsive barrier in the potential (Prieve & Ruckenstein, 1980).

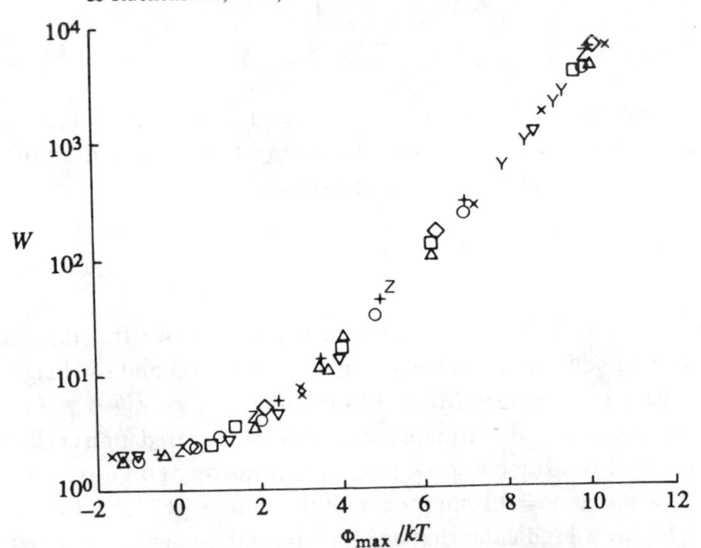

Several sensitive techniques now available can detect doublet formation for limited ranges of particle sizes: electronic counters, $0.5\,\mu m \leq 2a$ (Matthews & Rhodes, 1970); a stopped-flow spectrophotometer modified to reject forward scattering, $2a \leq 0.5\,\mu m$ (Lichtenbelt, Pathmananoharan, & Wiersema, 1974); and low-angle light-scattering instruments, $0.2 \leq 2a \leq 1.0\,\mu m$ (Lips & Willis, 1973).

The spectrophotometer measures the turbidity τ at a particular wavelength λ. At dilute concentrations,

$$\tau = c_1 n + c_2 n_2, \tag{8.6.1}$$

with n_2 the number density of doublets. The extinction cross-sections c_1 and c_2 can be calculated accurately from the Rayleigh–Gans–Debye theory, which assumes $8\pi a/\lambda \ll 1$, for $a \leq 0.2\,\mu m$. At short times when $n \sim n_0$ and $n_2 \ll n_0$, the time rate of change of the turbidity,

$$\frac{d\tau}{dt} = (\tfrac{1}{2}c_2 - c_1)J, \tag{8.6.2}$$

determines the flux J.

Low-angle light-scattering instruments operate at angles $\theta = 1$–$2°$. Provided $4\pi a\theta/\lambda < 1$ rad, the Rayleigh–Gans–Debye theory predicts the scattering intensity for a k-fold aggregate as $I_k = k^2 I_1$, with I_1 the scattering from a single particle (Lips & Willis, 1973). Consequently, the time derivative of the total intensity $I = nI_1 + n_2 I_2 + \ldots$ has initial value

$$\begin{aligned} \frac{dI}{dt} &= I_1 J \\ &= 2\frac{I_0}{t_p}, \end{aligned} \tag{8.6.3}$$

with $I_0 = n_0 I_1$.

Measurements of the rate of rapid flocculation of monodisperse polystyrene latices by these techniques (Table 8.2) confirm the effect of hydrodynamic interactions in retarding the process, since $W_\infty > 1$. In addition, comparison with Fig. 8.4 provides estimates for A_{eff}/kT, though the weak dependence on the Hamaker constant and the sensitivity to particle size introduce an uncertainty of ± 15 per cent or more. The values fall within a factor of two to three of the $\approx 2.5kT$ expected from the theory (§5.6). As noted earlier, retardation could affect the rate of Brownian flocculation for spheres larger than 0.05–0.10 μm. Indeed, the theoretical curve in Fig. 8.7, constructed with the approximations (5.9.3) and (5.10.1) for the retarded dispersion potential, provides clear evidence for the effect

Table 8.2. *Results of rapid-flocculation experiments with polystyrene latices*

	$2a/\mu m$	W_∞	A_{eff}/kT
Matthews & Rhodes (1970)	0.714	1.83	1.5
Lips & Willis (1973)	0.207	1.78	1.9
	0.357	1.89	1.2
	0.500	1.82	1.7
Lichtenbelt, Pathmananoharan, & Wiersema (1974)	0.091	1.67	2.7
	0.109	1.82	1.5
	0.176	1.59	4.1
	0.234	2.00	0.7
	0.357	1.96	0.7
Feke & Schowalter (1983)	0.675	1.87	1.1

Fig. 8.7. Stability ratio for rapid flocculation of polystyrene latices in water as function of sphere radius: ——, prediction from (8.4.7) with retarded dispersion potential from (5.9.3) and (5.10.1); ○, data from Table 8.2.

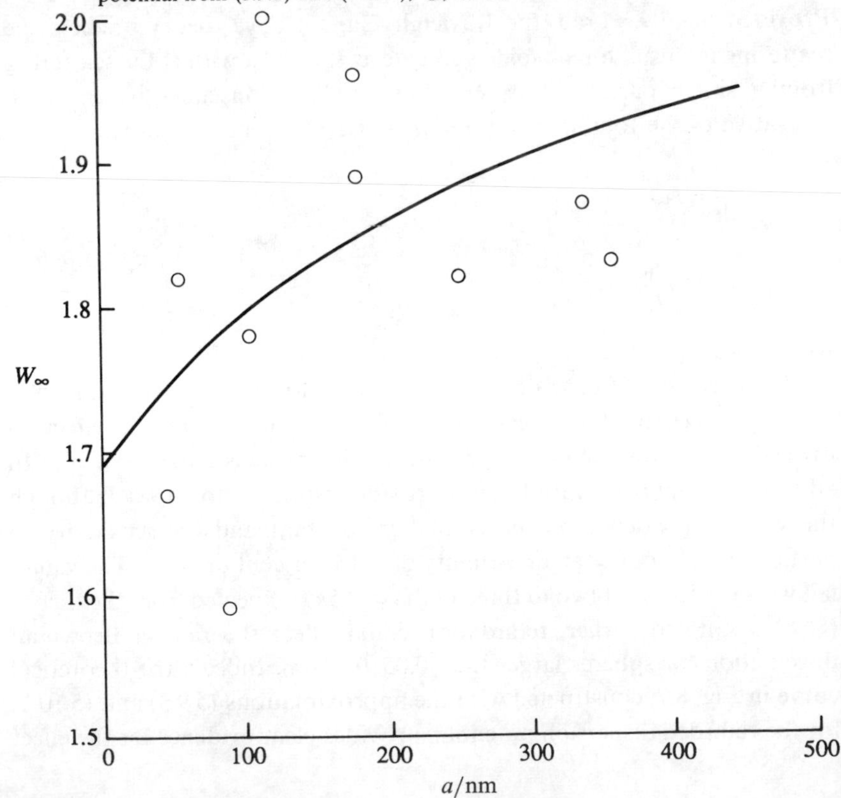

of retardation. The prediction changes ≈ 15 per cent for $0 \le a \le 0.4\,\mu m$ and the data points now scatter about the curve.

There exist few measurements of doublet formation rates under conditions of slow flocculation (Ottewill & Shaw, 1966; Lips & Willis, 1973; Zeichner & Schowalter, 1979) and the stability ratios extracted from the data compare poorly with theoretical predictions. Figure 8.8(a) illustrates the discrepancy common to the three sets of data for polystyrene latices, using the measured ionic strength, particle size, and surface potential and values for the Hamaker constant bracketing that expected from the theory.

The lack of agreement may stem from the extreme sensitivity of the rate to the magnitude of the repulsion. The repulsive barrier Φ_{max}/kT represents the difference between two large potentials, so a relatively small error in either translates into a large error in the predicted stability ratio. For example, with the linearized Derjaguin approximation and constant potential boundary conditions, an uncertainty of $\Delta\psi_s/\psi_s$ in the surface potential produces

$$\frac{\Delta\Phi_{max}}{kT} \approx 2\pi \ln 2 \frac{\varepsilon\varepsilon_0 \psi_s^2 a}{kT} \frac{\Delta\psi_s}{\psi_s}$$

$$\approx 290 \frac{\Delta\psi_s}{\psi_s}$$

for the conditions of Fig. 8.8(a). Accurate predictions of the stability ratio, therefore, would require knowledge of the surface potential within ± 1 per cent. In addition, the continuum theory could contain $O(1)$ errors at the small separations, $r - 2a \approx 1/\kappa \le 1\text{nm}$, characteristic of Φ_{max}. Hence quantitative predictions may lie beyond the capabilities of the theory.

The work of Ottewill & Shaw (1966) has received particular attention because the measured stability ratios deviate from the expected dependence on particle size. The data (Fig. 8.8(b, c)) actually differ from the theoretical expectations in two ways (Prieve & Ruckenstein, 1980). First, extrapolation of the linear portion of the curve to $W/W_\infty = 1$ determines an n_{crit} which depends on particle size, contrary to (8.2.3). Second, the slope at low ionic strengths, $\mathrm{d}\ln W/\mathrm{d}\ln[Ba(NO_3)_2] \approx \mathrm{d}\ln W/\mathrm{d}\ln\kappa$, varies little with particle size, whereas (8.2.2) and (8.5.4) lead one to expect a monotonic increase. Application of the theory to systems such as these with divalent counterions and surface charges spaced more than a Debye length apart is questionable, but whether the observed differences in the stability are fundamental or the consequence of these features of the experiments remains to be established.

Wiese & Healy (1970) and others since have suggested that the slow-flocculation experiments actually detect secondary minimum flocculation.

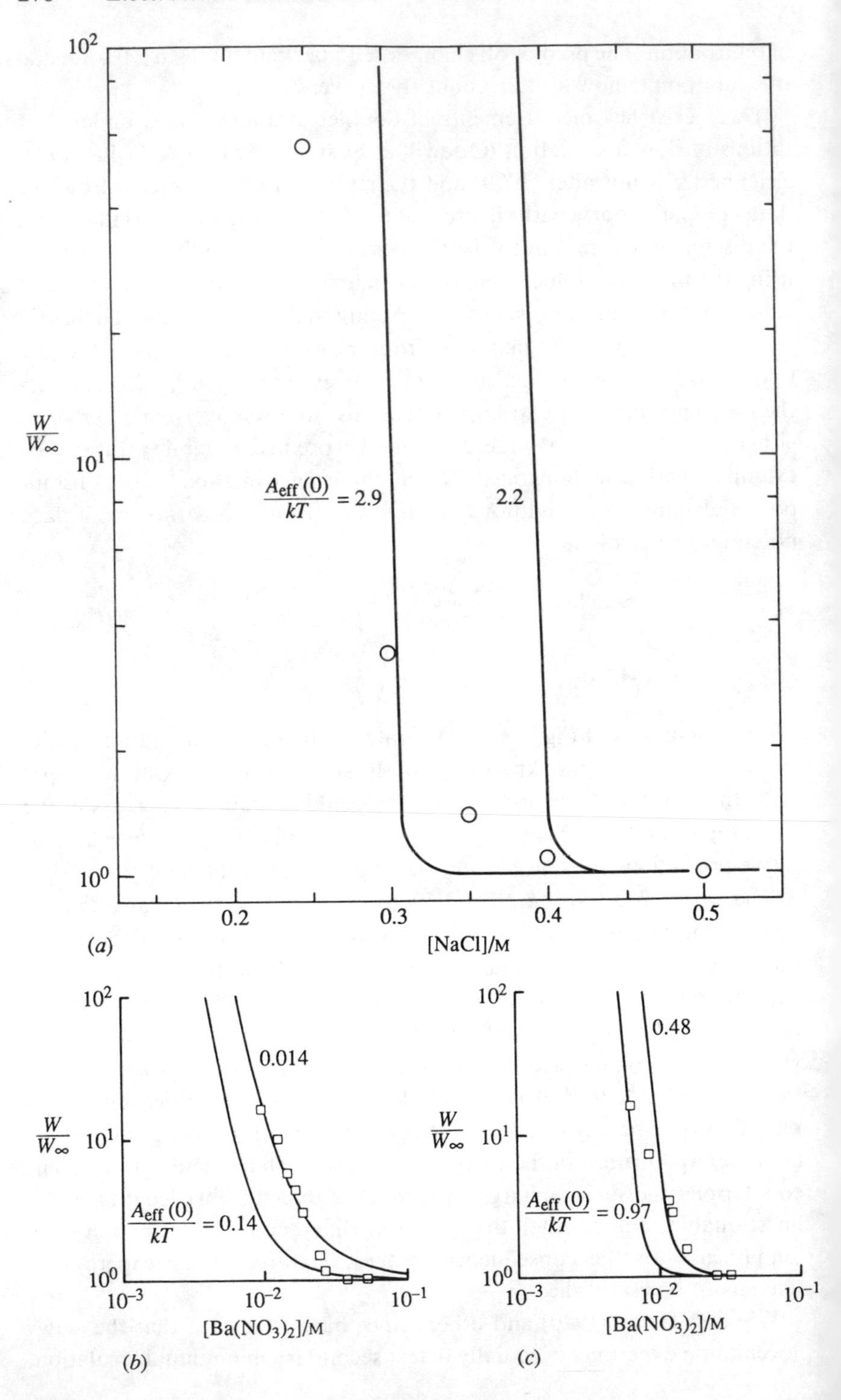

10²
W/W∞
10¹
10⁰
Aeff (0)/kT = 2.9
2.2
0.2
0.3
0.4
0.5
(a)
[NaCl]/M
10²
W/W∞
10¹
10⁰
0.014
Aeff (0)/kT = 0.14
10⁻³
10⁻²
10⁻¹
[Ba(NO₃)₂]/M
(b)
10²
W/W∞
10¹
10⁰
0.48
Aeff (0)/kT = 0.97
10⁻³
10⁻²
10⁻¹
[Ba(NO₃)₂]/M
(c)

Fig. 8.8. Comparison of measured and predicted stability ratios for Brownian flocculation of polystyrene latices: (*a*) $2a=0.50\,\mu m$ and $\psi_s=37\text{–}40\,mV$ (Zeichner & Schowalter, 1979). (*b*) $2a=0.10\,\mu m$ and $q=3.2\times10^{-3}\,C/m^2$. (*c*) $2a=0.24\,\mu m$ and $q=4.5\times10^{-3}\,C/m^2$. (*b*) and (*c*): data of Ottewill & Shaw, 1966 and calculations of Prieve & Ruckenstein, 1980.

They note that the depth of the secondary minimum increases as the salt concentration approaches the critical flocculation concentration. Thus a range of conditions can exist for which $-\Phi_{sec}/kT\approx2\text{–}5$, while $(\Phi_{max}-\Phi_{sec})/kT\geq5$. Consequently, pairs would accumulate in the secondary minimum quickly but would diffuse over the repulsive barrier relatively slowly. The experimental techniques would not distinguish between the two modes of flocculation, and, therefore, would indicate a stability ratio considerably lower than predicted. Thus far, the difficulty in either accounting unambiguously for this mode of flocculation in the theory or performing experiments that detect only primary minimum flocculation has frustrated attempts to reconcile the discrepancy between theory and experiment.

8.7 Growth and structure of large flocs

From the results for the rate of doublet formation, one can construct population balances to describe the growth in dilute suspensions of aggregates containing k spheres and having radius a_k. For Brownian flocculation the growth process is controlled by the collision rate between two aggregates (Smoluchowski, 1917),

$$J_{ij}=\frac{2kT}{3\mu}(a_i+a_j)\left(\frac{1}{a_i}+\frac{1}{a_j}\right)\frac{n_i n_j}{W_{ij}},\tag{8.7.1}$$

with n_i and n_j the corresponding number densities and W_{ij} the stability ratio. The conservation equation follows as

$$\frac{dn_k}{dt}=\tfrac{1}{2}\sum_{\substack{i=1\\ j=k-i}}^{k-1}J_{ij}-\sum_{i=1}^{\infty}J_{ki}\tag{8.7.2}$$

with $n_1(0)=n_0$ and $n_k(0)=0$ for $k>1$. The difficulties lie in relating a_k to k and in accounting for any size dependence of W_{ij}.

Smoluchowski (1917) obtained an analytical solution to (8.7.2) by assuming that (i) collisions between clusters of approximately equal size dominate such that

$$(a_i+a_j)\left(\frac{1}{a_i}+\frac{1}{a_j}\right)=4\tag{8.7.3}$$

and (ii) $W_{ij} = W$ independent of size. Then

$$n_k(t) = n_0 \frac{(t/t_p)^{k-1}}{(1+t/t_p)^{k+1}}, \tag{8.7.4}$$

with $t_p = 3\mu W/4n_0 kT$, and the total number density

$$n_{tot} = \sum_{i=1}^{\infty} n_i$$

decreases with time according to

$$\frac{n_0}{n_{tot}} = 1 + \frac{t}{t_p}. \tag{8.7.5}$$

Measurements of aggregate size distributions as functions of time for polystyrene spheres with $a = 0.49$ µm in 1.25 M KCl solution, obtained with an electronic counter by Higashitani & Matsuno (1979), provide a direct test of the theory. The rate of doublet formation yields $W = 1.74$,

Fig. 8.9. Brownian flocculation detected with a Coulter counter for polystyrene latices, with $2a = 0.974$ µm and $W = 1.74$, ○● (Higashitani & Matsuno, 1979), compared with predictions from (8.7.4): (*a*) Aggregates with $1 \leq k \leq 10$. (*b*) Aggregates with $k > 10$.

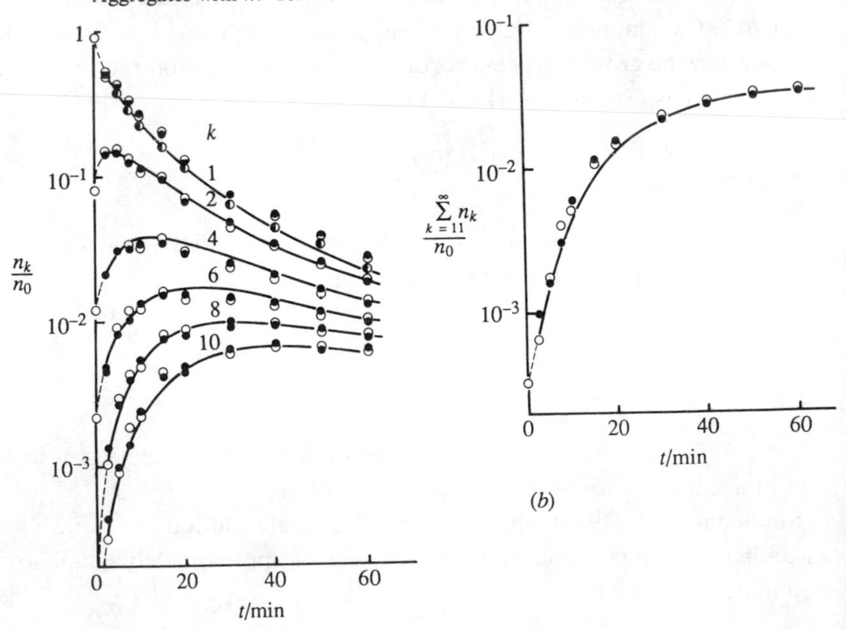

corresponding to $A_{eff}/kT = 2.2$ in satisfactory accord with the results cited in Table 8.2. As demonstrated by the solid lines in Fig. 8.9, the theory predicts quite accurately the growth of larger aggregates for rapid flocculation.

Although apparently successful in describing aggregate growth, this approach yields no information about their configuration. Properties that depend on the floc size have often been calculated by relating the degree of aggregation to the radius through

$$k = (a_k/a)^D \tag{8.7.6}$$

with $D = 3$, corresponding to a spherical aggregate of equal solid volume, generally assumed. Recent theoretical (e.g. Witten & Sander, 1981; Meakin, 1983; Jullien, Kolb, & Botet, 1984) and experimental (e.g. Schaefer, *et al.*, 1984; Weitz, Lin, & Huang, 1986) work reveals instead highly branched aggregates (Fig. 8.10). In many ways these resemble fractals satisfying (8.7.6) but with $D < 3$ (Feder, 1988) and with a self-similar structure exhibiting heterogeneities on lengths scales ranging from a to a_k.

Individual aggregates with fractal structure scatter light at intermediate wavenumbers q, $1/a > q > 1/a_k$, according to (Schaefer *et al.*, 1984):

$$\begin{aligned} I_k(q) &= \frac{k^2 I_1}{(qa_k)^D} \\ &= \frac{k I_1}{(qa)^D}, \end{aligned} \tag{8.7.7}$$

with I_1 the scattering intensity for a single sphere. A dispersion of such aggregates is characterized by a scattering intensity,

$$\begin{aligned} I(q) &= \sum_{k=1}^{\infty} n_k I_k(q) \\ &= \frac{I_0}{(qa)^D}, \end{aligned} \tag{8.7.8}$$

which is invariant during flocculation, i.e. independent of aggregate size, but exhibits a slope of $-D$ on a plot of $\ln I$ vs. $\ln q$. Also, from (8.7.6) the average number of particles in an aggregate, M, is related to the average radius R through

$$\begin{aligned} M &= \frac{1}{n_{tot}} \sum_{k=1}^{\infty} k n_k = \frac{1}{n_{tot}} \sum_{k=1}^{\infty} \left(\frac{a_k}{a}\right)^D n_k \\ &\equiv (R/a)^D. \end{aligned} \tag{8.7.9}$$

Fig. 8.10. Transmission electron micrographs of (*a*) gold sols with $a = 7.2 \pm 0.8$ nm (Weitz & Huang, 1984) and (*b*) silica sols with $a \approx 2.7$ nm (Schaefer *et al.*, 1984) flocculated by Brownian motion, showing the similarity of the structure for different cluster sizes.

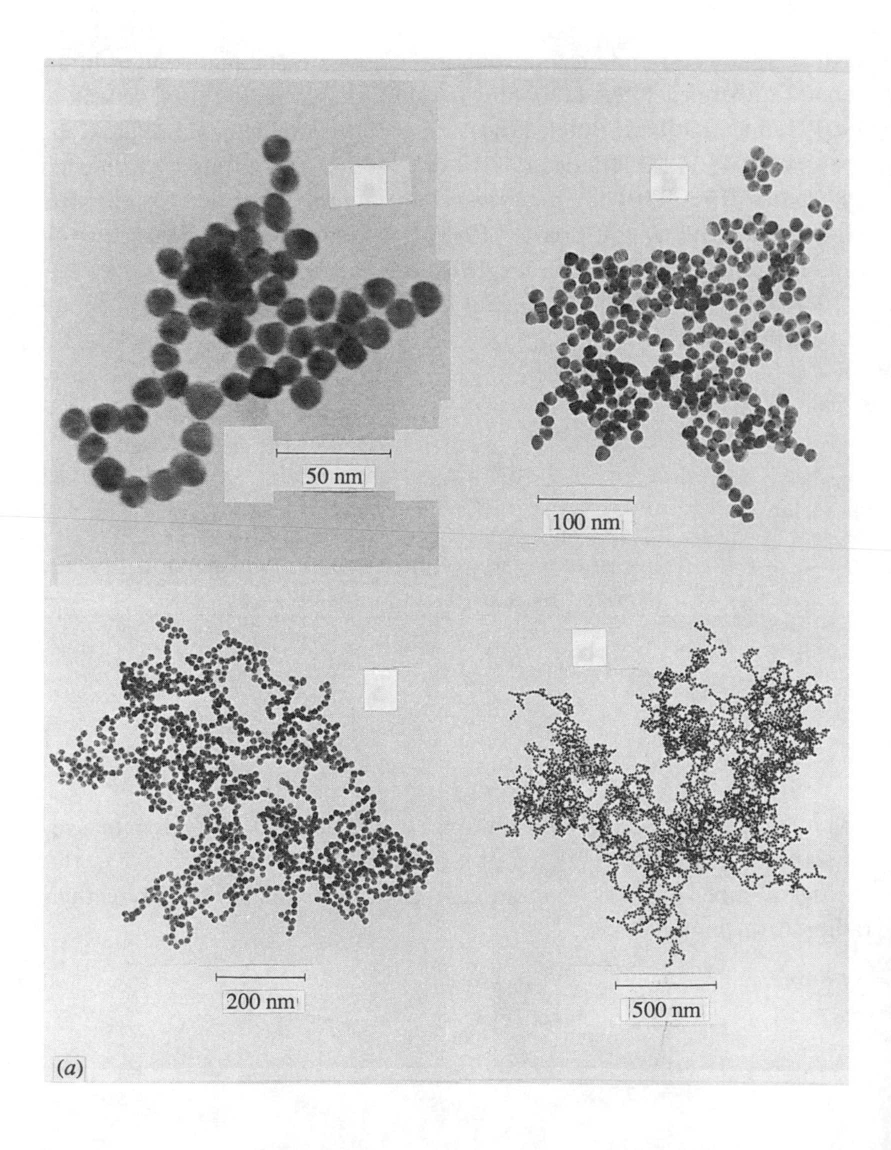

(b)
0.25 μm

The fractal dimension D reflects the internal structure of the flocs and depends on the mode of aggregation. For rapid, irreversible Brownian flocculation with no subsequent rearrangement, the computer simulations determine $D=2.5$ if the flocs grow by adding one particle at a time, but $D=1.75$ if cluster–cluster aggregation predominates. Any process that allows particles to penetrate further into the floc before sticking will increase D. For example, in slow flocculation, interacting clusters sample many configurations before sticking, resulting in $D\approx 2.0$. Only if internal rearrangements occur readily will D approach 3.0.

The scattering experiments of Weitz and co-workers (Dimon *et al.*, 1986), with aqueous dispersions of gold spheres, and of Schaefer *et al.* (1984) with aqueous silica dispersions, illustrate the phenomena nicely. Both are electrostatically stabilized, so flocculation is induced by reducing the surface charge, via displacing adsorbed ionic groups from the gold and altering the pH for the silica, as well as increasing the ionic strength for the latter. With complete removal of the surface charge, the gold aggregates appear to satisfy both the relationship between mass and size (8.7.9) and the power law decay of the scattering intensity (8.7.8) with $D\approx 1.7$–1.8 (Fig. 8.11). The theoretical results noted above confirm the process as rapid flocculation or diffusion-limited aggregation in which cluster–cluster interactions dominate. The silica aggregates, and the gold aggregates resulting from only partial removal of the surface charge, also exhibit the power law decay in the scattering intensity but with $D\approx 2.0$–2.2 (Fig. 8.12) in conformity with the predictions for slow flocculation or reaction-limited aggregation.

Complementary studies of the kinetics of these aggregation processes have employed photon correlation spectroscopy to follow the hydrodynamic radius of the flocs as a function of time. The distribution of floc sizes produces an autocorrelation function that comprises a sum of exponentials (§3.4). Expanding in powers of the delay time τ as

$$\ln F_s(\tau) = -\Gamma\tau + \ldots$$

provides a measure of the diffusion coefficient through the first cumulant

$$\Gamma(q) = \frac{\sum_{k=1}^{\infty} n_k I_k(q) D_k q^2}{\sum_{k=1}^{\infty} n_k I_k(q)}. \tag{8.7.10}$$

For measurements at wavenumbers $1/a > q > 1/a_k$, the intensity varies according to (8.7.7), while

$$D_k = D_0 k^{-1/D}$$

Fig. 8.11. Characterization of gold sols flocculated by Brownian motion after suppression of the electrostatic repulsion: (*a*) Floc mass as a function of floc radius indicating $M \sim R^{1.7 \pm 0.1}$ (Weitz & Huang, 1984). (*b*) Intensity as a function of wavenumber for light and neutron scattering indicating $I \sim q^{-1.8}$ for $R^{-1} \leq q \leq a^{-1}$ (Dimon *et al.*, 1986).

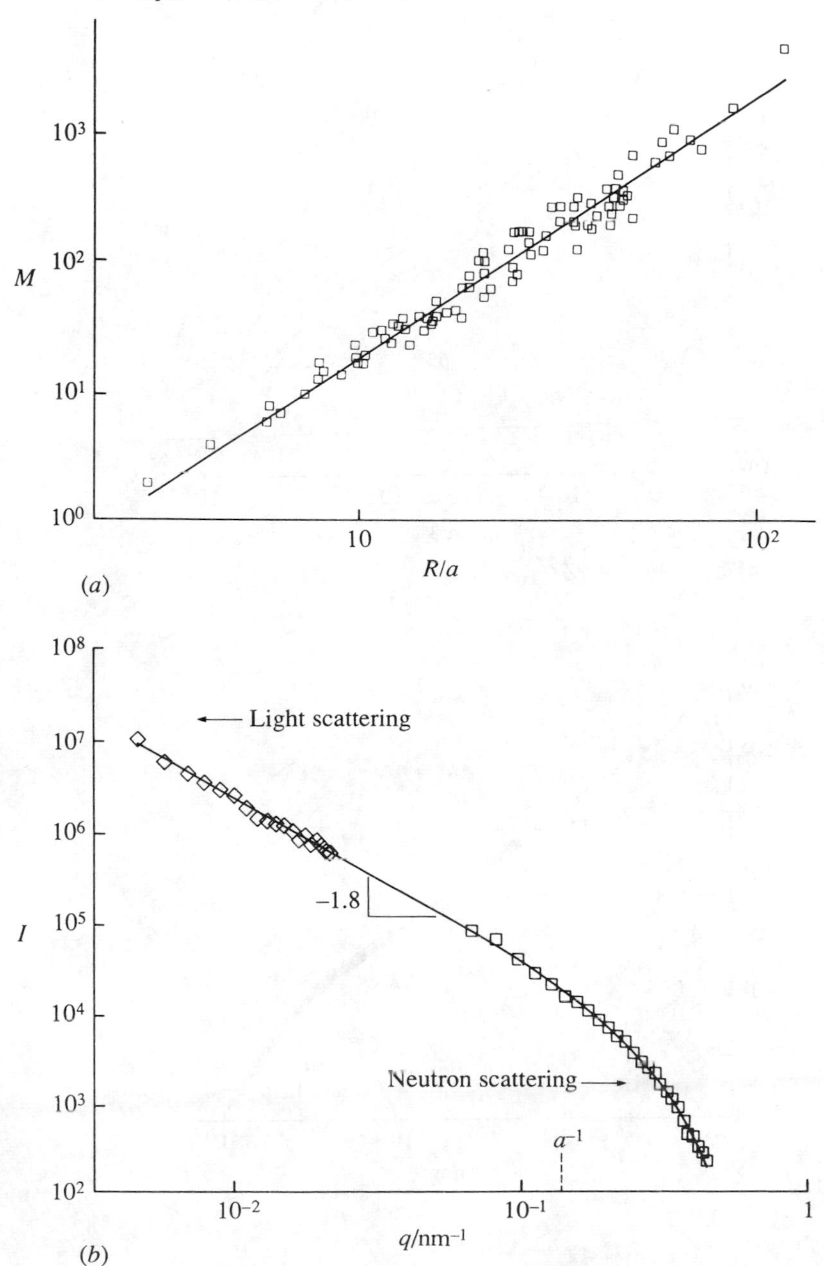

Fig. 8.12. Intensities as a function of wavenumber for slow Brownian flocculation: (*a*) X-ray scattering from gold sols (Dimon *et al.*, 1986). (*b*) X-ray and light scattering from silica sols (Schaefer *et al.*, 1984).

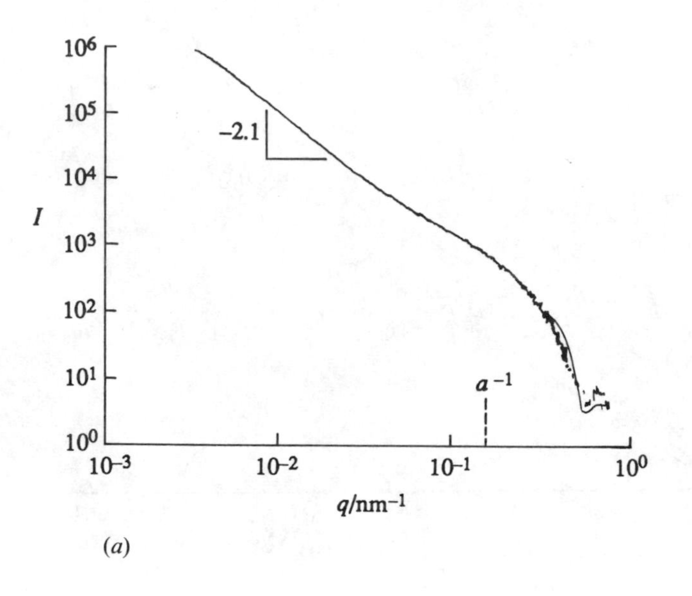

(*a*)

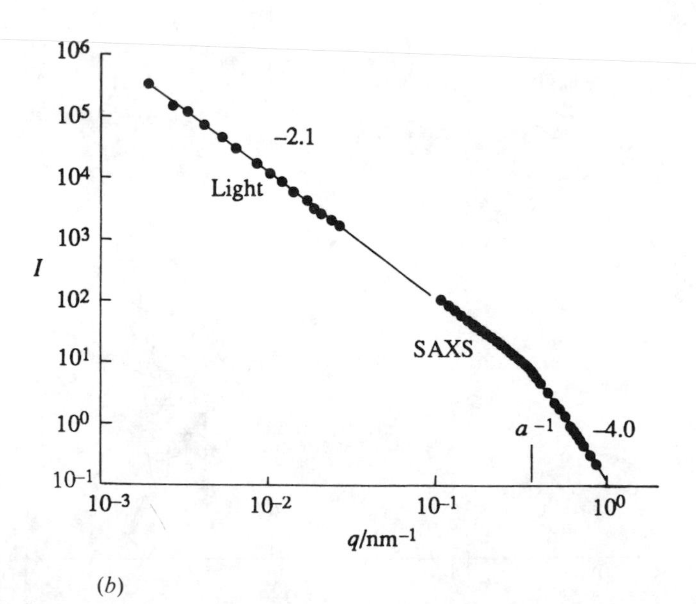

(*b*)

with $D_0 = kT/6\pi\mu a$. Substitution into (8.7.10) along with the Smoluchowski result (8.7.4) and expansion for $t/t_p \gg 1$ eventually leads to

$$\Gamma(q) = \frac{q^2 D_0}{(t/t_p)^{1/D}} \int_0^\infty z^{1-1/D} e^{-z} \, dz$$
$$\equiv \frac{kT}{6\pi\mu R} q^2, \qquad (8.7.11)$$

so that the effective hydrodynamic radius R increases with time according to

$$\frac{R}{a} = \frac{(t/t_p)^{1/D}}{\Gamma_*(2-1/D)} \qquad (8.7.12)$$

with Γ_* the gamma function. Thus a plot of $\ln(R/a)$ vs. $\ln(t/t_p)$ should become a straight line with slope $1/D$ at long time.

Radii detected by photon correlation spectroscopy for flocculation of the gold sols after complete removal of the surface charges and of polystyrene latices at high ionic strengths are displayed in Fig. 8.13. The range of sizes corresponds to $1 \leq Rq \leq 8$ for the gold and $1.5 \leq Rq \leq 30$ for the polystyrene, but the constancy of Γ/q^2 within this range was not demonstrated in either case. Nonetheless, each set of data very quickly attains an asymptote corresponding to $D \approx 1.6$–1.7, in accord with the static scattering and the mass–radius relationship for the gold sols. The horizontal shift between two sets of data for the polystyrene latices reflects the difference in ionic strength, and, therefore, stability ratio.

For slow flocculation with $W \gg 1$, the kinetics detected through the scattering experiment fail to follow (8.7.12). Instead, R increases exponentially with time (e.g. Weitz, Lin, & Huang, 1986). Theories incorporating stability ratios, W_{ij}, which decrease with increasing aggregate size appear capable of explaining the phenomena but lie beyond the scope of our treatment here.

The results described above establish that the simple description of the kinetics leading to (8.7.4), together with knowledge of the fractal structure of the aggregates, suffices for predicting the rate of rapid Brownian flocculation in dilute suspensions. For slow flocculation, however, the experiments indicate more compact fractals and quite different kinetics. A comparably simple, but quantitative, description of the kinetics remains to be developed. In addition, the nature of the transition between the slow and fast regimes and the effects of rearrangements within the aggregates have yet to be fully defined.

Fig. 8.13. Measurements of the kinetics of Brownian flocculation via photon correlation spectroscopy, showing that the average hydrodynamic radius increases with time as $R \sim t^{1/D}$: (*a*) Gold sols with $D \approx 1.7$ (Weitz & Huang, 1984). (*b*) Polystyrene latices with $a = 0.07\,\mu m$ in a glycerol-water mixture at $\phi = 10^{-5}$ (○, ●) and 5.0×10^{-5} (□, ■), with $I = 0.4$ M NaCl (hollow symbols) and 1.0 M NaCl (filled symbols) indicating $D \approx 1.6$ (Sonntag & Russel, 1986).

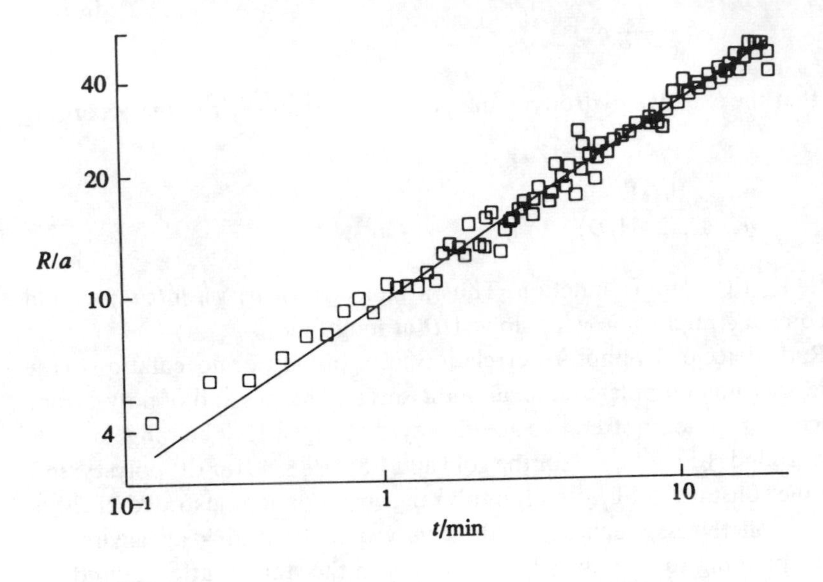

(*a*)

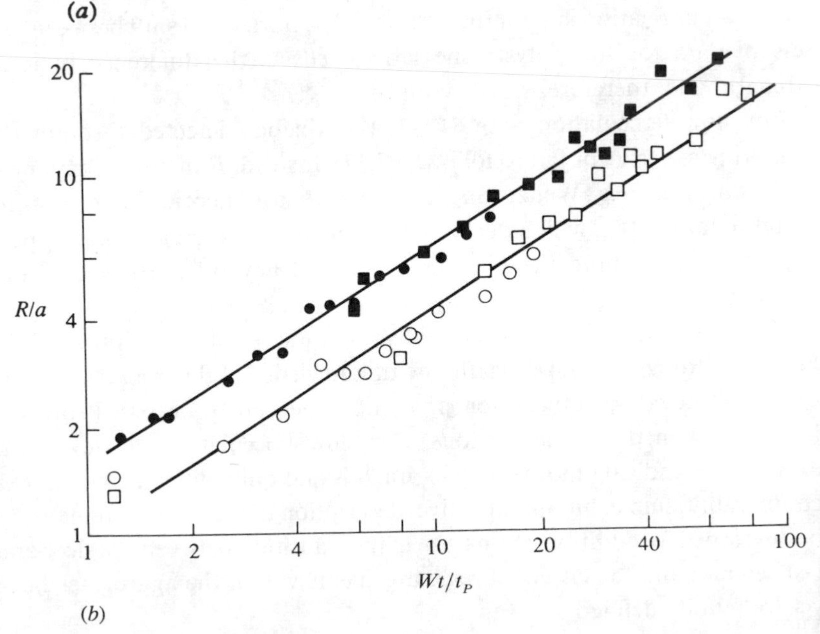

(*b*)

8.8 Doublet formation in shear flows

A shear flow can promote flocculation of colloidal particles by two modes, either increasing the rate in an unstable dispersion or mechanically destabilizing a dispersion resistant to Brownian flocculation. In the former situation, the degree of enhancement depends primarily on the rate of collisions due to shear relative to those from diffusion alone, as reflected by a Peclet number. The importance of this mode of shear flocculation derives from a number of applications, as in waste-water treatment processes. To achieve the second mode, viscous stresses forcing particles together must overcome the repulsive interparticle force responsible for kinetic stability. Warren (1975) demonstrated the value of selective destabilization of mineral particles by shear for enhancing the efficiency of flotation processes. Of course, when these viscous stresses exceed the attractive force at contact, existing flocs will break up. Indeed, the dispersion of any dry powder, such as a pigment, into a liquid requires the rupture of aggregates by mechanical forces (Parfitt, 1981). These flocculation and breakup processes depend on N_f, N_r, $a\kappa$, and δ/a, which characterize the relative magnitudes and ranges of the forces.

As with Brownian flocculation, we focus on doublet formation and breakup as described by the pair conservation equation (8.3.8). With flow, the pair probability loses the spherical symmetry of the Brownian case, precluding a general analytical solution. Thus for unstable dispersions we consider first $Pe \ll 1$, such that shear only slightly perturbs the Brownian flocculation process, and then $Pe \gg 1$, where the flow dominates with a lesser effect of diffusion. The straightforward generalization to larger aggregates is presented but not pursued, since experiments testing the results have not yet been performed. The questions of mechanical stability are addressed in §8.9.

Diffusion-dominated flocculation: $Pe \ll 1$

In this limit one might expect to calculate the enhancement due to shear through a regular perturbation expansion,

$$P_2 = n^2(p_0 + Pe\, p_1 + \ldots),$$

with p_0 representing the pair density for Brownian flocculation. Substituting into (8.3.8) and grouping terms of like order in Pe gives

$$\nabla_r \cdot \omega kT \cdot \left(\nabla_r p_0 + p_0 \nabla_r \frac{\Phi}{kT}\right) + Pe \nabla_r \cdot \omega kT \cdot \left(\nabla_r p_1 + p_1 \nabla_r \frac{\Phi}{kT}\right)$$

$$= Pe \nabla_r \cdot p_0 \mathbf{u}, \tag{8.8.1}$$

with $\mathbf{u} = \mathbf{U}/Pe$. Then setting the $O(1)$- and $O(Pe)$-terms individually to zero would determine p_0 and p_1, with the former reducing to (8.4.6).

Unfortunately, examination of the far-field behavior reveals a problem. For $r \to \infty$,

$$\mathbf{U} \sim \nabla \mathbf{u}_\infty \cdot \mathbf{r},$$

$$p_0 \sim 1 - \frac{2a}{Wr},$$

$$\boldsymbol{\omega} kT \sim \frac{kT}{3\pi\mu a}\boldsymbol{\delta},$$

so that

$$\frac{Pe \nabla_\mathbf{r} \cdot p_0 \mathbf{u}}{\nabla_\mathbf{r} \cdot \boldsymbol{\omega} kT \cdot \nabla_\mathbf{r} p_0} \sim Pe\left(\frac{r}{2a}\right)^2.$$

Clearly, convection becomes comparable to diffusion at separations $r/2a = O(Pe^{-1/2})$, even though $Pe \ll 1$. Consequently, the $Pe \to 0$ limit is singular; that is, convection is negligible in an inner region where $r/2a = O(1)$ but always becomes significant in an outer region where $r/2a = O(Pe^{-1/2})$. To accommodate this behavior, asymptotic solutions must be constructed in each region individually and then matched in an intermediate region where $1 \ll r/2a \ll Pe^{-1/2}$ (van de Ven & Mason, 1977).

The first step involves rescaling the separation in the outer region as $Pe^{1/2}\mathbf{r}/a = (X, Y, Z)$ to balance the convection and diffusion terms to first order. At such separations both hydrodynamic interactions and the interparticle force are negligible. For simple shear flow with $\mathbf{u}_\infty = (Y, 0, 0)$ (Fig. 8.14), the rescaled equation

$$Y\frac{\partial P_2}{\partial X} = \frac{\partial^2 P_2}{\partial X^2} + \frac{\partial^2 P_2}{\partial Y^2} + \frac{\partial^2 P_2}{\partial Z^2}, \tag{8.8.2}$$

with

$$P_2 \to n^2 \qquad \text{as} \qquad R^2 = X^2 + Y^2 + Z^2 \to \infty,$$

can be solved via a three-dimensional Fourier transform. In the intermediate region, this solution, $P_2 = P_{\text{out}}$, has the limiting form

$$\lim_{R \to 0} P_{\text{out}} = n^2 + \frac{C}{4\pi}\left(\frac{1}{R} - 0.257\right) + O(Pe). \tag{8.8.3}$$

In the inner region, (8.8.1) demonstrates that convection enters only at $O(Pe)$. The solution valid to $O(Pe^{1/2})$, which satisfies the equation and $P_2=0$ at $r=2a$, is

$$P_{\text{in}}=A\exp(-\Phi/kT)\int_{2a}^{r}\frac{\exp(\Phi/kT)}{r^2G(r)}\,dr+O(Pe), \tag{8.8.4}$$

so the distribution of pairs remains isotropic, i.e., spherically symmetric.

Matching the two solutions asymptotically in the intermediate region requires (van Dyke, 1975, §5.6–7)

$$\lim_{R\to 0} P_{\text{out}}(r/aPe^{1/2})=\lim_{r\to\infty} P_{\text{in}}(r),$$

i.e. the inner limit of the outer solution, converted to inner variables, must be identical to the outer limit of the inner solution, with terms of $O(Pe)$ neglected. Application to (8.8.3) and (8.8.4) determines

$$A=\frac{2an^2}{W}\left(1+0.257\frac{Pe^{1/2}}{W}\right),$$
$$C=-\frac{8\pi an^2}{W}. \tag{8.8.5}$$

Fig. 8.14. Coordinate system for interaction of two spheres in a simple shear flow.

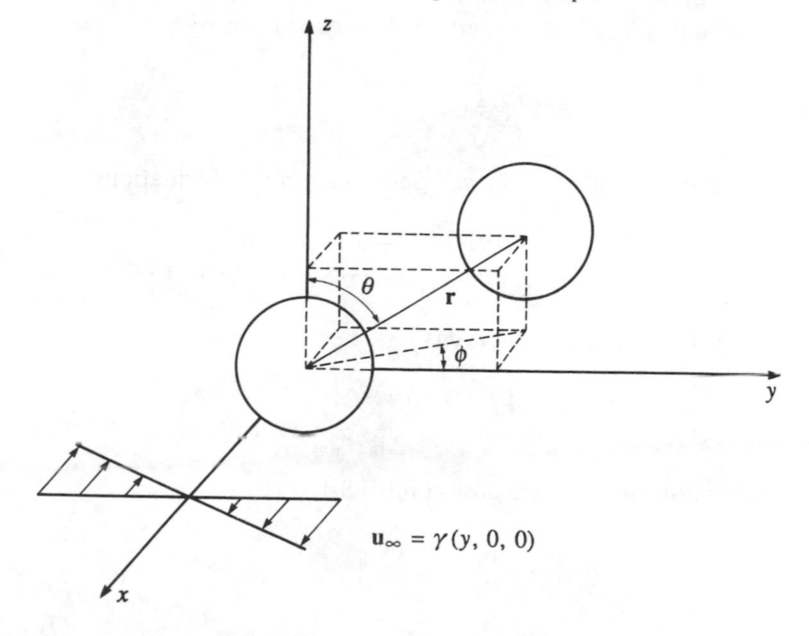

Substitution of A into (8.8.4) leads to the conclusion that convection increases the number of pairs in the inner region.

Calculation of the flux from (8.4.3) with the inner solution yields

$$J=\frac{8kTn^2}{3\mu W}\left(1+0.257\frac{Pe^{1/2}}{W}\right), \tag{8.8.6}$$

so that convection enhances the flux simply by increasing the number of pairs in the inner region. The simplicity of the result, valid for arbitrary W, derives from the fact that convection directly enters only the outer region where interactions are negligible.

Convection-dominated flocculation: $Pe \gg 1$

In strong flows the character of the phenomenon changes substantially as particles largely follow the fluid streamlines, deviating significantly only owing to hydrodynamic interactions and interparticle forces. The early work of Smoluchowski (1917) and more recent efforts (van de Ven & Mason, 1976; Zeichner & Schowalter, 1977) capitalized on the deterministic nature of the process in the absence of Brownian motion, calculating trajectories of interacting particles to identify those leading to doublet formation. Subsequently, Feke & Schowalter (1983) generalized the analysis to include weak Brownian motion at large but finite Pe. Here we outline their approach for systems with no repulsion, i.e. $N_r=0$.

For $Pe \gg 1$, a regular perturbation expansion with

$$P_2=n^2\left(p_0+\frac{p_1}{Pe}+\ldots\right) \tag{8.8.7}$$

succeeds in generating a well-posed hierarchy of equations:

$$\begin{aligned} &O(1) && \nabla_r \cdot p_0(\mathbf{U}-\boldsymbol{\omega}\cdot\nabla_r\Phi)=0 \\ &O(Pe^{-1}) && \nabla_r \cdot p_1(\mathbf{U}-\boldsymbol{\omega}\cdot\nabla_r\Phi)=\nabla_r\cdot\boldsymbol{\omega}kT\cdot\nabla_r p_0, \end{aligned} \tag{8.8.8}$$

with the boundary conditions

$$\begin{aligned} &p_0=p_1=0 && \text{at } r=2a \\ &p_0\to 1 \text{ and } p_1\to 0 && \text{as } r\to\infty \text{ upstream.} \end{aligned}$$

Substitution of the expansion into (8.4.3) results in

$$J=a^3n^2\gamma\left(j_0+\frac{j_1}{Pe}+\ldots\right), \tag{8.8.9}$$

with

$$a^3\gamma j_0 = -\int_{r=2a} (\mathbf{U} - \boldsymbol{\omega}\cdot\nabla_r\Phi)\cdot\mathbf{n}p_0\,\mathrm{d}\mathbf{r},$$

$$a^3\gamma j_1 = -\int_{r=2a} \{(\mathbf{U} - \boldsymbol{\omega}\cdot\nabla_r\Phi)p_1 - \boldsymbol{\omega}kT\cdot\nabla_r p_0\}\cdot\mathbf{n}\,\mathrm{d}\mathbf{r}.$$

The equations (8.8.8) comprise first-order partial differential equations which can be integrated along characteristics defined for simple shear flow by

$$\begin{aligned}
\frac{\mathrm{d}r}{\mathrm{d}t} &= \gamma r(1-A)\sin^2\theta\sin\phi\cos\phi - \frac{G}{3\pi\mu a}\frac{\mathrm{d}\Phi}{\mathrm{d}r},\\
\frac{\mathrm{d}\phi}{\mathrm{d}t} &= -\gamma\{\sin^2\phi + \tfrac{1}{2}B(\cos^2\phi - \sin^2\phi)\},\\
\frac{\mathrm{d}\theta}{\mathrm{d}t} &= \gamma(1-B)\sin\theta\cos\theta\sin\phi\cos\phi,
\end{aligned} \tag{8.8.10}$$

with A, B, and G the hydrodynamic functions described in §§2.8 and 2.9. These equations determine the trajectories of interacting spheres, generalizing (2.9.2, 17) to include the interparticle forces. In addition, (8.8.8) dictate that p_0 and p_1 vary along the characteristics according to

$$\begin{aligned}
\frac{\mathrm{d}p_0}{\mathrm{d}t} &= -p_0\left\{\gamma S\sin^2\theta\sin\phi\cos\phi \quad \frac{\mathrm{d}}{\mathrm{d}r}\left(\frac{G}{3\pi\mu a}\frac{\mathrm{d}\Phi}{\mathrm{d}r}\right)\right\},\\
\frac{\mathrm{d}p_1}{\mathrm{d}t} &= -p_1\left\{\gamma S\sin^2\theta\sin\phi\cos\phi - \frac{\mathrm{d}}{\mathrm{d}r}\left(\frac{G}{3\pi\mu a}\frac{\mathrm{d}\Phi}{\mathrm{d}r}\right)\right\}\\
&\quad + \nabla_r\cdot\boldsymbol{\omega}kT\cdot\nabla_r p_0,
\end{aligned} \tag{8.8.11}$$

with $S = 3(A-B) + r\mathrm{d}A/\mathrm{d}r$. The bracketed terms represent the non-zero divergence of the relative velocity of the pair, effectively crowding the trajectories and increasing the pair density. The $O(Pe)$-diffusion down probability gradients established by the convection process also alters p_1. Calculation of the trajectories and p_0 involves straightforward integration beginning far upstream where $p_0 = 1$. Determining p_1 along a particular trajectory, however, requires knowledge of adjacent values of p_0 in order to evaluate the diffusion term.

As shown for Brownian flocculation, an analysis neglecting long-range interactions provides a useful baseline for understanding the effects of hydrodynamic and dispersion forces. Then the trajectories simply follow the undisturbed fluid streamlines until the spheres collide, so that $\mathbf{U}=\gamma(r\sin\theta\sin\phi, 0, 0)$ and $p_0=1$, except in a zone downstream with $y^2+z^2=4a^2$ for $x>0$ and $y>0$ or $x<0$ and $y<0$. The flux in a simple shear flow, calculated as

$$a^3n^2\gamma j_0 = -2n^2\int_{\pi/2}^{\pi}\int_0^{\pi}\mathbf{U}\cdot\mathbf{n}(2a)^2\sin\theta\,\mathrm{d}\theta\,\mathrm{d}\phi = \tfrac{32}{3}a^3n^2\gamma, \qquad (8.8.12)$$

increases linearly with the shear rate and is proportional to ϕn.

As already noted in §2.9, hydrodynamic interactions displace spheres from the linear trajectories derived from this simplistic analysis. In addition, an attraction converts the closed trajectories that are due to hydrodynamic interactions in a simple shear flow (Fig. 2.13) into spirals that eventually lead to doublet formation. Figure 8.15 depicts trajectories in the vertical plane $z=0$ without repulsion. In Region 1, the trajectories are

Fig. 8.15. Trajectories in the $z=0$ plane for two spheres in a simple shear flow, with $N_r=0$ and $N_f=1$ (Feke & Schowalter, 1983): 1, open trajectories extending from upstream to downstream infinity; 2, trajectories leading to collision after crossing the plane of shear and reversing direction; 3, trajectories leading to collision before crossing the plane of shear.

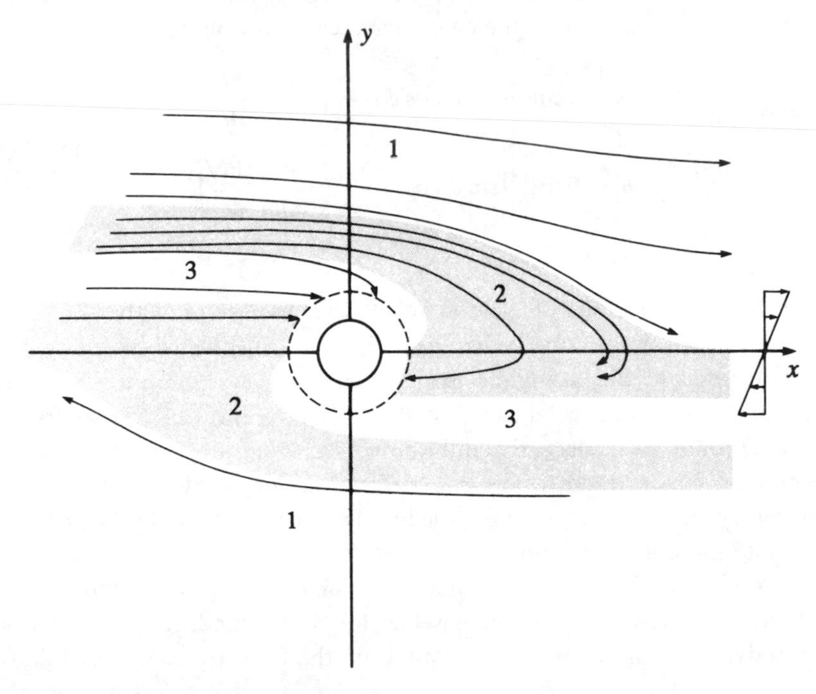

open, extending from upstream infinity to downstream infinity. Those in Region 2 are displaced across the plane of shear downstream by hydrodynamic interactions, reversing the direction of motion and providing a second opportunity for the attraction to form a doublet. Trajectories in Region 3 lead to doublet formation before crossing the plane of shear. While $p_0 = 1$ in the portions of Regions 1, 2, and 3, far upstream, this is not true in the part of Region 2 across the plane of shear.

The effects of these interactions on the rate of doublet formation for $N_r = 0$ and $Pe = \infty$ are reflected in j_0 calculated with the non-retarded Hamaker form for the dispersion force (Fig. 8.16). Only for $N_f \approx 0.5$ is $j_0 \approx \frac{32}{3}$, as expected from (8.8.12). In general, j_0 decreases with increasing N_f as the viscous forces become stronger relative to the attractive interparticle

Fig. 8.16. Dimensionless rate of doublet formation for spheres in a simple shear flow at $Pe = \infty$, calculated with the non-retarded Hamaker form for the dispersion potential and no repulsion (Feke & Schowalter, 1983).

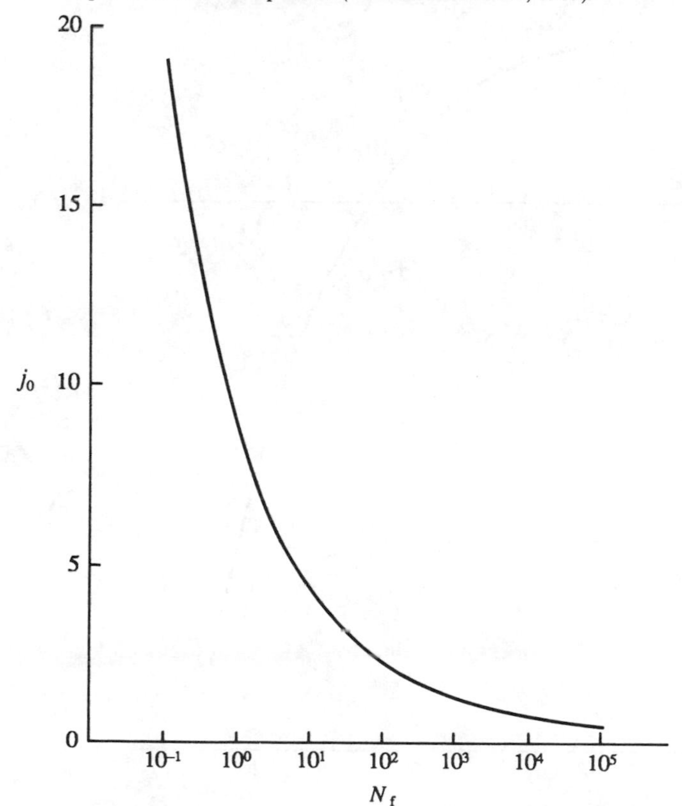

potential. This means that J increases slower than linearly with increasing shear rate, i.e. viscous interactions retard the formation of doublets as in Brownian flocculation.

These results demonstrate two quantitative distinctions between Brownian and shear flocculation. The first is the difference in time scales given by

$$\frac{t_{\mathrm{p}}^{\mathrm{Br}}}{t_{\mathrm{p}}^{\mathrm{sh}}}=\frac{4}{3\pi}\,Pe\,\frac{W}{W^{\mathrm{sh}}},$$

Fig. 8.17. $O(1/Pe)$ correction to the rate of doublet formation for spheres in a simple shear flow calculated with the same interparticle potential as in Fig. 8.16 (Feke & Schowalter, 1983).

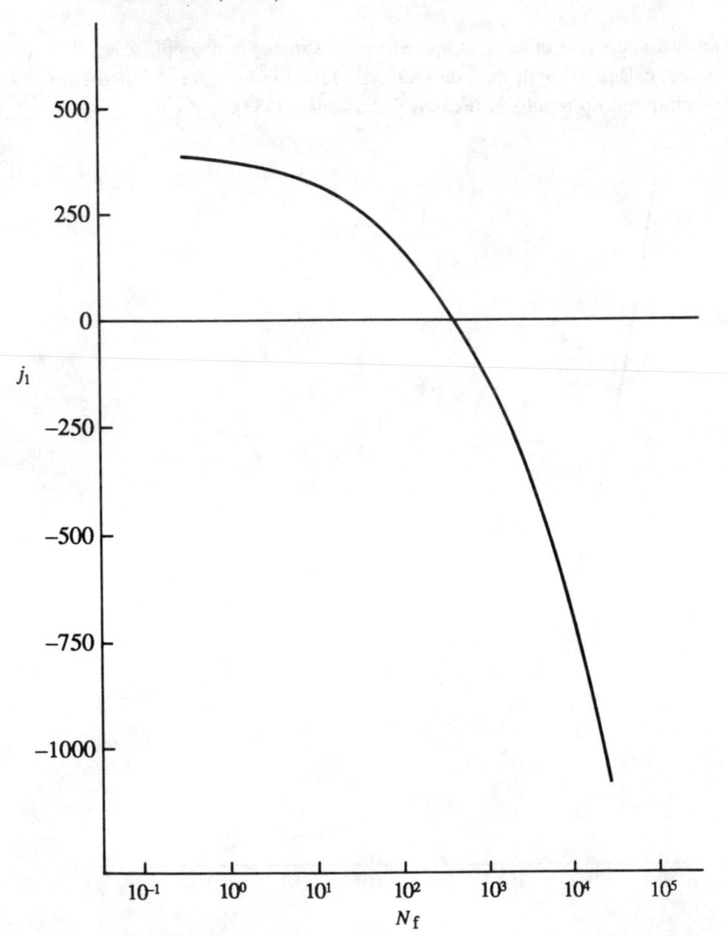

with $W^{sh} \equiv 32/3j_0$. For $A_{eff}/kT = 2$, the ratio of the time scales is about 15 at $Pe = 10^2$ and 500 at $Pe = 10^4$. The second is the greater sensitivity to the effective Hamaker constant. Recall that W_∞ varies less than a factor of two with a hundredfold change in A_{eff}/kT (Fig. 8.4). Over a similar range of N_f, 0.1–10, j_0 and $1/W^{sh}$ decrease by a factor of five. In addition, retardation can affect shear flocculation more, since particles are generally larger and the attraction acts at large distances in regions of weak flow.

The results for j_1 (Fig. 8.17) indicate, rather surprisingly, that a little Brownian motion can either enhance or reduce the rate of flocculation, depending on the spatial variations in p_0. With $N_f = 1$, diffusion from Region 3 (Fig. 8.15), where $p_0 > 1$, into Region 1, where flow separates the particles, is unlikely; so diffusion toward contact, where $p_0 = 0$, dominates and the result is $j_1 > 0$. For larger N_f, however, Region 3 diminishes in size, thereby increasing the gradient, driving diffusion away from contact, and diminishing j_1. The magnitudes of j_0 and j_1 suggest a significant effect of

Fig. 8.18. Rate of doublet formation for spheres as a function of Pe, showing the results from the perturbation expansions in the $Pe \ll 1$ (8.8.6) and $Pe \gg 1$ (8.8.9) limits for $A_{eff}/kT = 2$ (——) and 20 (- - -) and no repulsion ($N_r = 0$).

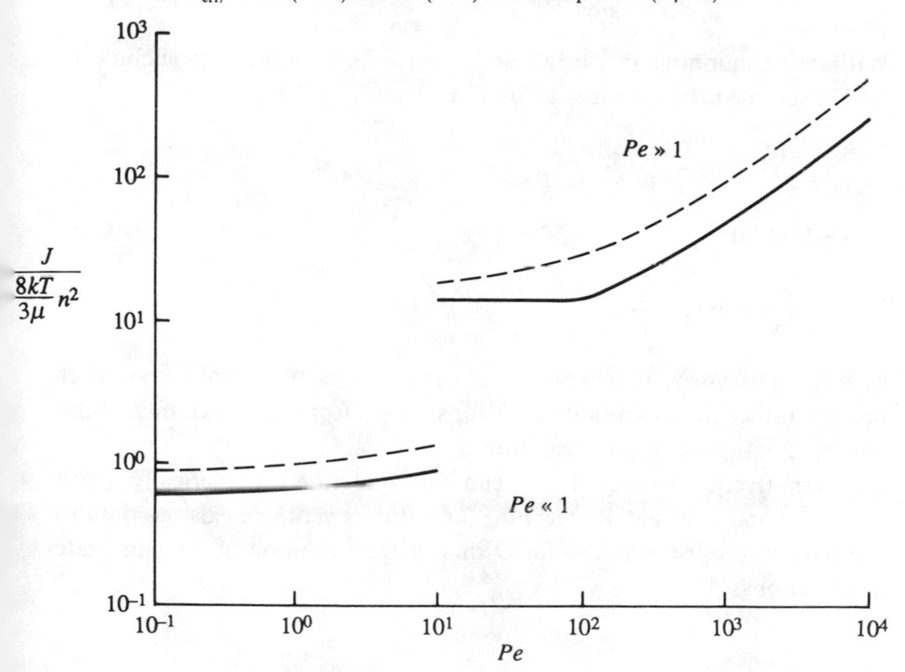

Brownian motion even at substantial Peclet numbers. However, the expansion remains valid only if

$$\frac{j_1}{Pe\, j_0} = \frac{2kT j_1}{A_{\rm eff} N_{\rm f}\, j_0} \ll 1,$$

which requires $A_{\rm eff}/kT > 50$ for $Pe = 10$ and $A_{\rm eff}/kT > 1$ when $Pe = 10^3$–10^4.

These asymptotic results reflect a complex coupling between convection and diffusion at finite values of the Peclet number. The rate of doublet formation depends on the detailed form of the pair probability density resulting from the combination of shear, interparticle forces, and Brownian motion. For $Pe \ll 1$ shear always increases the rate, but for $Pe \gg 1$ diffusion can either enhance or retard the process for $N_{\rm r} = 0$. In the intermediate range, $1 < Pe < 100$, the situation is uncertain. Extension of the asymptotic results beyond their ranges of validity (Fig. 8.18) reveals no region of overlap and no clear means of interpolating between them.

Growth of large aggregates in shear

The population balances (8.7.2) apply equally well to shear flocculation, but require the generalized form of the collision rate for unequal spheres,

$$J_{ij} = \tfrac{4}{3}(a_i + a_j)^3 n_i n_j \gamma / W_{ij}^{\rm sh}. \tag{8.8.13}$$

With the assumptions $W_{ij}^{\rm sh} = W^{\rm sh}$ and $a_i + a_j = 2a_i$ for $a_i > a_j$, Smoluchowski (1917) summed the balances to obtain

$$\frac{{\rm d}n_{\rm tot}}{{\rm d}t} = -\frac{4}{\pi}\frac{\phi\gamma n_{\rm tot}}{W^{\rm sh}},$$

which leads to (8.8.14)

$$\frac{n_{\rm tot}}{n_0} = \exp\left(-\frac{t}{t_{\rm p}}\right),$$

with $t_{\rm p} = \pi W^{\rm sh}/4\phi\gamma$. This exponential decay, characteristic of a first-order process rather than two-body collisions, arises from the invariance of the volume fraction during flocculation.

Alternatively, the equations can be integrated numerically, with $a_k = ak^{1/D}$ and constant $W^{\rm sh}$ as in (8.7.6). But as yet there exist no data to establish appropriate values for D or test the evolution of the aggregate distributions.

8.9 Criteria for mechanical stability

Although the effect of electrostatic repulsion on the initial rate of shear flocculation can be calculated by the approach described above, few results are available. Instead, attention has focused on the limits of stability in the absence of Brownian motion, i.e. the conditions for which $j_0 \to 0$. In a simple shear flow, this requires only consideration of a pair initially aligned along the y-axis with $x = -\infty$. If this trajectory does not lead to the formation of a doublet, then the dispersion is stable to shear.

The question of stability hinges on the magnitude of the viscous force F_{vis}, pushing particles together on the upstream portion of the trajectory and pulling them apart downstream, relative to the interparticle force. The interparticle potential, e.g. Figs 8.1 and 8.2, typically has three extrema in the force: $-(\mathrm{d}\Phi/\mathrm{d}r)_{min} < 0$, coincident with the primary minimum; $-(\mathrm{d}\Phi/\mathrm{d}r)_{max} > 0$, just beyond the primary maximum; and $-(\mathrm{d}\Phi/\mathrm{d}r)_{sec} < 0$, associated with the secondary minimum. Particles remain dispersed if the viscous force exceeds all three in magnitude or if $(\mathrm{d}\Phi/\mathrm{d}r)_{sec} < F_{vis} < (-\mathrm{d}\Phi/\mathrm{d}r)_{max}$. Primary minimum aggregation can occur for $(-\mathrm{d}\Phi/\mathrm{d}r)_{max} < F_{vis} < (\mathrm{d}\Phi/\mathrm{d}r)_{min}$ and secondary minimum aggregation for $F_{vis} < (\mathrm{d}\Phi/\mathrm{d}r)_{sec} < (-\mathrm{d}\Phi/\mathrm{d}r)_{max}$.

Numerical integration of the trajectory equations (van de Ven & Mason, 1976; Zeichner & Schowalter, 1977) produces stability diagrams defining the state of a dispersion as a function of the dimensionless groups characterizing the relative strengths of the interparticle and viscous forces. With the linearized, thin-double-layer approximation for the repulsion (4.10.10) and the non-retarded Hamaker form for the attraction (5.2.15), N_r, N_f, $a\kappa$, and δ/a determine the behavior (Fig. 8.19). With weak repulsion, i.e. small N_r, there is no repulsive barrier and primary minimum aggregation comprises the rest state. Only if the viscous force exceeds the attraction at contact, e.g. for $N_f > 10^5$ if $\delta/a < 10^{-3}$, is dispersion achieved. At sufficiently large values of N_r, a repulsive barrier exists but secondary minimum aggregation persists in the absence of flow. Increasing the shear rate, and thereby N_f, first extracts particles from the secondary minimum, then pushes them over the repulsive barrier into the primary minimum, and finally returns them to the dispersed state.

Reasonable analytical approximations for these stability boundaries can be derived by balancing the interparticle force $-\mathrm{d}\Phi/\mathrm{d}r$ against the radial component of the viscous force such that $\mathrm{d}r/\mathrm{d}t = 0$ in (8.8.10). For simple shear, the extrema in the viscous force for spheres correspond to $\sin\theta = 1$ and $\sin\phi\cos\phi = \pm\frac{1}{2}$, while near contact $r \approx 2a$ and $(1-A)/G \approx 2$. Thus the

Fig. 8.19. Stability diagram for two spheres in a simple shear flow with $Pe = \infty$, based on the linearized Derjaguin approximation, with constant-potential boundary conditions for the electrostatic repulsion and the non-retarded Hamaker form for the attraction (Zeichner & Schowalter, 1977).

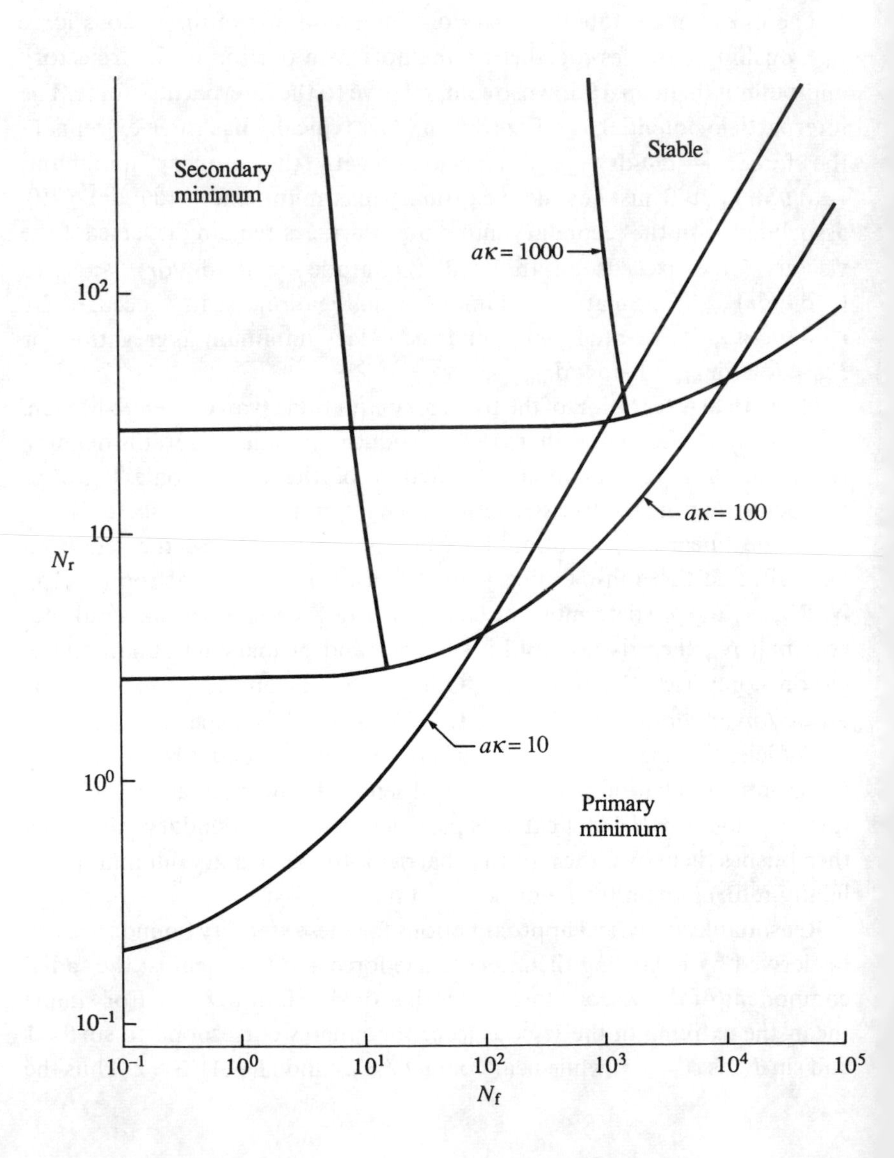

stability boundaries should correspond to

$$-\frac{\mathrm{d}\Phi}{\mathrm{d}r} = \pm 6\pi\mu a^2\gamma, \tag{8.9.1}$$

with the sign depending upon whether the formation or breakup of a doublet is of interest. The positions of the extrema in the interparticle force are determined by $\mathrm{d}^2\Phi/\mathrm{d}r^2 = 0$, or

$$\frac{\exp[\kappa(r-2a)]\{1+\exp[-\kappa(r-2a)]\}^2}{[\kappa(r-2a)]^3} = 12\pi\frac{N_\mathrm{r}}{a\kappa}. \tag{8.9.2}$$

Analytical approximations are possible for $12\pi N_\mathrm{r}/a\kappa \gg 1$:

(i) The maximum attraction in the secondary minimum, where $\kappa(r-2a) \gg 1$, occurs at

$$\kappa(r-2a) \sim \ln\left\{\frac{12\pi N_\mathrm{r}}{a\kappa}\left[\ln\frac{12\pi N_\mathrm{r}}{a\kappa}\left(\ln\frac{12\pi N_\mathrm{r}}{a\kappa}\ldots\right)^3\right]^3\right\}. \tag{8.9.3}$$

Substitution into (8.9.1) with the explicit form for Φ fixes the stability boundary between secondary minimum flocculation and free pairs as

$$N_\mathrm{f} \sim \frac{(a\kappa)^2}{12\left(\ln\frac{12\pi N_\mathrm{r}}{a\kappa}\ldots\right)^2}. \tag{8.9.4}$$

(ii) The maximum repulsion appears at

$$\kappa(r-2a) \sim \left(\frac{a\kappa}{3\pi N_\mathrm{r}}\right)^{1/3} \ll 1, \tag{8.9.5}$$

yielding the boundary between free pairs and primary minimum flocculation as

$$N_\mathrm{f} \sim \pi a\kappa N_\mathrm{r}\left[1 - 0.118\left(\frac{a\kappa}{N_\mathrm{r}}\right)^{1/3}\right]. \tag{8.9.6}$$

(iii) The repulsive barrier disappears when (8.9.2) ceases to have a solution, or

$$\frac{N_\mathrm{r}}{a\kappa} < 0.0215, \tag{8.9.7}$$

which defines the boundary between primary and secondary minimum flocculation at rest.

(iv) To pull a pair out of the primary minimum the viscous force must exceed the attractive force at the minimum separation δ, which translates into

$$N_f = \frac{1}{12}\left(\frac{a}{\delta}\right)^2\left[1 - 12\pi a\kappa N_r\left(\frac{\delta}{a}\right)^2\right]. \tag{8.9.8}$$

These asymptotic results actually represent the stability diagram quite well (Fig. 8.20), and provide a means for treating other values of $a\kappa$ and δ/a. The discrepancies arise from the asymptotic estimation of the extrema and the fact that these force balances comprise a pseudo-steady approximation. In reality, the vorticity of the flow field rotates the pair, limiting the time available for the viscous forces to overcome the interparticle force. For flows without vorticity, such as uniaxial extension, the force balance gives the exact result (Zeichner & Schowalter, 1977).

To illustrate magnitudes, consider 1.0 μm radius polystyrene spheres in water with an ionic strength of 3×10^{-3} M and $e\psi_s/kT = 1$ so that $a\kappa = 10^2$ and $N_r = 42$. For surfaces coated by a surfactant layer, $\delta = 1$ nm might be reasonable. Then the shear rates corresponding to the stability boundaries are as follows:

Fig. 8.20. Comparison of the approximate and exact stability boundaries for $a\kappa = 10^2$ as in Fig. 8.19: – – –, (8.9.4); ···, (8.9.7); – – –, (8.9.6); –●–, (8.9.8).

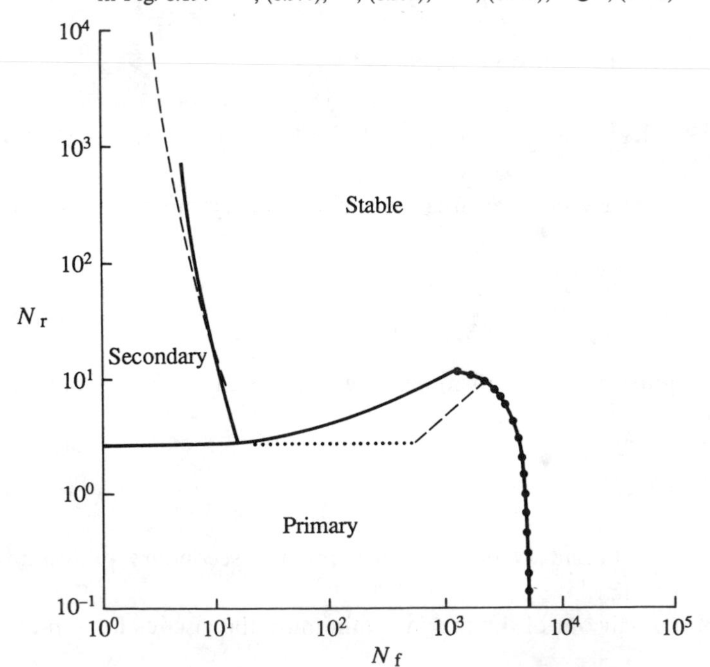

	γ/s^{-1}
Secondary minimum→stable	5
Stable→primary minimum	6 100
Primary minimum→stable	38 000

Clearly, separating two particles from the primary minimum requires extreme conditions.

8.10 Experimental studies of shear flocculation

Several groups have measured the rate of shear flocculation in the absence of repulsion, but, as with the Brownian case, have actually detected the formation of aggregates larger than doublets. However, Feke & Schowalter (1985), sheared samples of 0.675 μm diameter polystyrene spheres in 0.6 M NaCl solutions at constant rate in a concentric cylinder apparatus for times less than $0.25t_p$ and then diluted for low-angle light-scattering measurements of the degree of aggregation. The resulting data clearly represent doublet formation rates and best test the theory described above.

Figure 8.21 summarizes the results as the stability ratio W^{sh} plotted against the Peclet number. The solid curve is the prediction of the asymptotic expansion (8.8.9) for $Pe \gg 1$ based on $N_r = 0$ and $A_{eff}/kT = 1.2$, as deduced from the value obtained for W_∞ in Brownian flocculation experiments. The broken-line curve is the prediction for $A_{eff}/kT = 2.5$, as derived from the continuum theory (§5.6) for polystyrene in water at high ionic strengths. The difference clearly lies outside the range of experimental error.

At the experimental conditions, the validity of both the non-retarded Hamaker form for the dispersion force and the asymptotic expansion appear somewhat questionable. The critical portion of the limiting trajectory, separating Region 1 from Region 2 (Fig. 8.15), corresponds to the point where the viscous force pulling the spheres apart just equals the attraction, i.e.

$$\frac{aA_{eff}}{12(r-2a)^2} = 6\pi\mu a^2\gamma, \tag{8.10.1}$$

or

$$\left(\frac{r-2a}{a}\right)_{crit} = 0.29N_f^{-1/2}.$$

For $A_{eff}/kT = 2.5$ and the shear rates of the experiments, $(r-2a)_{crit} = 19$–53 nm, for which Fig. 5.6 indicates significant retardation of the dispersion force, with $0.6 \leqslant (A_{eff}(r-2a)/kT)_{crit} \leqslant 1.5$. In addition, the attraction is significant at much longer range in the regions of weak flow near the plane of shear. Thus a smaller apparent value for the Hamaker constant is not surprising. Note also that the magnitude of the $O(Pe^{-1})$ correction relative to the $O(1)$ term in the theory ranges from 4.3 at the lowest shear rate to 0.41 at the highest. This means that the accuracy of the asymptotic expansion is assured only at the high end of the range. Nonetheless, the theory remains in reasonable accord with the data.

The only direct observation of mechanical stability is van de Ven & Mason's (1977) study of electrostatically stabilized 1.0 μm radius polystyrene latices in their traveling microtube apparatus. With an ionic strength of 10^{-3} M and $\psi_s = -41$ mV, a shear rate of $\gamma \approx 16\,s^{-1}$ pulled particles out of the secondary minimum, while 24–36 s^{-1} produced flocculation into the primary minimum, in qualitative accord with the predictions. The quantitative comparison is less satisfying. These

Fig. 8.21. Comparison of the stability ratio predicted by the two-term expansion (8.8.9), with $N_r = 0$ and $A_{eff}/kT = 1.2$ (——) or 2.5 (- - -), with the values measured (I-symbols) for polystyrene latices with $2a = 0.675$ μm (Feke & Schowalter, 1985).

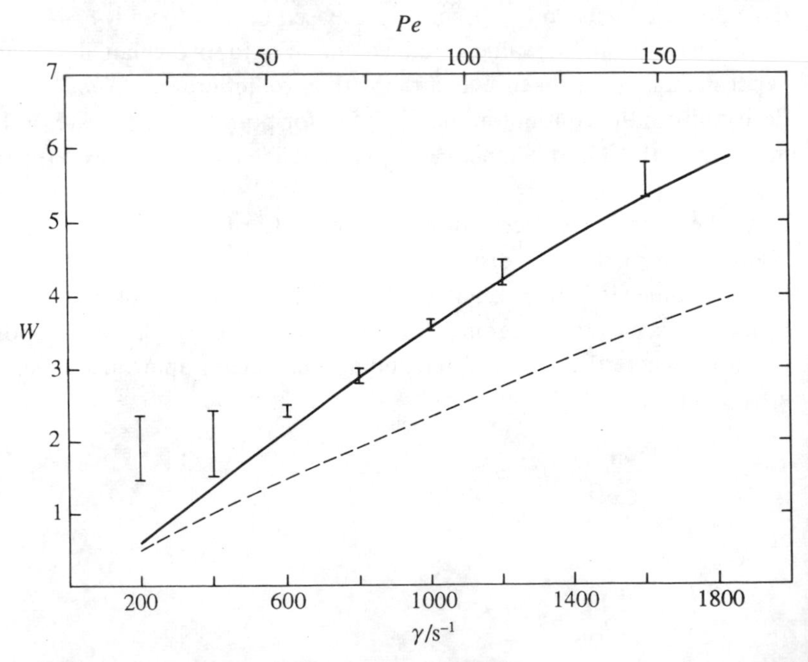

conditions correspond to $N_r = 112$ ($A_{eff}/kT = 2.5$) and $a\kappa = 10^2$, for which the observed values of N_f at the secondary minimum-stable and stable-primary minimum boundaries, 29 and 42–64, respectively, lie well away from the predicted boundaries in Fig. 8.20.

8.11 Summary

Analyses of doublet formation still form the cornerstone for the understanding of electrostatic stabilization. The theory now includes rigorous treatments of hydrodynamic interactions and the interparticle potentials for spheres. Predictions of the rate for rapid flocculation in both quiescent ($Pe = 0$) and sheared ($Pe \gg 1$) suspensions agree with measurements, if allowance is made for retardation of the dispersion potential. With significant repulsion, rates are much slower. Predictions are essentially complete for flocculation dominated by Brownian motion ($Pe \ll 1$), but limited for shear flocculation, mainly dealing with mechanical stability. Both agree qualitatively with the available data but are not quantitative. Under these conditions, the magnitude of the electrostatic and dispersion forces are large but the difference is small. Hence small uncertainties in the parameters characterizing these forces, or in the underlying theory, lead to large uncertainties in the predicted rates.

The growth of larger flocs has received renewed attention with the development of simple techniques for characterizing them both theoretically, as fractals, and experimentally, through scattering of radiation. The kinetics attributed to Smoluchowski accurately describe rapid Brownian flocculation, but fail when repulsions slow the process. Computer simulations and scattering experiments reveal fundamental changes in structure with mode of flocculation, providing a guide for the next generation of theory. Comparable studies of growth under shear have not been performed.

References

Adler, P. M. (1981). Heterocoagulation in shear flow. *J. Colloid Interface Sci.* **83**, 106–15.

Derjaguin, B. V. (1966). Colloid Stability. *Disc. Far. Soc.* **42**, 317–21.

Derjaguin, B. V. & Landau, L. (1941). Theory of the stability of strongly charged lyophobic sols and the adhesion of strongly charged particles in solutions of electrolytes. *Acta Physicochim. URSS* **14**, 633–62.

Derjaguin, B. V. & Muller, V. M. (1967). Slow coagulation of hydrophobic colloids. *Dokl. Akad. Nauk SSSR* **176**, 738–41.

Dimon, P., Sinha, S. K., Weitz, D. A., Safinya, C. R., Smith, G. S., Varady, W. A. & Lindsay, H. M. (1986). Structure of aggregated gold colloids. *Phys. Rev. Lett.* **57**, 595–8.

Feder, J. (1988). *Fractals.* Plenum Press.

Feke, D. L. & Schowalter, W. R. (1983). The effect of Brownian diffusion on shear-induced coagulation of colloidal dispersions. *J. Fluid Mech.* **133**, 17–35.

Feke, D. L. & Schowalter, W. R. (1985). The influence of Brownian diffusion on binary flow-induced collision rates in colloidal dispersions. *J. Colloid Interface Sci.* **106**, 203–14.

Freundlich, H. (1910). Die Bedeutung der Adsorption bei der Fallung der Suspensions kolloide. *Z. Physik. Chem.* **73**, 385–423.

Fuchs, N. (1934). Über der stabilität und aufladung der aerosole. *Z. Phys.* **89**, 736–43.

Hardy, W. B. (1900). A preliminary investigation of the conditions which determine the stability of irreversible hydrosols. *Proc. Roy. Soc. Lond.* **66**, 110–25.

Higashitani, K. & Matsuno, Y. (1979). Rapid Brownian coagulation of colloidal dispersions. *J. Chem. Eng. Japan* **12**, 460–5.

Hogg, R., Healy, T. W. & Fuerstenau, D. W. (1966). Mutual coagulation of colloidal dispersions. *Far. Soc. Trans.* **62**, 1638–51.

Honig, E. P., Roeberson, G. J. & Wiersema, P. H. (1971). Effect of hydrodynamic interaction on the coagulation rate of hydrophobic colloids. *J. Colloid Interface Sci.* **36**, 97–109.

Hough, D. B. & White, L. R. (1980). The calculation of Hamaker constants from Lifshitz theory with applications to wetting phenomena. *Adv. Colloid Interface Sci.* **14**, 3–41.

Hunter, R. J. (1987). *Foundations of Colloid Science.* Clarendon Press.

Jullien, R., Kolb, M. & Botet, R. (1984). Chemically limited versus diffusion limited aggregation. *J. Phys.* (*Paris*) *Lett.* **45**, L977–81.

LaMer, V. K. (1964). Coagulation of colloids. *J. Colloid Sci.* **19**, 291–2.

Lichtenbelt, J. W. Th., Pathmamanoharan, C. & Wiersema, P. H. (1974). Rapid coagulation of polystyrene latex in a stopped-flow spectrophotometer. *J. Colloid Interface Sci.* **49**, 281–5.

Lips, A. & Willis, W. E. (1973). Low angle light scattering technique for the study of coagulation. *J. Chem. Soc. Far. Trans. I* **69**, 1226–36.

Matthews, B. A. & Rhodes, C. T. (1970). Studies of the coagulation kinetics of mixed suspensions. *J. Colloid Interface Sci.* **32**, 332–8.

McQuarrie, D. A. (1976). *Statistical Mechanics.* Harper & Row.

Meakin, P. (1983). The Vold–Sutherland and Eden models of cluster formation. *J. Colloid Interface Sci.* **96**, 415–24.

Napper, D. H. (1983). *Polymeric Stabilization of Colloidal Dispersions.* Academic Press.

Ottewill, R. H. & Shaw, J. N. (1966). Stability of monodisperse polystyrene latex dispersions of various sizes. *Disc. Far. Soc.* **42**, 154–63.

Parfitt, G. D. (ed.) (1981). *Dispersion of Powders in Liquids*, 3rd edn. Applied Science Publ.
Parsegian, V. A. (1975). Long range van der Waals forces. In *Physical Chemistry: Enriching Topics in Colloid and Surface Science* (eds H. van Olphen and K. J. Mysels), pp. 27–72. Theorex.
Prieve, D. C. & Ruckenstein, E. (1980). Role of surface chemistry in primary and secondary coagulation and heterocoagulation. *J. Colloid Interface Sci.* **73**, 539–55.
Reerinck, H. & Overbeek, J. Th. G. (1954). The rate of coagulation as a measure of the stability of silver iodide sols. *Disc. Far. Soc.* **18**, 74–84.
Schaefer, D. W., Martin, J. E., Wiltzius, P. & Cannell, D. S. (1984). Fractal geometry of colloidal aggregates. *Phys. Rev. Lett.* **52**, 2371–4.
Schulze, H. (1882). Schwefelarsen im wässeriger Lösung. *J. Prakt. Chem.* **25**, 431–52. (1883). Antimontrisulfid im wässeriger Lösung. **27**, 320–32.
Smoluchowski, M. von (1917) Versuch einer mathematischen Theorie der Koagulationkinetik kolloider Lösungen. *Z. Phys. Chem.* **92**, 129–68.
Sonntag, R. & Russel, W. B. (1986). Structure and breakup of flocs subjected to fluid stresses. *J. Colloid Interface Sci.* **113**, 399–413.
Spielman, L. A. (1970). Viscous interactions in Brownian coagulation. *J. Colloid Interface Sci.* **33**, 562–71.
van de Ven, T. G. M. & Mason, S. G. (1976). The microrheology of colloidal dispersions. IV. Pairs of interacting spheres in a shear flow. *J. Colloid Interface Sci.* **57**, 505–16. V. Primary and secondary doublets of spheres in shear flow. *J. Colloid Interface Sci.* **57**, 517–34.
van de Ven, T. G. M. & Mason, S. G. (1977). The microrheology of colloidal dispersions. VIII. The effect of shear on perikinetic doublet formation. *Colloid Poly. Sci.* **255**, 794–804.
van Dyke, M. (1975). *Perturbation Methods in Fluid Mechanics.* Parabolic Press.
Verwey, E. J. W. & Overbeek, J. Th. G. (1948). *Theory of the Stability of Lyophobic Colloids.* Elsevier.
Warren, L. J. (1975). Shear flocculation of ultrafine scheelite in sodium oleate solutions. *J. Colloid Interface Sci.* **50**, 307–18.
Weitz, D. A. & Huang, J. S. (1984). Self-similar structures and the kinetics of aggregation of gold colloids. In *Kinetics of Aggregation and Gelation* (eds P. Family and D. P. Landau), p. 19. Elsevier.
Weitz, D. A., Lin, M. Y. & Huang, J. S. (1986). Fractals and scaling in kinetic colloid aggregation. In *Complex and Supramolecular Fluids* (eds S. A. Safron and N. A. Clark), pp. 509–49. Wiley-Interscience.
Wiese, G. R. & Healy, T. W. (1970). Effect of particle size on colloid stability. *Far. Soc. Trans.* **66**, 490–9.
Witten, T. A., Jr. & Sander, L. M. (1981). Diffusion limited aggregation, a kinetic critical phenomenon. *Phys. Rev. Lett.* **47**, 1400–3.

Zeichner, G. R. & Schowalter, W. R. (1977). Use of trajectory analysis to study stability of colloidal dispersions in flow fields. *AIChE J.* **23**, 243–54.

Zeichner, G. R. & Schowalter, W. R. (1979). Effects of hydrodynamic and colloidal forces on the coagulation of dispersions. *J. Colloid Interface Sci.* **71**, 237–53.

Problems

1 Construct the interaction potential between spheres in aqueous solution by superimposing the non-linear Derjaguin approximation for the electrostatic repulsion (Problem 4.8) and the modified Hamaker form (5.10.1) with (5.9.3) for the dispersion attraction. Then, for silica spheres of radius 75 nm with a surface potential of 50 mV, determine the ionic strengths for which (i) $\Phi_{max}/kT=10$ and (ii) $\Phi_{max}/kT=0$ and plot the interaction potentials as functions of the dimensionless separation.

2 Evaluate the stability ratio, with suitable approximations for the electrostatic and dispersion potentials and either the exact G or an approximation constructed from the near- and far-field forms, to demonstrate
(i) the effect of $a\kappa$ on W with all other parameters fixed and
(ii) the effect of retardation on W_∞ as reflected by the dimensionless relaxation frequency $n_0(\bar{n}_0^2+n_0^2)^{1/2}a\omega/c$ (cf. 5.9.3).

3 The solution (8.7.4) and (8.7.5) for the growth of large flocs due to Brownian collisions depends on the assumption (8.7.3). Plot the predictions for $n_k(t)$ from this solution as a function of k for several t/t_p and the actual form of $(a_i+a_j)(1/a_i+1/a_j)$ as a function of i/j for $D=1.6$. From these results ascertain
(i) possible reasons for the success of the approximation and
(ii) how the exact $n_k(t)$, obtained by integrating (8.7.2) numerically, might deviate from the values predicted by (8.7.4). You might support your assertions with numerical results for short times such that $n_k\approx 0$ for $k\geq 10$.

4 For silica spheres with $a=1\,\mu m$ and $\phi=10^{-5}$ in water containing 1 M NaCl calculate as well as possible from the available theory the rate of doublet formation at shear rates from 0 to 4000 s^{-1}. Express the results in terms of the time scale $t_p=n_0/J_0$.

5 Calculate trajectories for two equal spheres with $\varepsilon_0\varepsilon\psi_s^2 a/A_{eff}(0)=100$ interacting in the plane $\theta=\pi/2$ for $a\kappa=10$ and $\mu a^2\gamma/\epsilon_0\epsilon\psi_s^2=10^{-2}$, 10^{-1}, 1, and 10. Use the superposition approximation for the electrostatic repulsion, the non-retarded Hamaker form for the

dispersion attraction, and the far-field forms for the hydrodynamic functions. Explain the nature of the trajectories as quantitatively as possible in terms of the magnitudes of the various forces.

6 The dispersion of powders into liquids requires that aggregates be ruptured by mechanical stresses. Assess the difficulty of this for 0.5 μm TiO_2 spheres coated with 1 nm surfactant layers in water (cf. Problem 5.5) by calculating the shear rate required to rupture a doublet.

9

POLYMERIC STABILIZATION

9.1 Introduction

As suggested in Chapter 6, the adsorption or anchoring of polymer onto the surface of colloidal particles provides an alternate means of imparting stability. Indeed, polymeric stabilization was exploited by the ancient Egyptians as early as 2500 BC (Napper, 1983, §2.1). They formulated inks by dispersing carbon black particles in a solution of naturally occurring polymer such as casein or gum arabic. Adsorption of the polymer onto the carbon black maintained the dispersion and also allowed redispersion after drying.

Several reasons persist for using polymeric, instead of electrostatic, stabilization. In some aqueous systems, electroviscous effects and the accompanying sensitivity to electrolyte concentration may be undesirable. In non-aqueous solvents with low dielectric constants, and surface charge densities typically one to two orders of magnitude smaller than in water, electrostatic repulsion frequently does not suffice. In addition, polymeric stabilization can be more robust than the electrostatic mode, providing stability for a longer time and at higher solids concentrations. When flocculation or phase separation does occur, it is normally reversible, i.e. a suitable change in the solvent conditions will redisperse the particles spontaneously.

Napper's (1983, §2.4) review of early studies indicates the evolution in the nature of the polymers used for this purpose. Work in the nineteenth century, and the early part of the twentieth, dealt with aqueous systems and employed biopolymers that were generally globular, crosslinked, or highly branched. The recognition that linear polymers that freely interpenetrate in bulk solution can impart similar stability apparently dates to Rehbinder, Lagutina, & Wenstrom (1930), Verwey & de Boer (1938), and Heller &

Pugh (1954). In recent years, block or graft copolymers have emerged as the most effective stabilizers (e.g. Barrett, 1975). These combine a soluble, non-adsorbing polymer with a nominally insoluble polymer intended to adsorb irreversibly onto the particle. In some cases, the same effect can be achieved without the insoluble block by covalently bonding the soluble polymer directly to the particle (Laible & Hamann, 1980).

The requirements for polymeric stabilization follow from the interaction potentials discussed in previous chapters. First, the polymeric layer must prevent particles from approaching one another close enough for the dispersion forces to become significant. This implies a minimum thickness, dependent on the particle size and the magnitude of the Hamaker constant. For thicker layers the polymer-induced forces control the interparticle potential. If the interaction between the polymeric layers is purely repulsive then the dispersion will be stable.

For adsorbed homopolymers, the results of §6.5 demonstrate the interaction potential to be entirely repulsive only for strongly adsorbing polymer at full coverage in a good solvent. However, the repulsion reflects a constrained equilibrium for the polymer in the gap. Macromolecules squeezed between surfaces at separations less than the coil dimension have free energy greater than those in the bulk solution. This drives slow desorption, reducing the repulsion and inducing slow flocculation, or ageing, in a concentrated dispersion. Thus homopolymers are not entirely satisfactory.

With terminally anchored chains, on the other hand, good solvent conditions always yield a repulsion. Hence the most effective stabilizers tend to be copolymers, e.g. diblock or comb, and chains grafted directly onto particles (Fig. 9.1). Advances in polymer synthesis continue to expand the variety of individual polymers available in these forms.

The current understanding of polymeric stabilization evolved largely from a series of experiments by Napper and co-workers identifying the critical flocculation point, generally a well-defined temperature, pressure,

Fig. 9.1. Stabilizers attached to surfaces in the form of (*a*) diblock and (*b*) comb copolymers and (*c*) by grafting.

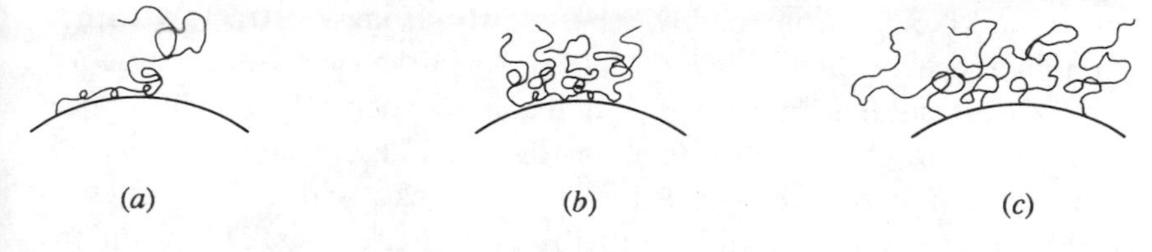

or composition of a mixed solvent at which the particles flocculate. Subsequently, theory has advanced to the point where the effects of particle size and type and the nature, molecular weight, and surface coverage of the stabilizing polymer can be rationalized at least qualitatively. In the following, we first consider the expectations based on the interparticle potentials discussed in previous chapters and then compare them with the experimental results. A more detailed survey of the subject is available in the monograph by Napper (1983).

Our treatment is limited to an assessment of stability in terms of the interaction potential, analogous to §8.2 for electrostatic stabilization. This seems appropriate, since elastic compression of anchored chains produces a repulsion that dominates the dispersion attraction at small separations. Thus the total interaction potential exhibits no minimum at contact and the colloidal suspension is thermodynamically stable for sufficiently thick layers in good solvents if the chains do not desorb. Instability is associated with a minimum, at a separation comparable to the layer thickness, due to either dispersion forces acting across thin layers or polymer–polymer interactions in poor solvents for thick layers.

Our treatment characterizes the polymer and the associated forces in terms of the segment length l, the number of segments per chain N, the segment–segment excluded volume v, the physical volume of a segment w, and the number of chains attached per unit area n_p. The relationship between these parameters and the properties of specific polymers was discussed in §6.2. The mean-field nature of the theory, with respect to both segment–segment interactions and the density profile in the layer, suggests that the predictions may not be quantitative, but may still convey the dominant features.

9.2 Criteria for stability

This section combines the earlier results for the dispersion potential between spheres (§§5.9, 5.10) with predictions for the interaction between terminally anchored layers from the mean-field theory of §6.3. The objective is to identify conditions producing an attractive minimum just sufficient to cause either flocculation or phase separation. An initially dispersed system flocculates when the interparticle potential changes from purely repulsive to strongly attractive with a minimum deeper than $-5kT$ to $-10kT$, as in the electrostatic situation (§8.2). The resulting critical flocculation point is independent of the particle concentration. Phase separation, on the other hand, requires a weak attraction of $-1kT$ to $-5kT$ to allow particles within aggregates to condense into the configuration with the lowest free energy. The conditions for the onset of phase separation vary with particle concentration as observed for dispersion forces acting across thin polymer layers by Long, Osmond & Vincent (1973).

In this section we address the question of stability in steps, first identifying the layer thickness necessary to render the dispersion forces insignificant, then illustrating the emergence of an attractive minimum in the potential with decreasing solvent quality, and finally noting the variation of the critical excluded volume parameter with the surface coverage. The next section demonstrates that these trends explain, at least qualitatively, the features of the phenomena identified experimentally.

Interaction potential between spheres with polymer layers

In §6.3 we obtained the interaction potential between planar layers of terminally anchored polymer by minimizing the free energy of interacting layers, each assumed to have a uniform segment density. The dimensionless layer thickness, $\alpha = L/N^{1/2}l$, obtained from the minimization, varies with the dimensionless separation, $H = h/N^{1/2}l$; the surface coverage, $\phi_p = Nn_p w^{1/2}/l$; and the excluded-volume parameter, $N^{1/2}v/l^3$. At large separations, $\alpha \to \alpha_0$, corresponding to an isolated layer. For good solvents $\alpha_0 > 1$ and for sufficiently poor solvents $\alpha_0 < 1$.

The interactions between layers at separations $H \leq 2\alpha_0$ depend on the values of the excluded volume parameter and the surface coverage (Fig. 9.2). In a theta solvent ($v = 0$) at low coverages ($\phi_p < 1$), the layers freely interpenetrate; only when the chains are compressed between the surface for $H \leq \alpha_0$ does the layer thickness change and the interaction potential become repulsive. At higher surface coverages ($\phi_p > 1$) or in better solvents ($v > 0$), interpenetration becomes unfavorable, producing repulsion for $H \leq 2\alpha_0$. In some cases, the chains are compressed without interpenetration for $\alpha_0 \leq H \leq 2\alpha_0$. In poor solvents ($v < 0$) at low coverages, interpenetration is favorable, causing extension of the chains, i.e. $\alpha \geq \alpha_0$, for $\alpha_0 \leq H \leq 2\alpha_0$ and an attractive interaction potential. With increasing coverage, poorer solvent quality is required to produce the same effect. Under all conditions, the elastic compression of chains for $H < \alpha_0$ produces a strongly repulsive potential.

We convert these predictions for flat plates into interaction potentials between equal spheres through the Derjaguin approximation (5.7.2), giving the total potential as

$$\Phi = \pi a \int_{r-2a}^{\infty} \Phi_{fp}\, dh - \tfrac{1}{6} A_{eff}(r - 2a)\left(\frac{2a^2}{r^2 - 4a^2} + \frac{2a^2}{r^2} + \ln\frac{r^2 - 4a^2}{r^2}\right), \tag{9.2.1}$$

with Φ_{fp} from (6.3.18) and A_{eff} from (5.9.3). When expressed in dimensionless form, Φ/kT depends on the separation $(r-2a)/N^{1/2}l$ and the parameters characterizing the two interaction potentials, expressed as the surface coverage, ϕ_{p}; the excluded volume, $N^{1/2}v/l^3$; the ratio of particle and segment sizes, a/l; the non-retarded Hamaker constant, $A_{\mathrm{eff}}(0)/kT$; and the dimensionless frequency characterizing retardation, $\Omega = n_0(\bar{n}_0^2 + n_0^2)^{1/2} a\omega/c$ from (5.9.3). This treatment accounts for dispersion forces between the polymer layers only through the excluded-volume

Fig. 9.2. Modes of interaction between polymer layers according to the mean-field theory (§6.3) for (*a*) low coverage at theta conditions, (*b*) moderate coverage in a good solvent, and (*c*) low coverage in a poor solvent.

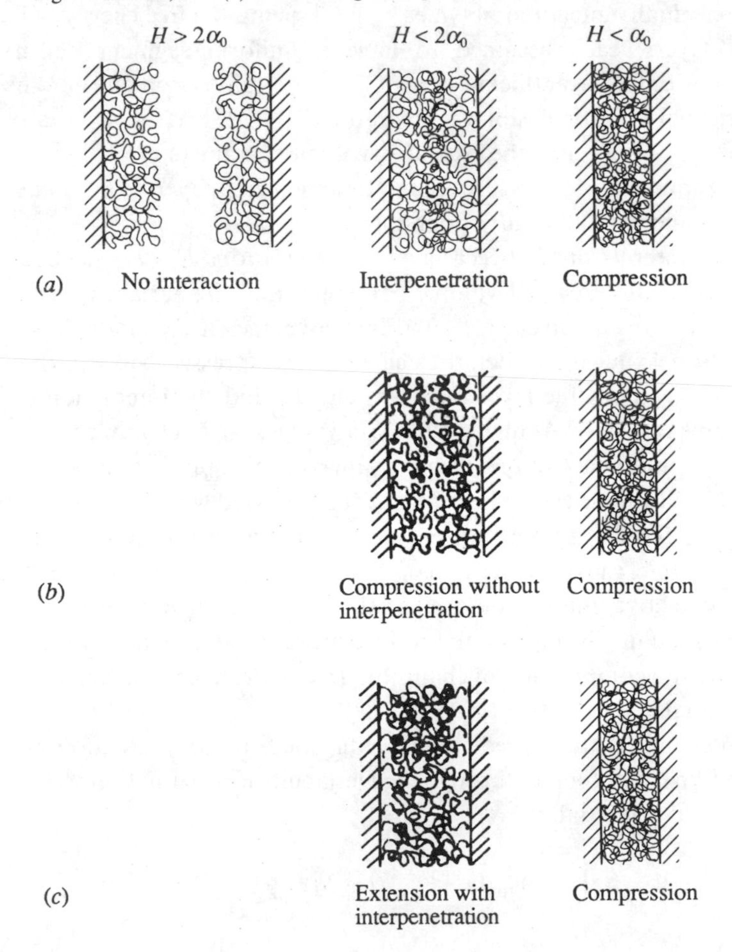

parameter. Longer-range polymer–polymer interactions, which must become important as the polymer approaches bulk density, are neglected.

Stability with respect to dispersion forces

To illustrate the minimum layer thickness necessary to render the dispersion attraction negligible, we evaluate (9.2.1) for polystyrene latices in water bearing a polymer layer with $\phi_p = 1.0$ and $N^{1/2}v/l^3 = 0$, i.e. full coverage and theta conditions. Setting $a/l = 200$ then leaves the molecular weight N as the sole remaining parameter. Figure 9.3 shows the diminishing attraction with increasing chain length and, hence, increasing layer thickness. Note that minima occur at $r - 2a \approx 2N^{1/2}l \ll 2a$, corresponding to twice the thickness of the individual layers. For these conditions, dispersion forces become insignificant, i.e. $-\Phi_{min}/kT \leq 0.1$, for $N \geq 250$ or $N^{1/2}l/a \geq 1/12.5$. Flocculation or phase separation requires $-\Phi_{min}/kT \geq 2$–10, so that $N \geq 16$ or $N^{1/2}l/a \geq 1/50$ suffices for stability.

This criterion for the minimum layer thickness can be generalized, since the minimum remains at $r - 2a \approx 2N^{1/2}l \ll 2a$, independent of the molecular

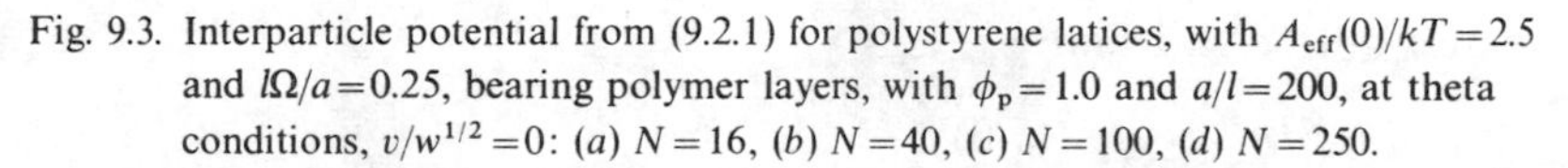

Fig. 9.3. Interparticle potential from (9.2.1) for polystyrene latices, with $A_{eff}(0)/kT = 2.5$ and $l\Omega/a = 0.25$, bearing polymer layers, with $\phi_p = 1.0$ and $a/l = 200$, at theta conditions, $v/w^{1/2} = 0$: (*a*) $N = 16$, (*b*) $N = 40$, (*c*) $N = 100$, (*d*) $N = 250$.

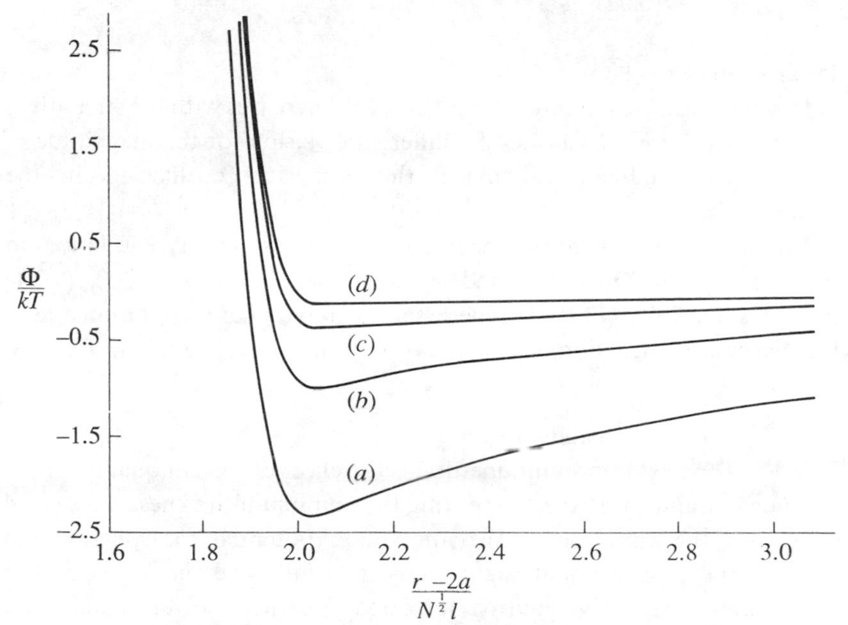

weight. While the layers do not interact at larger separations, slight interaction produces a strong repulsion, so the minimum corresponds to the dispersion potential evaluated at $r-2a \approx 2N^{1/2}l$

$$-\frac{\Phi_{\min}}{kT} \approx \frac{A_{\text{eff}}(2N^{1/2}l)}{24kT} \frac{a}{N^{1/2}l} \tag{9.2.2}$$

for $N^{1/2}l/a \ll 1$. Then, setting $-\Phi_{\min}/kT \leq 2$, for example, and using the approximation from §5.9,

$$A_{\text{eff}}(2N^{1/2}l) = A_{\text{eff}}(0)\left[1+\left(\frac{\pi\Omega}{2\sqrt{2}}\frac{N^{1/2}l}{a}\right)^{3/2}\right]^{-2/3}, \tag{9.2.3}$$

to account for retardation provides an explicit criterion,

$$\frac{N^{1/2}l}{a} \geq \frac{0.57}{\Omega}\left\{\left[1+0.014\left(\frac{\Omega A_{\text{eff}}(0)}{kT}\right)^{3/2}\right]^{1/2}-1\right\}^{2/3}. \tag{9.2.4}$$

The result simplifies to

$$\frac{Nl}{a} \gtrsim 0.019\frac{a}{l\Omega}\frac{A_{\text{eff}}(0)}{kT},$$

for $\Omega A_{\text{eff}}(0)/kT \gg 1$ and (9.2.5)

$$\frac{N^{1/2}l}{a} \geq 0.021\frac{A_{\text{eff}}(0)}{kT}$$

for $\Omega A_{\text{eff}}(0)/kT \ll 1$.

In either case the minimum layer thickness increases with the strength of the dispersion forces. Choice of a different criterion for the magnitude of $-\Phi_{\min}/kT$ would merely change the numerical coefficients in the expressions.

For an illustration of the magnitudes, consider polystyrene latices in salt water with $l\Omega/a=0.25$ and $A_{\text{eff}}(0)/kT=2.5$. For $a/l=200$, (9.2.4) indicates that $N^{1/2}l/a \gtrsim 0.031$ is necessary to avoid phase separation due to the dispersion attraction, in accord with the numerical results in Fig. 9.3.

Critical flocculation point

The next important question is the effect of solvent quality, or the excluded volume, for layers exceeding this minimum thickness. Figure 9.4 illustrates the variation in the interaction potential for spheres with $a/l=200$ bearing layers at slightly less than full coverage, $\phi_p=0.5$. The interaction is strongly repulsive for good and theta solvent conditions,

$N^{1/2}v/l^3 \geq 0$. Indeed, the repulsion persists to somewhat poorer than theta conditions, but a strong attraction appears for $N^{1/2}v/l^3 < -0.5$. Thus, for $N \gg 1$, the transition from a stable dispersion to a flocculated one should occur near the theta point over a narrow range of $-v/l^3 \ll 1$, referred to as the critical flocculation point.

The origin of this abrupt transition is evident from the forms of the interaction potential between flat plates and the Derjaguin approximation. Combining (6.3.18) with (9.2.1) and expressing the separation in dimensionless form leads to

$$\frac{\Phi}{kT} = \pi \frac{a}{l} \phi_p F\left(\frac{r-2a}{N^{1/2}l}, \frac{N^{1/2}v}{l^3}, \phi_p\right), \tag{9.2.6}$$

with the function F representing the dimensionless free energy difference between flat plates integrated over the gap between the spheres. Having $N \gg 1$ amplifies the effect of excluded volume and the factor of $a/l \gg 1$ magnifies any attractive minimum in the flat-plate potential. For smaller particles or shorter polymer chains, the transition would be more gradual.

Near the theta point, the excluded-volume parameter can be related to the temperature or composition of the polymer solution (e.g. Napper, 1983,

Fig. 9.4. Interparticle potential from (9.2.1) for spheres with $a/l = 200$ bearing polymer layers, with $\phi_p = 0.5$ at $N^{1/2}v/w^{1/2} = 2.24$ (*a*), 0.0 (*b*), -0.45 (*c*), -0.67 (*d*).

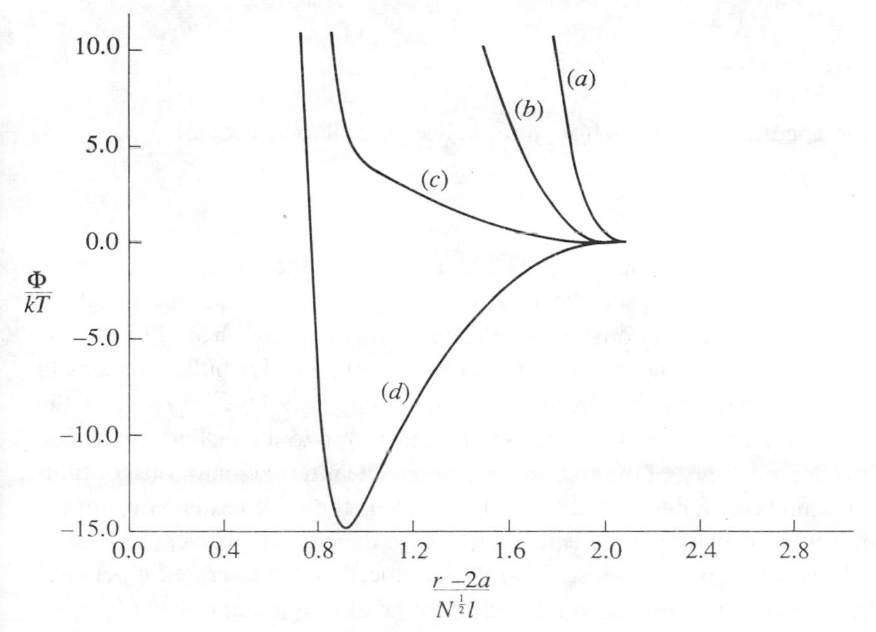

§3.2.7). For example, if temperature is the controlling variable, expanding the excluded volume in a Taylor series about the theta temperature, θ, leads to

$$\frac{v}{w^{1/2}}=\psi_1\left(1-\frac{\theta}{T}\right), \tag{9.2.7}$$

with $\psi_1=-\mathrm{d}(v/w^{1/2})/\mathrm{d}(\theta/T)\,|_{T=\theta}$ having either sign. At the conditions of Fig. 9.4, the transition from stability to flocculation would occur over a temperature range $\Delta T/T_f \approx 0.1/N^{1/2}$ centered about $T_f \approx \theta(1+0.5/N^{1/2})$ for $\psi_1=1.0$. For $N=500$ and $\theta=300$ K, this leads to $T_f \approx 306.7$ K and $\Delta T \approx 1.4$ K. This range becomes narrower and T_f moves closer to θ as a/l and N increase.

The transition, though abrupt, occurs for non-zero v. This displacement of the predicted flocculation point from the theta point reflects the contribution to the free energy (6.3.10) from the higher-order term, $Nwn^2/6$, representing three-body interactions. In fact, the expression for the free energy (6.2.25) suggests that two layers should attract only when interpenetration decreases the chemical potential, i.e.

$$-Nvn>\tfrac{1}{2}Nwn^2.$$

Since the layer thickness is approximately $N^{1/2}l$ and the relevant segment density in the region of overlap is twice that of an individual layer or

$$n\approx 2\frac{N^{1/2}n_p}{l},$$

the condition for attraction, and, hence, flocculation becomes

$$-N^{3/2}\frac{n_p v}{l}>\phi_p^2. \tag{9.2.8}$$

Evaluation of the potential (9.2.1) to identify the numerical value of $N^{1/2}v/w^{1/2}$ at which the potential becomes attractive yields the results in Fig. 9.5. The continuous line conforms to that expected from (9.2.8). The broken line identifies the phase boundary from §6.2 for bulk solutions in poor solvents at the nominal concentration within the layer, $\phi_p/N^{1/2}$. In the two-phase region above the phase boundary, horizontal tie lines determine the compositions of the coexisting phases. The intersection of the continuous and broken lines indicates that decreasing the solvent quality produces flocculation before phase separation within the individual layers for $\phi_p \lesssim 2$. The consequences of phase separation induced by the interaction between two layers have only been addressed for adsorbing homopolymer (Ingersent, Klein, & Pincus, 1987) to date.

In summary, the theory from §6.3 predicts three important qualitative features for interactions between spheres bearing polymer layers:

(i) a minimum layer thickness necessary to mask dispersion forces (9.2.4),
(ii) an abrupt transition from stability to flocculation at slightly worse than theta conditions (Fig. 9.4), and
(iii) the possibility of stability at significantly worse than theta conditions for dense polymer layers (Fig. 9.5).

The second and third points refer to layers that are thick enough for dispersion forces to be insignificant.

9.3 Measurements of the critical flocculation point

The experimental characterization of polymeric stabilization has focused primarily on determining the critical flocculation point. Since the

Fig. 9.5. Values of the dimensionless excluded volume, $-N^{1/2}v/w^{1/2}$, corresponding to the critical flocculation point (——) for colloidal particles with surface coverage $\phi_p = Nw^{1/2}n_p/l$ and the phase boundary (- - -) for polymer solutions at concentration $\phi_p/N^{1/2} = N^{1/2}w^{1/2}n_pl$.

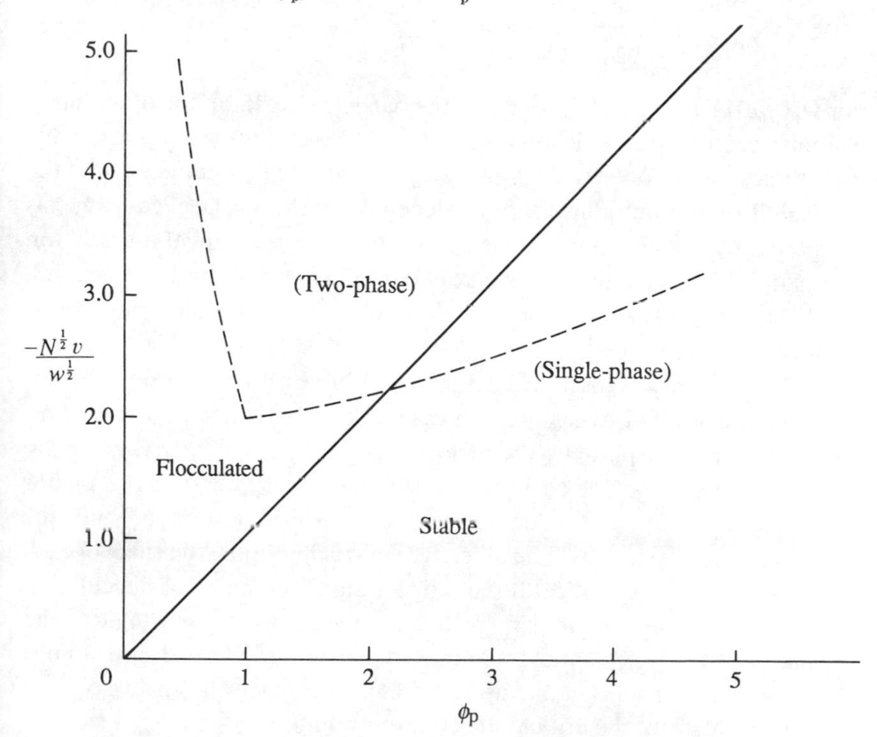

polymer–polymer interactions generally control the interparticle potential, the experiment involves modifying the solvent to change the segment–segment excluded volume. For many solutions, varying the temperature suffices. As noted earlier, $v/w^{1/2} = \psi_1(1-\theta/T)$ near the theta temperature, but ψ_1 can be either positive or negative. Thus, the transition from stable to flocculated states accompanies a decrease in temperature if $\psi_1 > 0$ and an increase in temperature if $\psi_1 < 0$. Often, however, the theta temperature lies outside the range of temperatures readily accessible in the laboratory. Then, the addition of a non-solvent provides a convenient alternative. Thus the critical flocculation point may correspond to either a temperature or a fraction of non-solvent added to the original good solvent. Napper (1983, Ch. 7) provides a thorough discussion of these thermodynamic factors.

The relationship between the critical flocculation point and the theta point, together with the dependence on the parameters characterizing the system, provides evidence about the dominant forces acting between the particles. For example, flocculation or phase separation in good solvents must be caused by dispersion forces. The critical flocculation point then corresponds to the layer thickness, $\alpha_0 N^{1/2} l$, at which $\Phi(2\alpha_0 N^{1/2} l)/kT \approx -2$, e.g.

$$\alpha_0(\phi_p, z) = \left(\frac{\sqrt{2}}{24\pi}\frac{A_{\text{eff}}(0)}{\Omega kT}\right)^{1/2} \frac{a}{N^{1/2} l}, \tag{9.3.1}$$

for $\Omega A_{\text{eff}}(0)/kT \gg 1$, and depends on the particle size. If, on the other hand, stability persists to significantly worse than theta conditions, repulsion due to interactions among three or more segments must be significant, and the critical flocculation point should depend on the surface coverage as suggested by Fig. 9.5. A close correlation between the critical flocculation point and the theta point, for a range of particle sizes and surface coverages, establishes excluded volume effects as dominant, so that only the solvent quality matters.

In the experiments discussed below, the soluble polymer responsible for stability was grafted either onto an insoluble polymer chain, the anchor, or directly onto the particle. The properties that characterize the system are the nature and size of the particles, the stabilizing polymer, and the anchor polymer; the nature of the solvent; the surface coverage; and the concentration of particles. The objective was to determine the effect of each characteristic on the critical flocculation point. The onset of flocculation was detected by techniques similar to those described for sensing doublet formation with electrostatically stabilized dispersions (§8.6). For example, an increase in turbidity indicates the presence of flocs and the associated conditions identify the critical flocculation point.

Table 9.1. *Surface coverage*

a/nm	$m\Big/\dfrac{\text{kg}}{\text{mol}}$	n_p/nm^{-2}	Nl^2n_p
poly(12-hydroxystearic acid) on PMMA (Walbridge, 1975)			
53	1.5	0.33	3.4
69	1.5	0.29	2.9
70	1.6	0.33	3.6
110	1.6	0.26	2.9
250	1.5	0.32	3.3
polystyrene grafted on silica (Edwards *et al.*, 1984)			
52	13.8	0.055	3.3
52	25.6	0.077	8.6
86	25.6	0.038	4.2
86	25.6	0.083	9.3
86	25.6	0.172	19
299	34.4	0.077	12
poly(dimethyl siloxane) grafted on silica (Edwards *et al.*, 1984)			
130	6.7	0.88	22
130	72.4	0.12	32

The polymer stabilizers used in these experiments were poly(12-hydroxystearic acid) and poly(oxyethylene). Solution properties of these polymers are summarized in Table 6.1. Note that even the flexible polymer poly(oxyethylene) has about four bonds, and, hence, two monomers per segment and $w^{1/2}/l^3 \approx 0.4$.

The poly(12-hydroxystearic acid) chains were terminally grafted at multiple points along the backbone of an insoluble anchor polymer such as poly(methyl methacrylate). Adsorption of the resulting comb copolymer onto the particle surface produces $Nl^2n_p \approx 3$–4 at full coverage (Table 9.1). The poly(oxyethylene) was formed into diblock copolymers, e.g. with poly(vinyl acetate). The scant data available on other block copolymers (e.g. Walbridge, 1975) indicates somewhat lower surface coverages of $Nl^2n_p \approx 1$, but little is known about the actual magnitude or the variation with the molecular weight of either block.

Napper's early experiments (Napper, 1968, 1970) demonstrated that optimum stabilization with block copolymers requires strong anchoring, full surface coverage, and sufficiently high molecular weights. Then the nature of the particles and anchoring block do not affect the critical

Table 9.2. *Effect of particle composition* (*Napper, 1968, 1970*)

polymer/fluid	θ/K	particle	T/K
poly(oxyethylene)/ 0.39 M $MgSO_4$	315±3	poly(vinyl acetate)	318±2
		poly(methyl acrylate)	320±2
		polystyrene	323
	c_θ		c
poly(12-hydroxy stearic acid)/ *n*-heptane	0.39±0.01 ethanol	poly(vinyl acetate)	0.395±0.005
		poly(methyl methylacrylate)	0.39±0.01

Effect of anchor polymer (*Napper, 1968*)

polymer/fluid	θ/K	anchor	T/K
poly(oxyethylene)/ 0.39 M $MgSO_4$	315±3	poly(vinyl acetate)	318±2
		poly(methyl acrylate)	314±2
		poly(methyl methacrylate)	320±2
		polystyrene	314±3
		poly(vinyl stearate)	322±3

flocculation point. The data reproduced in Table 9.2 illustrate the evidence. For poly(12-hydroxystearic acid) in *n*-heptane, the volume fraction of non-solvent, ethanol, required for flocculation (c) corresponded to that at theta conditions (c_θ) for two different latices. For poly(oxyethylene) in an electrolyte solution, the critical flocculation temperature (T) deviated somewhat, but not substantially, from the theta temperature for a selection of polymer latices and anchor groups. The implication is that characteristics of the anchor and the particle surface affect the stability only when (i) adsorption is weak, allowing desorption from the gap between particles, or (ii) the surface is incompletely covered, permitting the polymer to move laterally out of the gap. Either enables closer approach of the surfaces, exposing the particles to stronger dispersion forces that depend on the bulk composition of the particle. Associated experiments with partially coated particles confirmed that flocculation then occurs in good solvents at a temperature that depends on the coverage.

Given strong anchoring and full coverage, the next question concerns the minimum layer thickness necessary to provide stability against flocculation caused by the dispersion attraction. Two experiments appear to confirm the criterion set out in the previous section. Figure 9.6 depicts data (Napper, 1968) for poly(oxyethylene) of molecular weights ranging from 1 to 10^3 kg/mol attached to latices with radii from 15 to 115 nm in 0.39 M $MgSO_4$ solutions. These dispersions flocculate upon heating, but the critical flocculation temperature lies close to the theta temperature of 315 ± 3 K in all cases. Conversion of the temperatures to excluded volumes

Fig. 9.6. Data from Napper (1968) for the critical flocculation point for poly(oxyethylene) of molecular weights 1–10^3 kg/mol attached to latices with $a = 15$–115 nm in 0.39 M $MgSO_4$ solutions: (*a*) critical flocculation temperature T and (*b*) dimensionless excluded-volume parameter calculated with $\psi_1 = -1.0$. The broken lines and the error bars represent the uncertainties.

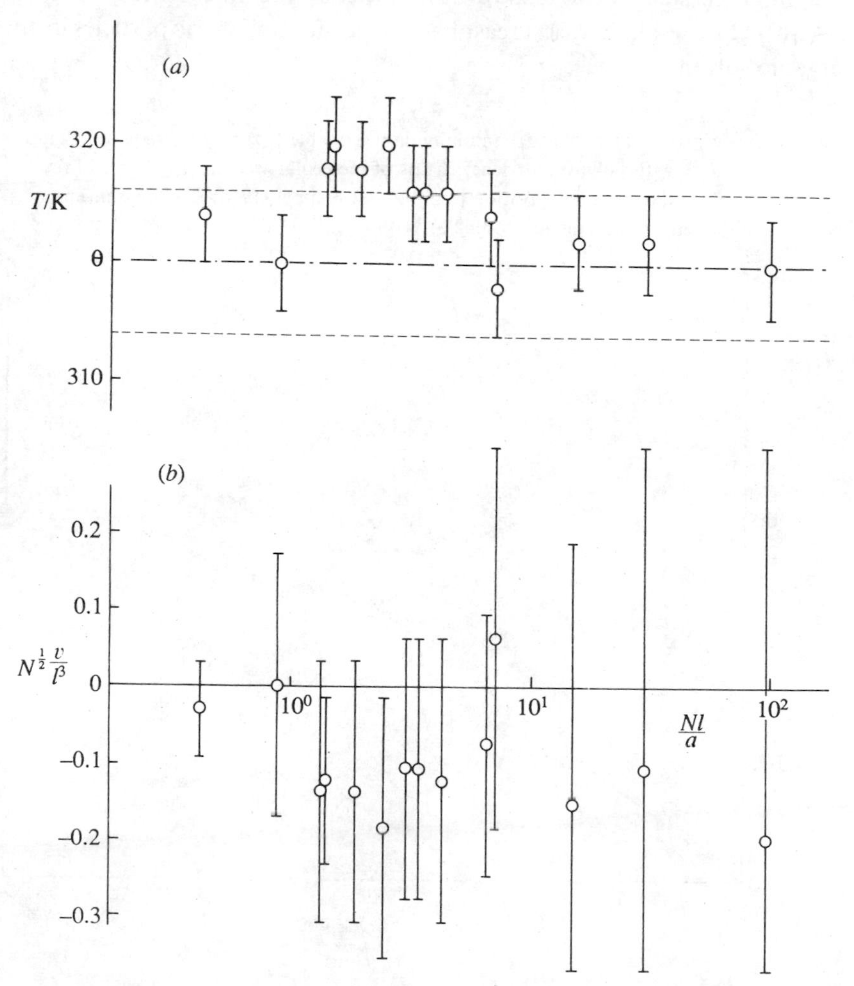

using the molecular parameters in Table 6.1 and $\psi_1 = -1.0$ produces consistently negative, but small, values. Thus dispersion forces appear to play no role, consistent with the fact that all values of Nl/a exceed the minimum layer thickness of ≈ 0.2 expected from (9.2.5).

Smitham & Napper (1976) dispersed latices bearing poly(oxyethylene) chains in melts of identical chains. This produces a uniform number density of segments, eliminating all but the elastic contributions to the free energy of the interacting layers. Stability then is determined solely by the balance between the elastic compression of chains and the dispersion forces. A plot of the maximum stable and minimum unstable particle sizes (Fig. 9.7) suggests a linear correlation between particle size and polymer molecular weight, consistent with (9.2.5). The slope of the line corresponds to $aA_{\text{eff}}(0)/\Omega lkT \approx 4$ in (9.2.5), a reasonable value for polymeric particles in an organic solvent.

Fig. 9.7. The maximum stable (●) and minimum unstable (○) particle radii a for latex spheres with poly(oxyethylene) chains of molecular weight M dispersed in a melt of the same composition (Smitham & Napper, 1976). The broken line represents the prediction of (9.2.5), with $aA_{\text{eff}}(0)/\Omega lkT = 4$.

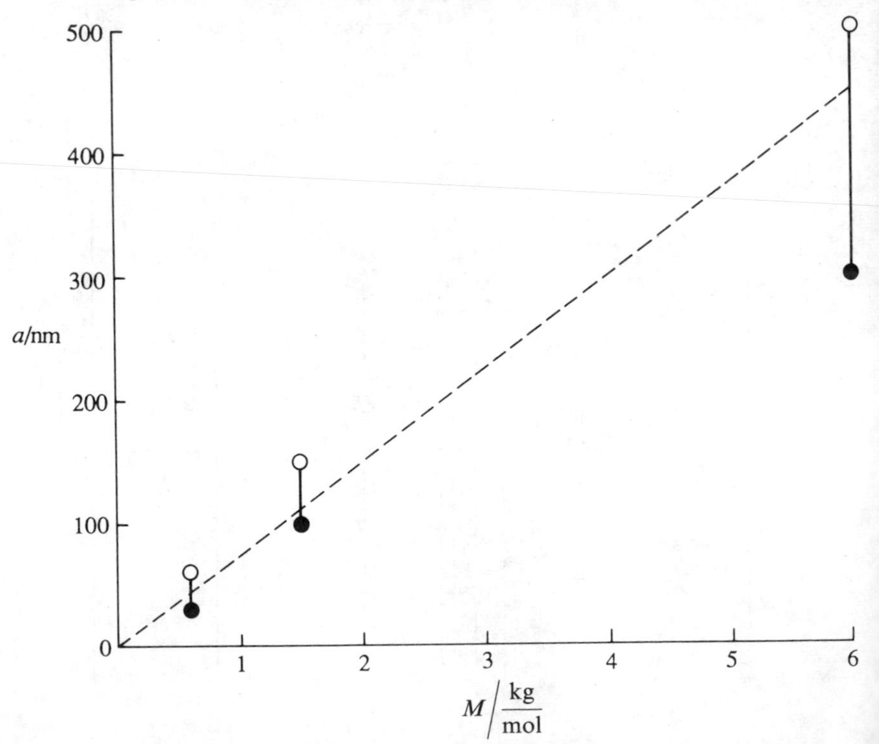

Table 9.3. *Dependence of critical flocculation point on surface coverage*

polymer/fluid	ϕ_p	$-N^{1/2}\dfrac{v}{w^{1/2}}$
poly(oxyethylene)/0.39 M $MgSO_4$	≈ 0.4	0.1 ± 0.08
poly(dimethyl siloxane)/	12	0.7 ± 0.2
bromocyclohexane	18	3.3 ± 1.0

The most striking feature of data on critical flocculation points is the close correlation with the theta point of the stabilizing polymer in solution, independent of other characteristics of the system. This observation, originally stressed by Napper (1968), has been confirmed by others for a wide variety of block and comb copolymers of the type illustrated above. The theory presented in §9.2 predicts flocculation at $-N^{1/2}v/w^{1/2} \approx \phi_p$. The data of Fig. 9.6, for the poly(oxyethylene) block copolymers expected to have $Nl^2 n_p \approx 1$, suggests $-N^{1/2}v/w^{1/2} = 0.11 \pm 0.08$, in reasonable accord with the value of $\phi_p = Nl^2 n_p w^{1/2}/l \approx 0.4$ expected.

Recent experiments with denser layers provide additional data on the effect of ϕ_p. Edwards *et al.* (1984) grafted polystyrene and poly(dimethyl siloxane) chains onto silica particles at much higher surface coverages than were attainable with block copolymers, e.g. $Nl^2 n_p = 3$–30 (Table 9.1). Subsequent measurements of critical flocculation temperatures revealed stability at considerably worse than theta conditions and a significant dependence on the volume fraction of particles (Fig. 9.8). Together with the data for poly(oxyethylene) these establish a significant correlation between the approximate values of the excluded volume at the critical flocculation point and the surface coverage (Table 9.3). Thus, the observed stability in poor solvents and flocculation at values of $-N^{1/2}v/w^{1/2}$ that increase with ϕ_p agree qualitatively with the expectations from Fig. 9.5. However, the dependence on the volume fraction of particles suggests a more gradual transition from stability to flocculation at the higher coverages, akin to the equilibrium phase separations discussed in the next chapter.

The quantitative discrepancy between the observations and the critical value of the excluded volume parameter, $-N^{1/2}v/w^{1/2} \approx \phi_p$, expected from the theory of §6.3 could arise from several sources. For example, the attraction responsible for flocculation derives from segment–segment interactions at lower segment densities than characteristic of the bulk of the layer. Proper accounting for the variation in segment density, however, requires a more sophisticated approach such as the self-consistent field theory or the equivalent lattice theory (e.g. Scheutjens & Fleer, 1985). In addition, dispersion forces between the polymer layers could play a role at these higher densities. Identification of the correct explanation for the quantitative difference between the observations and the predictions awaits data for better-characterized adsorbed layers and calculations with the more complete theories.

In summary, extensive observations show a robust correlation between the critical flocculation point and the theta point – independent of the size and type of the particle, anchor polymer, and stabilizing polymer; the particle concentration; and the solvent – provided the stabilizer layer

Fig. 9.8. The critical flocculation temperatures, T, for silica particles dispersed in bromocyclohexane at volume fraction ϕ with grafted poly(dimethyl siloxane) chains of molecular weights 72.4 kg/mol (○) and 6.7 kg/mol (∗) (Edwards *et al.*, 1984).

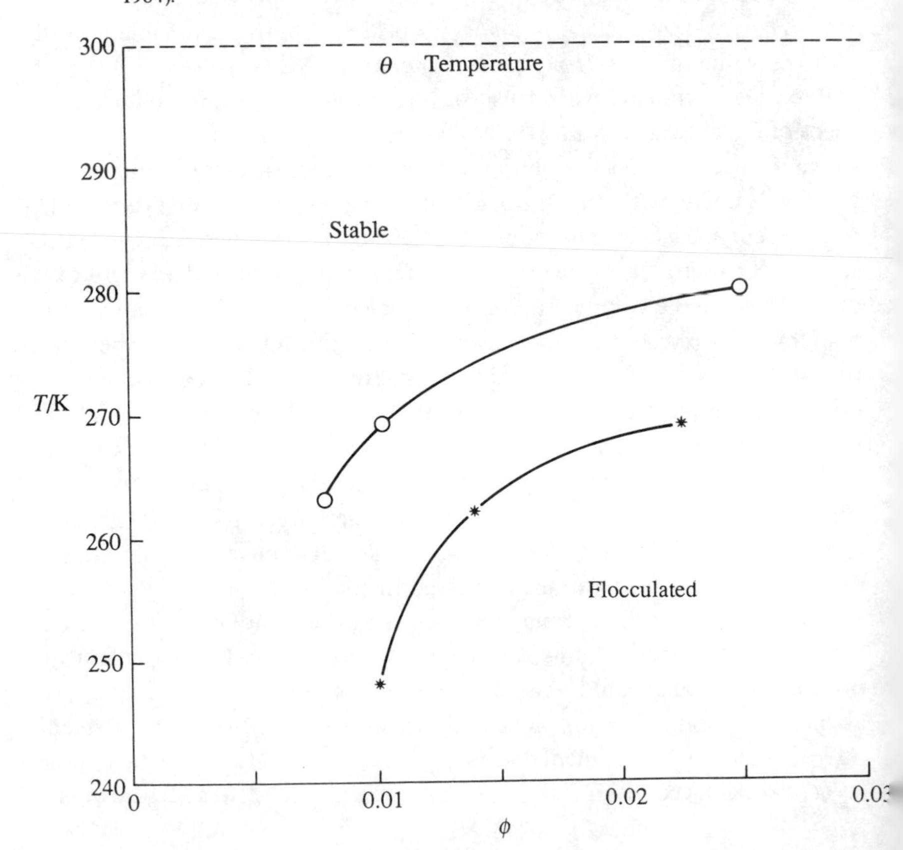

(i) is anchored strongly,
(ii) fully covers the particle, and
(iii) exceeds a minimum thickness.

At less than full coverage, i.e. $\phi_p < 1$, or for layers of less than a minimum thickness, dispersion forces cause flocculation in good solvents. At substantially higher surface coverages, i.e. $\phi_p \gg 1$, recent data demonstrate stability in poor solvents.

9.4 Summary

The efficacy of stabilization of colloidal particles with copolymers firmly adsorbed at full coverage is well established. For layers exceeding a minimum thickness, dispersion forces are unimportant. Then a well-defined critical flocculation point exists, and, at moderate surface coverages, is practically indistinguishable from the theta point. For grafted chains at much higher coverages, stability at considerably worse than theta conditions is possible though not yet fully defined. At a qualitative level, theory and experiment agree on these features of the phenomena.

Several important fundamental questions have received no attention here and little in the literature:

(i) the dependence of the surface coverage on the structure of the block copolymers and the effect on stability in poor solvents;
(ii) the effect of curvature on polymeric layers attached to small particles;
(iii) the design and synthesis of stabilizers for specific particle-solvent combinations, and
(iv) dispersion forces between polymer layers.

References

Barrett, K. E. J. (ed.) (1975). *Dispersion Polymerization in Organic Media*. Wiley. London.

Edwards, J., Lenon, S., Toussaint, A. F. & Vincent, B. (1984). Preparation and stability of polymer grafted silica dispersions. In *Polymer Adsorption and Dispersion Stability* (eds E. D. Goddard and B. Vincent). 'ACS Symp. Series' **240**, pp. 281–96.

Heller, W. & Pugh, T. L. (1954). Steric protection of hydrophobic colloidal particles by adsorption of flexible macromolecules. *J. Chem. Phys.* **22**, 1778.

Ingersent, K., Klein, J. & Pincus, P. (1986). Interactions between surfaces with adsorbed polymers: Poor solvents 2 – calculations and comparison with experiment. *Macromolecules* **19**, 1374–81.

Laible, R. & Hamann, K. (1980). Formation of chemically bonded polymer layers on oxide surfaces and their role in colloidal stability. *Adv. Colloid Interface Sci.* **13**, 65–99.

Long, J. A., Osmond, S. W. J. & Vincent, B. (1973). The equilibrium aspects of weak flocculation. *J. Colloid Interface Sci.* **42**, 545–53.

Napper, D. H. (1968). Flocculation studies of non-aqueous sterically stabilized dispersions of polymer. *Trans. Far. Soc.* **64**, 1701–11.

Napper, D. H. (1970). Flocculation studies of sterically stabilized dispersions. *J. Colloid Interface Sci.* **32**, 106–14.

Napper, D. H. (1983). *Polymeric Stabilization of Colloidal Dispersions.* Academic Press.

Rehbinder, P., Lagutina, L. & Wenstrom, E. (1930). Stabilizing action of interface-active materials on suspensions of hydrophobe and hydrophil powders in water and non-aqueous dispersion agents. I. *Z. Phys. Chem. A* **146**, 63–78.

Scheutjens, J. M. H. M. & Fleer, G. J. (1985). Interactions between two adsorbed polymer layers. *Macromolecules* **18**, 1882–1900.

Smitham, J. B. & Napper, D. H. (1976). Elastic steric stabilization in polymer melts. *J. Colloid Interface Sci.* **54**, 467–70.

Verwey, E. J. W. & de Boer, J. H. (1938). Dilatancy. *Rec. Trav. Chim.* **57**, 383–9.

Walbridge, D. J. (1975). The design and synthesis of dispersants for dispersion polymerization in organic media. In *Dispersion Polymerization in Organic Media* (ed. K. E. J. Barrett), pp. 45–114. Wiley.

Problems

1 Since terminally anchored layers extend further from the surface in good solvents, lower molecular weight chains suffice to negate the effect of dispersion forces. Derive the analogue to (9.2.4) for good solvent conditions with $z = N^{3/2} n_p v / l \gg 1$ and determine the critical molecular weight for poly(oxyethylene) grafted to polystyrene latices with $a = 0.1\ \mu m$ in 0.39 M $MgSO_4$ solutions at 300 K.

2 Evaluate the molecular weight required for stability at theta conditions for:
(i) polystyrene grafted at $\phi_p = 1$ and 10 to silica spheres with $a = 50$ nm in cyclohexane;
(ii) poly(oxyethylene) grafted at $\phi_p = 10$ to polystyrene latices with $a = 20$ and 500 nm in aqueous solutions.

3 For a fixed contour length does a terminally anchored chain with a flexible or a relatively stiff backbone impart better stability? Support your answer with arguments based on the theories here and in Chapter 6.

10

EQUILIBRIUM PHASE BEHAVIOR

10.1 **Introduction**

While much attention has been focused on stability toward irreversible flocculation as described in Chapters 8 and 9, reversible phase separations also occur in colloidal systems. The distinction between equilibrium phase behavior and non-equilibrium states is one of time scales, that for relaxation toward equilibrium compared with the observation time. For an aggregated dispersion the relaxation time can be estimated from the rate of doublet breakup, equivalent to diffusion over an energy barrier (Chandrasekhar, 1943). This time scale takes the form

$$\frac{6\pi\mu a^3}{kT}\exp\left(-\frac{\Phi_{min}}{kT}\right), \tag{10.1.1}$$

with $-\Phi_{min}$ the magnitude of the minimum in the potential. If $-\Phi_{min}/kT > 20$ for spheres with $a = 0.1\,\mu m$ in water then equilibrium cannot be expected on time scales of normal interest, i.e. days to months (Table 10.1). Hence, primary minimum flocculation arising from dispersion forces (Chapter 8) and destabilization of polymer-coated particles in poor solvents (Chapter 9), for which attractions are strong, are intrinsically non-equilibrium processes. The attraction induced by dissolved non-adsorbing polymer (§6.4) or by dispersion forces acting across a thin polymer layer (§9.3), which vary more gradually with the solution chemistry, can provide attractive potentials with $-\Phi_{min}/kT = 1$–3 and equilibrium behavior.

In this and subsequent treatments of the macroscopic properties of colloidal suspensions, we speak of a dispersion of particles in a fluid, containing electrolyte, polymer, and, perhaps, other species, as a one-component system as polymer solutions were treated in Chapter 6. Hence, appropriate state variables are the temperature and the concentration of

Table 10.1. *Time scale for doublet breakup,* $\frac{6\pi\mu a^3}{kT}\exp\left(\frac{-\Phi_{min}}{kT}\right)$

$a/\mu m$	$\frac{-\Phi_{min}}{kT}=2$	10	20
0.1	0.034 s	100 s	26 days
1.0	34 s	1.2 days	70 years

particles, with the other species accounted for solely through the interparticle potential. A phase then consists of a homogeneous volume of dispersion, containing many particles, characterized by a pressure that depends on concentration and temperature. This represents the osmotic pressure acting across a membrane permeable to the fluid and all dissolved species, but not to the particles. The true thermodynamic pressure is indeterminant, since the system is assumed to be incompressible.

An additional important feature is the structure or spatial configuration of the particles. Disordered or fluid phases have local structure, i.e. correlations among the relative positions of particles, that persist over distances of the order of the particle size or the range of the interparticle potential. Ordered or solid phases have a periodic structure of effectively infinite extent. Two coexisting phases must have equal Gibbs free energies and pressures, even though the concentrations and, perhaps, the structures differ.

Phase transitions observed in colloidal systems divide into two classes:

(i) disorder–order transitions driven by entropic effects in systems dominated by repulsive interparticle potentials and
(ii) fluid–fluid or fluid–solid transitions caused by weak attractions.

The first class corresponds to the liquid–solid or freezing transition in molecular systems and is distinguished by the small difference in density between the two coexisting phases and the crystalline order of the denser phase. The second, akin to gas–liquid and gas–solid transitions, involves $O(1)$ density differences.

One of the most striking and earliest examples of the former is the occurrence of ordered phases, known as Schiller layers, for rod-like particles, noted by Zocher (1925) and explained by Onsager (1949) as the consequence of configurational constraints in the crowded disordered phase. The more subtle, disorder–order transition for hard spheres was predicted quantitatively in the 1950s (e.g. Wood & Jacobsen, 1957; Alder & Wainwright, 1957, 1959) but not quantified experimentally until the work of Hachisu & Kobayashi (1974). Monodisperse, charged spheres develop similar crystalline order at low ionic strengths, owing to long-range electrostatic repulsions. Since the first report by Luck, Klier, & Wesslau (1963), a multitude of experiments, beginning with Hiltner & Krieger (1969) and Hachisu, Kobayashi, & Kose (1973), have probed the periodic microstructure, phase behavior, and rheology of these intriguing systems. Since the range of concentrations for which two phases coexist is rather narrow, attention has focused primarily on the structure and properties of the ordered solid, and, more recently, the dynamics of the phase transition and the possibility of metastable glassy states (e.g. Pusey & van Megen, 1987).

Equilibrium phase transitions due to dispersion forces have been observed for particles large enough to have significant secondary minima (e.g. Efremov, 1976) or smaller particles stabilized by thin steric layers (Long, Osmond, & Vincent, 1973). In addition, as noted in §6.4, dissolved, non-adsorbing polymer can produce weak attractions between colloidal particles. Although observed (Bondy, 1939) and predicted (Asakura & Oosawa, 1954, 1958) much earlier, equilibrium phase separations induced by non-adsorbing polymer have been identified clearly only recently. Experiments with a variety of systems, demonstrated to have minimal adsorption, have now established the coexistence of equilibrium phases, one dense and one dilute in particles, above a critical polymer concentration (e.g. Cowell, Li-In-On, & Vincent, 1978; Vincent, Luckham, & Waite, 1980; Sperry, Hopfenberg, & Thomas, 1981; de Hek & Vrij, 1981). Although the nature of the repulsive portion of the interparticle potential differs, the phase transition in each case arises from the osmotic attraction due to free polymer.

The primary difficulty in predicting equilibrium phase behavior lies in the many-body interactions intrinsic to any condensed phase. Fortunately, the synthesis of several methods – integral equation approaches, perturbation theories, and computer simulations – now provides accurate predictions of the thermodynamic properties and phase behavior of dense fluids consisting of simple molecules. Furthermore, McMillan and Mayer and Kirkwood and Buff (Hill, 1960, Ch. 19) formulated the treatment of dispersions as pseudo-one-component systems with the colloid osmotic pressure Π replacing the normal pressure as described for polymer solutions in Chapter 6. Hence, for particles large with respect to the intervening molecules, the fluid can be treated as a homogeneous continuum whose properties enter only as parameters in the interaction potential as assumed in Chapters 4–6.

10.2 The statistical mechanical approach

This section reviews the necessary concepts from equilibrium statistical mechanics. Those interested in a more thorough treatment of the subject should consult any of several basic texts (Reed & Gubbins, 1973; McQuarrie, 1976; Hansen & McDonald, 1986) or reviews (Castillo, Rajagopalan, & Hirtzel, 1984: van Megen & Snook, 1984).

As noted in §6.2, the thermodynamic properties of a system can be evaluated from knowledge of the Helmholtz free energy as a function of volume and temperature through (6.2.9) and (6.2.11):

$$A = -kT \ln Q \tag{10.2.1}$$

with

$$Q = \frac{Z_N}{N!}$$

and

$$Z_N = \int \ldots \int \exp\left(-\frac{E_N}{kT}\right) \mathrm{d}\mathbf{r}_1 \ldots \mathrm{d}\mathbf{r}_N,$$

with E_N the internal energy of a volume V containing N particles in the configuration $(\mathbf{r}_1 \ldots, \mathbf{r}_N)$. In particular, the osmotic pressure and Gibbs free energy follow from

$$\Pi = -\left[\frac{\partial A}{\partial V}\right]_T$$

and

$$G = A + \Pi V. \tag{10.2.2}$$

Alternatively, specific thermodynamic functions can be related directly to the interaction potential and the equilibrium microstructure as characterized by P_N, the probability density corresponding to a particular configuration of particles. For example, the equation of state is obtained from (10.2.1) and (10.2.2) by assuming pairwise additive interaction potentials, such that

$$E_N = \sum_{i>j} \Phi(r_{ij}), \tag{10.2.3}$$

with $r_{ij}^2 = (\mathbf{r}_i - \mathbf{r}_j) \cdot (\mathbf{r}_i - \mathbf{r}_j)$; then differentiating (10.2.1) with respect to the volume and integrating over the positions of all but two particles determines

$$\Pi = nkT - \frac{2\pi}{3} n^2 \int_0^\infty r^3 g(r) \frac{\mathrm{d}\Phi}{\mathrm{d}r} \, \mathrm{d}r. \tag{10.2.4}$$

Here the equilibrium radial distribution function, defined by

$$n^2 g(r) = \frac{1}{(N-2)!} \int \cdots \int P_N \, \mathrm{d}\mathbf{r}_3 \ldots \mathrm{d}\mathbf{r}_N, \tag{10.2.5}$$

depends on the volume fraction, except in the dilute limit, due to interactions among three or more particles.

For coexisting phases, the Gibbs free energies and pressures must be equal to ensure both thermodynamic and mechanical equilibrium. Hence the calculation requires evaluating G and Π as functions of ϕ, for a particular interparticle potential Φ/kT. Then the intersection on a plot of G vs. Π identifies the coexistence point corresponding to that potential or temperature. A complete phase diagram is constructed by varying Φ/kT and represented by plotting the volume fractions ϕ of the coexisting phases as functions of a dimensionless temperature, e.g. $-kT/\Phi_{\min}$.

Further progress requires either Z_N or $g(r)$ for the interaction potential of interest. In neither case can the multidimensional integrals (10.2.1) or (10.2.5) be evaluated directly, so a variety of strategies have been devised, e.g. virial expansions, integral equations, Monte Carlo and molecular dynamics simulations, and perturbation methods. The first comprises a regular perturbation expansion in volume fraction $\phi \ll 1$, which is tractable to reasonably high order for simple potentials, but does not converge at higher volume fractions. The following section illustrates the results for pair interactions. Extensive results from the next three approaches listed define the behavior of systems with simple interaction potentials, such as hard spheres. Perturbation methods describe concentrated systems with more complex potentials by expanding about the hard sphere limit, capitalizing on results from other methods for hard spheres at concentrations from infinite dilution to closest packing. As demonstrated in subsequent sections, this approach is often appropriate for colloidal particles, e.g. when the electrostatic, dispersion, and polymer-induced forces are significant only at separations slightly greater than the particle diameter.

10.3 Equilibrium properties of dilute suspensions

At dilute concentrations, the probability of three-body and higher-order interactions is small, allowing the pair distribution (10.2.5) to be expressed as (Reed & Gubbins, 1973, §7.20)

$$\begin{aligned} P_2 &= n^2 g(r) \\ &= n^2 \exp\left(-\frac{\Phi(r)}{kT}\right)\{1+O(\phi)\}. \end{aligned} \tag{10.3.1}$$

Thus the equilibrium radial distribution function $g(r)$ is independent of concentration to a first approximation. The hard-sphere potential

$$\Phi(r) = \begin{cases} \infty & r<2a \\ 0 & 2a<r \end{cases} \tag{10.3.2}$$

produces $g(r)=H(r-2a)$, with H the Heaviside step function. Potentials with longer range alter the pair distribution function for $r>2a$, with $g>1$ for attractions and $g<1$ for repulsions. In either case, $g\to 1$ as $\Phi\to 0$ for $r\to\infty$.

Expansion of the pressure (10.2.4) in powers of the volume fraction as

$$\Pi = nkT(1+A_2\phi+\ldots) \tag{10.3.3}$$

identifies the second virial coefficient

$$A_2 = \tfrac{3}{2}\int_0^\infty \{1-g(s)\}s^2\,\mathrm{d}s \tag{10.3.4}$$

with $s=r/a$. Evaluation of A_2 requires only a straightforward numerical integration, but generally involves specification of a number of parameters. For example, the combination of electrostatic repulsion and dispersion attraction involves three dimensionless groups, $\varepsilon\varepsilon_0\psi_s^2 a/kT$, $a\kappa$, and $A_{\text{eff}}(0)/kT$, even without retardation. However, the macroscopic properties calculated in subsequent chapters often prove insensitive to the details of the interaction potential, motivating a simplified form incorporating both attraction and repulsion.

At high ionic strengths, the Debye length satisfies $a\kappa \gg 1$, making the range of the repulsion short relative to the particle size. Also, the dispersion potential exceeds kT only for $r-2a \ll a$. Thus, the electrostatic potential resembles a hard sphere repulsion, with range slightly greater than the actual particle diameter, and the dispersion potential can be approximated as a stickiness that retains pairs at small separations. Consequently, an adhesive excluded shell model, with the pair potential (Baxter, 1968)

$$\frac{\Phi(s)}{kT} = \begin{cases} \infty & s < s_0 \\ -\ln \dfrac{\delta(s-s_0)}{6\tau} & s = s_0 \\ 0 & s_0 > s \end{cases} \tag{10.3.5}$$

and the radial distribution function

$$g(s) = H(s-s_0) + \frac{\delta(s-s_0)}{6\tau}, \tag{10.3.6}$$

determined by the shell diameter s_0 and the stickiness $1/\tau$, should suffice. Substitution into (10.3.4) determines the second virial coefficient as

$$A_2 = \frac{s_0^3}{2} - \frac{s_0^2}{4\tau}. \tag{10.3.7}$$

Thus the excluded shell increases the pressure by forcing the spheres farther apart, but the doublets produced by the adhesion compensate. Indeed, $A_2 = 0$ for $s_0 = 1/2\tau$, producing a pseudo-ideal state analogous to the theta condition for polymer solutions.

The two parameters, s_0 and τ, can be related to those characterizing the actual potential by comparing (10.3.7) with the exact expression for the second virial coefficient (10.3.4). Defining a separation σ such that $\Phi(\sigma) = 0$ suggests setting

$$\frac{s_0^3}{2} = 4 + \tfrac{3}{2} \int_2^{\sigma} [1 - \exp(-\Phi/kT)] s^2 \, ds$$

and (10.3.8)

$$\frac{1}{\tau} = \frac{6}{s_0^2} \int_{\sigma}^{\infty} [\exp(-\Phi/kT) - 1] s^2 \, ds.$$

Numerical integration then provides accurate values of the two parameters. The relatively small effect on the second virial coefficient might be difficult to measure, but the transport coefficients differ significantly from those for hard spheres as described later.

10.4 Perturbation theory

This section outlines a perturbation theory (Barker & Henderson, 1967a, b) that formalizes the *ad hoc* approximation for dilute suspensions in the previous section and enables the prediction of the thermodynamic properties and equilibrium structure of dispersions over the full range of concentrations. The method provides the basis for predictions of the disorder–order transitions in latices at low ionic strengths (§10.6) and fluid–fluid and fluid–solid transitions induced by dissolved polymer (§10.7). The perturbation expansion employs as the reference state the hard-sphere dispersion whose properties are summarized in §10.5. Further discussions of this material and related approaches are available in McQuarrie (1976, §14.3) and Hansen & McDonald (1976, Ch. 6).

The approach requires resolving the interaction potential into the hard sphere repulsion, an additional repulsion of range $a\delta$, and a weak, longer range attraction as (Fig. 10.1)

$$\Phi(r) = \begin{cases} \infty & 2a > r \\ \Phi[(2+\delta\bar{s})a] & \sigma > r > a \\ \lambda\Phi^*(r) & r > \sigma, \end{cases} \tag{10.4.1}$$

with

$$\bar{s} = (s-2)/\delta,$$
$$\lambda = -\Phi_{\min}/kT,$$
$$\Phi^* = -kT\Phi/\Phi_{\min},$$
$$\Phi(\sigma) = 0.$$

Fig. 10.1. Form of interaction potential with soft repulsion for $2a \le r \le \sigma$ and weak attraction for $\sigma < r$ and effective hard-sphere diameter $2a \le d < \sigma$.

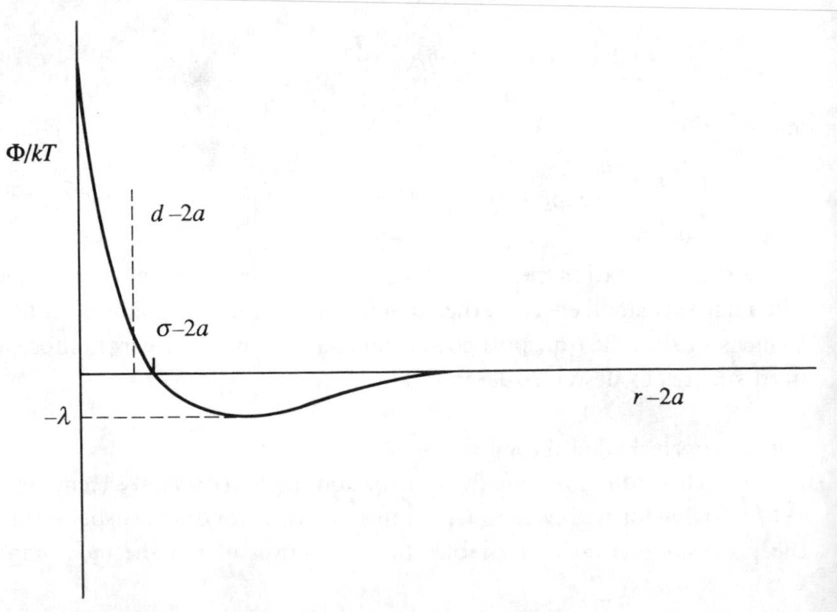

When $\delta \ll 1$ and $\lambda < 1$, the potential resembles that of a hard sphere with effective diameter $d > 2a$, plus a weak attraction, permitting expansion of the Helmholtz free energy about the value for hard spheres $A_{\mathrm{HS}}(n, d)$ as

$$A = A_{\mathrm{HS}} + \delta\left(\frac{\partial A}{\partial \delta}\right)_{\delta=\lambda=0} + \lambda\left(\frac{\partial A}{\partial \lambda}\right)_{\delta=\lambda=0} + \ldots . \tag{10.4.2}$$

For pairwise additive potentials the configuration integral, which determines the Helmholtz free energy, takes the form

$$Z_N = \int \ldots \int \exp\left[-\sum_{i>j} \Phi(r_{ij})/kT\right] \mathrm{d}\mathbf{x}_1 \ldots \mathrm{d}\mathbf{x}_N, \tag{10.4.3}$$

permitting the necessary derivatives to be expressed analytically.

Differentiation with respect to the parameter δ leads to

$$\frac{1}{N}\left(\frac{\partial \ln Z_N}{\partial \delta}\right)_{\delta=\lambda=0} = 2\pi n a d^2 g_{\mathrm{HS}}(d)\left\{d - 2a - \int_{2a}^{\sigma} [1 - \exp(-\Phi/kT)]\,\mathrm{d}r\right\} \tag{10.4.4}$$

Thus choosing the effective hard sphere diameter such that

$$d = 2a + \int_{2a}^{\sigma} [1 - \exp(-\Phi/kT)]\,\mathrm{d}r \tag{10.4.5}$$

eliminates the $O(\delta)$-term from the expansion. Evaluation of the derivative with respect to the magnitude of the attractive potential and substitution into (10.4.2) then gives

$$A = A_{\mathrm{HS}} + 2\pi N n \int_{\sigma}^{\infty} \Phi(r) g_{\mathrm{HS}}(r) r^2 \mathrm{d}r + O(\delta^2, \lambda^2, \delta\lambda). \tag{10.4.6}$$

Approximate evaluation of the $O(\lambda^2)$-term is also possible to test the convergence of the expansion (McQuarrie 1976, §14.3). Although an alternative approach for determining the effective diameter (Andersen, Chandler, & Weeks, 1971) can be somewhat more accurate at $O(\lambda)$, (10.4.6) suffices for our purposes.

Perturbation theories have been tested extensively for the square well and Lennard-Jones potentials (McQuarrie, 1976) and for the superposition of electrostatic and dispersion potentials (van Megen & Snook, 1984), with the results generally falling within ± 10 per cent of the values from computer simulations, even when $\lambda = O(1)$. The predictions require, of course, accurate and complete results for the hard sphere reference state as described in the next section.

10.5 Suspensions of hard spheres

Hard spheres comprise both the simplest colloidal system realizable in the laboratory and a convenient reference state for the perturbation theory discussed above. This section summarizes for later use the information available on the radial distribution function, the thermodynamic functions, and the phase behavior, and demonstrates the hard-sphere nature of some colloidal systems. Readers interested in the origins of either the theoretical or the experimental results presented should consult the original papers cited.

The radial distribution function for the fluid phase is generated most easily from the solution to the Percus–Yevick equation (Smith & Henderson, 1970),

$$g_{\mathrm{HS}}(s)=\sum_{j=1}^{\infty} H\left(\frac{s}{2}-j\right) g_j(s), \tag{10.5.1}$$

with

$$\frac{s}{2} g_j(s)=\frac{(-12\phi)^{j-1}}{(j-1)!} \sum_{i=0}^{2} \lim_{t \to t_i} \frac{\mathrm{d}^{j-1}}{\mathrm{d}t^{j-1}}\left\{(t-t_i)^j t\left(\frac{L(t)}{S(t)}\right)^j \exp\left[t\left(\frac{s}{2}-j\right)\right]\right\},$$

$$S(t)=(1-\phi)^2 t^3+6\phi(1-\phi)t^2+18\phi^2 t-12\phi(1+2\phi),$$

$$L(t)=\left(1+\frac{\phi}{2}\right)t+1+2\phi,$$

where $t_i(i=0, 1, \ldots)$ are the zeros of $S(t)$. The corresponding equation of state for the fluid derived by Carnahan & Starling (1969),

$$Z(\phi)=\frac{\Pi}{nkT}=\frac{1+\phi+\phi^2-\phi^3}{(1-\phi)^3}, \tag{10.5.2}$$

agrees with both the first seven virial coefficients and the molecular dynamics results. The function $Z(\phi)$ is known as the compressibility factor.

Figure 10.2 illustrates the structure produced in g_{HS} by many-body interactions at finite volume fractions. The local order represented by the maxima and minima in g reflect the difficulty of packing spheres at high densities, i.e. the limited number of configurations available. Such constraints increase the entropy or free energy of the system, and, therefore, the pressure as indicated by (10.5.2).

Silica spheres stabilized with a thin organophilic layer and dispersed in cyclohexane provide a model hard-sphere colloid (Vrij *et al.*, 1983; de Kruif, Jansen, & Vrij, 1987). The choice of a non-polar fluid with a refractive index close to that of the particles suppresses both the electrostatic and the dispersion forces. Comparison of the measured compressibilities with those predicted by (10.5.2) demonstrates their hard-sphere nature (Fig. 10.3).

Molecular dynamics and Monte Carlo simulations and statistical mechanical theories for hard spheres predict a transition from a disordered fluid for $\phi < 0.50$ to an ordered face-centered-cubic solid for $0.55 < \phi < 0.74$ (Alder & Wainwright, 1957; Wood & Jacobson, 1957; Alder, Hoover, & Young, 1968; Haymet & Oxtoby, 1986). Within the solid, the results from the simulations conform to (Hall, 1972)

$$Z(\phi) = 2.558 + 0.125\beta + 0.176\beta^2 - 1.053\beta^3 + 2.819\beta^4 - 2.922\beta^5 + 1.118\beta^6 + 3(4-\beta)/\beta, \tag{10.5.3}$$

with $\beta = 4(1-\phi/0.74)$, indicating that the pressure diverges near close packing as

$$\frac{\Pi}{nkT} \sim \frac{2.22}{0.74-\phi}. \tag{10.5.4}$$

Fig. 10.2. Radial distribution function for suspensions of hard spheres in the disordered state calculated from (10.5.1).

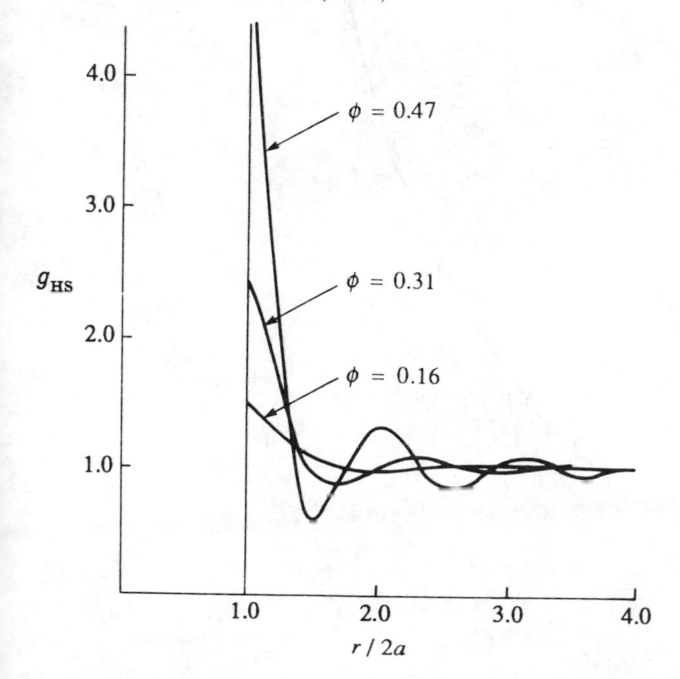

Kincaid & Weis (1977) correlated their numerical results for the radial distribution function of the solid in terms of spheres localized near the lattice sites in a face-centered-cubic crystal with the positions and widths of the peaks varying with volume fraction as illustrated in Fig. 10.4. The areas under the peaks reflect the number of neighbors in a particular spherical shell. Only for $\phi > 0.63$ does $g_{HS} \to 0$ between the peaks, while at lower volume fractions the radial distribution function still resembles that for the disordered fluid.

Although a face-centered-cubic solid phase appears at equilibrium for $\phi > 0.50$, disordered dispersions of hard spheres can persist for extended

Fig. 10.3. Osmotic compressibility of suspensions of hard spheres from the Carnahan–Starling equation (10.5.2) compared with measurements for organophilic silica spheres with $a = 11$ nm dispersed in cyclohexane (Vrij *et al.*, 1983).

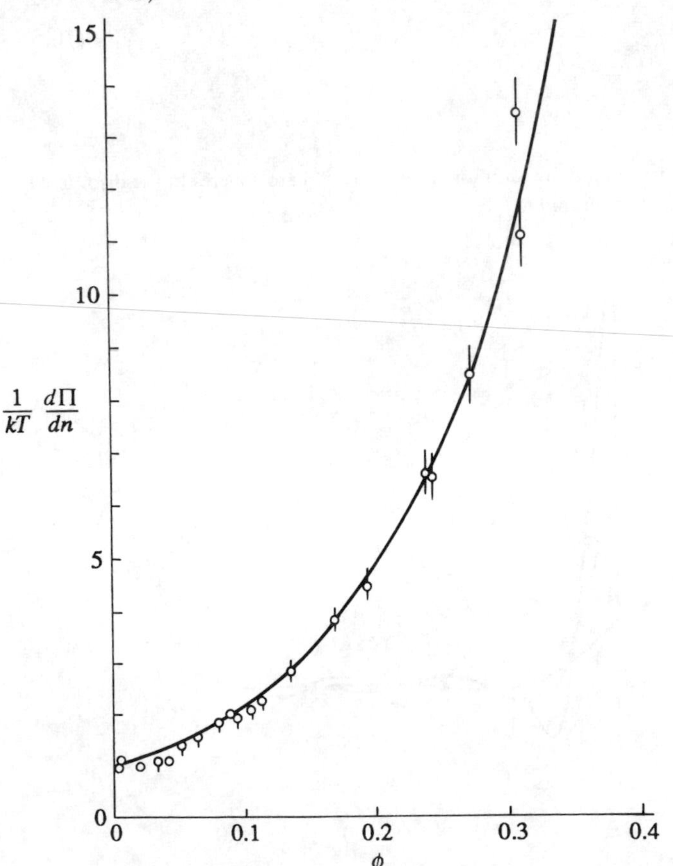

times (Woodcock, 1981; Pusey & van Megen, 1987). The simulations reveal an osmotic pressure that increases smoothly from the values predicted from (10.5.2) for $\phi \leq 0.50$ but diverges according to

$$\frac{\Pi}{nkT} \sim \frac{1.85}{0.64 - \phi}. \tag{10.5.5}$$

Note that the divergence occurs at random close packing $\phi \approx 0.63$–0.64, consistent with the idea that higher volume fractions require some degree of order.

Figure 10.5 displays the compressibility factor over the full range of concentrations, indicating the fluid and solid phases and the coexistence region. The other thermodynamic functions can be generated from the equation of state through the relationships (10.2.2).

The small difference in volume fractions between the two phases makes the disorder–order transition a subtle one, which is difficult to detect experimentally except by the iridescence of the crystalline solid. Nonetheless, Hachisu & Kobayashi (1974) did quantify the coexistence of two equilibrium phases with aqueous polystyrene latices that behave as hard spheres at high ionic strengths. The volume fractions, measured directly by

Fig. 10.4. Radial distribution function for suspensions of hard spheres in the ordered state calculated from the equations in Kincaid & Weis (1977).

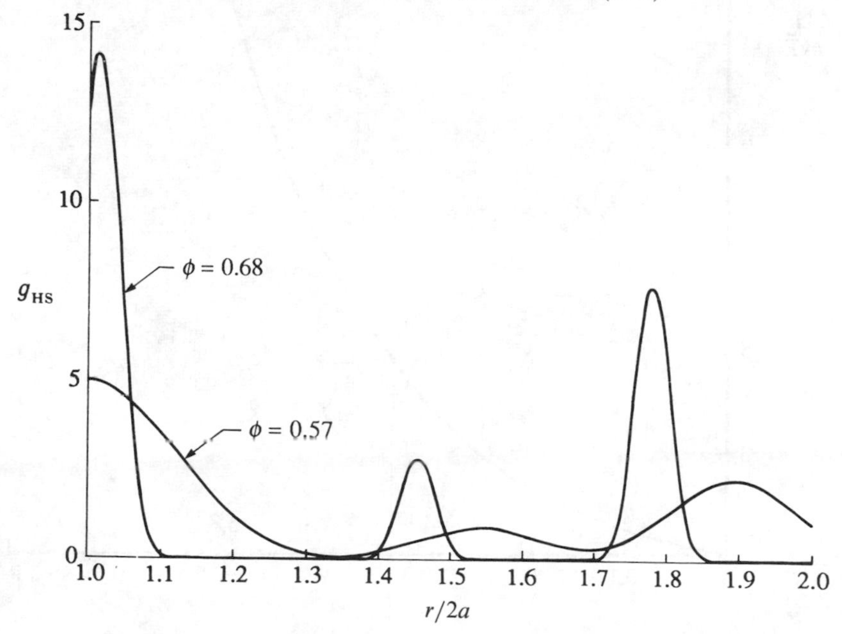

Fig. 10.5. Compressibility factor Π/nkT for suspensions of hard spheres including the fluid (10.5.2) and solid (10.5.3) curves and the coexistence region for $0.50<\phi<0.55$.

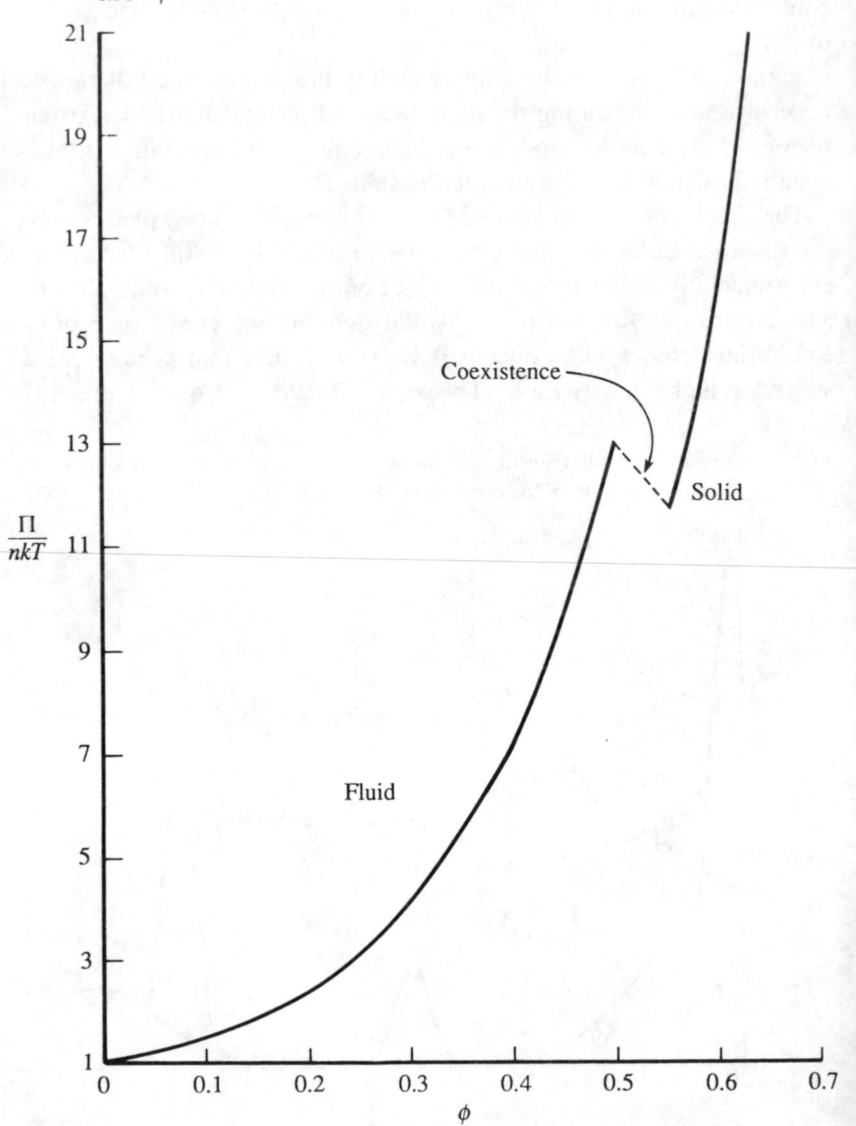

sampling the individual phases, accord well with the predictions of the hard-sphere theory indicated by the horizontal lines (Fig. 10.6). The silica dispersions also crystallize at high volume fractions (de Kruif, Jansen, & Vrij, 1987); indeed, the phenomenon provides a means for synthesizing opals (Darragh, Gaskin, & Sanders, 1976). Thus there exist colloidal systems that behave as hard spheres, and the theories predict their thermodynamic properties and phase behavior with reasonable accuracy.

10.6 Disorder–order transition for charged spheres

At low ionic strengths, the long-range electrostatic repulsion between charged spheres induces a phase transition at volume fractions as low as 10^{-3}, with characteristics similar to that for hard spheres: (i) The transition is accompanied by an iridescence linked by Hiltner & Krieger (1969) to the Bragg diffraction of visible light by the crystalline array of suspended particles. (ii) The suspension reverts to a disordered fluid with dilution (Hiltner & Krieger, 1969), or, equivalently, the addition of excess electrolyte (Hachisu, Kobayashi, & Kose, 1973). (iii) The region of fluid–solid coexistence is narrow.

Fig. 10.6. Volume fractions of coexisting fluid (○) and solid (●) phases for polystyrene latices with $a = 0.09\,\mu m$ in water, demonstrating the hard-sphere disorder–order transition (Hachisu & Kobayashi, 1974).

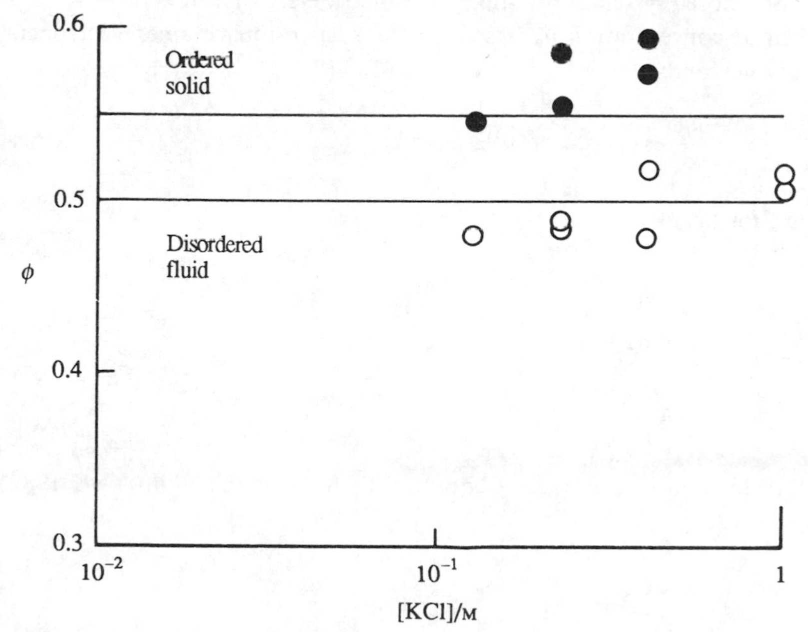

Such a disorder–order transition can be predicted by the perturbation theory in its simplest form, as suggested in the original papers of Hachisu and exploited more recently by others (Snook & van Megen, 1976; Barnes *et al.*, 1978; Beunen & White, 1981; Pieranski, 1983). Accurate calculation of the effective hard-sphere diameter from (10.4.5), however, requires proper treatment of electrostatic interactions in concentrated systems.

Recall that within the fluid Poisson's equation (4.6.1),

$$\nabla^2\psi = -\frac{e}{\varepsilon\varepsilon_0}\Sigma z^k n^k, \tag{10.6.1}$$

governs the electrostatic potential with the ion densities determined by the Boltzmann distributions

$$n^k = c^k \exp\left(-\frac{ez^k\psi}{kT}\right). \tag{10.6.2}$$

For moderately charged particles the exponential can be expanded about the mean potential in the fluid $\langle\psi\rangle$ as $\psi = \psi' + \langle\psi\rangle$, with $\langle\psi'\rangle = 0$, so that

$$n^k = c^k \exp\left(-\frac{ez^k\langle\psi\rangle}{kT}\right)\left(1 - \frac{ez^k\psi'}{kT} + \ldots\right). \tag{10.6.3}$$

For a closed system containing a symmetric electrolyte ($z^1 = -z^2 = z > 0$) added at concentration n_b based on the suspension volume, electroneutrality demands that

$$-\frac{3q\phi}{a} = (1-\phi)ez(\langle n^1\rangle - \langle n^2\rangle). \tag{10.6.4}$$

Thus, for negatively charged spheres,

$$\langle n^1\rangle = \frac{n_b - \dfrac{3q\phi}{aez}}{1-\phi}, \tag{10.6.5}$$

$$\langle n^2\rangle = \frac{n_b}{1-\phi}.$$

Substitution into Poisson's equation then yields (Chaikin *et al.*, 1982; Russel, 1983)

$$\nabla^2\psi' = \kappa^2\psi' + \frac{3q\phi}{\varepsilon\varepsilon_0 a(1-\phi)} \tag{10.6.6}$$

with

$$\kappa^2 = \frac{e^2}{\varepsilon\varepsilon_0 kT} \frac{2z^2 n_b - \dfrac{3qz\phi}{ae}}{1-\phi}.$$

The second term on the right-hand side of (10.6.6) represents a uniform density of counterions, while the modified Debye length accounts for the counterions and the reduction in the fluid volume due to the presence of particles. The former can be substantial, since $-q/ae \approx 10^{-3}$ M for $a = 0.1$ μm and $q = -0.01$ C/m^2. Thus, at low n_b the Debye length can be significantly shorter than expected from the added electrolyte alone.

This formulation provides the basis for constructing a pairwise additive description of the interactions between spheres in concentrated dispersions. The homogeneous solution to (10.6.6) coincides with that for pair interactions at infinite dilution, except for the modified Debye length, while the constant inhomogeneous solution does not affect the interaction potential. Hence the pair potential remains as presented in §4.10, except for the volume-fraction-dependent Debye length and the corresponding variation in either the surface charge or the potential. At these low ionic strengths, neither the primary nor the secondary minima are significant, allowing the dispersion potential to be neglected.

The strong repulsions characteristic of the ordered systems generally preclude substantial overlap of the double layers, making the superposition approximation

$$\frac{\Phi(r)}{kT} = \alpha \frac{\exp(-\kappa r)}{\kappa r}, \tag{10.6.7}$$

with $\alpha = 4\pi\varepsilon\varepsilon_0 \psi_s^2 a^2 \kappa \exp(2a\kappa)/kT$, appropriate. The pre-exponential factor is generally quite large, e.g. $\alpha = 1.2 \times 10^6$ for $a = 0.1$ μm, $\psi_s = 50$ mV, and $n_b = 10^{-4}$ M. Consequently, the exponential in (10.4.5) changes rapidly from zero at small separations to unity at large separations (Fig. 10.7), suggesting the approximation

$$1 - \exp\left(-\frac{\Phi(r)}{kT}\right) = \begin{cases} 1 & r < L \\ 0 & L < r \end{cases}$$

with the transition occurring about $\Phi(L)/kT \approx 1$ so that

$$L \sim \frac{1}{\kappa} \ln\{\alpha/\ln[\alpha/\ln(\alpha/\ldots)]\}.$$

The effective hard-sphere diameter then follows trivially from (10.4.5) as $d = L$.

For a purely repulsive potential application of the perturbation theory requires only the calculation of d. The thermodynamic properties such as the osmotic pressure then correspond to those for a hard-sphere suspension with $\phi_{\text{eff}} = \phi(d/2a)^3$. Consequently, the suspension remains a disordered fluid for

$$\phi < 0.50\left(\frac{2a}{d}\right)^3,$$

and becomes an ordered, face-centered cubic solid when

$$\phi > 0.55\left(\frac{2a}{d}\right)^3.$$

At intermediate volume fractions the two phases co-exist. The resulting phase diagrams, plotted as ϕ versus $a\kappa_0$ with $\kappa_0^2 = 2e^2z^2n_b/\varepsilon\varepsilon_0 kT$, depend only on the dimensionless surface potential or charge through

$$\frac{4\pi\varepsilon\varepsilon_0\psi_s^2 a}{kT} \quad \text{or} \quad \frac{4\pi a^3 q^2}{\varepsilon\varepsilon_0 kT}.$$

The predictions differ quantitatively according to how the surface charge and potential vary as the excess electrolyte changes, reflecting the role of surface chemistry (Beunen & White, 1981). But the qualitative features

Fig. 10.7. Boltzmann factor, $\exp(-\Phi(r)/kT)$, corresponding to electrostatic repulsion between spheres as a function of the dimensionless separation.

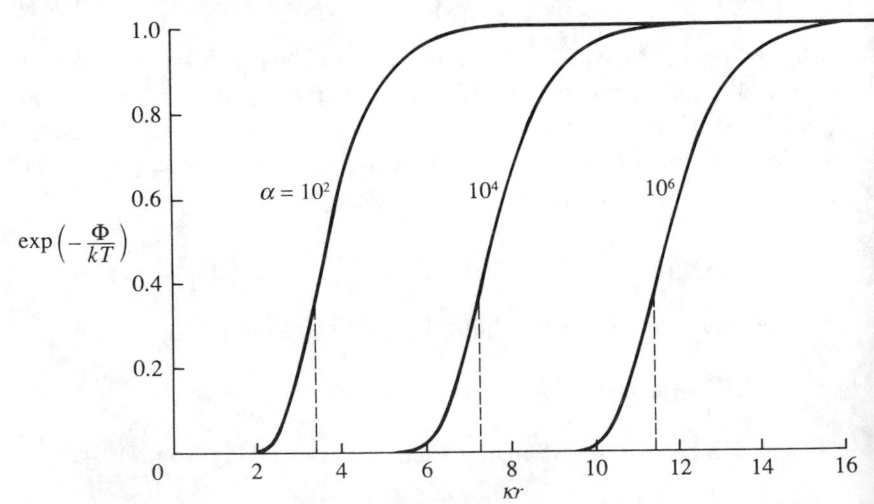

remain the same with the ordered phase at high volume fractions and/or low ionic strengths coexisting over a narrow range of volume fractions with the disordered phase.

As an illustration, we compare the predictions based on 5000 charges fixed on 0.1 μm radius spheres with the data of Hachisu & Kobayashi (1974) (Fig. 10.8). The corresponding charge density of $6.4 \times 10^{-3}\,\mu C/m^2$ falls in the range generally deduced for latices. Microscopic observations (Fig. 10.9) and scattering experiments (Williams & Crandall, 1974; Ackerson & Clark, 1981) confirm the face-centered-cubic structure for the ordered phase for $\phi > 0.01$.

At very low ionic strengths and low volume fractions, the perturbation theory errs because the $O(1/a\kappa)^2$ term in the expansion becomes significant. Also, the softness of the repulsion alters the crystal structure to body-centered-cubic (Williams & Crandall, 1974; Ackerson & Clark, 1981). Prediction of this structural transition at low ionic strengths and volume fractions requires a more sophisticated approach (e.g., Hone *et al.*, 1983; Shih, Aksay, & Kikuchi, 1987).

Fig. 10.8. Phase diagram for disorder–order transition for charged spheres in an electrolyte solution. Data of Hachisu, Kobayashi, & Kose (1973) for polystyrene latices with $a = 0.085\,\mu m$: open circles, disordered; half-filled circles, two-phase; filled circles, ordered; curves, predictions of phase boundaries from the perturbation theory for $a = 0.10\,\mu m$ and $4\pi a^2 q = 5000e$ (Russel, 1987).

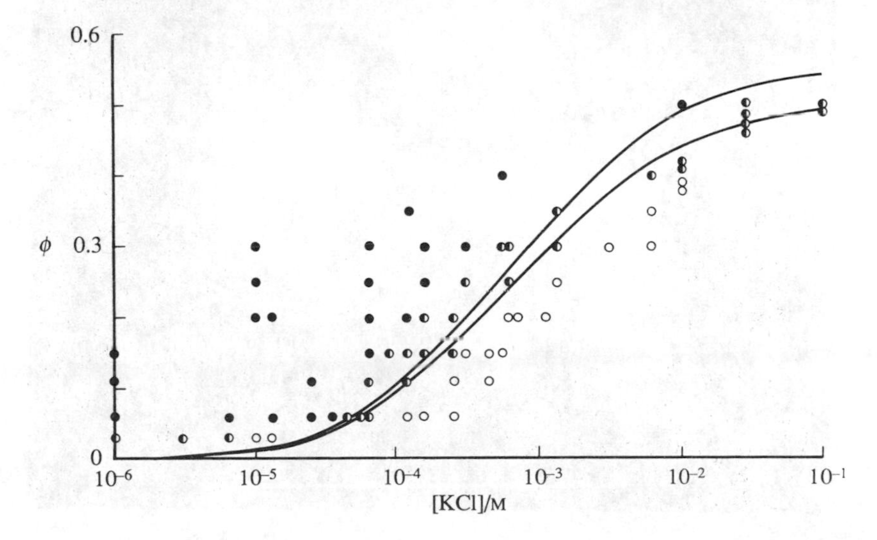

Fig. 10.9. Photomicrograph of ordered structure in deionized polystyrene latex with $2a = 0.33\,\mu\text{m}$ and $\phi = 0.01$ (Kose *et al.*, 1973).

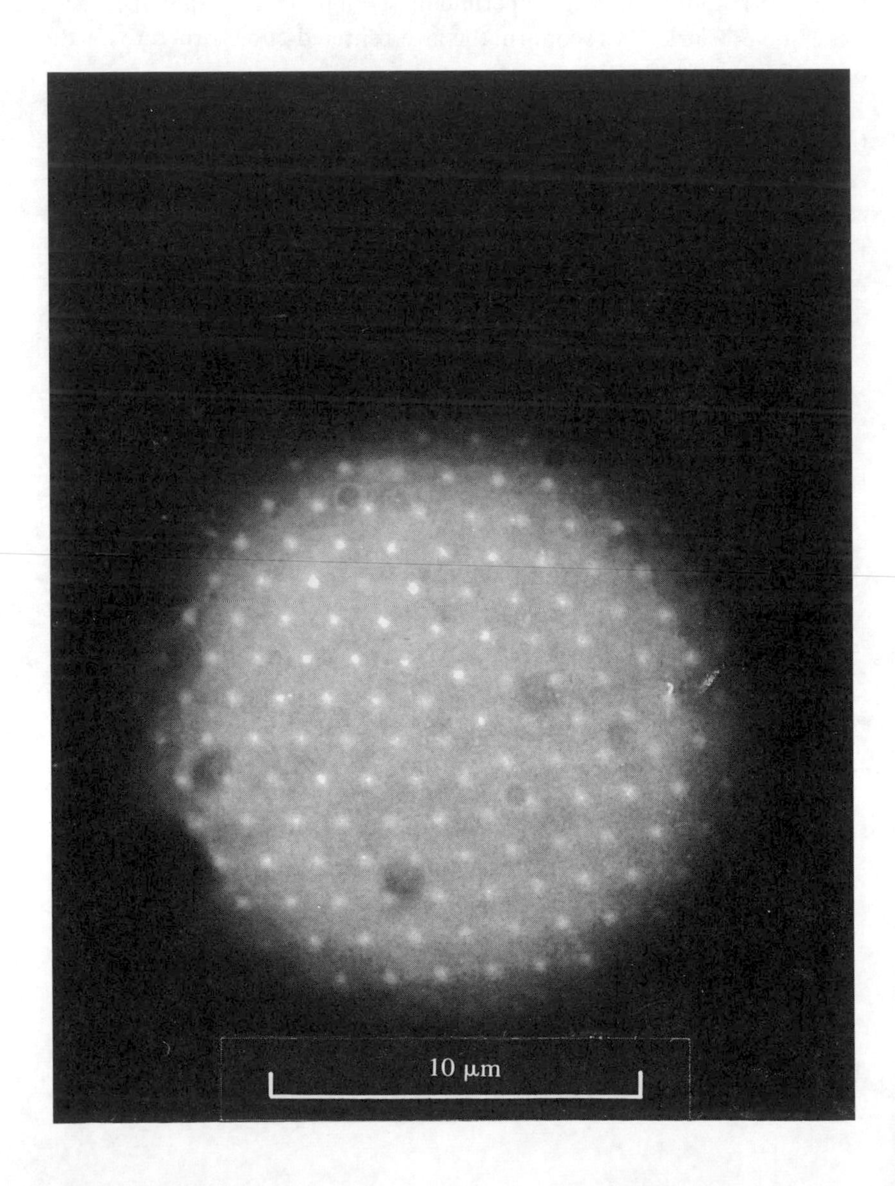

10.7 Phase transitions induced by dissolved polymer

A second class of phase transitions is produced in otherwise stable suspensions by the addition of soluble, non-adsorbing polymer. The observations are quite simple (Fig. 10.10). At low polymer concentrations, the suspension remains a single-phase fluid, with the particles uniformly dispersed. Above a critical polymer concentration, though, the system separates into a dense phase in equilibrium with dilute phase. Increasing the polymer concentration further depletes the dilute phase of particles and concentrates the dense phase. Frequently, however, there exists a second critical concentration above which the dispersion reverts to a single phase. The phenomena occur with silica particles dispersed in cyclohexane containing polystyrene at theta conditions (de Hek & Vrij, 1981), for electrostatically stabilized latices and a non-ionic, water-soluble polymer such as hydroxyethyl cellulose (Sperry, Hopfenberg, & Thomas, 1981), and for particles stabilized by grafted polymer chains in solutions of the same or different polymer (Cowell, Li-In-On, & Vincent, 1978; Vincent, Luckham, & Waite, 1980; Clarke & Vincent, 1981; Vincent, Clarke, Barnett, 1986; Vincent *et al.*, 1986).

Fig. 10.10. Schematic of the phase transition caused by the addition of soluble, non-adsorbing polymer to a stable dispersion showing a single phase at low polymer concentrations, coexisting phases at intermediate concentrations, and return to a single phase at high concentrations.

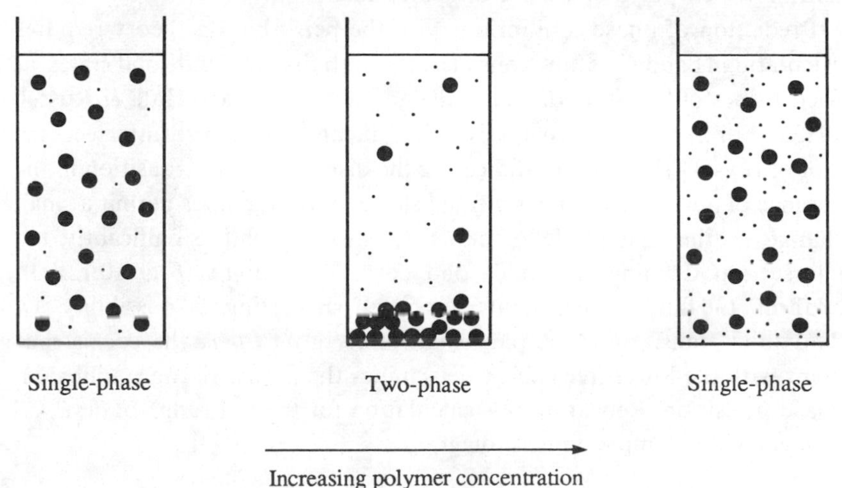

Application of the perturbation theory

Phase separation arises from the attraction induced by the exclusion of polymer from the gap between two particles at separations less than the macromolecular size as described in §6.4. Several of the authors mentioned above have correlated successfully the critical polymer concentration as a function of polymer molecular weight, particle size, and the solution chemistry with theories based simply on the magnitude of the attractive potential (6.4.13). These approaches firmly establish the mechanism but provide no information on the compositions or structure of the individual phases and generally apply to initially dilute suspensions.

The perturbation theory (§10.4), on the other hand, yields complete phase diagrams as a function of the characteristics of the polymer, the particles, and the solution. Specification of the total interparticle potential determines the effective hard-sphere diameter d from (10.4.5) and the perturbation due to the attraction from (10.4.6). For hard spheres $d=2a$, but electrostatic or polymeric repulsions generate $d/2a>1$. For dilute polymer solutions and particles without polymeric layers, (6.4.13) comprises the attractive component of the potential, with the thickness of the depletion layer equal to the radius of gyration r_g of the polymer and the osmotic pressure of the polymer solution P related to the concentration and molecular weight through the virial expansion (6.2.24). From these, the relevant thermodynamic functions – A/NkT, G/NkT, and $a^3\Pi/kT$ – can be calculated through (10.4.6) and (10.2.2) as functions of ϕ, a/r_g, $a\kappa$, $\epsilon\epsilon_0\psi_s^2 a/kT$, and ar_g^2P/kT, with the last measuring the magnitude of the attractive potential. Interaction between polymerically stabilized particles and dissolved polymer will be discussed later.

Prediction of phase equilibrium with the perturbation theory requires calculating Π and G as functions of ϕ, for both the fluid and solid states, at fixed values of the other dimensionless parameters (Gast, Hall, & Russel, 1983a). For hard spheres themselves the fluid and solid curves intersect on a plot of G vs. Π (Fig. 10.11), indicating the disorder–order transition in the absence of polymer. The nearly equal slopes of these curves permit a small change in the magnitude of either function to shift significantly the intersection defining the fluid–solid coexistence point (Fig. 10.12). In addition, the fluid curve also intersects itself, suggesting the possibility of a fluid–fluid transition. At a particular value of ar_g^2P/kT, the coexistence point with the lower free energy determines the nature of the equilibrium phase transition. Repeating this calculation for the full range of ar_g^2P/kT delineates the complete phase diagram.

Fig. 10.11. Gibbs free energy versus pressure for a suspension of hard spheres, showing the intersection of the fluid and solid curves corresponding to the disorder-order transition with $4\pi a^3 n_0/3 = 0.74$ (Gast, Hall, & Russel, 1983a).

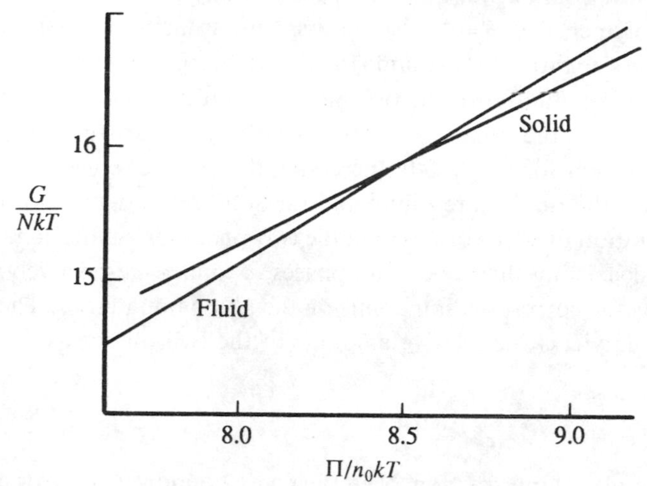

Fig. 10.12. Gibbs free energy versus pressure for hard spheres in an ideal polymer solution, with $a/r_g = 2.5$ and $4\pi a r_g^2 P/3kT = 1.0$, showing intersections of the fluid and solid curves and the fluid curve with itself.

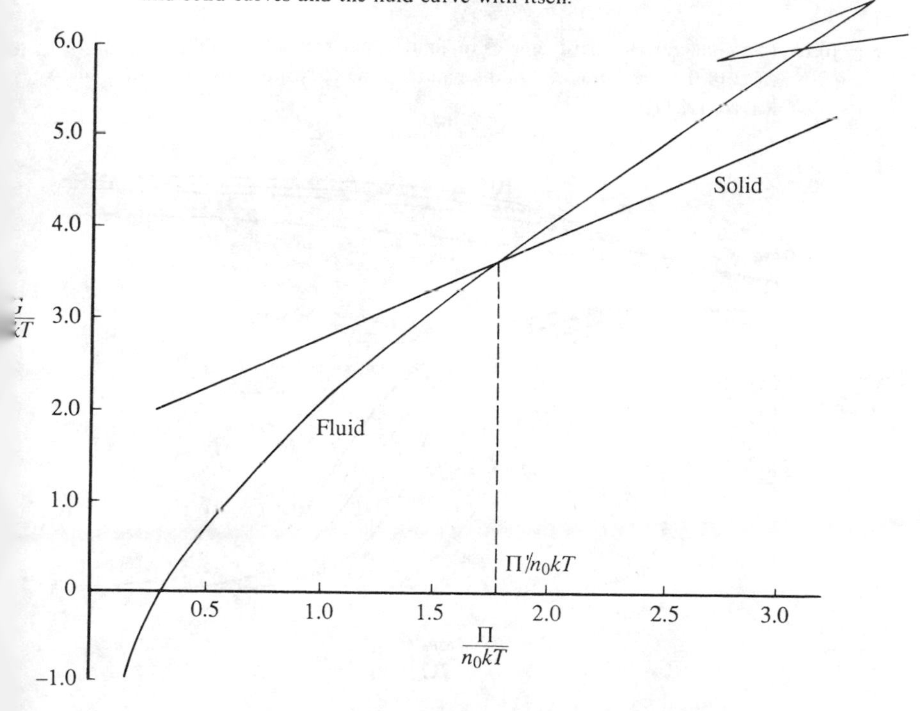

Hard spheres in ideal polymer solutions

For a suspension of hard spheres in a polymer solution at theta conditions, the phase boundaries (Fig. 10.13) depend only on the size ratio, a/r_g. Without polymer, i.e. $P=0$, the hard-sphere transition persists, independent of the size ratio. At a volume fraction $\phi<0.50$, the dispersion remains a single-phase fluid until the polymer concentration exceeds the value corresponding to the phase boundary, whereupon coexisting fluid and solid phases appear for $a/r_g > 3.0$. Increasing the polymer concentration further makes the fluid more dilute and the solid more dense. At a polymer concentration or value of P within the two-phase region the compositions of the coexisting fluid and solid phases, ϕ_f and ϕ_s respectively, are determined by the corresponding points on the phase boundaries. The lever arm rule then yields the relative amounts of the two phases as

$$\frac{\text{solid}}{\text{fluid}}=\frac{\phi-\phi_f}{\phi_s-\phi},$$

where ϕ is the overall volume fraction. Near the phase boundary, where the magnitude of the attractive potential is $1kT$ to $3kT$, except at very low volume fractions, the two phases should equilibrate easily.

Fig. 10.13. Phase diagram for hard spheres in an ideal polymer solution, illustrating the effect of the size ratio a/r_g on the nature of the coexisting phases (Gast, Hall, & Russel, 1983a).

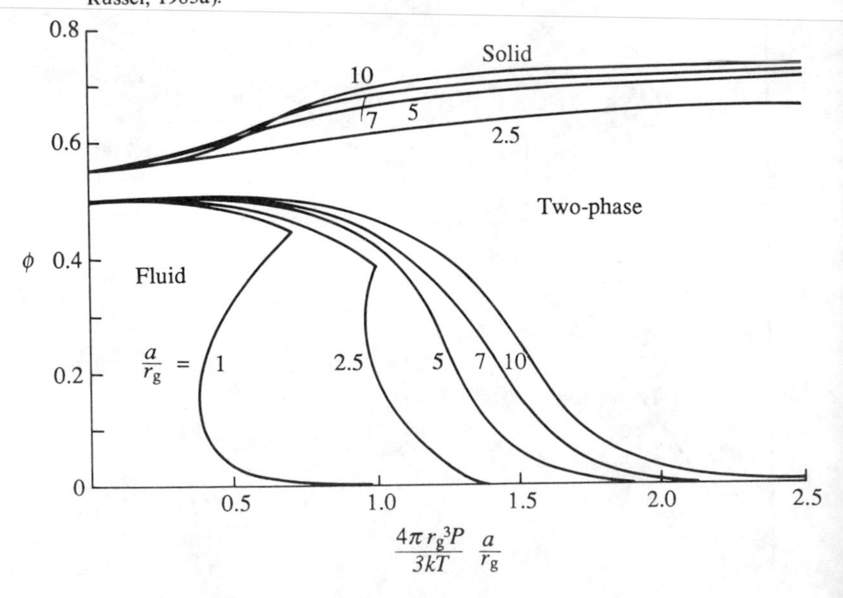

For $a/r_g < 3.0$, two fluid phases can coexist, producing a phase diagram with a critical point and a triple point. Then, for $0 < \phi < \phi_{triple}$ the addition of polymer first produces the fluid–fluid transition, followed by a reversion to a fluid–solid coexistence beyond the triple point.

Electrostatically stabilized dispersions

For aqueous suspensions, the phase diagrams remain qualitatively similar to those for hard spheres (Gast, Hall, & Russel, 1983b). Since the theory accommodates the electrostatic repulsion through the effective hard-sphere diameter, the disorder–order transition moves to lower-volume fractions, as discussed in §10.6, and the phase boundaries depend on $\varepsilon\varepsilon_0\psi_s^2 a/kT$ and $a\kappa$ as well as a/r_g.

The non-ideality of the polymer solutions affects both the concentration dependence of the osmotic pressure and the depletion layer thickness, though the latter is difficult to quantify. The magnitude of $ar_g^2 P/kT$ no longer reflects the attractive minimum in the potential because of the electrostatic repulsion, but the actual minimum remains in the range $2kT$ to $3kT$ along the phase boundary, as observed by Sperry (1982).

Several sets of observations for the addition of hydroxyethyl cellulose or dextran to aqueous latices illustrate the phenomena and the capabilities of the theory (Sperry, Hopfenberg, & Thomas, 1981; Sperry, 1984; Gast, Russel, & Hall, 1986; Patel & Russel, 1989). The polymer concentration required to induce phase separation decreases with increasing molecular weight, ionic strength, and particle size. In each case the trends simply reflect variations in the attractive minimum in the total interaction potential, and agree quantitatively with the theory.

Sperry (1984) also observed the morphologies of the phase-separated dispersions for a range of particle sizes and polymer concentrations (Fig. 10.14). The top row of photos corresponds to polymer concentrations just into the two-phase region. For $a/r_g \leq 1.52$, the dense phase resembles droplets with a smooth interface, characteristic of a fluid phase with an interfacial tension. For $a/r_g \geq 3.37$, however, the interface is irregular on a scale large relative to the individual particles, suggesting a solid phase incapable of flow. These observations agree qualitatively with the prediction of a transition from fluid–solid to fluid–fluid coexistence about $a/r_g \approx 3$. The polymer concentrations at the phase boundary conform reasonably well with the predictions for $a/r_g > 1$, but the theory fails for smaller particles (Fig. 10.15). When the polymer concentration is increased for $a/r_g = 0.97$ and 1.52, the dense phase begins to resemble a solid, as would be expected from Fig. 10.13 as the triple point is passed.

Observed phase diagrams, though limited, generally conform to those predicted (e.g. Fig. 10.16). The agreement is quantitative for the fluid–solid transition at moderate polymer concentrations and the dense phase shows iridescence. Further into the two-phase region the iridescence takes longer to appear and the volume fractions can fall below those expected, suggesting a failure to equilibrate completely due to the deeper attractive minimum. The theory also predicts the form of the fluid–fluid phase boundary accurately, demonstrating its capability for anticipating the structure of the phases.

The perturbation theory does have limitations, though. Problems arise with small particles for which the repulsion is relatively soft, causing the effective hard sphere diameter to exceed the actual particle diameter significantly. Then since $\lambda = O(1)$, the $O(\delta\lambda)$- and $O(\delta^2)$-errors in (10.4.5) degrade the predictions of both the fluid–solid and the fluid–fluid

Fig. 10.14. Photomicrograph showing the structure of the dense phase as a function of the particle diameter and the hydroxyethyl cellulose concentration for $\phi = 0.10$, $M_n = 306$ kg/mol and 0.023 M electrolyte (Sperry, 1984). The top row corresponds to conditions just across the phase boundary into the two-phase region.

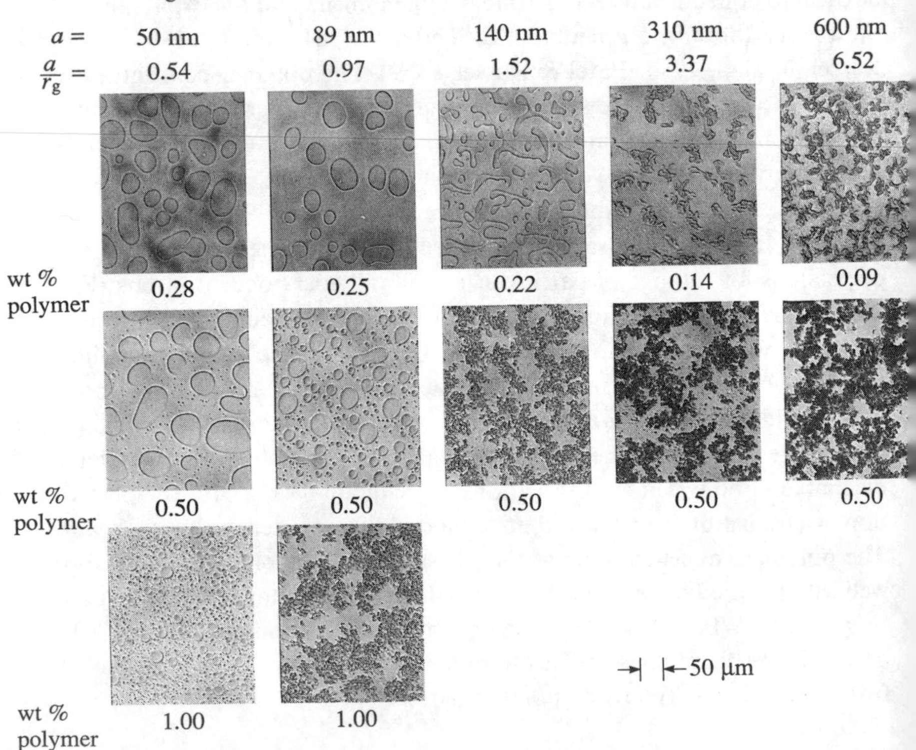

Fig. 10.15. Concentration of hydroxyethyl cellulose necessary to phase separate aqueous polystyrene latices as a function of particle size with $\phi = 0.10$, $M_n = 306$ kg/mol, and 0.023 M electrolyte: I-symbols represent the data of Sperry (1984); unbroken and broken lines are, respectively, the fluid–solid and fluid–fluid predictions of Gast, Russel, & Hall (1986).

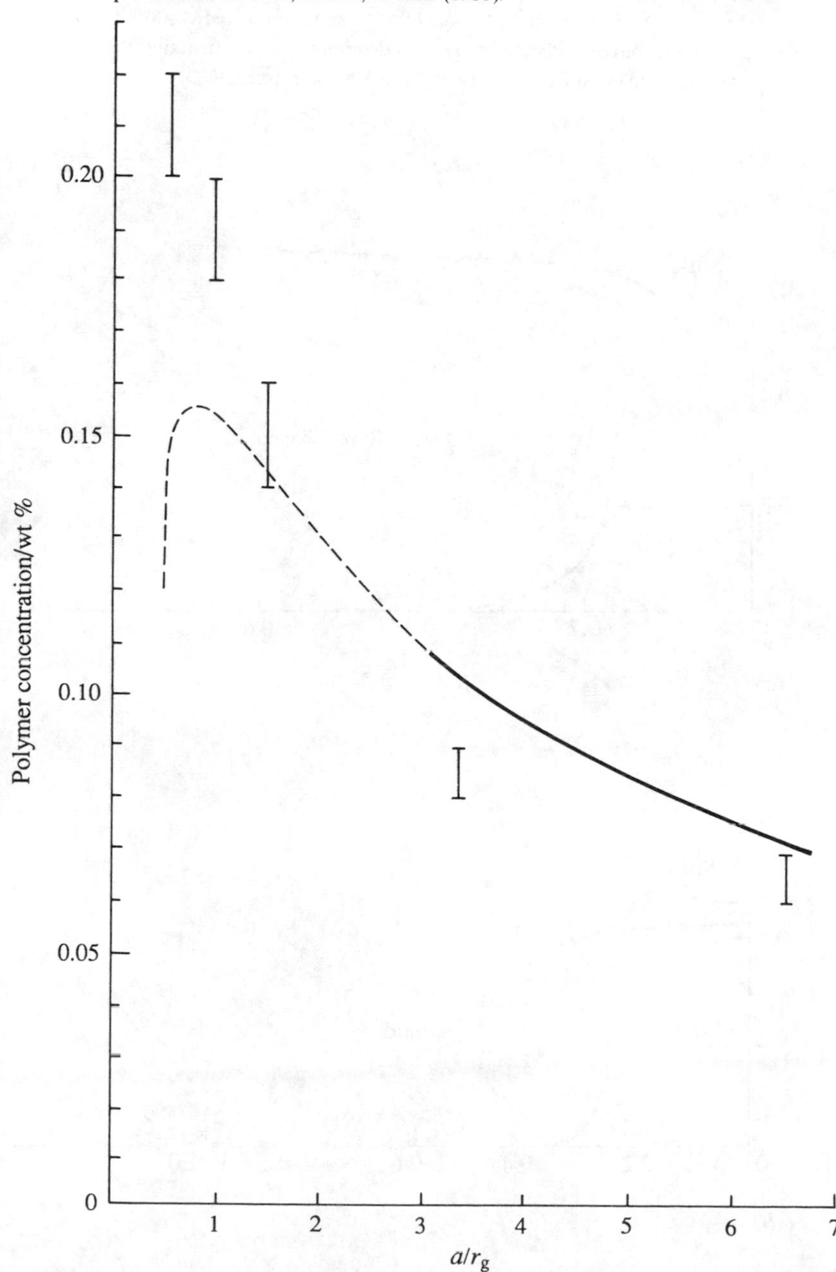

Fig. 10.16. Phase diagrams for aqueous polystyrene latices at 0.06 M NaCl, with dextran ($M_w = 733 \pm 100$ kg/mol, $M_n = 300 \pm 50$ kg/mol, $r_g = 16 \pm 3$ nm) added to induce separation (Patel and Russel, 1989): (*a*) $2a = 0.24 \pm 0.02$ µm and $a/r_g = 6.9$; ○ denote measured compositions; '⊢—⊣' indicates brackets on phase boundary for $\phi = 0.2$; curves show prediction from theory of Gast, Hall, & Russel (1983). (*b*) $2a = 0.059 \pm 0.003$ µm and $a/r_g = 1.9$; ○ denote measured compositions; broken-line curves indicate boundaries determined experimentally; unbroken ones are prediction from theory of Gast, Hall & Russel (1983).

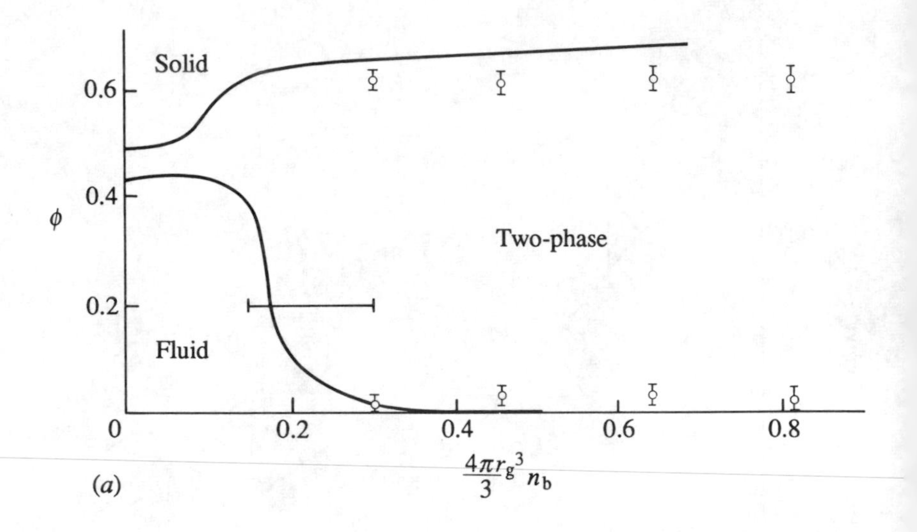

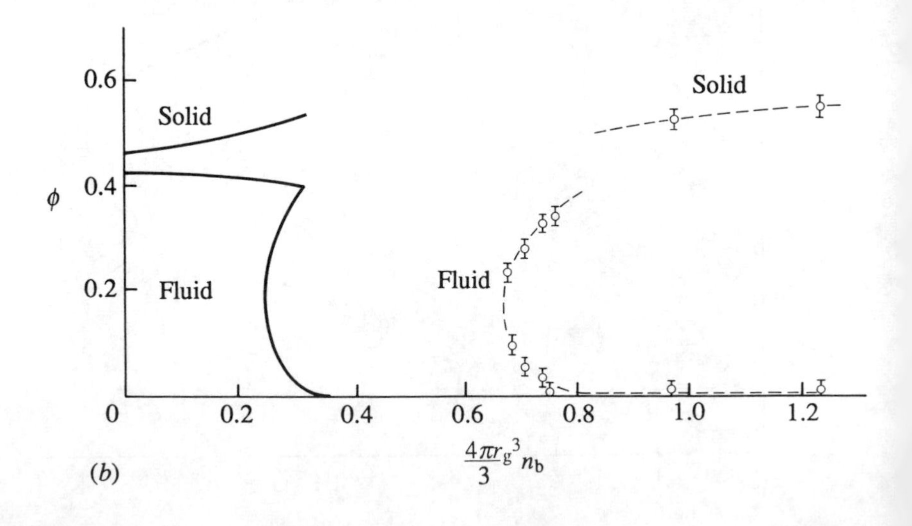

transitions. In addition, the assumption of pairwise additive potentials becomes suspect when $a/r_g = O(1)$.

Polymerically stabilized dispersions

With polymerically stabilized particles, the phenomena are similar but more subtle, owing to the interaction of the dissolved polymer with the stabilizing layer. Extensive data on carefully characterized dilute dispersions in the papers of Vincent and co-workers cited above establish the range of polymer concentrations over which particles aggregate. As noted in recent work (Edwards *et al.*, 1984), the onset of aggregation precedes phase separation, so these concentrations differ from those along the phase boundary but exhibit similar trends. For systems with chemically similar dissolved and anchored polymer, the relevant parameters are the size and volume fraction of the particles, the molecular weight and solution concentration, n_b, of the dissolved polymer, the molecular weight and surface density of chains, n_p, of the anchored polymer, and the excluded-volume parameter, v.

The results in Fig. 10.17, for silica spheres at volume fraction ϕ bearing a layer of polystyrene immersed in solutions of polystyrene in toluene at concentrations n_b, illustrate the phenomena. The data points and associated curve represent a stability boundary, distinguishing polymer

Fig. 10.17. Stability diagram for silica spheres ($a = 115\,\mu m$) with a grafted polystyrene layer of molecular weight 7.5 kg/mol in a solution of polystyrene in toluene (molecular weight 31 kg/mol) (Edwards *et al.*, 1984).

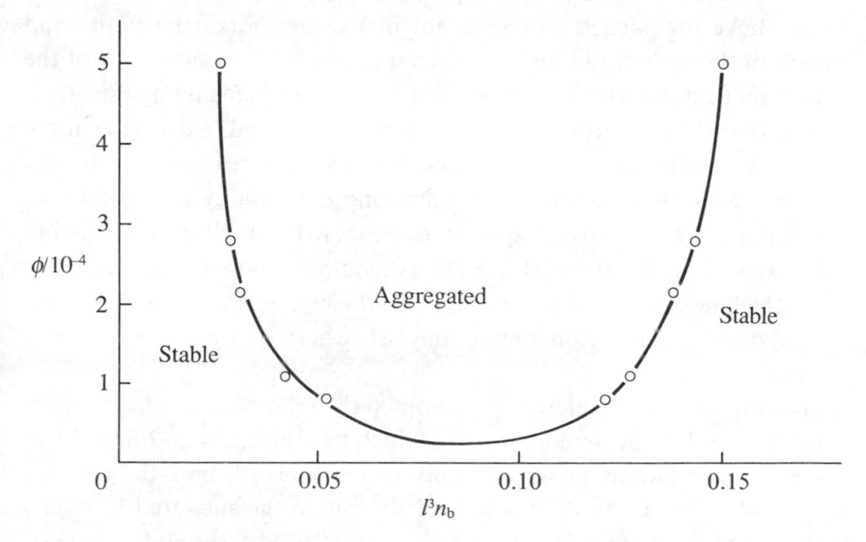

concentrations producing no detectable aggregation from those at which aggregation occurred. For a particular volume fraction, increasing the concentration of polystyrene first induces aggregation, i.e. destabilizes the dispersion, at the concentration corresponding to the left-hand boundary. With increasing concentration, the degree of aggregation initially increases, but then passes through a maximum and decreases effectively to zero at the right-hand boundary. Beyond that point, the dispersion is completely stable. The stability boundary resembles a phase boundary but differs quantitatively, since some aggregation occurs within the single phase region.

The left-hand branch of the stability boundary, reflecting destabilization, shifts to lower polymer concentrations with increasing particle size or increasing molecular weight of the dissolved polymer. The phase boundary for the hard sphere and electrostatically stabilized systems exhibits the same trends. Increasing the molecular weight of the grafted polymer shifts this boundary to higher polymer concentrations (Clarke & Vincent, 1981), as does increasing the Debye length for the electrostatically stabilized situation. These features of the phenomena can be understood and correlated through the superposition of a repulsive interaction between the steric layers and an attraction due to depletion of free polymer from the gap. Relatively simple models for the penetration of free polymer into the grafted layer provide reasonable predictions for the onset of aggregation (Vincent, Luckham, & Waite, 1980; Vincent *et al.*, 1986) and the associated phase boundary (Rao & Ruckenstein, 1985).

Two features of the aggregation phenomena, however, differ from those noted above for particles without polymer layers. First, the right-hand branch of the stability boundary, corresponding to restabilization of the dispersion at polymer concentrations above those inducing aggregation, implies that the depletion layer disappears. The second, a decrease in the polymer concentration required to destabilize with increasing graft density (Vincent *et al.*, 1986), implies that enhancing the stability with respect to dispersion forces or solvent quality decreases the stability to depletion flocculation. Understanding these effects requires a more detailed analysis of the thickness of the adsorbed layer and the concentration profile of the free polymer resulting from interactions between the free and the anchored polymer.

Recently, Gast & Leibler (1986) adopted the self-consistent field theory to determine that the depletion layer, which has thickness $\pi l/2(6vn_b)^{1/2}$ in concentrated solutions, penetrates a distance $\pi l/2(6vn_a)^{1/2}$ into the polymer layer. The segment density $n_a=(6n_p^2/vl^2)^{1/3}$ within the anchored layer can be extracted from (6.3.12). These conclusions lead to a simple approximation for the minimum in the interaction potential between flat plates exposed to a concentrated polymer solution,

$$\frac{\Phi_{\min}}{kT} = -\frac{\pi}{2} ln_b \left(\frac{vn_b}{6}\right)^{1/2} \left[1 - \left(\frac{n_b}{(6n_p^2/vl^2)^{1/3}}\right)^{1/2}\right]. \tag{10.7.1}$$

The prefactor corresponds to the attractive minimum for a flat plate without a polymer layer at the same conditions (6.4.11).

This result can be generalized through (6.4.13) into the interaction potential between spheres large with respect to both the depletion and polymeric layers,

$$\frac{\Phi_{\min}}{kT} = -\frac{\pi^3}{24} al^2 n_b \left[1 - 0.74 \left(\frac{lv^{1/2} n_b^{3/2}}{n_p}\right)^{1/3}\right]^2.$$

Adopting the criterion that $-\Phi_{\min}/kT = 2$ along the stability or phase boundary then provides two polymer concentrations $l^3 n_b$ for each value of the dimensionless surface density $l^{7/2} n_p / v^{1/2}$ and the particle size a/l. The surface density controls the penetration of free polymer into the polymer layer, preventing penetration and thereby enhancing depletion at low bulk concentrations, but allowing penetration and eliminating depletion when the bulk concentration approaches n_a. Figure 10.18 depicts the results for

Fig. 10.18. Polymer concentrations corresponding to $\Phi_{\min}/kT = -2$ calculated from (10.7.2), showing the effect of surface density of anchored chains, $(l^3/v)^{1/2} l^2 n_p$, and the particle size, a/l.

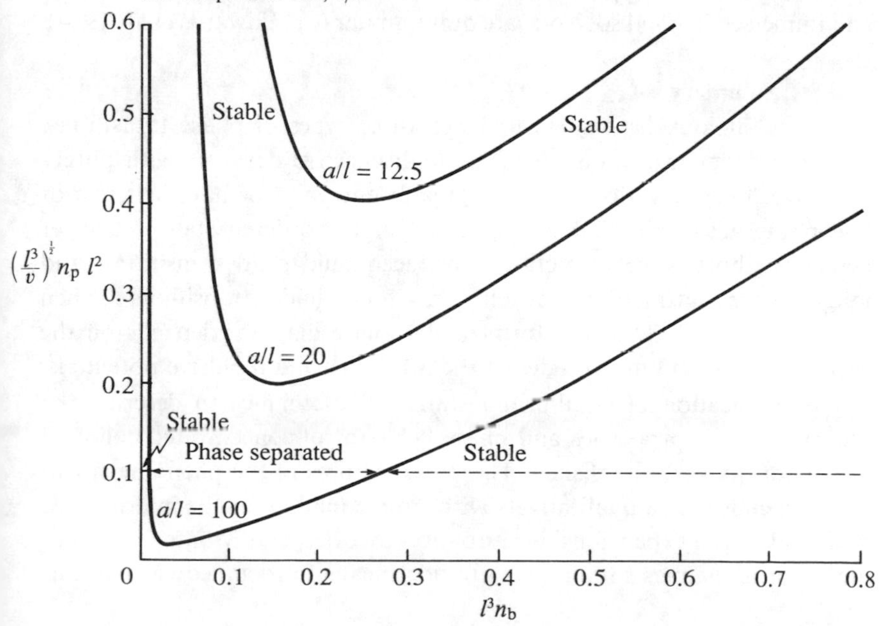

several values of a/l. For example, a dispersion with $a/l = 100$ and $(l^3/v)^{1/2} n_p l^2 = 0.1$ should be stable for polymer solution concentrations $l^3 n_b$ less than 0.01 or greater than 0.27, but phase separated at intermediate values.

The volume fraction dependence of the experimental results arises from entropic effects not included in this analysis. Nonetheless, the theory explains the mechanism of restabilization and the effect of graft density on destabilization noted above. Also, the shifts in the left-hand branch of the curves with decreasing particle size and solvent quality explain the observed increase in the polymer concentration required to destabilize the dispersion.

The results described in this section establish the basic features of the phase behavior of dispersions of hard, charged, and polymerically stabilized spheres containing dissolved polymer and the predictive capability of statistical mechanics, specifically the perturbation theory. Two fluid phases, analogous to a gas and a liquid, and an ordered solid phase appear as a consequence of the weak attraction induced by the dissolved polymer. The nature of the phases depends primarily on the size ratio of the two species and their compositions on the parameters characterizing the magnitude of the attractive minimum in the potential. Equilibrium phases are observed near the phase boundary where the attraction is $1kT$–$3kT$. Observations of phase transitions caused by other weak attractions, such as dispersion forces acting between particles stabilized by thin steric layers and immersed in good solvents, are quite similar (e.g. Edwards *et al.*, 1984).

10.8 Summary

Colloidal dispersions undergo three types of phase transitions. Disorder–order transitions occur at high effective densities with purely repulsive potentials, driven by the greater entropy (for hard spheres) or lower internal energy (for soft spheres) of the ordered state. At lower densities, short-range attractions produced fluid–solid transitions, and longer-range attractions result in fluid–fluid transitions when $-\Phi_{min}/kT \approx 1$–3. Thus the form of the phase diagram depends on the relative ranges and magnitudes of the attractive and repulsive potentials.

The application of equilibrium statistical mechanics to describe the thermodynamic properties and phase behavior of concentrated colloidal dispersions is well established. The basic features of the phenomena can often be understood qualitatively with simple models for the interparticle potential. Virial expansions, perturbation theories, and Monte Carlo and molecular dynamics simulations provide means for detailed calculations with specific potentials. The theoretical efforts have been tested and, in many cases, motivated by experiments with a variety of model dispersions in which the interactions can be carefully varied. The treatment here has addressed only pseudo-one-component systems at equilibrium. Many interesting questions remain with respect to the dynamics of the phase transitions described above and the behavior of truly multicomponent systems.

References

Ackerson, B. J. & Clark, N. A. (1981). Shear induced melting. *Phys. Rev. Lett.* **46**, 123–6.

Alder, B. J. & Wainwright, T. E. (1957). Phase transition for hard sphere fluid. *J. Chem. Phys.* **27**, 1208–9. (1959). Studies in molecular dynamics. *J. Chem. Phys.* **31**, 459–66.

Alder, B. J., Hoover, W. G. & Young, D. A. (1968). Studies in molecular dynamics. V. High density equation of state for hard disks and spheres. *J. Chem. Phys.* **49**, 3688–96.

Andersen, H. C., Chandler, D. & Weeks, J. D. (1971). Relationship between the hard sphere fluid and fluids with realistic repulsive forces. *Phys. Rev. A* **4**, 1597–607.

Asakura, S. & Oosawa, F. (1954). Interaction between two bodies immersed in a solution of macromolecules. *J. Chem. Phys.* **22**, 1255–6.

Asakura, S. & Oosawa, F. (1958). Interactions between particles suspended in solutions of macromolecules. *J. Polym. Sci.* **33**, 183–92.

Barker, J. A. & Henderson, D. (1967a). Perturbation theory and equation of state for fluids: I. The square well potential. *J. Chem. Phys.* **47**, 2856–61.

Barker, J. A. & Henderson, D. (1967b). Perturbation theory and equation of state for fluids: II. A successful theory of liquids. *J. Chem. Phys.* **47**, 4714–21.

Barnes, C. J., Chan, D. Y. C., Everett, D. H. & Yates, D. E. (1978). Disorder/order transitions in concentrated electrocratic dispersions. *J. Chem. Soc. Far. Trans. II* **74**, 136–48.

Baxter, R. J. (1968). Percus–Yevick equation for hard spheres with surface adhesion. *J. Chem. Phys.* **49**, 2770–4.

Beunen, J. A. & White, L. R. (1981). The disorder–order transition in latex dispersions. *Colloids and Surfaces* **3**, 371–90.

Bondy, C. (1939) Creaming of rubber latex. *Trans. Far. Soc.* **35**, 1093–108.

Carnahan, N. F. & Starling, K. E. (1969). Equation of state for nonattracting rigid spheres. *J. Chem. Phys.* **51**, 635–6.

Castillo, C. A., Rajagopalan, R. & Hirtzel, C. S. (1984). Statistical mechanics of colloidal dispersions. *Rev. Chem. Eng.* **2**, 237–348.

Chaikin, P. M., Pincus, P., Alexander, S. & Hone, D. (1982). BCC–FCC, melting, and re-entrant transitions in colloidal crystals. *J. Colloid Interface Sci.* **89**, 555–62.

Chandrasekhar, S. (1943). Stochastic problems in physics and astronomy. *Rev. Mod. Phys.* **15**, 1–89.

Clarke, J. & Vincent, B. (1981). Nonaqueous silica dispersions stabilized by terminally anchored polystyrene. The effect of added polymer. *J. Colloid Interface Sci.* **82**, 208–16.

Cowell, C., Li-In-On, R. & Vincent, B. (1978). Reversible flocculation of sterically stabilized dispersions. *J. Chem. Soc. Far. Trans. I* **74**, 337–47.

Darragh, P. J., Gaskin, A. J. & Sanders, J. V. (1976). Opals. *Scientific American* **234** (4), 84–95.

de Hek, H. & Vrij, A. (1981). Interactions in mixtures of colloidal silica spheres and polystyrene molecules in cyclohexane. I. Phase separations. *J. Colloid Interface Sci.* **84**, 409–22.

de Kruif, C. G., Jansen, J. W. & Vrij, A. (1987). Sterically stabilized silica colloids as a model supramolecular fluid. In *Complex and Supramolecular Fluids* (ed. S. A. Safran and N. A. Clark), pp. 315–43. Wiley-Interscience.

Edwards, J., Everett, D. H., O'Sullivan, T., Pangalou, I. & Vincent, B. (1984). Phase separation in model colloidal dispersions. *J. Chem. Soc. Far. Trans. I* **80**, 2599–607.

Efremov, I. F. (1976). Periodic colloidal structures. In *Colloid and Surface Science* (ed. E. Matijevic), **8**, 85–192.

Gast, A. P. & Leibler, L. (1986). Interactions of sterically stabilized particles suspended in a polymer solution. *Macromolecules* **19**, 686–91.

Gast, A. P., Hall, C. K. & Russel, W. B. (1983a). Polymer induced phase separations in non-aqueous colloidal suspensions. *J. Colloid Interface Sci.* **96**, 251–67.

Gast, A. P., Hall, C. K. & Russel, W. B. (1983b). Phase separations induced in aqueous colloidal suspensions by dissolved polymer. *Far. Disc.* **76**, 189–201.

Gast, A. P., Russel, W. B. & Hall, C. K. (1986). An experimental and theoretical study of phase transitions in the polystyrene latex and hydroxyethylcellulose system. *J. Colloid Interface Sci.* **109**, 161–71.

Hachisu, S. & Kobayashi, Y. (1974). Kirkwood-Alder transition in monodisperse latexes. II. Aqueous latexes at high electrolyte concentration. *J. Colloid Interface Sci.* **46**, 470–6.

Hachisu, S., Kobayashi, Y. & Kose, A. (1973). Phase separation in monodisperse latices. *J. Colloid Interface Sci.* **42**, 342–8.

Hall, K. R. (1972). Another hard sphere equation of state. *J. Chem. Phys.* **57**, 2252–4.

Hansen, J. P. & McDonald, I. R. (1986). *Theory of Simple Liquids*, 2nd Edition. Academic Press.

Haymet, A. D. J. & Oxtoby, D. W. (1986). A molecular theory for freezing: Comparison of theories and results for hard spheres. *J. Chem. Phys.* **84**, 1769–77.

Hill, T. L. (1960). *Introduction to Statistical Thermodynamics*. Addison-Wesley.

Hiltner, A. & Krieger, I. M. (1969). Diffraction of light by ordered suspensions. *J. Phys. Chem.* **73**, 2386–9.

Hone, D., Alexander, S., Chaikin, P. M. & Pincus, P. (1983). The phase diagram of charged colloidal suspensions. *J. Chem. Phys.* **79**, 1474–9.

Kincaid, J. M. & Weis, J. J. (1977). Radial distribution function of a hard-sphere solid. *Mol. Phys.* **34**, 931–8.

Kose, A., Ozaka, M., Takano, K., Kobayashi, Y. & Hachisu, S. (1973). Direct observation of ordered latex suspension by metallurgical microscope. *J. Colloid Interface Sci.* **44**, 330–8.

Long, J. A., Osmond, S. W. J. & Vincent, B. (1973). The equilibrium aspects of weak flocculation. *J. Colloid Interface Sci.* **42**, 545–53.

Luck, W., Klier, M. & Wesslau, H. (1963). Crystallization of macromolecular subunits. *Naturwiss.* **50**, 485–94.

McQuarrie, D. A. (1976). *Statistical Mechanics*. Harper & Row.

Onsager, L. (1949). The effects of shapes on the interactions of colloidal particles. *Ann. N.Y. Acad. Sci.* **51**, 627–59.

Patel, P. D. & Russel, W. B. (1989). An experimental study of aqueous suspensions containing dissolved polymer: A. Phase separation. *J. Colloid Interface Sci.* **131**, 192–200.

Pieranski, P. (1983). Colloidal crystals. *Contemp. Phys.* **24**, 25–73.

Pusey, P. N. & van Megen, W. (1987). Properties of concentrated suspensions of slightly soft colloidal spheres. In *Complex and Supramolecular Fluids* (ed. S. A. Safran and N. A. Clark), pp. 673–98. Wiley-Interscience.

Rao, I. V. & Ruckenstein, E. (1985). Phase behavior of mixtures of sterically stabilized colloidal dispersions and free polymer. *J. Colloid Interface Sci.* **108**, 389–402.

Reed, T. M. & Gubbins, K. E. (1973). *Applied Statistical Mechanics*. McGraw-Hill.

Russel, W. B. (1983). Effects of interactions between particles on the rheology of dispersions. In *Advanced Seminar on the Theory of Multiphase Flow* (ed. R. E. Meyer), pp. 1–34. Academic Press.

Russel, W. B. (1987). *Dynamics of Colloidal Systems*. University of Wisconsin Press.

Shih, W. Y., Aksay, I. A. & Kikuchi, R. (1987). Phase diagrams of charged colloidal particles. *J. Chem. Phys.* **86**, 5127–32.

Smith, W. R. & Henderson, D. (1970). Analytical representation of the Percus–Yevick hard sphere radial distribution function. *Mol. Phys.* **19**, 411–15.

Snook, I. & van Megen, W. (1976). Predictions of ordered and disordered states in colloidal dispersions. *J. Chem. Soc. Far. Trans. II* **72**, 216–23.

Sperry, P. R. (1982). A simple quantitative model for the volume restriction flocculation of latex by water soluble polymers. *J. Colloid Interface Sci.* **87**, 375–84.

Sperry, P. R. (1984). Morphology and mechanism in latex flocculated by volume restriction. *J. Colloid Interface Sci.* **99**, 97–108.

Sperry, P. R., Hopfenberg, H. B. & Thomas, N. L. (1981). Flocculation of latex by water-soluble polymers: Experimental confirmation of a non-bridging, nonadsorptive, volume restriction mechanism. *J. Colloid Interface Sci.* **82**, 62–76.

van Megen, W. & Snook, I. (1984). Equilibrium properties of dispersions. *Adv. Colloid Interface Sci.* **21**, 119–94.

Vincent, B., Clarke, J. & Barnett, K. G. (1986). The flocculation of non-aqueous sterically stabilised latex dispersions in the presence of free polymer. *Colloids and Surfaces* **17**, 51–65.

Vincent, B., Luckham, P. F. & Waite, F. A. (1980). The effect of free polymer on the stability of sterically stabilized dispersions. *J. Colloid Interface Sci.* **73**, 508–21.

Vincent, B., Edwards, J., Emmett, S. & Jones, A. (1986). Depletion flocculation in dispersions of sterically stabilized particles (soft spheres). *Colloids and Surfaces* **18**, 261–81.

Vrij, A., Jansen, J. W., Dhont, J. K. G., Pathmamanoharan, C., Kops-Werkhoven, M. M. & Fijnaut, H. M. (1983). Light scattering of colloidal dispersions in non-polar solvents at finite concentrations. Silica spheres as model particles for hard-sphere interactions. *Far. Dis.* **76**, 19–36.

Williams, R. & Crandall, R. S. (1974). The structure of crystallized suspensions of polystyrene spheres. *Phys. Lett* **48A**, 225–6.

Wood, W. W. & Jacobsen, J. D. (1957). Preliminary results from a recalculation of the Monte Carlo equation of state of hard spheres. *J. Chem. Phys.* **27**, 1207–8.

Woodcock, L. V. (1981). Glass transition in the hard sphere model and Kauzman's paradox. *Ann. New York Acad. Sci.* **371**, 274–98.

Zocher, H. (1925). Spontaneous structure formation in sols: A new kind of anisotropic liquid media. *Z. Anorg. Allg. Chem.* **147**, 91–110.

Problems

1 Calculate the second virial coefficient A_2 from (10.3.4) and deduce the corresponding values for the excluded shell diameter s_0 and the stickiness $1/\tau$ for spheres of radius a with
(i) a non-retarded dispersion attraction acting across dense grafted lyophilic layers of thickness L and
(ii) a retarded dispersion attraction plus an electrostatic repulsion.
In each case explore the effect of the relevant dimensionless groups on the result, using typical values for the dimensional parameters to

identify appropriate magnitudes. Set $\Phi = \infty$ for $r < r_{max}$ to account for the inaccessibility of the primary minimum on the time scales of interest.

2 Calculate the effective hard sphere diameter d (10.4.5) for the same conditions chosen for Problem 1(ii) and explain any differences between s_0 and d/a.

3 (i) Show that for hard spheres (10.2.4) can be reduced to

$$\Pi_{HS} = nkT[1 + 4\phi g_{HS}(2a)].$$

(ii) For particles that interact through a hard sphere repulsion plus a weak short-range attraction derive, via the perturbation theory, the Helmholtz free energy A in terms of the second virial coefficient A_2 for the actual system and A_{HS} and Π_{HS} for the hard-sphere reference state.

4 A single-phase system becomes unstable and particles precipitate when $d\Pi/d\phi = 0$, a phenomenon known as spinodal decomposition. Use the results of Problem 10.3 to show that for hard spheres with weak attractions

$$\frac{d \ln Z_{HS}}{d \ln \phi} = \frac{A_2}{4 - A_2} + O(4 - A_2)$$

defines the spinodal curve. Plot ϕ vs. $1/(4 - A_2) \approx -kT/\Phi_{min}$ which plays the role of the dimensionless temperature in this case.

5 Dialysis of charged latices against water at a known ionic strength provides a well-defined reference state with $\psi = 0$ and $n^k = n_b^k$. Use this condition together with electroneutrality (10.6.4) to determine the average potential $\langle \psi \rangle$ and the effective Debye length $1/\kappa$ for a concentrated dispersion. Comment on the differences between these results and those for a closed system in (10.6.6).

6 With the results of Problem 10.5 calculate the phase boundaries corresponding to the disorder–order transition for latices with fixed surface charges such that $4\pi a^3 q^2/\varepsilon_0 \varepsilon kT = 10^4$.

7 The addition of non-adsorbing polymer induces phase separation when $-\Phi_{min}/kT = 2$–3. Formulate a rule of thumb for the critical polymer concentration in an aqueous system by combining the linear superposition approximation for the electrostatic repulsion and (6.4.13) for the attraction. Predict the dependence on (i) a/r_g for the conditions of Fig. 10.15 and (ii) $a\kappa$ for $a/r = 6.5$. Assume ideal polymer solutions and $e\psi_s/kT = 3$ in both cases.

11

PARTICLE CAPTURE

11.1 Introduction

In the preceding chapters, fundamental aspects of colloid behavior have been emphasized. Now we are ready to apply this knowledge to processes involving suspensions. Here we investigate the capture of small particles by stationary collector units, one aspect of filtration technology.

Elementary considerations show that a strong attractive force is necessary if freely suspended particles are to come together, because at close separations viscous resistance increases dramatically. Since the interparticle force derives from the combination of electrostatic and dispersion forces, capture is particularly sensitive to the balance between colloidal and hydrodynamic forces. Several mechanisms contribute to particle capture and retention. Inertia is the dominant factor when fast-moving particles impact on a stationary object, whereas geometry and proximity govern the interception of slow-moving particles. The capture of submicron particles is influenced enormously by interparticle forces and Brownian motion. All these aspects are treated here, but technological issues are ignored. For example, a persistent problem encountered in the filtration of small particles is buildup of a deposit. Our treatment deals with the behavior of clean collector units to emphasize basic colloidal phenomena.

Aerosols have received the most study by a wide margin and many comprehensive reviews exist, e.g. Hidy & Brock (1970), Davies (1973), Friedlander (1977), and Kirsch & Stechkina (1978). Ives (1975) and Tien & Payatakes (1979) present broad reviews of liquid filtration; Spielman (1977) concentrates on small-scale processes in liquids. Hydrosol behavior differs from aerosol behavior in several ways. Electrical forces are important in either case, but screening by the diffuse charge layer reduces the interaction distance considerably in polar liquids. Moreover, in gases the continuum approximation often begins to break down at length scales where colloidal forces are just coming into play. Liquids behave as continua down to length scales of a few angstroms, much smaller than the interaction lengths for forces relevant to the capture process. Capture from liquids will be emphasized here in keeping with the themes set out in Chapter 1.

This chapter deals with microscale phenomena and describes how attractive and repulsive forces combine with other colloidal phenomena governing the capture of small particles by much larger objects. The focus is on situations where the theory can be tested experimentally. We begin by defining the capture efficiency of an individual collector and calculate the rate at which particles are intercepted by a spherical object.[†] Next, a scale analysis is used to classify other processes. This leads to a discussion of capture with non-Brownian particles, including interception, impaction, and the role of colloidal forces. Brownian motion is an important transport mechanism with smaller particles and their capture is described before we turn to the experimental studies.

Capture efficiency and the filter coefficient

A connection between microscopic behavior and overall filter performance can be made via the filter coefficient and capture efficiency, which encompass colloidal transport processes through a single parameter. To fix ideas, imagine a control volume $\Delta x\,\Delta y\,\Delta z$ within an array of collectors; see Fig. 11.1. Interstitial fluid movement is characterized by a

[†] Throughout our presentation, it is assumed that a permanent 'bond' is formed when the separation between the surfaces of particle and collector is a few angstroms. Such a bond could arise from the strong dispersion force, but we will not delve further into its nature. Suffice it to say that the bond is taken to be strong enough to prevent re-entrainment.

Fig. 11.1. Schematic diagram of a packed-bed filter array.

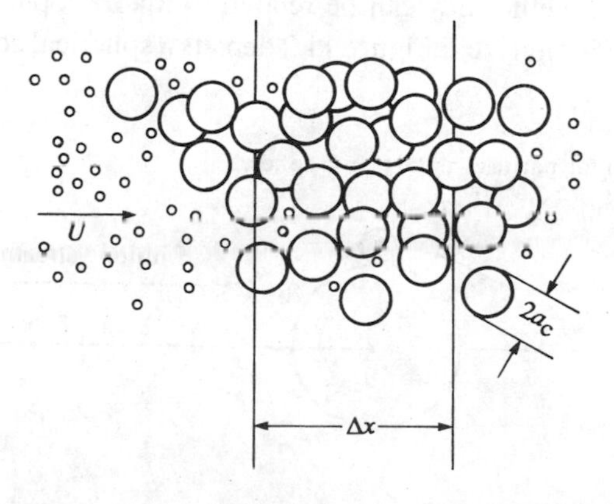

velocity, U, a length scale for the obstacles, a_c, and a length scale for suspended particles, a. Balancing the movement of particles across the boundaries of $\Delta x\,\Delta y\,\Delta z$ against the average flux to the collector units (per unit volume), J, yields

$$-U\frac{\mathrm{d}n}{\mathrm{d}x}=J, \tag{11.1.1}$$

as $\Delta x \to 0$ for a one-dimensional system; n is the average particle number density. The filter coefficient, λ, is defined as the proportionality factor between flux, local concentration, and velocity, viz.,

$$J=\lambda n U. \tag{11.1.2}$$

Accordingly, with a constant filter coefficient, the average particle concentration decreases exponentially with distance, i.e.,

$$n=n_0\exp(-\lambda x). \tag{11.1.3}$$

The filter coefficient is related to the capture efficiency, η, as follows. Imagine a cylinder parallel to the mean flow formed by lines passing through the periphery of a single collector. A certain fraction of the particles moving towards the collector inside this cylinder will be captured; this fraction is the capture efficiency or capture cross-section. Thus

$$J=\lambda n U=\eta n A_c U N. \tag{11.1.4}$$

Here N stands for the number of collectors per unit volume and A_c is the projected area of a collector. The capture efficiency, or capture cross-section, like the filter coefficient, is widely used in the correlation and interpretation of experimental data. In some situations the capture cross-section can be calculated from first principles, making it possible to compare theory with experiments.

To see how the capture efficiency can be related to microscopic events, capture by interception is analyzed. Figure 11.2 depicts a spherical collector

Fig. 11.2. Definition sketch for particle capture on a sphere.

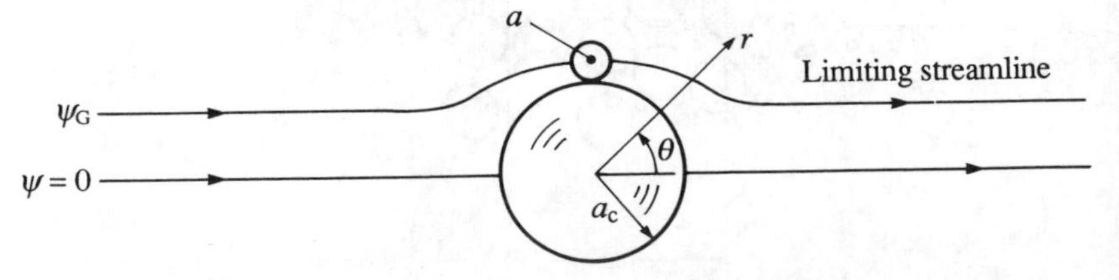

immersed in a uniform flow containing small suspended particles. If viscous interactions, inertia, buoyancy, and interparticle forces are ignored, suspended particles follow fluid streamlines, so the flux of particles to the surface can be calculated directly from the stream function (see Chapter 2). For slow, axisymmetric flow past a sphere this is

$$\psi = a_c^2 U(\tfrac{1}{2}r^2 - \tfrac{3}{4}r + \tfrac{1}{4}r^{-1})\sin^2\theta, \tag{11.1.5}$$

with r denoting the radial position scaled on a_c; θ denotes the angle from the rear stagnation point. The local fluid velocity can be calculated from derivatives of the stream function (cf. Batchelor, 1967) and is tangent to a streamline. Thus, a stream tube formed by rotating a given streamline (where ψ is constant) around an axis of symmetry always encloses the same fluid, as does the annular region between two such stream tubes. It was shown (§2.3) that the volumetric flow through the region between two axisymmetric stream tubes is 2π times the difference in the value of the stream function between the two surfaces, thus the flux of particles through any annular stream tube is

$$2\pi(\psi_2 - \psi_1)n.$$

Two stream surfaces are involved in calculating the flux of particles onto our spherical collector. One consists of the axis of symmetry and the surface of the collector; ψ is set to zero here. The other is formed by the envelope of grazing streamlines. A suspended spherical particle is deemed to be intercepted by the collector surface if its center passes within one particle radius of the surface. Thus, all particles within the stream tube defined by the grazing streamlines are captured. On the grazing streamline the stream function has the value

$$\psi_G = \tfrac{1}{2}a_c^2 U[(1+\alpha)^2 - \tfrac{3}{2}(1+\alpha) + \tfrac{1}{2}(1+\alpha)^{-1})], \quad \alpha = a/a_c, \tag{11.1.6}$$

and the flux of particles to the surface of the sphere is $2\pi\psi_G n$. The capture cross-section for this situation is

$$\eta = (1+\alpha)^2 - \tfrac{3}{2}(1+\alpha) + \tfrac{1}{2}(1+\alpha)^{-1} \approx \tfrac{3}{2}\alpha^2. \tag{11.1.7}$$

Interception efficiencies are yardsticks against which other effects are measured; Table 11.1 lists values for cylinders and spheres for reference. Note that here it is assumed that the spherical test particle moves undisturbed along a grazing streamline, so viscous effects on the scale of the test particle are ignored. Note also that interception efficiencies for viscous flows are much smaller than those for inviscid flows, owing to the structure of the velocity field adjacent to a no-slip boundary.

Table 11.1. *Interception efficiencies for various geometries* ($a/a_c \ll 1$)

Collector shape	Flow field	η
sphere	potential	$3\dfrac{a}{a_c}$
sphere	Stokes	$\dfrac{3}{2}\left(\dfrac{a}{a_c}\right)^2$
cylinder	potential	$2\dfrac{a}{a_c}$
cylinder	Stokes	$\dfrac{1}{\ln(3.7\nu/a_c U)}\dfrac{3}{2}\left(\dfrac{a}{a_c}\right)^2$

Methodologies to adapt calculations for isolated collectors to arrays of identical collectors are semi-theoretical, at best, since the detailed structure of the flow influences the efficiency of individual collectors. Cell models and other effective medium theories are used to characterize the flow structure through a model-specific parameter, which accounts for the influence of the array on the single-collector efficiency. For example, the interception efficiency for a spherical collector in an array can be written as

$$\eta = \frac{3}{2}\left(\frac{a}{a_c}\right)^2 A_F, \qquad \frac{a}{a_c} \ll 1, \tag{11.1.8}$$

where A_F is the dimensionless, structure-specific parameter. Spielman's 1977 review lists sources for these parameters. The influence of the array is not considered here, since it pertains more to fluid mechanics than to colloidal phenomena.

The program for the rest of this chapter is to include various colloidal effects in the calculation of the capture efficiency for an isolated collector and compare the theory with experimental results. The first task is to sort out the various phenomena.

Scale analysis

To identify factors that dominate particle capture, consideration is given to inertia, viscous drag, Brownian motion, sedimentation, dispersion forces based on van der Waals attraction, and electrostatic repulsion. The dimensionless numbers summarized in Table 11.2 follow upon examining forces and time scales for the various processes.

Table 11.2. *Dimensionless groups*[a]

Reynolds number	$Re \equiv \frac{\rho a_c U}{\mu}$
Stokes number	$St \equiv \frac{2}{9}\left(\frac{a}{a_c}\right)^2 \frac{\rho_p}{\rho} Re$
Peclet number	$Pe \equiv \frac{6\pi\mu a a_c U}{kT}$
sedimentation number	$N_G \equiv \frac{2}{9} \frac{a^2 \Delta\rho g}{\mu U}$
attraction number	$N_A \equiv \frac{A_{eff}}{9\pi\mu a^2 U}\left(\frac{a_c}{a}\right)^2$
repulsion number	$N_R \equiv \frac{3\varepsilon\varepsilon_0 \zeta\zeta_c a}{2A_{eff}}$

[a]Numerical factors are added to conform to conventions adopted for spherical particles.

A time scale for flow is a_c/U, so the inertial force per unit volume in the fluid is $O(\rho U^2/a_c)$. The corresponding viscous force is $O(\mu U/a_c^2)$ and the ratio of the two forces is the Reynolds number characterizing motion around a collector. In most situations studied here the Reynolds number is small, so inertial effects in the fluid are negligible.

Particle inertia, Brownian motion, and sedimentation transport particles across streamlines into the neighborhood of a collector where colloidal forces are important. The relative importance of these factors can be established by comparing characteristic times for the different transport processes with the $O(a_c/U)$ time a particle spends in the neighborhood of a collector.

First, the inertial force on a particle as it is carried around a collector is $O(a^3 \rho_p U^2/a_c)$. Combining this force with the mobility yields the velocity at which inertia forces particles across streamlines and shows that the characteristic time for the particle to travel distance of $O(a_c)$ is $O(\mu a_c^2/\rho_p a^2 U^2)$. The ratio of the time spent in the neighborhood of the collector to this time is $O(\rho_p a^2 U/\mu a_c)$. When numerical factors are inserted to conform with the conventional definition, this defines the Stokes number, *St*.

Brownian motion causes particles to wander close to the collector in a time scale set by the Brownian velocity, and, since it is $O(kT/a a_c \mu)$, the

Table 11.3. *Particle capture parameters*

particle radius, a	arbitrary
fluid velocity, U	arbitrary
collector radius radius, a_c	0.5 mm
fluid viscosity, μ	10^{-3} Pa s
fluid density, ρ	10^3 kg/m^3
particle density, ρ_p	2×10^3 kg/m^3
kT	4×10^{-21} J

Brownian time scale is $O(aa_c^2\mu/kT)$. The ratio of this scale to the convective time scale defines the Peclet number, *Pe*. Similarly, the sedimentation velocity is $O(\Delta\rho a^2 g/\mu)$, which yields a characteristic time of $O(a_c\mu/\Delta\rho a^2 g)$. The ratio of the convective time scale to this one produces the sedimentation number, N_G.

Different velocity and time scales are appropriate close to the particle surface where viscous and colloidal forces come into play. The characteristic fluid velocity is smaller, i.e., $O(aU/a_c)$, owing to the proximity of the surface, so the time spent in the neighborhood of collector, the dwell time, is $O(a_c^2/Ua)$. The importance of interparticle attraction is assessed by calculating the time required to capture a particle once it has been moved close to the surface and comparing this with the dwell time. A particle mobility of $O(1/\mu a)$ and a dispersion force of $O(A_{eff}/a)$ (see Chapter 5) produce a velocity towards the surface of $O(A_{eff}/\mu a^2)$. Therefore, a representative capture time, the time required to move a distance of $O(a)$, is $O(\mu a^3/A_{eff})$. The ratio of the dwell time to the capture time is called the attraction or adhesion number, N_A. Small values imply that the particle moves beyond the range of the attractive force before capture can occur. Electrostatic forces that act to thwart capture are $O(\varepsilon\varepsilon_0\zeta\zeta_c)$; ζ and ζ_c are the ζ-potentials of the particle and collector surface, respectively. The ratio of repulsive to attractive forces is called the repulsion number, N_R.

The daunting list in Table 11.2 emphasizes that many processes influence particle capture. However, for extreme values of certain dimensionless groups, a dominant mechanism may be singled out. For example, the Stokes, Peclet, and sedimentation numbers characterize processes that move particles into close proximity with the collector where colloidal forces come into play. Figure 11.3 depicts circumstances were inertia, convection, convection and sedimentation, or diffusion are the dominant mechanisms under conditions delineated in Table 11.3. The reasoning leading to Fig. 11.3 is as follows. When the Stokes number is greater than unity, inertia is pre-eminent. By using data from Table 11.3, a velocity–particle size relation can be derived to identify the region denoted 'inertia' in which $St > 1$. For a Stokes number smaller than unity, other mechanisms take over. Thus, when fluid velocity and particle size are such that $Pe < 1$, diffusion is the principal transport mechanism. Similarly, when $Pe > 1$, convection or, if $Pe > 1$ and $N_G > 1$, convection and sedimentation provide the impetus.

Fig. 11.3. Classification of dominant mechanisms using the parameters assigned in Table 11.3.

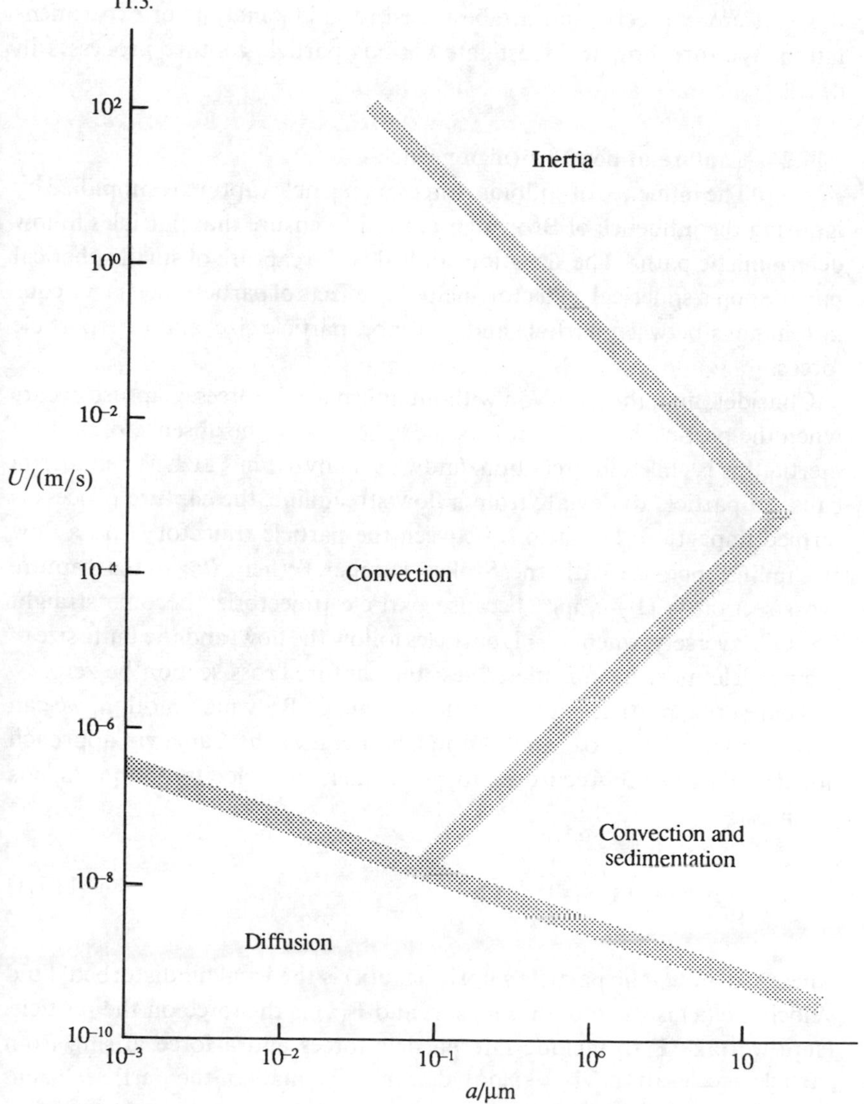

These criteria were used to construct the other boundaries. The figure emphasizes that under 'ordinary conditions', i.e., where the fluid velocity is a centimeter per second or so, convection is the principal mechanism transporting colloidal-size particles to the neighborhood of the collector surface. Inertia and sedimentation are effective only with larger particles and diffusion on the length scale of a collector is too slow. Diffusion may, however, become important within a thin boundary layer near the collector surface, as we will see in §11.3. Nevertheless, categorizations of this sort are imprecise and are best used to guide analysis or experimentation. We turn now to investigate various particle capture processes in detail.

11.2 Capture of non-Brownian particles

The influence of colloidal forces on particle capture is simplified by ignoring the influence of Brownian motion to ensure that particles follow deterministic paths. The situation studied is the capture of small spherical particles on a spherical collector, including effects of particle inertia, viscous interactions between particle and collector, particle size, and interparticle forces.

Consider, first, the situation without interparticle forces. Capture occurs when the particle surface contacts the collector. In the absence of particle inertia this is called interception, and was analyzed in §11.1. When inertia causes a particle to deviate from a flow streamline, the capture process is termed impaction. Deviations between the particle trajectory and a flow streamline increase with the Stokes number. When $St \gg 1$, the capture cross-section is $(1+a/a_c)^2$, because particle trajectories become straight lines. Conversely, when $St \ll 1$, particles follow the flow, and the finite size of the particle must be considered lest the capture cross-section be zero.

To describe particle capture in the absence of Brownian motion, we can adopt either the approach set out in Chapter 2 or the Langevin approach noted in Chapter 3. According to the former, the velocity of a particle is given by

$$\frac{\mathrm{d}}{\mathrm{d}t}\mathbf{x} = \mathbf{u}(\mathbf{x}) + \boldsymbol{\omega}(\mathbf{x}) \cdot \mathbf{F}(\mathbf{x}), \tag{11.2.1}$$

where $\mathbf{x}$ denotes the particle's position, $\mathbf{u}(\mathbf{x})$ is the local 'undisturbed' fluid velocity, $\boldsymbol{\omega}(\mathbf{x})$ is the mobility tensor, and $\mathbf{F}(\mathbf{x})$ is the force on the particle. Here we take $\mathbf{F}$ to include interparticle forces and a force arising from particle acceleration, viz. $-m\mathrm{d}^2\mathbf{x}/\mathrm{d}t^2$; m is the mass of the particle. Then, given expressions for the velocity, mobility, and interparticle force, the equation can be integrated to furnish a particle trajectory. Two situations will be analyzed: inertial impaction, and capture with attractive and repulsive forces.

Inertial capture

It is interesting that there is an inertial threshold below which capture is impossible unless interparticle forces are present or finite size particles are allowed. To see this, we take the mobility tensor to be a constant equal to the Stokes mobility for a sphere, $1/6\pi\mu a$, and omit the interparticle forces from (11.2.1), yielding

$$\frac{m}{6\pi\mu a}\frac{\mathrm{d}^2}{\mathrm{d}t^2}\mathbf{x}+\frac{\mathrm{d}}{\mathrm{d}t}\mathbf{x}=\mathbf{u}(\mathbf{x}). \tag{11.2.2}$$

Scaling velocities on the free-stream velocity, U, lengths on the characteristic dimension of the collector, a_c, and time on a_c/U gives

$$St\frac{\mathrm{d}^2}{\mathrm{d}t^2}\mathbf{x}+\frac{\mathrm{d}}{\mathrm{d}t}\mathbf{x}=\mathbf{u}(\mathbf{x}). \tag{11.2.3}$$

Given a flow field and initial conditions for the particle, integration of this equation furnishes a trajectory and from a family of trajectories the capture efficiency can be calculated as a function of the Stokes number.

To illustrate the threshold effect, attention is focused on point particles, the stagnation streamline, and for simplicity, potential flow towards a sphere (§2.3). Close to the sphere

$$St\frac{\mathrm{d}^2y}{\mathrm{d}t^2}+\frac{\mathrm{d}y}{\mathrm{d}t}=-3y, \tag{11.2.4}$$

where y measures distances upstream from the surface. The solution of this equation is the sum of two exponentials,

$$\mathrm{e}^{\lambda_1 t} \quad \text{and} \quad \mathrm{e}^{\lambda_2 t},$$

with

$$St\,\lambda^2+\lambda+3=0. \tag{11.2.5}$$

For $St=0$, there is one real (negative) root and separation between the particle and the surface decreases exponentially with time. However, the (point) particle never reaches the surface since particle velocity and fluid velocity coincide. Similarly, for $0<St<\frac{1}{12}$, particle inertia fails to overcome viscous forces. For $St>\frac{1}{12}$, inertia causes the roots of the characteristic equation to be complex. This is interpreted as evidence of capture, since negative values for the separation distance correspond to locations inside the collector. A more complete analysis shows that particles just off the stagnation streamline veer away and then intersect the collector for $St \geq \frac{1}{12}$ (Michael & Norey, 1969).

Calculation of the capture efficiency as a function of the Stokes number requires detailed analysis; Fig. 11.4 shows a portion of the results for a spherical collector. Note the threshold value of the Stokes number and the slow asymptotic approach to a collection of efficiency of unity. Capture efficiencies for the potential flow field around a sphere exceed those for viscous flow, because streamlines for the former are packed closer to the surface. Because of their relevance to aerosol filtration, inertial efficiencies have been calculated for several other shapes (Langmuir & Blodgett, 1946).

Fig. 11.4. Inertial capture efficiency for a spherical collector with potential flow (Michael & Norey, 1969).

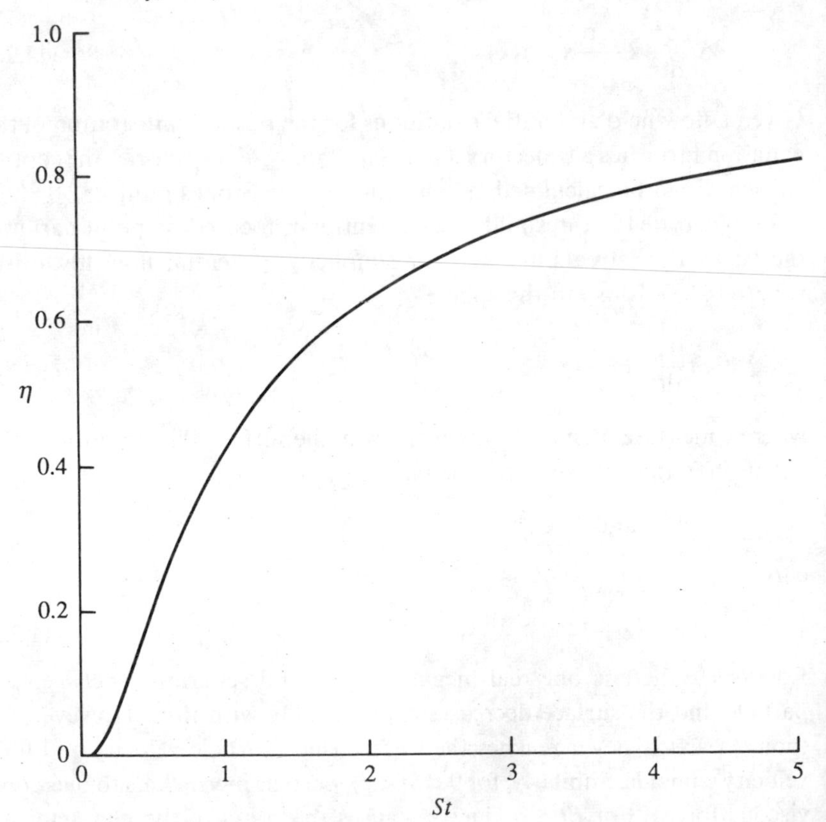

Earlier, we noted that the force required to bring the particle and collector into close proximity is inversely proportional to the film thickness when the separation is small. Hence, if particle and collector are to come into contact, the particle must be moved onto the surface by a stronger force. Clearly, inertial effects are not sufficient at low Stokes numbers, as evidenced by the fact that the Stokes number must exceed a certain value for capture of something as simple as a point-particle. It follows that capture of colloidal particles will require an attractive force between particle and collector to overcome the strong effect of viscosity at close separations. These situations are analyzed next.

Capture with attractive forces

Here a trajectory analysis is appropriate with exact expressions for hydrodynamic interactions and the attractive force based on the Hamaker potential (Chapter 5). The path of an individual (finite-size) particle, found by solving

$$\frac{\mathrm{d}}{\mathrm{d}t}\mathbf{x}=\mathbf{u}(x)+\boldsymbol{\omega}\cdot\mathbf{F}(\mathbf{x}), \tag{11.2.6}$$

is depicted in Fig. 11.5. For axisymmetric flow past a sphere, a single-trajectory equation can be formed by eliminating time as a variable from (11.2.6) Spielman & Goren (1970). Then

$$\frac{3}{2}(H+1)\frac{F_3(H)}{F_1(H)}\sin\theta\frac{\mathrm{d}H}{\mathrm{d}\theta}=-\frac{3}{2}(H+1)^2F_2(H)\cos\theta+\frac{N_{\mathrm{A}}}{H^2(H+2)^2}. \tag{11.2.7}$$

Here, H denotes the dimensionless separation $(r-a_c-a)/a$, and θ is the angle measured from the rear stagnation point (Fig. 11.5). The F-functions account for hydrodynamic interactions between the particle and collector: $F_1(H)$ pertains to changes in the mobility, $F_2(H)$ to the force exerted by flow along the line of centers between particle and collector, and $F_3(H)$ to

Fig. 11.5. Definition sketch for particle capture on a sphere due to dispersion attraction.

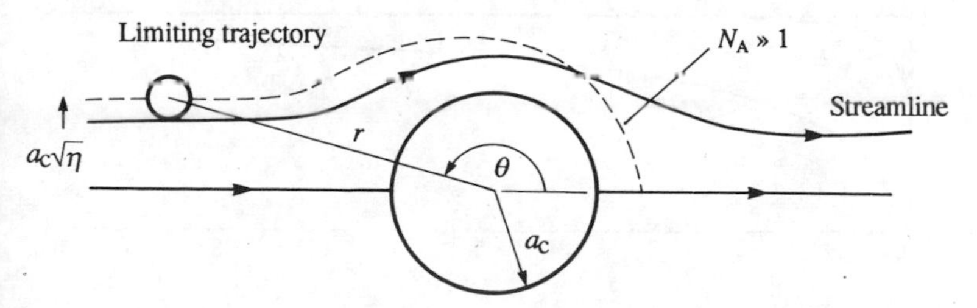

the force exerted by flow perpendicular to the line of centers. These expressions depend on the geometry of the collector, except when the particle–collector separation is small where the collector surface can be regarded as a plane. The form of each F-function is shown in Fig. 11.6; Spielman & Fitzpatrick (1973a) give explicit formulas for their asymptotic behavior. Numerical solutions of (11.2.7) yield trajectories from which capture efficiencies can be calculated. Results for spherical, cylindrical, and rotating disc collectors, including the influence of gravitational settling, are available (Spielman & Fitzpatrick, 1973a).

Figure 11.7 shows capture efficiencies for a spherical collector without sedimentation. Results for two special cases are shown, along with those from a numerical solution of (11.2.7). The asymptotic solution for strong attraction is a good approximation over a wide range of the attraction number whereas the solution for pure interception is not. This can be understood by examining the asymptotic solution in detail.

If attraction is strong then the limiting trajectory lies far from the collector, outside the region where viscous interaction occurs. In this

Fig. 11.6. Hydrodynamic interaction functions, showing how viscous effects are altered by the proximity of the collector (Spielman & Fitzpatrick, 1973a).

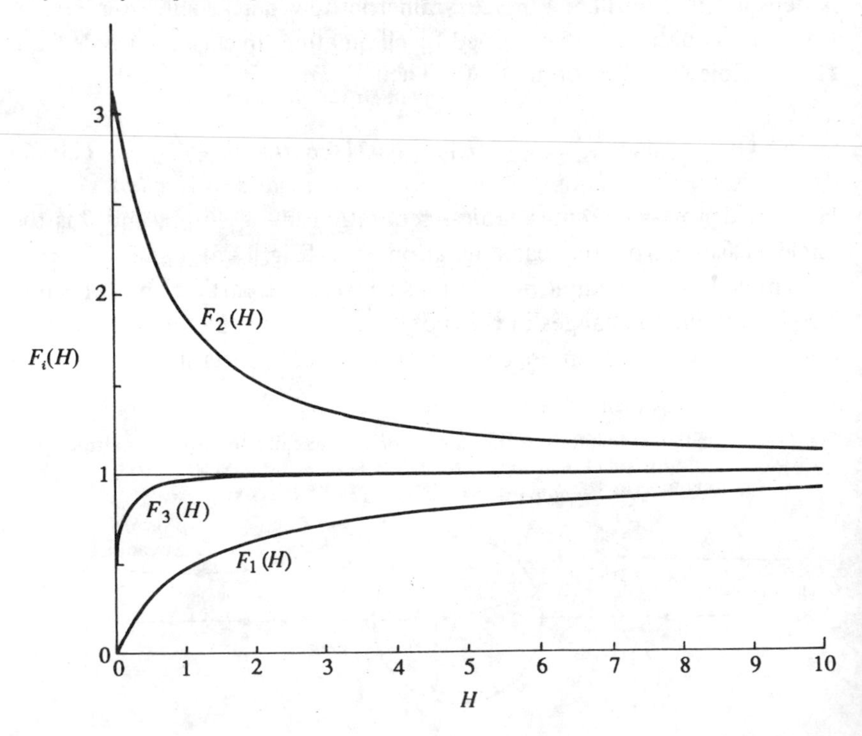

situation, the capture efficiency can be calculated as follows. Inside the limiting trajectory, all particle paths intersect the spherical collector; the limiting trajectory intersects the rear stagnation line ($\theta=0$), since the attractive force is large. If we trace this trajectory back upstream then it will become coincident with a streamline, and, once the value of the stream function on this streamline is known, the capture cross-section can be calculated in the fashion used with the interception, cf. (11.1.6) and (11.1.7). Thus

$$\eta = 2\frac{\psi_1}{a_c^2 U}. \tag{11.2.8}$$

To establish the value of the stream function on the limiting streamline, the trajectory equation can be analyzed. Since $a/a_c \ll 1$ and

$$r - 1 = \frac{a}{a_c}(H+1), \tag{11.2.9}$$

we can have $H \gg 1$ and $r-1 \ll 1$. Therefore, the stream function can be approximated by

$$\psi \approx \tfrac{3}{4} a_c^2 U\,(a/a_c)^2 (H+1)^2 \sin^2\theta, \qquad \frac{a}{a_c} \ll 1, \tag{11.2.10}$$

Fig. 11.7. Capture efficiencies for a spherical collector. Solid line from a numerical solution of (11.2.7) (Spielman & Fitzpatrick, 1973a).

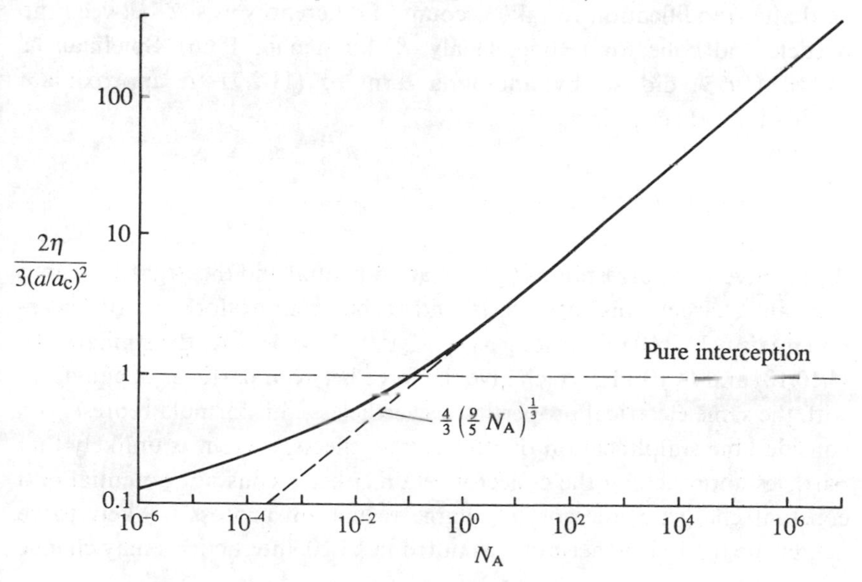

and the capture efficiency follows from the limiting value of $(H+1)^2 \sin^2 \theta$ as $\theta \to \pi$, i.e.,

$$\eta = \lim_{\theta \to \pi} [\tfrac{3}{2}(a/a_c)^2 (H+1)^2 \sin^2 \theta]. \tag{11.2.11}$$

To find the limiting value of $(H+1)^2 \sin^2 \theta$, we integrate (11.2.7) upstream for situations where $N_A \gg 1$, yielding (Spielman & Goren, 1970)

$$\eta = 2(a/a_c)^2 (\tfrac{9}{5} N_A)^{1/3}. \tag{11.2.12}$$

The asymptotic formula is in good agreement with the numerical solution when $N_A \gg 1$. This follows from the fact that here the limiting trajectory lies well away from the collector and Stokes' law represents viscous effects accurately. When N_A is $O(1)$ or smaller, the limiting trajectory is close to the collector where the attractive force must contend with effects due to the proximity of the particle and the collector. The underestimate in the capture rate stems from their omission. Since both interparticle attraction and the alteration to the viscous drag are omitted in the pure interception calculation, it is hardly surprising that this result is a poor approximation for the entire range of N_A.

Capture with electrostatic repulsion

Electric charge affects the force on a colloidal particle, and, to account for electrostatic interactions, formulas given in Chapter 4 can be used after modification to take account of differences in size between the particle and collector (Hogg, Healy, & Furstenau, 1966). Spielman & Cukor (1973) did so by adding a term to (11.2.7) to approximate electrostatic effects, viz.

$$N_A N_R \kappa a \frac{e^{-\kappa a H}}{1 \pm e^{-\kappa a H}};$$

the positive sign corresponds to constant potential and the negative sign to constant charge. This formula is said to be accurate for $\kappa a > 10$ and ζ-potentials below 60 mV as long as $(\zeta - \zeta_c)^2 / |\zeta \zeta_c| \ll 1$. Note the similarity to (4.10.10) and (4.10.11), which give the force between particles of equal size with the same electrical properties. Nevertheless, this formula represents a considerable simplification of electrostatic effects, since it is unlikely that particles approaching the collector retain either a constant potential or a constant charge. Some sort of charge regulation process is likely to be nearer the truth. Furthermore, as noted in §4.10, interactions may change from attraction to repulsion as the separation changes. Even with the simplified formula, capture behavior is complicated by the addition of two new dimensionless groups, N_R and κa.

Several possibilities can be identified: no capture, capture in either the primary or secondary minimum (see Chapter 8), and a combined mode where particles are captured in both potential wells. It is possible to decide which sort of behavior exists and prepare stability diagrams without integrating trajectory equations by examining the balance of forces at particle stagnation points (Spielman & Cukor, 1973). Figure 11.8 shows results for interactions at constant potential. In the region marked 'no collection', the repulsive barrier thwarts capture, whereas flow carries some particles over the repulsive barrier in the zone marked 'primary minimum

Fig. 11.8. Stability diagram for a spherical collector showing the influence of electrostatic repulsion at constant potential (Spielman & Cukor, 1973).

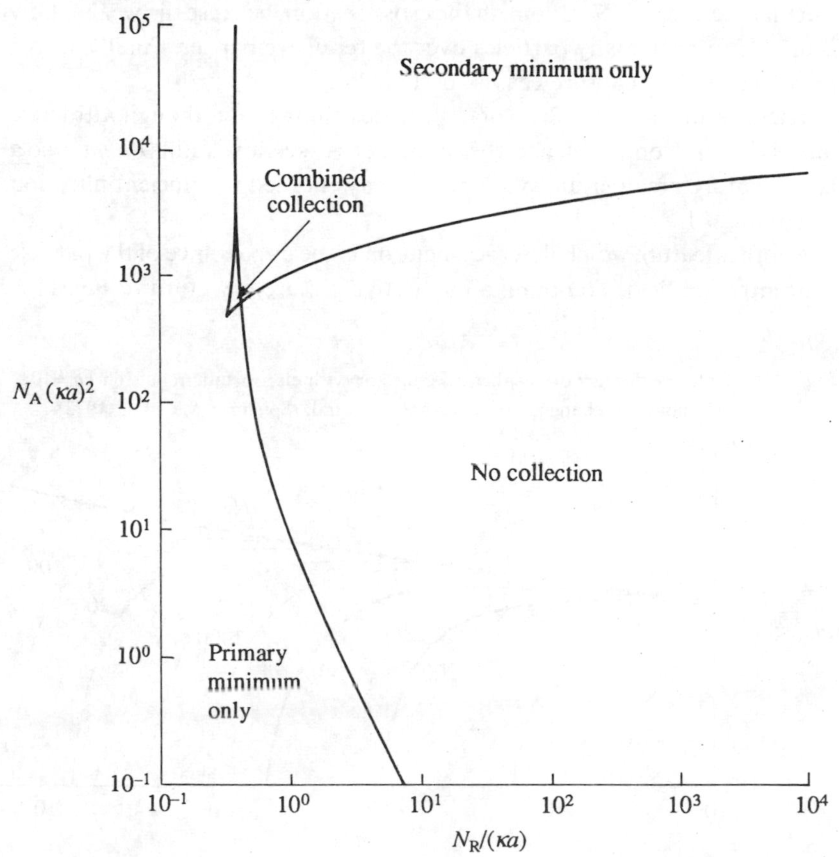

only'. Particle capture in the secondary minimum results when the flow is too weak to carry particles through the secondary well. Capture by the combined mechanisms occurs in a small region of the parameter space.

Capture efficiencies predicted from numerical integration of the trajectory equation are shown in Fig. 11.9. Recall that N_R represents the ratio of repulsive to attractive forces and N_A stands for the ratio of the dwell time (the time spent in the neighborhood of the collector) to the time necessary for capture, cf. §11.1. Thus $N_A \ll 1$ corresponds to either a weak attractive force or a strong flow. A constant value of the repulsion number fixes the attractive and repulsive interparticle forces, so an increase in N_A corresponds to slower flow. The shape of the curves on the diagram can be explained as follows. Take $N_R = 10^2$ and $N_A = 10^{-5}$, with interactions at constant change; the capture cross-section is lower than for $N_R = 0$, owing to the repulsive force. Now, as the flow strength is diminished, N_A increases and the capture cross-section increases somewhat until $N_A \approx 5 \times 10^{-5}$. Further increases in N_A diminish the cross-section, because the weaker flow is unable to push many particles over the repulsive barrier. Finally, as N_A increases further, capture ceases; cf. Fig. 11.8.

These results for non-Brownian particles show that although attractive forces have a strong influence, the capture cross-section is still $O(a^2/a_c^2)$, so a large capture area per unit volume is always needed for efficient filtration when $a/a_c \ll 1$.

Another feature which deserves mention is the appearance of the particle concentration field. To obtain a qualitative picture we return to Fig. 11.2,

Fig. 11.9. Capture efficiency on a spherical collector with electrostatic repulsion $\kappa a = 10$; ——, constant change; - - - constant potential, (Spielman & Cukor, 1973).

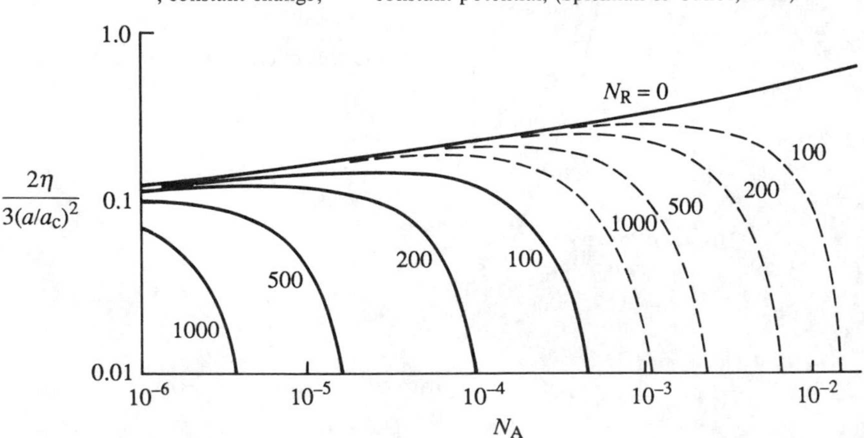

which depicts particle capture by interception. The particle number density is uniform and equal to the density far upstream, except within: (i) a layer whose thickness is $O(a)$ where particles touch the collector surface and (ii) the particle-free shadow downstream of the collector. This shadow has a circular cross-section whose radius asymptotes to $a_c\sqrt{\eta}$ far downstream. These features are due to the assumption of an infinitely strong, zero-range attractive force between particles and collector. A different situation exists when the dispersion forces produce a long-range attraction. Here, the downstream shadow disappears and the only concentration jump is at the collector surface. The next section illustrates how random Brownian motion of the suspended particles smooths the concentration field.

11.3 Capture of Brownian particles

Brownian motion produces diffuse particle trajectories, making the analysis in §11.2 inappropriate. A conservation equation may be derived by modifying, somewhat, the definition of the pair probability discussed in Chapter 8. Using P to denote the conditional probability of finding a Brownian particle at $\mathbf{x}$, given a second (immobile) particle at the origin, leads to a conservation equation for P, viz.

$$\nabla\cdot\{[\mathbf{u}(\mathbf{x})+\boldsymbol{\omega}(\mathbf{x})\cdot\mathbf{F}(\mathbf{x})]P\}=\nabla\cdot[\mathbf{D}(\mathbf{x})\cdot\nabla P] \tag{11.3.1}$$

Here, $\mathbf{F}$ stands for the central force between the particles, $\mathbf{D}=\boldsymbol{\omega}kT$ is the diffusivity tensor, and $\boldsymbol{\omega}$ is the mobility tensor. Boundary conditions for a spherical collector and a spherical particle are:

$$\begin{aligned} P&=0 \quad \text{at} \quad x=a+a_c,\\ P&\to n \quad \text{as} \quad x\to\infty. \end{aligned} \tag{11.3.2}$$

The flux to the surface of the collector,

$$J=-n\int_{S>S_c}\{P[\mathbf{u}(\mathbf{x})+\boldsymbol{\omega}(\mathbf{x})\cdot\mathbf{F}(\mathbf{x})]\cdot\mathbf{n}-\mathbf{n}\cdot\mathbf{D}(\mathbf{x})\cdot\nabla P\}\,\mathrm{d}S, \tag{11.3.3}$$

with S denoting a surface enclosing the spherical collector, determines the capture efficiency as

$$\eta=J/n\pi a_c^2 U. \tag{11.3.4}$$

Note the close similarity to the mathematical model used to describe Brownian flocculation kinetics in Chapter 8.

Special cases provide insight into the general characteristics of capture in the presence of Brownian motion. For example, if the size of the suspended particles and the interparticle force is ignored, results from the theory of heat and mass transfer may be taken over directly. With a constant (scalar) particle diffusivity, (11.3.1) becomes

$$\nabla \cdot \{\mathbf{u}(\mathbf{x})P\} = D_0 \nabla^2 P, \tag{11.3.5}$$

the convective diffusion equation. With this equation, we can begin to resolve details of the sharp transitions encountered earlier. When $a_c U \gg D_0$, the transition from $P=n$ to $P=0$ takes place within a thin layer adjacent to the surface. Balancing diffusion towards the surface with convection tangent to the surface identifies the boundary layer thickness, δ, as $O[a_c(D_0/a_c U)^{1/3}]$. Note, however, that by treating the particles as mathematical 'points', events on the particle scale, a, are omitted, so $[D_0 a_c^2/Ua^3]^{1/3}$ must be large. The flux to the collector surface calculated from (11.3.3) is $O(na_c^2 D_0/\delta)$, so the capture cross-section is $O(D_0/\delta U)$, i.e.,

$$\eta \propto (D_0/aU_c)^{2/3} \equiv Pe^{-2/3}. \tag{11.3.6}$$

The proportionality constant must be calculated by solving the partial differential equation. Natanson (1957b) and Friedlander (1977) describe capture by cylinders; for spheres, analyses by Lighthill (1950) and Levich (1962) yield

$$\eta = 2.52 Pe^{-2/3}. \tag{11.3.7}$$

To account for interparticle forces, finite particle size, and hydrodynamic effects on $\mathbf{D}(\mathbf{x})$, equation (11.3.1) must be integrated numerically. Extensive results are available for spherical (Prieve & Ruckenstein, 1974) and cylinderical collectors (Adamczyk & van de Ven, 1981). One set of results for spherical collectors is shown as Fig. 11.10, along with asymptotic results calculated from the Lighthill–Levich formula (11.3.7) and trajectory equations; cf. §11.2. These results show the trajectory formula to be a good approximation at high Peclet numbers, where the diffusion boundary layer is much thinner than the range of the attractive force. Deviations occur when the boundary layer thickness becomes comparable with the effective range of the attractive force. The Lighthill–Levich formula supplies an adequate approximation over an intermediate range of Peclet numbers. At high Peclet numbers, the diffusion boundary layer is too thin to encompass the range of the attractive force, so the boundary layer theory underestimates the capture rate. At low Peclet numbers, particles are captured from a layer much thicker than that subsumed in the Lighthill–Levich formula, which errs on the low side.

Prieve & Ruckenstein (1974) delineated regions where certain processes dominate by adopting a subjective criterion: a transport process was judged dominant if the error incurred by neglecting other processes was less than 10 per cent. Figure 11.11 shows one region where convection and diffusion control the rate because the diffusional boundary layer is thicker than the range of the attractive force, a second region where attractive forces dominate because the boundary layer is thinner, and a transition region.

Although Prieve & Ruckenstein (1974) did not report calculations for electrostatic repulsion, they did investigate the influence of gravitational settling, which can have a strong effect when forced convection is weak. When $Pe \ll 1$, it is convenient to measure the strength of sedimentation by comparing the $O(4\pi a^3 \Delta\rho g/3)$ gravitational body force with the $O(kT/a)$ Brownian force. Their ratio,

$$N_G^* \equiv \frac{4\pi a^4 \Delta\rho g}{3kT},$$

affects capture cross-sections, as shown in Fig. 11.12. As expected, the influence of sedimentation is substantial.

Fig. 11.10. Capture efficiency for the Brownian capture of spherical particles on a spherical collector with attraction and $a_c/a = 1000$. Dotted lines calculated for non-Brownian particles from a trajectory analysis; dashed line, the Lighthill–Levich formula (Prieve & Ruckenstein, 1974).

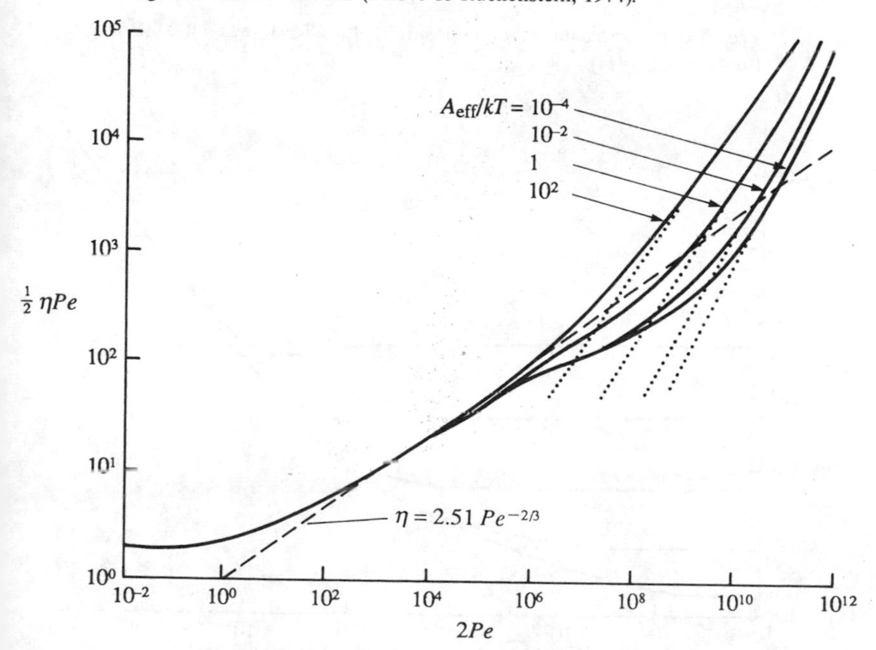

Fig. 11.11. Diagram showing regions where different processes dominate capture (Prieve & Ruckenstein, 1974).

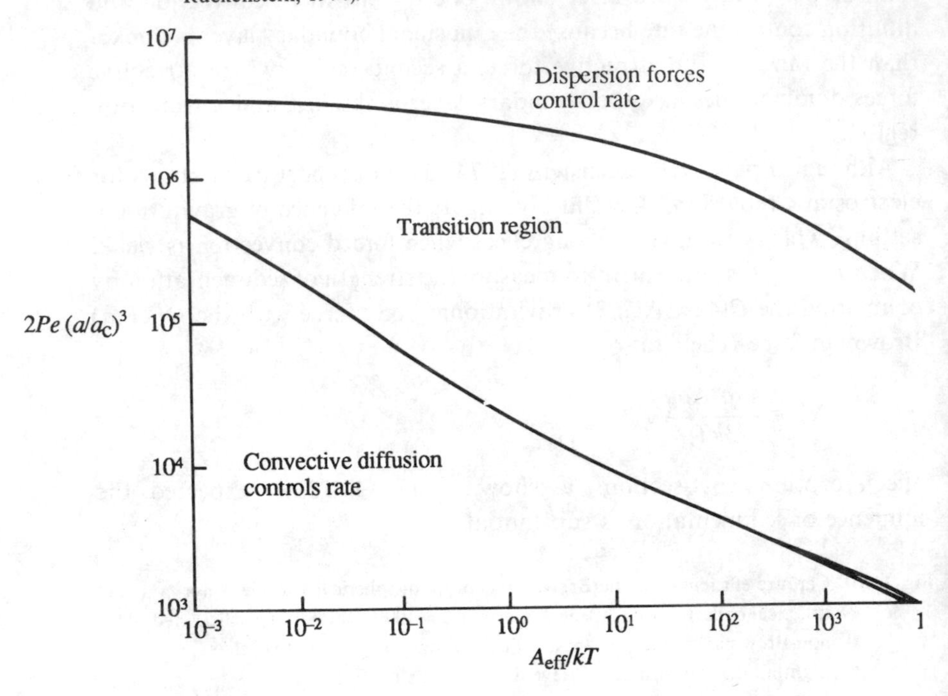

Fig. 11.12. The effect of sedimentation on deposition; $A_{eff}/kT = 1$, $a_c/a = 1000$ (Prieve & Ruckenstein, 1974).

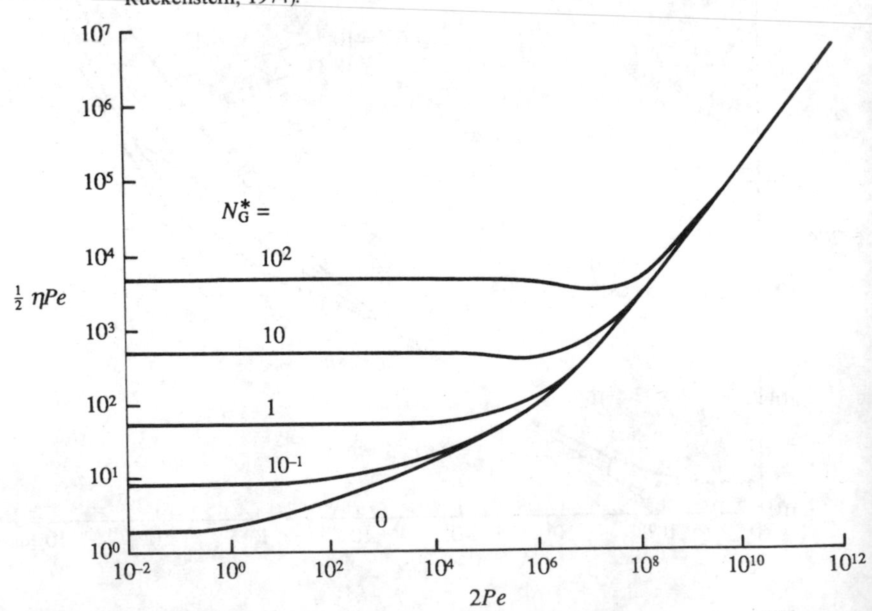

It is tedious to calculate capture rates when all the transport mechanisms are treated simultaneously, so it is worth considering whether adding individual process rates produces a useful approximation. Yao, Habibian & O'Melia (1971) concluded that adding individual rates from sedimentation, convective diffusion of 'point' particles, and attraction does yield an adequate approximation. However, their results were obtained with a model that ignored hydrodynamic interactions between particle and collector. Prieve & Ruckenstein (1974) made a similar comparison using their exact model. The points on Fig. 11.13 show results from rigorous calculations; the solid line comes from adding individual contributions from Brownian motion, attraction, and sedimentation. Although the agreement is excellent here, the notion of additivity should be treated cautiously, since its entire domain of validity is yet to be established.

11.4 Experimental measurements

Here the discussion will focus on comparisons between theory and experiments for rotating disc collectors and packed beds. In dealing with capture on a rotating disc, the flow field is known and agreement between theory and experiment is good under conditions where electrostatic

Fig. 11.13. Diagram showing the agreement between the exact solution and the additivity rule for $a_c/a = 10^4$, $A_{eff}/kT = 10^4$. Points are from the exact solution; solid line from adding contributions from separate calculations of individual effects (Prieve & Ruckenstein, 1974).

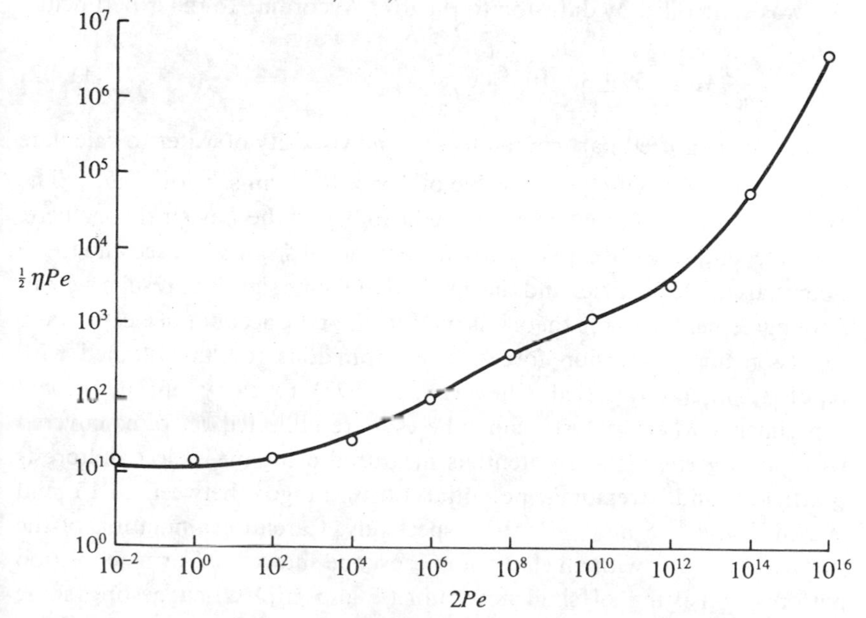

repulsion and hydrodynamic interactions between particle and collector are negligible. When electrostatic repulsion is present the agreement is less satisfactory, but there appears to be a plausible explanation for the discrepancy. With packed beds, the flow fields are largely unknown and various approximations must be introduced (cf. §11.1). The quantitative agreement between theory and experiment is satisfactory in the absence of electrostatic repulsion. In its presence, the agreement is qualitative at best.

Experiments with rotating discs

This configuration is used because the flow field is known and the thickness of the diffusion boundary layer is constant over most of the disc. With small particles and no repulsive interactions between disc and particles, the convective diffusion equation (11.3.5) can be solved exactly. The flux per unit area of the disc, j, is

$$\frac{j}{n\omega^{1/2}} = 0.62\frac{D_0^{2/3}}{\nu^{1/6}} \tag{11.4.1}$$

where ω is the rotation rate, D_0 is the particle diffusivity, and n is the concentration of particles far from the surface (Levich, 1962). Hull & Kitchener (1969) measured the rate of deposition of small polystyrene particles ($a = 0.154\,\mu\text{m}$) from dilute suspensions onto a rotating disc covered with a smooth polyvinylpyridine copolymer film. The ζ-potentials of the particles and film were -70 mV and $+72$ mV, respectively, and the flux was controlled by diffusion to the disc. According to the experiments,

$$\frac{j}{n\omega^{1/2}} = 7.59 \pm 3 \times 10^{-6}\ \text{cm/s}^{1/2}. \tag{11.4.2}$$

Using the measured particle radius and the viscosity of water to calculate the particle diffusivity gives a value of 7.66×10^{-6} cm/s$^{1/2}$ for $j/n\omega^{1/2}$. The excellent agreement confirms the applicability of the Levich theory here.

In a second series of experiments, a different substrate was used and the ζ-potentials of the particles and the disc had the same sign. The results were in poor agreement with the theory as modified to take account of electrostatic repulsion and dispersion forces. The anomalous results obtained with repulsive interactions led Clint *et al.* (1973) to carry out additional experiments where polystyrene spheres were collected on discs covered with polystyrene films. ζ-potentials measured using microelectrophoresis (particles) and streaming potential (disc) ranged between -15 and -25 mV and -5 and -7 mV, respectively. Careful examination of the surface of the disc with an electron microscope showed uniform deposition patterns consisting of singlets, doublets, and triplets, suggesting some coagulation in the bulk. Good agreement with the theory was obtained at high ionic strengths by allowing for coagulation in the bulk and the capture of aggregates by the surface. At low ionic strengths, where repulsion is strongest, agreement between theory and experiment was poor. This was attributed to the acute sensitivity of the capture rate to small changes in the surface potential. Upon making small adjustments (≈ 1 mV) to the values of the ζ-potential of the film used in the theory, good agreement was obtained. This sensitivity was also identified by Ruckenstein & Prieve (1973).

Experiments with packed beds

Experiments with beds of spheres have been used to establish the applicability of theories based on a single collector to collector arrays where several transport processes operate and the flow field is not known in detail. Here filter coefficients are measured instead of single-particle capture efficiencies and a flow structure parameter used. Considering the number of experimental variables, scatter in the data comes as no surprise. However, it is possible to distinguish between alternative theories in most cases.

An extensive study of non-Brownian particles was carried out by Spielman & Fitzpatrick (1973b) to test theories for attraction dominated capture. Several hundred experiments were done wherein latex particles were filtered from aqueous suspensions using beds of glass beads. Particle diameters ranged from 0.7 μm to 21 μm; bead diameters from 0.1 mm to 4 mm; and superficial velocities between 0.01 and 1.0 cm/s. Electrolyte type and ionic strength were varied to control electrostatic repulsion and, to avoid interference from particle deposits, the glass beads were cleaned before each experiment. Filter coefficients were calculated from particle concentrations measured with a Coulter counter.

Figure 11.14 shows data for conditions where the electrostatic repulsion and sedimentation are negligible. Note the appearance of the flow structure parameter, A_F, mentioned in §11.1. Although there is a fair amount of scatter, the data cluster about the line derived from the theory given by Spielman & Fitzpatrick (1973a).

To indicate when electrostatic repulsion is important, these authors used a semi-theoretical criterion, viz.

$$\Gamma \equiv \frac{4\kappa a N_R N_A}{(1.94 + 4N_G)} \frac{e^{-x}}{1+e^{-x}} > 0.2, \tag{11.4.3}$$

where

$$x = \frac{2N_A \kappa a}{3(1.94 + 4N_G)}.$$

Fig. 11.14. Filter coefficients for non-Brownian particles in packed beds; electrostatic repulsion and sedimentation are negligible (Spielman & Fitzpatrick, 1973b).

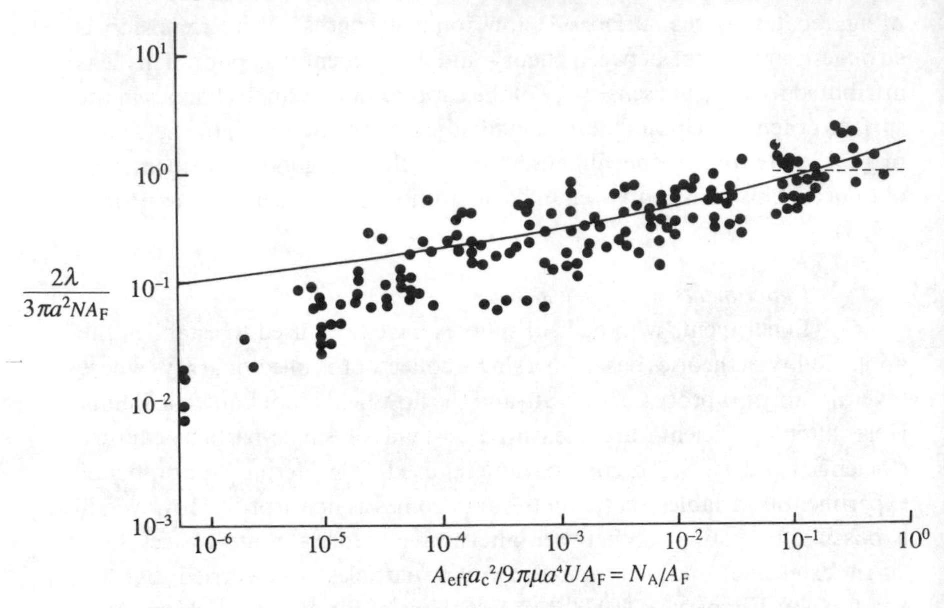

Fig. 11.15. Filter coefficients for non-Brownian particles in packed beds with sedimentation (Spielman & Fitzpatrick, 1973b).

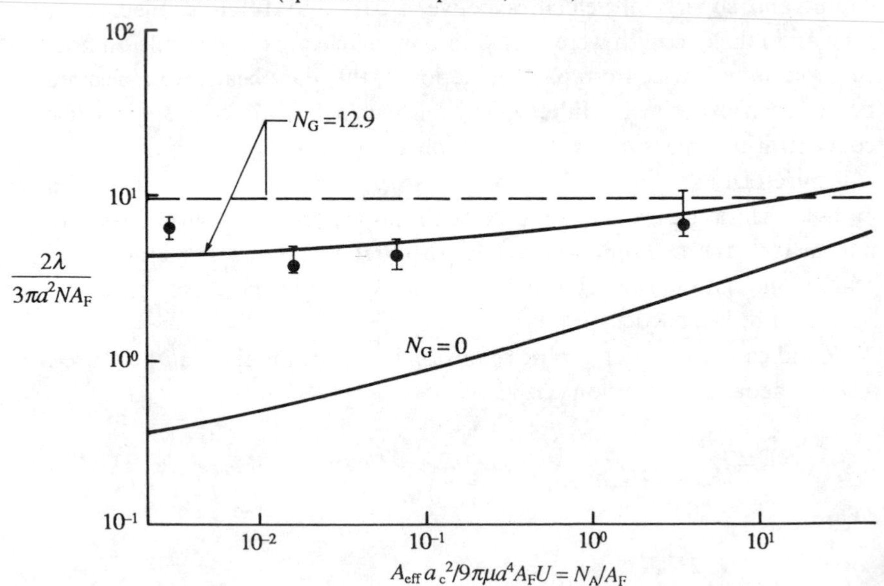

Data taken under conditions where this criterion was satisfied deviated significantly from the rest, indicating that repulsion was present. However, these data show considerable scatter and could not be correlated with Γ. This is not surprising, considering the findings by Clint *et al.* (1973), discussed previously.

Another aspect covered in Fitzpatrick & Spielman's experimental work was sedimentation. Figure 11.15 shows filter coefficients measured under conditions where sedimentation aided capture. The solid line labelled $N_G = 12.9$ shows results of an exact solution of (11.3.1) where hydrodynamic interactions between the particles and the collector were taken into account, whereas the broken line is the theory without these interactions. These results support the contention that hydrodynamic interactions are important under practical conditions.

Reports of studies involving the filtration of Brownian particles are sparse. Nevertheless, those extant agree with the formulas adapted from the theory of heat and mass transfer. For example, Yao, Habibian, & O'Melia (1971) were able to bring the Lighthill–Levich theory into reasonable agreement with their experimental data by adjusting the flow structure parameter.

11.5 Summary

Particle capture from low speed flows has been surveyed in this chapter to illustrate the influence of colloidal forces. Dispersion forces and electrostatic repulsion influence the capture rate strongly, and, in configurations where the flow field is well known, agreement between theory and experiment is adequate. Capture rates turn out to be extraordinarily sensitive to small differences between repulsive surface potentials. A similar sensitivity was found in the study of coagulation, which points up the need to understand the electrostatics fully for quantitative application of the theory. Under conditions encountered in filtration practice, the theory is not fully developed, owing largely to ambiguities connected with the structure of the flow fields.

References

Adamczyk, A. & van de Ven, T. G. M. (1981). Deposition of Brownian particles onto cylindrical collectors. *J. Colloid Interface Sci.* **84**, 497–518.

Batchelor, G. K. (1967). *An Introduction to Fluid Mechanics.* Cambridge University Press.

Clint, G. E., Clint, J. H., Corkhill, J. M. & Walker, T. (1973). Deposition of latex particles onto a planar surface. *J. Colloid Interface Sci.* **44**, 121–32.

Davies, C. N. (1973). *Air Filtration.* Academic Press.

Friedlander, S. K. (1977). *Smoke, Dust and Haze.* Wiley.

Hidy, G. M. & Brock, J. R. (1970). *The Dynamics of Aerocolloidal Systems*, vol. 1. Pergamon Press.

Hogg, R., Healy, T. W. & Fuerstenau, D. W. (1966). Mutual coagulation of colloidal dispersions. *Trans. Faraday Soc.* **62**, 1638–51.

Hull, M. & Kitchener, J. A. (1969). Interaction of spherical colloidal particles with planar surfaces. *Trans. Faraday Soc.* **65**, 3093–104.

Ives, K. J. (ed.) (1975). *The Scientific Basis of Filtration. NATO Advanced Study Institute Series E: Applied Sciences*, vol. 2. Noordhoff.

Kirsch, A. A. & Stechkina, I. B. (1978). The theory of aerosol filtration in *Fundamentals of Aerosol Science* (ed. D. Shaw). Wiley.

Langmuir, I. & Blodgett, K. B. (1946). A mathematical investigation of water drop trajectories. Army Air Forces Tech. Report No. 5418.

Levich, V. G. (1962). *Physicochemical Hydrodynamics.* Prentice-Hall.

Lighthill, M. J. (1950). Contributions to the theory of heat transfer through a laminar boundary layer. *Proc. Roy. Soc. Lond. A* **202**, 359–77.

Michael, D. H. & Norey, P. W. (1969). Particle collision efficiencies for a sphere. *J. Fluid Mech.* **37**, 565–75.

Natanson, G. L. (1957a). Diffusion precipitation of aerosols on a streamlined cylinder for small capture coefficients. *Dokl. Akad. Nauk SSSR* **112**, 100–3.

Natanson, G. L. (1957b). Deposition of aerosol particles from gas streams flowing around a cylinder. *Dokl. Akad. Nauk SSSR* **112**, 696–8.

Prieve, D. C. & Ruckenstein, E. (1974). Effect of London forces upon the rate of deposition of Brownian particles. *AIChE J.* **20**, 1178–87.

Ruckenstein, E. & Prieve, D. C. (1973). Rate of deposition of Brownian particles under the action of London and double-layer forces. *J. Chem. Soc. Faraday Trans. II* **69**, 1522–36.

Spielman, L. A. (1977). Particle capture from low speed flows. *Ann. Rev. of Fluid Mech.* **9**, 297–319.

Spielman, L. A. & Goren, S. L. (1970). Capture of small particles by London forces from low speed flows. *Enviro. Sci. Technol.* **4**, 135–40. (Corrections in *Enviro. Sci. Technol.* **5**, 254, 1971.)

Spielman, L. A. & Cukor, P. M. (1973). Deposition of non-Brownian particles under colloidal forces. *J. Colloid Interface Sci.* **43**, 51–65.

Spielman, L. A. & Fitzpatrick, J. A. (1973a). Theory for particle collection under London and gravity forces. *J. Colloid Interface Sci.* **42**, 607–23.

Spielman, L. A. & Fitzpatrick, J. A. (1973b). Filtration of latex suspensions through beds of glass spheres. *J. Colloid Interface Sci.* **43**, 350–69.

Tien, C. & Payatakes, A. C. (1979). Advances in deep bed filtration. *AIChE J.* **25**, 735–59.

Yao, K. M., Habibian, M. T. & O'Melia, C. R. (1971). Water & waste water filtration: Concepts and applications. *Enviro. Sci. Technol.* **5**, 1105–12.

Problems

1 Derive the formula for the interception efficiency analogous to (11.1.7) for a spherical collector in a potential flow.

2 The threshold for inertial capture in a potential flow follows from (11.2.3). Derive the counterpart for viscous flow and find the critical Stokes number.

3 For $N_A \gg 1$, analytical solutions of (11.2.7) are possible. Show that the equation simplifies to

$$Z^5 \sin\theta \frac{\mathrm{d}Z}{\mathrm{d}\theta} + Z^6 \cos\theta = \frac{2}{3} N_A,$$

where $Z \equiv 1 + H$ for Z and $N_A \gg 1$ and derive (11.2.12).

4 Simplify (11.3.5) for $Pe \gg 1$ by stretching the radial distance and derive the boundary layer equation for the concentration field. Show that there is a similarity transformation that leads to (11.3.7).

5 The advantage of the rotating-disc configuration is that the convective diffusion equation reduces to an ordinary differential equation since the concentration field depends only on distance from the disc.

Expressions for the velocity field are given in Schlichting's book.[†] Use (11.3.1) with this velocity field to derive the conservation equation for the capture of Brownian particles on a rotating disc. Show that the equation can be integrated to give (11.4.1) for situations where attractive and repulsive interactions between the disc and the particles are negligible. Can the differential equation be solved in closed form when the interactions are not negligible?

[†] H. Schlichting (1960). *Boundary Layer Theory*, 4th edn, Ch. 5. McGraw-Hill.

12

SEDIMENTATION

12.1 Introduction

The sedimentation of colloidal particles is important both in technology and in the laboratory. Gravity settlers, thickeners, or clarifiers commonly remove particles from waste streams issuing from a variety of processes. These generally operate as continuous processes that split the feed into two product streams, one the clear fluid and the other a sludge. Successful design requires knowledge of the sedimentation velocity of the particles over the relevant range of volume fractions and the role of interparticle forces in determining the structure of the dense sludge. Centrifugation provides a means of enhancing the driving force for commercial-scale operations, as well as concentrating or analyzing dispersions in the laboratory.

Despite their longstanding use, much remains to be understood about the details of processes which convert dilute dispersions into dense sediments. The key issues appear to be

(i) the variation of the settling velocity with volume fraction and interparticle potential,
(ii) the role of forces transmitted by interparticle potentials, and
(iii) the formulation of macroscopic models to predict the evolution of volume fraction as a function of position and time.

As with other colloidal phenomena, the complexity arises from the importance of a variety of interparticle forces and the fact that many systems of interest tend to be flocculated.

In this chapter we address all three points, but only for stable dispersions of spheres. At infinite dilution, a small sphere with density differing from that of the surrounding liquid by $\Delta\rho$ moves at the Stokes velocity,

$$U_0 = \frac{2a^2 \Delta\rho\, g}{9\mu},$$

determined by the balance between the gravitational force and the viscous drag (§2.6.5). Inertial effects remain unimportant, provided the Reynolds number is less than unity, i.e.

$$Re = \frac{\rho a U_0}{\mu} = \frac{2a^3 \rho\, \Delta\rho\, g}{9\mu^2} < 1.$$

Hereafter, we assume this to be satisfied, which for $\Delta\rho/\rho \approx 1$ in water requires $a < 75\,\mu\text{m}$.

At finite concentrations, hydrodynamic interactions can either increase or decrease the sedimentation velocity relative to U_0, depending on the nature of the non-hydrodynamic interactions. For purely repulsive interaction potentials, settling is hindered. Attractions, however, induce aggregation, causing two or more particles to settle as an effectively larger entity and, thereby, increasing the velocity. In both cases the net effect reflects the spatial distribution of particles, or the microstructure of the suspension, which depends on the range and the magnitude of the potential.

In the absence of an external field, the balance between the interparticle potential and Brownian motion completely determines the spatial distribution of particles in the suspension. But any relative velocity induced by an external field perturbs this microstructure. For example, for two particles with Stokes settling velocities differing by ΔU_0, the perturbation is characterized by a Peclet number for sedimentation,

$$Pe = \frac{2a\, \Delta U_0}{D_0} = \frac{8\pi a^4 \Delta\rho g}{3kT} \frac{\Delta U_0}{U_0}.$$

For aqueous systems with $\Delta\rho/\rho \approx 1$ and $\Delta U_0/U_0 \approx 1$, $Pe \approx 1$ for $1\,\mu\text{m}$ diameter particles. For $Pe \ll 1$, the perturbation is small, but for $Pe \gg 1$ the hydrodynamic forces dominate Brownian motion and determine the microstructure. Hence the concentration dependence of the sedimentation velocity for macroscopic particles can differ substantially from that for colloidal particles.

In a settling process, particles eventually accumulate at the bottom of the vessel, forming the sediment. Clearly, the boundary must exert a force to balance the weight of the particles. The transmission of this force upward through the dispersion represents the macroscopic consequence of thermodynamic interactions among particles in a non-uniform environment. The force acting on an individual particle is proportional to the gradient in its chemical potential, and, consequently, the gradient in the volume fraction. When balanced against gravity, this thermodynamic force determines the volume fraction profile within the sediment.

The following sections first address the dependence of the sedimentation velocity on the volume fraction and interparticle potential for both monodisperse and multimodal size distributions. Section 12.2 contains the formulation of the ensemble averages required to predict the average velocity from knowledge of interactions in dilute dispersions. The results are then described and compared with experimental data in §§12.3 and 12.4. The final two sections confront the questions of the thermodynamic force and macroscopic description of batch settling, for both incompressible sediments and stable but compressible ones.

12.2 Ensemble average velocities

The sedimentation velocity of particles in a concentrated suspension represents an average over all configurations of all particles. A prediction a priori, therefore, requires three elements: the velocities of particles for each configuration, the probability of observing an individual configuration, and a valid means of averaging. Since exact treatments of the hydrodynamic interactions exist only for two spheres, we restrict our attention to dilute concentrations for which pair interactions dominate. Then the pair distribution function characterizing the configurational probability can be determined from (8.3.8). Ensemble averaging (e.g., §3.2) is required to obtain the desired bulk property, but must be modified to accommodate the long-range hydrodynamic interactions.

Consider a representative volume V containing N equal spheres enclosed within a larger container of characteristic dimension L. The velocity at a point $\mathbf{x}$ within the volume is $\mathbf{U}$ if $\mathbf{x}$ resides within a particle and $\mathbf{u}$ otherwise. Both the particle and fluid velocities depend on the configuration of all N spheres. The respective ensemble averages, $\langle \mathbf{U} \rangle$ and $\langle \mathbf{u} \rangle$, will be independent of the size and geometry of the container only if $L^3 \gg V \gg a^3/\phi$ so that $N \gg 1$.

At dilute concentrations, it is tempting to construct a virial expansion valid to $O(\phi)$ by expressing the velocity of the ith sphere through a pairwise additive approximation based on (2.8.1),

$$\begin{aligned} \mathbf{U}_i &= \mathbf{U}_0 + \sum_{\substack{j=1 \\ \neq i}}^{N} \left[\boldsymbol{\omega}_{11}(\mathbf{r}_{ij}) - \frac{\boldsymbol{\delta}}{6\pi\mu a} + \boldsymbol{\omega}_{12}(\mathbf{r}_{ij}) \right] \cdot \mathbf{F}_j \\ &\equiv \mathbf{U}_0 + \sum_{\substack{j=1 \\ \neq i}}^{N} \mathbf{U}_{ij}(\mathbf{r}_{ij}), \end{aligned} \tag{12.2.1}$$

with $\mathbf{r}_{ij} = \mathbf{x}_i - \mathbf{x}_j$. Ensemble averaging then yields

$$\begin{aligned} \langle \mathbf{U} \rangle &= \frac{1}{N!} \int P_N \mathbf{U}_1 \, d\mathbf{x}_1 \ldots d\mathbf{x}_N \\ &= \mathbf{U}_0 + n \int \mathbf{U}_{12}(\mathbf{r}) p(\mathbf{r}) \, d\mathbf{r} + O(\phi^2). \end{aligned} \tag{12.2.2}$$

Unfortunately $\boldsymbol{\omega}_{12}$ decays as $1/r$, while the pair distribution function $P_2(\mathbf{r}) = n^2 p(\mathbf{r})$ asymptotes to a constant, resulting in a divergent integral. Batchelor (1972) first resolved the problem and calculated the $O(\phi)$-correction to the Stokes velocity for hard spheres. Here we follow the approach of O'Brien (1979) as adapted by Glendinning & Russel (1982), one of several that have confirmed Batchelor's result.

The velocity at a point $\mathbf{x}$ in the fluid within the representative volume, expressed from (2.6.1) in terms of integrals over the surfaces of the particles $A_i (i = 1, \ldots, N)$ and a bounding surface S (Fig. 12.1) as described by O'Brien (1979), is given by

$$\mathbf{u}(\mathbf{x}) = -\int_{S + \Sigma A_i} \{ \mathbf{I} \cdot \boldsymbol{\sigma} \cdot \mathbf{n} - \mathbf{n} \cdot (\mathbf{u} \cdot \mathbf{J}) \} \, d\mathbf{x}'. \tag{12.2.3}$$

$\mathbf{I}$ is the fundamental solution to the Stokes equations given by (2.5.16) for a point force and $\mathbf{J} = -3\mathbf{xxx}/4\pi x^5$ is the corresponding solution for a force dipole as deduced from (2.5.20); $\mathbf{n}$ is the unit normal directed into the fluid. The integrals over the A_i account for interactions among particles within the representative volume, while the integral over S represents the effect of particles outside. Strictly speaking, S should deivate from the surface enclosing V in order to bypass particles and remain in the fluid. However, in the dilute limit taking S to coincide with the surface of V introduces an error of only $O(\phi^2)$. Note that (12.2.3) comprises only a formal solution, since $\mathbf{u}$ depends on the unknown velocities and stresses on all the surfaces.

The velocity of the ith sphere follows from the local fluid velocity as

$$\mathbf{U}_i = \frac{1}{4\pi a^2} \int_{A_i} \mathbf{u}\, d\mathbf{x}. \tag{12.2.4}$$

Expanding $\mathbf{I}$ and $\mathbf{J}$ in Taylor series about the center of the ith sphere and employing the dilute approximation (12.2.1) for interactions with spheres within the representative volume yields

$$\mathbf{U}_i = \mathbf{U}_0 + \sum_{\substack{j=1 \\ \neq i}}^{N} \mathbf{U}_{ij} - \int_S \{\mathbf{I}\cdot\boldsymbol{\sigma}\cdot\mathbf{n} - \mathbf{n}\cdot(\mathbf{u}\cdot\mathbf{J}) + \frac{a^2}{6}\nabla^2\mathbf{I}\cdot\boldsymbol{\sigma}\cdot\mathbf{n}\}\, d\mathbf{x}'. \tag{12.2.5}$$

Fig. 12.1. Schematic of a representative volume V with surface S containing N particles at positions $\mathbf{x}_i$ with surfaces A_i. The unit normal $\mathbf{n}$ points into the fluid from S and all the A_i.

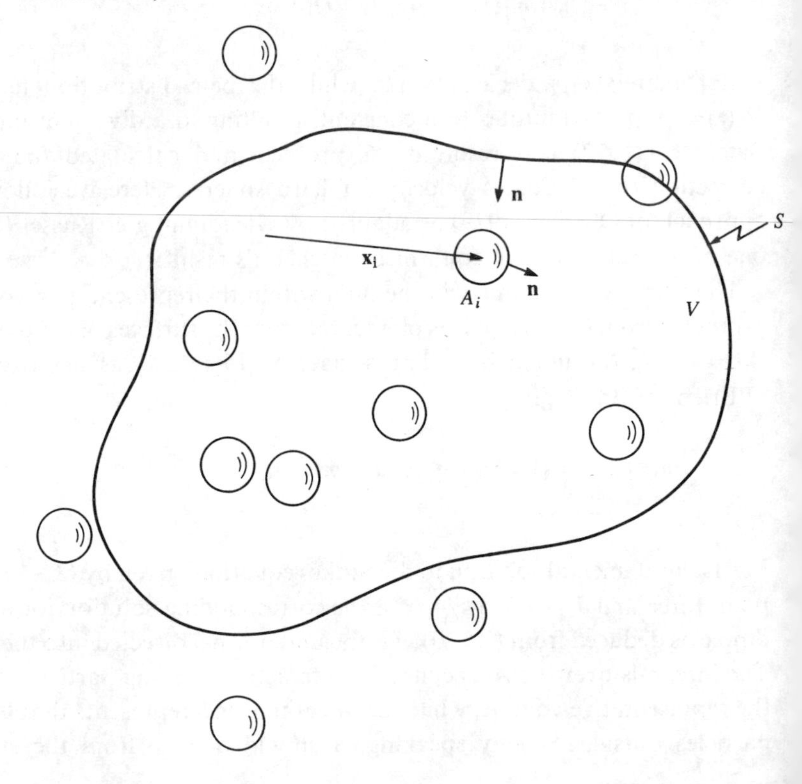

Similarly, the fluid velocity can be written as

$$\mathbf{u}=\sum_{j=1}^{N}\mathbf{u}_j-\int_S\{\mathbf{I}\cdot\boldsymbol{\sigma}\cdot\mathbf{n}-\mathbf{n}\cdot(\mathbf{u}\cdot\mathbf{J})\}\,d\mathbf{x}',$$

with (12.2.6)

$$\mathbf{u}_j=6\pi\mu a\mathbf{U}_0\cdot\left(1+\frac{a^2}{6}\nabla^2\right)\mathbf{I}$$

representing the disturbance velocity generated by an isolated sphere.

For particles settling in an otherwise quiescent suspension, the average velocity of the suspension is zero, so

$$\mathbf{0}=\phi\langle\mathbf{U}\rangle+(1-\phi)\langle\mathbf{u}\rangle,$$

with (12.2.7)

$$\langle\mathbf{u}\rangle=\frac{1}{(N-1)!}\int\mathbf{u}\,P^*_{N-1}\,d\mathbf{x}_2\,\ldots\,d\mathbf{x}_N,$$

where P^*_{N-1} specifies the probability of a particular configuration of the $N-1$ spheres given that $\mathbf{x}_1$ lies in the fluid. Thus $\langle\mathbf{u}\rangle=-\phi\langle\mathbf{U}\rangle/(1-\phi)$ represents the fluid velocity necessary to compensate for the volume flux of particles. The quantity of interest, the sedimentation velocity, is the difference between the ensemble averages of the particle and suspension velocities. Subtracting (12.2.7) from (12.2.5) leaves

$$\langle\mathbf{U}\rangle=\mathbf{U}_0(1-\phi)+\phi(\langle\mathbf{U}_{12}\rangle-\langle\mathbf{u}_1\rangle)-\frac{a^2}{6}\int_S\nabla^2\mathbf{I}\cdot\langle\boldsymbol{\sigma}\rangle\cdot\mathbf{n}\,d\mathbf{x}+O(\phi^2).\tag{12.2.8}$$

The ensemble average has been commuted with the integral, since **I** and **n** do not vary with the configuration of the particles.

Next, the bulk stress $\langle\boldsymbol{\sigma}\rangle$ must be derived from the microscopic momentum equation, which can be written as

$$\mathbf{0}=\nabla\cdot\boldsymbol{\sigma}+\begin{cases}\dfrac{9\mu}{2a^2}\mathbf{U}_0 & \text{within particles}\\ \mathbf{0} & \text{in the fluid.}\end{cases}\tag{12.2.9}$$

Straightforward ensemble averaging produces the macroscopic analogue

$$0=\nabla\cdot\langle\boldsymbol{\sigma}\rangle+\frac{9\mu}{2a^2}\phi\mathbf{U}_0\tag{12.2.10}$$

and integration determines

$$\langle \boldsymbol{\sigma} \rangle = -\frac{9\mu}{2a^2}\,\phi \mathbf{U}_0 \cdot \mathbf{x}\,\boldsymbol{\delta}. \tag{12.2.11}$$

This bulk stress simply accounts for the additional hydrostatic pressure due to the presence of the particles.

Substitution into (12.2.8) and use of the divergence theorem to convert the surface integral into a volume integral yields the sedimentation velocity as

$$\langle \mathbf{U} \rangle = \mathbf{U}_0 - \phi \mathbf{U}_0 - 6\pi\mu a n \mathbf{U}_0 \cdot \int_{r>a} \left(1 + \frac{a^2}{6}\nabla^2\right)\mathbf{I}(1-p)\,\mathrm{d}\mathbf{r} + \tfrac{1}{2}\phi \mathbf{U}_0$$

$$+\,6\pi\mu a n \mathbf{U}_0 \cdot \int_{r>2a} \left\{ \boldsymbol{\omega}_{11} - \frac{\boldsymbol{\delta}}{6\pi\mu a} + \boldsymbol{\omega}_{12} - \left(1 + \frac{a^2}{6}\nabla^2\right)\mathbf{I} \right\} p\,\mathrm{d}\mathbf{r} + O(\phi^2). \tag{12.2.12}$$

The terms in (12.2.12) have clear physical significance. At infinite dilution only the first, the *Stokes velocity*, remains. The second and third terms comprise the effect of '*backflow*', the reverse flow of fluid necessary to compensate for the volumetric flux of particles plus the associated fluid. The fourth term arises from the additional gradient in the *hydrostatic pressure.* The final integral, containing the mobilities, accounts for the *near-field hydrodynamic interactions.*

This formal expression, and the results presented earlier for the hydrodynamic mobilities and the pair distribution function, enable exact calculations of the sedimentation velocity to $O(\phi)$ in monodisperse suspensions. Furthermore, the generalization to multimodal dispersions is available (§12.4), and the extension to higher concentrations is possible, though limited by incomplete knowledge of the hydrodynamic interactions (e.g. Beenakker & Mazur, 1983).

12.3 Monodisperse suspensions of spheres

For two spheres of equal sizes and densities, the relative velocity

$$\Delta \mathbf{U} = (\boldsymbol{\omega}_{22} - \boldsymbol{\omega}_{21} - \boldsymbol{\omega}_{11} + \boldsymbol{\omega}_{12}) \cdot 6\pi\mu a \mathbf{U}_0$$

is identically zero, since $\boldsymbol{\omega}_{11} = \boldsymbol{\omega}_{22}$ and $\boldsymbol{\omega}_{12} = \boldsymbol{\omega}_{21}$. Thus at dilute concentrations the structure remains at equilibrium with $p(\mathbf{r}) = g(r) = \exp$

$(-\Phi/kT)$. With the explicit forms (2.8.4) for the mobilities and the spherical symmetry of $p(\mathbf{r})$, the expression for the sedimentation velocity reduces to

$$\langle \mathbf{U} \rangle = \mathbf{U}_0\left(1-\phi\left\{5+3\int_2^\infty (1-g)s\,\mathrm{d}s - \right.\right.$$

$$\left.\left. -\int_2^\infty [A_{11}+A_{12}+2(B_{11}+B_{12})-3(1+s^{-1})]gs^2\,\mathrm{d}s\right\}+O(\phi^2)\right) \tag{12.3.1}$$

$$\equiv \mathbf{U}_0\{1+K_2\phi+O(\phi^2)\},$$

with $s=r/a$.

For hard spheres with $g=H(s-2)$, the result (Batchelor, 1972)

$$\frac{U}{U_0}=1-6.55\phi+O(\phi^2) \tag{12.3.2}$$

includes an $O(\phi)$-correction that consists of a relatively large negative contribution from the backflow (-5.5ϕ), a small positive effect from the pressure gradient $(+0.5\phi)$, and the hindrance due to the near field hydrodynamics (-1.55ϕ).

Experimental confirmation of this prediction requires demonstrating the hard sphere nature of the particle, e.g. $A_2=4$ (§10.3), as well as measuring K_2. For fd bacteriophage DNA, a compact spherical macromolecule with a hydrodynamic radius of 31.8 nm, Newman *et al.* (1974) measured a second virial coefficient of 4.0 ± 2.0 for the osmotic pressure and an $O(\phi)$-coefficient of -6.7 ± 0.8 for sedimentation. For the silica spheres shown to behave thermodynamically as hard spheres in §10.5, the corresponding value for sedimentation is -6.6 ± 0.6 (Kops-Werkhoven *et al.*, 1982).

The forms of the individual contributions to (12.3.1) suggest the effect of interparticle potentials. Attraction increases the population of nearby particles, thereby reducing the retardation from backflow while enhancing the modest contribution from near-field hydrodynamic interactions. Since the former is larger, attractions decrease in magnitude the negative $O(\phi)$-coefficient, increasing the sedimentation velocity. Conversely, a repulsion of longer range than the hard-sphere diameter causes greater retardation via backflow while reducing the near-field hydrodynamic interactions, thereby slowing sedimentation (Batchelor, 1972; Reed & Anderson, 1980; Batchelor & Wen, 1982).

As discussed in §10.3, the adhesive excluded-shell potential provides a useful model for demonstrating the combined effects of electrostatic repulsion and attraction into a secondary minimum. The radial distribution function

$$g(s)=H(s-s_0)+\frac{\delta(s-s_0)}{6\tau}, \tag{12.3.3}$$

together with the far-field approximations (2.8.9) for the mobilities yield

$$\frac{U}{U_0}=1-\left(K_0-\frac{K_1}{\tau}\right)\phi+O(\phi^2),$$

with (12.3.4)

$$K_0=\frac{3}{2}s_0^2-1+\frac{15}{4s_0}-\frac{9}{8s_0^3}-\frac{107}{20s_0^5}+O(s_0^{-7}),$$

$$K_1=\frac{1}{2}s_0-\frac{5}{8s_0^2}+\frac{16}{9s_0^4}+\frac{107}{24s_0^6}+O(s_0^{-7}).$$

Both K_0 and K_1 are positive and increase monotonically with the excluded-shell radius. The two parameters, s_0 and $1/\tau$, also affect the second virial coefficient according to (10.3.7). Indeed, plotting K_2 vs. A_2 (Fig. 12.2)

Fig. 12.2. The $O(\phi)$-coefficient in the sedimentation velocity, K_2, as a function of the second virial coefficient in the osmotic pressure, A_2: ——, predictions for the adhesive excluded shell potential from (12.3.4) and (10.3.7); ●, data for silica spheres with $a=32.5\pm0.2$ nm in toluene from Table 12.1. (Jansen, de Kruif, & Vrij, 1986).

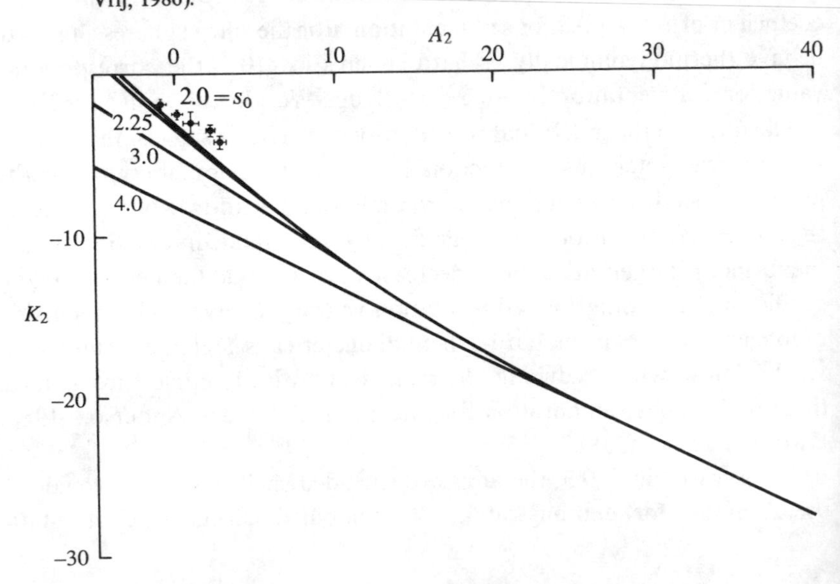

Table 12.1. *Sedimentation of dilute suspensions with weak attractions*

(a) Silica spheres in toluene (Jansen, de Kruif & Vrij, 1986); $a = 32.5 \pm 0.2$ nm, $\Delta\rho = 890 \pm 10$ kg/m^3.

T/°C	A_2	K_2
25.0	2.86	-4.25 ± 0.39
20.0	2.19	-3.53 ± 0.31
14.8	1.02	-3.03 ± 0.68
12.1	0.11	-2.43 ± 0.15
10.0	-0.81	-1.89 ± 0.14

(b) Polystyrene latices in water

	Buscall *et al.* (1982)	Cheng & Schachman (1955)
a	1.55 μm	0.13 μm
I	10^{-3} M	10^{-1} M
ψ_0 (assumed)	-25 mV	-75 mV
$a\kappa$	155	130
s_0	2.04	2.04
$1/\tau$	0.23	2.22
A_2	4.0	1.9
K_2 (experiment)	-5.4 ± 0.1	-5.1
(theory)	-6.5	-5.0

produces a single curve for weak attractions ($1/\tau \leq 1.5$) or short-range repulsions ($s_0 - 2 \leq 0.25$). Under these conditions, which probably include most aqueous systems with significant secondary minima (Batchelor & Wen, 1982) and non-aqueous dispersions with interactions across thin polymer layers, the detailed form of the potential is unimportant.

Since the organophilic silica dispersions are stabilized by octadecyl chains grafted to their surfaces, weak attractions can be induced by exchanging the good solvent cyclohexane for a poor one such as toluene. Jansen, de Kruif, & Vrij (1986) measured the osmotic compressibility by light scattering and the sedimentation velocity with an analytical ultracentrifuge for particles of 32.5 nm radius in toluene over a range of temperatures. The values extracted for A_2 and K_2 indicate significant attractive interactions (Table 12.1) and the correlation between the two parameters closely resembles that predicted (Fig. 12.2).

Some data also exist for well-characterized aqueous systems. For example, Cheng & Schachman (1955) detected $K_2 = -5.06$ for 0.13 μm radius latices in 10^{-1} M NaCl and Buscall *et al.* (1982) found $K_2 = -5.4$ for

1.55 μm radius latices in 10^{-3} M NaCl. The values for s_0 and $1/\tau$ deduced with reasonable values for the unknown ψ_0 and the expected dispersion potential (§5.9, 5.10) generate the $O(\phi)$-coefficients listed in Table 12.1. In both cases A_2 and $-K_2$ lie at or below the values for hard spheres, indicating that the populations of doublets produced by the secondary minima outweigh the depletion of pairs at smaller separations due to the electrostatic repulsion.

At finite concentrations, the experimental results reported by Buscall *et al.* (1982) for the aqueous latices and de Kruif, Jansen, & Vrij (1987) for silica spheres in cyclohexane reflect the concentration dependence in the absence of either strong attractions or long range repulsions (Fig. 12.3). More accurate correlations can be constructed, but the simple expression,

$$\frac{U}{U_0}=(1-\phi)^{-K_2}, \tag{12.3.5}$$

with the exponent conforming to the $O(\phi)$ coefficient in the dilute limit, characterizes the behavior at both the dilute and the concentrated ends of

Fig. 12.3. The dimensionless sedimentation velocity as a function of volume fraction: ●, data for polystyrene spheres with $a=1.55\,\mu$m in 10^{-3} M NaCl solution (Buscall *et al.*, 1982); ○, ▽, data for silica spheres with $a=71\pm2$ nm and $a=34.6\pm0.5$ nm in cyclohexane (de Kruif, Jansen, & Vrij, 1987): ···, (12.3.5) with $K_2=-6.6$; ‐ ‐ ‐, (12.3.5) with $K_2=-5.4$; ‐·‐, (12.3.5) with $K_2=-4.7$.

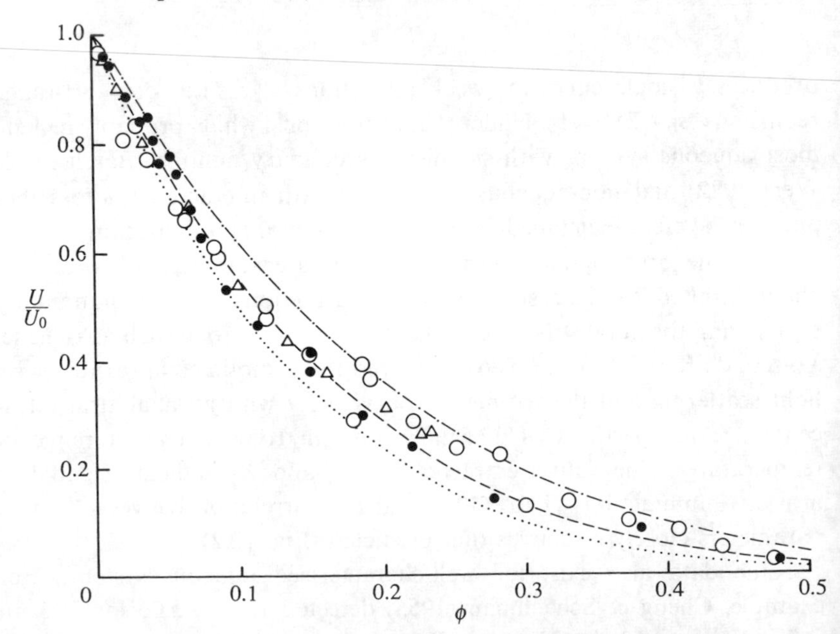

the range. Thus, K_2 equals -6.6 for the silica dispersions and -5.4 for the aqueous latices (the experimental values). Both dispersions settle much like hard spheres; for stronger attractions or long-range repulsions, the volume fraction dependence might be quite different.

We began this section by noting that for dilute monodisperse spheres, sedimentation does not perturb the equilibrium microstructure. For higher-volume fractions or non-uniform particles, however, non-equilibrium effects can enter. Either three-body interactions or slight variations in particle size and/or density would generate a non-zero $\Delta\mathbf{U}$, thereby perturbing the pair distribution and altering the sedimentation velocity. Indeed, the volume fraction dependence of the sedimentation velocity for macroscopic spheres, represented by the third curve for $K_2 = -4.7$ in Fig. 12.3, differs from that for colloidal spheres in qualitative accord with the dilute theory for unequal spheres described in the following section.

12.4 Polydisperse suspensions of spheres

Spheres of unequal sizes or densities settle at different velocities, producing a relative velocity for an interacting pair with magnitude scaling on

$$\Delta\mathbf{U}_0 = \mathbf{U}_{0j} - \mathbf{U}_{0i} = (\gamma\lambda^2 - 1)\mathbf{U}_0$$

with $\gamma = \Delta\rho_j/\Delta\rho_i$, $\lambda = a_j/a_i$, and $U_0 = U_{0i}$. Consequently, the pair distribution function will depart from equilibrium even in dilute suspensions. Generalization of the result of the previous section yields the average velocity for spheres of radius a_i, density difference $\Delta\rho_i$, and volume fraction ϕ_i interacting with spheres of $m-1$ other radii or densities as (Batchelor & Wen, 1982)

$$\langle\mathbf{U}_i\rangle - \langle\mathbf{u}\rangle = \mathbf{U}_{0i}\left(1 + \sum_{j=1}^{m} K_{ij}\phi_j + O(\phi)\right),$$

with (12.4.1)

$$K_{ij} = -\gamma(\lambda^2 + 3\lambda + 1) + \left(\frac{1+\lambda}{2\lambda}\right)^3 \int_2^\infty \left\{(A_{11} + 2B_{11} - 3)p_{ij} + 2\gamma\frac{\lambda^3}{1+\lambda}\left[(A_{12} + 2B_{12})p_{ij} - \frac{3}{s}\right]\right\} s^2\, ds,$$

where $n^2 p_{ij}$ is the distribution function for the ij pair and $s = 2r/(a_i + a_j)$. The mobilities for unequal spheres are discussed in §2.8.

The pair conservation equation governing p_{ij}, a generalization of (8.3.8), accounts for relative motion due to the gravitational force, the interparticle potential, and Brownian motion. The variation of the coefficients with separation renders the equation difficult to solve analytically in any general sense, but semi-analytical solutions are possible for small and large values of the Peclet number (Batchelor & Wen, 1982). In the former case, a regular perturbation expansion suffices to determine the small departure from equilibrium. In the latter, Brownian motion is negligible, reducing the conservation equation to a first-order partial-differential equation that can be integrated along trajectories. The results define the asymptotic behavior of the sedimentation velocity for small and large particles. In this section we restrict our attention to the high Peclet number limit to demonstrate the effect of particle size on the behavior of nearly monodisperse suspensions and some qualitative features of bimodal systems.

For sufficiently large Peclet numbers, the interaction potential as well as Brownian motion can be neglected, reducing the steady-state conservation equation to

$$\nabla \cdot p_{ij}\mathbf{U}_{ij} = 0,$$

with

$$\mathbf{U}_{ij} = (\gamma\lambda^2 - 1)\mathbf{U}_0 \cdot \left[L\frac{\mathbf{rr}}{r^2} + M\left(\boldsymbol{\delta} - \frac{\mathbf{rr}}{r^2}\right)\right] \tag{12.4.2}$$

$$L = A_{ii} - \frac{2\lambda}{1+\lambda}A_{ij}, \quad M = B_{ii} - \frac{2\lambda}{1+\lambda}B_{ij},$$

from (2.8.2). The boundary condition

$$p_{ij} \to 1 \qquad \text{as} \qquad r_{ij} \to \infty$$

reflects the absence of long range structure.

For trajectories which start and end at infinite separation, the pair conservation equation can be integrated to determine (Batchelor, 1982)

$$p_{ij} = \exp \int_s^{\infty} \left(\frac{2(L-M)}{sL} + \frac{1}{L}\frac{\mathrm{d}L}{\mathrm{d}s} \right) \mathrm{d}s \tag{12.4.3}$$

Since p_{ij} depends only on the scalar separation, the hydrodynamic interactions preserve spherical symmetry but cause pairs to accumulate near contact. For equal spheres at infinite Peclet number, (12.4.3) yields different forms for p_{ij} at $\gamma = \lambda = 1$ according to how this limit is approached. The non-equilibrium distribution functions for both $\lambda = 1$, with $\gamma \to 1$, and $\gamma = 1$, with $\lambda \to 1$ (Fig. 12.4), resemble the equilibrium pair distribution resulting from a long range attractive potential, rather than $p_{ij} = H(s-2)$, which is characteristic of perfectly monodisperse hard spheres. Closed trajectories, which appear for small ranges of λ when the smaller sphere is more dense than the larger one, are not considered here.

For sufficiently large particles, even small variations in either density or radius lead to either

$$Pe = \frac{2aU_0}{D_0}(\lambda - 1) \gg 1 \qquad \text{for } \gamma = 1$$

or

$$Pe = \frac{aU_0}{D_0}(\gamma - 1) \gg 1 \qquad \text{for } \lambda = 1.$$

For example, for glass beads of $10^2\,\mu$m radius in water, a 1 per cent variation in radius would produce a Peclet number of 10^7. Thus the volume-fraction dependence of the sedimentation velocity of large particles will not conform to that for a monodisperse suspension.

Substitution of (12.4.3) for the pair distribution into (12.4.1) determines the K_{ij}. The contributions can be separated into the far-field effects of backflow and the mean pressure gradient, the viscous drag due to a passive

Fig. 12.4. The dimensionless pair distribution function for $Pe = \infty$ for unequal spheres at several ratios of radii, λ, and density differences, γ, (Batchelor & Wen, 1982). Note the distinct difference between p_{ij} for $\lambda \to 1$ with $\gamma = 1$ and the value for identical spheres, $p_{ij} = 1$.

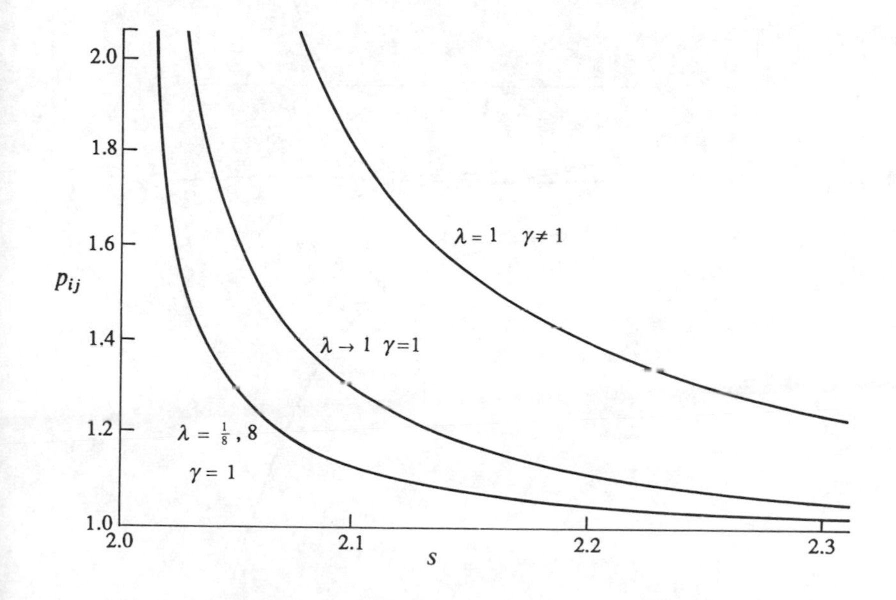

second sphere, and the velocity induced in the ith sphere by the motion of the jth. If the jth sphere is small, $\lambda \ll 1$, then only the viscous drag and backflow are significant, leaving

$$K_{ij} = -\tfrac{5}{2} - \gamma + O(\lambda). \tag{12.4.4}$$

Thus, the larger particle effectively settles through a continuum, with relative viscosity increased by 2.5ϕ and density altered by $\gamma\phi$. For the smaller sphere, only the far-field term contributes, giving

$$K_{ji} = -\frac{1}{\gamma}\left(1 + \frac{3}{\lambda} + \frac{1}{\lambda^2}\right) + O(\lambda). \tag{12.4.5}$$

So the smaller sphere is simply convected by the velocity and pressure field generated by the larger. Between these limits, the $O(\phi)$ coefficient varies with λ, as shown in Fig. 12.5. The open circle about $\lambda = \gamma = 1$ reflects the

Fig. 12.5. The coefficient characterizing the dependence of the sedimentation velocity of spheres of type i on the volume fraction of spheres of type j as a function of the size ratio $\lambda = a_j/a_i$ and the relative density difference $\gamma = \Delta\rho_j/\Delta\rho_i$ for $Pe = \infty$: ——, predictions of Batchelor & Wen (1982); ●, data of Davis & Birdsell (1988) for mixtures of glass beads ($\lambda = 0.52$, $\gamma = 1.0$) and glass and acrylic beads ($\lambda = 0.99$, $\gamma = 0.11$).

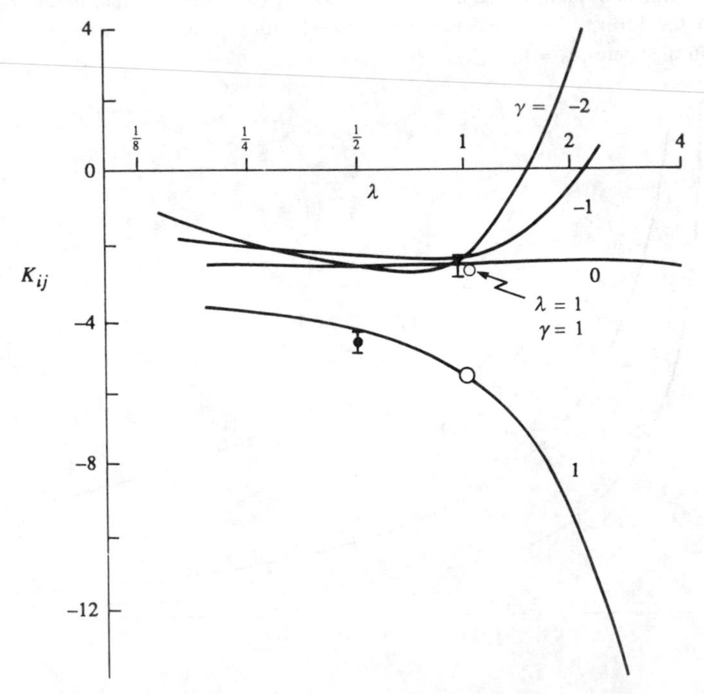

singular nature of the point. The two data points from Davis & Birdsell (1988) for glass and acrylic spheres with $Pe \approx 10^7$ and $\gamma = 0.11$ with $\lambda = 1.0$ and $\gamma = 1.0$ with $\lambda = \frac{1}{2}$ agree with the predictions.

Despite the complicated form of Fig. 12.5, several aspects of the results are surprisingly simple and help to explain observations in the literature:

(i) For $\gamma = 0$, $K_{ij} = -2.5$ for all size ratios, indicating that neutrally buoyant spheres simply increase the effective viscosity of the medium. Recent experiments by Mondy, Graham, & Jensen (1986) find this prevails for heavy spheres settling in concentrated suspensions of neutrally buoyant spheres as well.

(ii) For equal sizes, K_{ij} depends weakly on the density ratio,

$$K_{ij} = -2.52 - 0.13\gamma, \tag{12.4.6}$$

except near $\gamma = 1$, where it jumps from -2.65 to -5.62.

(iii) When $\lambda \gg 1$ and $\gamma < 0$, $K_{ij} > 0$, indicating the backflow of the rising second species, which enhances the settling of the first, outweighs the near-field hydrodynamic interactions.

The different values for $\gamma \approx 1$ and $\lambda \approx 1$ provide some rationale for the difference between the settling velocities for large 'monodisperse' glass beads (Fig. 12.6), with

$$\frac{U}{U_0} = (1-\phi)^{4.65}, \tag{12.4.7}$$

and the result noted earlier for monodisperse colloidal spheres (Fig. 12.3). For example, suppose one mixes equal parts of spheres with radii and density differences (1) a and $\Delta\rho$, (2) $a(1+\varepsilon)$ and $\Delta\rho$, and (3) a and $\Delta\rho(1+\varepsilon)$ to obtain a total volume fraction of ϕ. Then the average settling velocity would be

$$\begin{aligned} \frac{U}{U_0} &= \frac{U_1 + U_2 + U_3}{3U_0} \\ &= 1 + \tfrac{1}{9}\phi(K_{11} + K_{22} + K_{33} + 2K_{12} + 2K_{13} + 2K_{23}). \end{aligned} \tag{12.4.8}$$

With $K_{11} = K_{22} = K_{33} = -6.55$, $K_{12} = -5.62$, $K_{13} = -2.65$, and $K_{23} = -(5.62 + 2.65)/2 = -4.13$, this yields $K_1 = -4.92$, close to the value extracted from the data.

Though still limited, the theoretical and experimental results suggest a qualitative distinction between large and small particles and the effects of bimodal distributions. Some further information is available elsewhere (e.g. Davis & Acrivos, 1985). We now return our attention to monodisperse systems to examine the settling process.

Fig. 12.6. Dimensionless sedimentation velocity as a function of volume fraction for glass beads ($a = 0.35$ mm) in glycerine–water mixtures ($\mu = 0.001$–0.390 N s/m²) such that $Re < 0.01$ and $aU_0/D_0 > 10^{11}$ (Hanratty & Bandukwala, 1957). The solid curve corresponds to (12.3.5), with $K_2 = -4.65$.

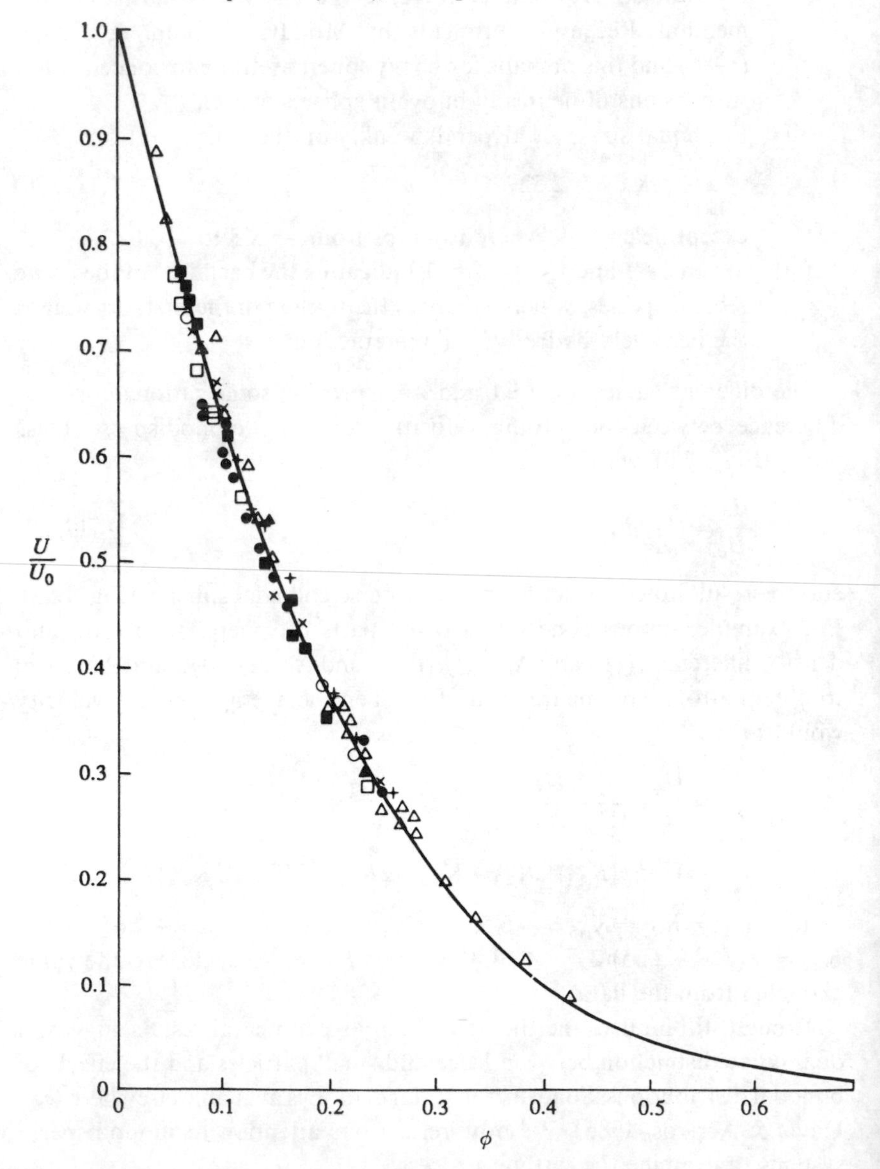

12.5 Theory of batch settling

As a dispersion settles in a closed container, the concentration necessarily becomes non-uniform (Fig. 12.7). A clear layer, devoid of particles, forms at the top and a sediment at the bottom. In addition, concentration gradients can appear in the region above the sediment. Kynch (1952) first formulated the theory capable of describing these phenomena for incompressible sediments. Here we generalize his treatment of monodisperse suspensions to account for compressibility imparted by the Brownian motion of and interactions among particles in the sediment.

In the absense of an external field, a spatially varying concentration represents a non-equilibrium state for a dispersion. Hence both particles and fluid molecules experience forces, proportional to the gradients in their respective chemical potentials (e.g. Batchelor, 1976),

$$\begin{aligned} \mathbf{F}_p &= -\nabla\mu_p, \\ \mathbf{F}_s &= -\nabla\mu_s. \end{aligned} \tag{12.5.1}$$

But a uniform body force, $-\mathbf{F}_s/l^3$ per unit volume, with l^3 the volume of a fluid molecule, acting on particles and fluid alike produces no relative motion between the two species. Hence the molecules can be treated as force-free if the effective force acting on each particle is written as

$$\begin{aligned} \mathbf{F} &= \mathbf{F}_p - \frac{4\pi a^3}{3l^3}\mathbf{F}_s \\ &= -\nabla\left(\mu_p - \frac{4\pi a^3}{3l^3}\mu_s\right). \end{aligned} \tag{12.5.2}$$

Combining the Gibbs–Duhem relation,

$$n\nabla\mu_p + n_s\nabla\mu_s = 0, \tag{12.5.3}$$

with n and n_s the number densities of the particles and fluid molecules, respectively, and the volume-filling constraint,

$$\frac{4\pi a^3}{3}n + l^3 n_s = 1, \tag{12.5.4}$$

then yields

$$\begin{aligned} \mathbf{F} &= \frac{1}{nl^3}\nabla\mu_s \\ &= -\frac{1}{n}\nabla\Pi. \end{aligned} \tag{12.5.5}$$

The final form for the thermodynamic force arising from the non-uniform concentration follows from the relationship (6.2.22) between Π, the osmotic pressure, and the chemical potential of the fluid.

With (12.5.5), the flux of particles due to the combined effects of gravity and the thermodynamic force can be written as

$$\phi\frac{K(\phi)}{6\pi\mu a}\left(\frac{4\pi a^3}{3}\Delta\rho\mathbf{g}+\mathbf{F}\right)=K(\phi)\left(\phi\mathbf{U}_0-D_0\frac{\mathrm{d}}{\mathrm{d}\phi}[\phi Z(\phi)]\nabla\phi\right), \quad (12.5.6)$$

with $Z(\phi)=\Pi/nkT$, the compressibility factor; $K(\phi)=U/U_0$, the sedimentation coefficient; and $D_0=kT/6\pi\mu a$, the Stokes–Einstein diffusion coefficient. Note that $K(\phi)/6\pi\mu a$ characterizes the mobility of particles in a concentrated dispersion. The conservation equation for a one-dimensional settling process then takes the form (Davis & Russel, 1988)

$$\frac{\partial\phi}{\partial t}+U_0\frac{\partial}{\partial x}\phi K(\phi)=D_0\frac{\partial}{\partial x}\left[K(\phi)\frac{\mathrm{d}}{\mathrm{d}\phi}[\phi Z(\phi)]\frac{\partial\phi}{\partial x}\right], \quad (12.5.7)$$

with $\phi(x,0)=\phi_0$ and no flux boundary conditions at the top and bottom of the vessel, $x=0, h$.

Rendering the conservation equation dimensionless by scaling time on h/U_0 and the position on h identifies a single dimensionless group

$$Pe^{-1}=\frac{D_0}{hU_0}=\frac{3kT}{4\pi a^3\Delta\rho gh},$$

which multiplies the term on the right-hand side. This quantity, an inverse Peclet number for the macroscopic settling process, gauges the magnitude of the thermodynamic force relative to the gravitational force for a dilute dispersion. Settling in a conventional sense, i.e., to form a recognizable sediment, requires a large Peclet number. For example, for spheres of 1.0 μm radius with $\Delta\rho/\rho=1$ in a 1 m column of water the value is 10^7.

For large Peclet numbers the thermodynamic force is clearly negligible wherever $\phi Z(\phi)$ and the dimensionless gradient of the volume fraction are $O(1)$. Then, recasting the conservation equation as the quasi-linear first-order partial-differential equation

$$\frac{\partial\phi}{\partial t}+U_0\frac{\mathrm{d}}{\mathrm{d}\phi}[\phi K(\phi)]\frac{\partial\phi}{\partial x}=0 \quad (12.5.8)$$

leads to the general solution (cf. Rhee, Aris, & Amundson, 1987, §2)

$$\phi(x,t)=f(x-ut)$$

with (12.5.9)

$$u=U_0\frac{\mathrm{d}}{\mathrm{d}\phi}\phi K(\phi).$$

Consequently, in this free-settling region, ϕ is constant along characteristic curves with slopes $dx/dt = u$. The initial or boundary condition, where a characteristic intersects a coordinate axis, determines the value of ϕ.

The bottom of the container, however, comprises an impenetrable boundary at which the solution fails because the particle flux must vanish, even though $K(\phi)$ does not. Here particles accumulate until the volume fraction approaches closest packing. Since the compressibility factor diverges as $Z \sim (\phi_m - \phi)^{-1}$, as discussed in §10.5 (Woodcock, 1981), the thermodynamic force balances gravity only when $\phi_m - \phi = O(Pe^{-1})$. Then the time derivative becomes $O(Pe^{-1})$ smaller than the other terms. Invoking the pseudo-steady approximation, integrating the conservation

Fig. 12.7. Schematic for a batch settling process in a container of height h.

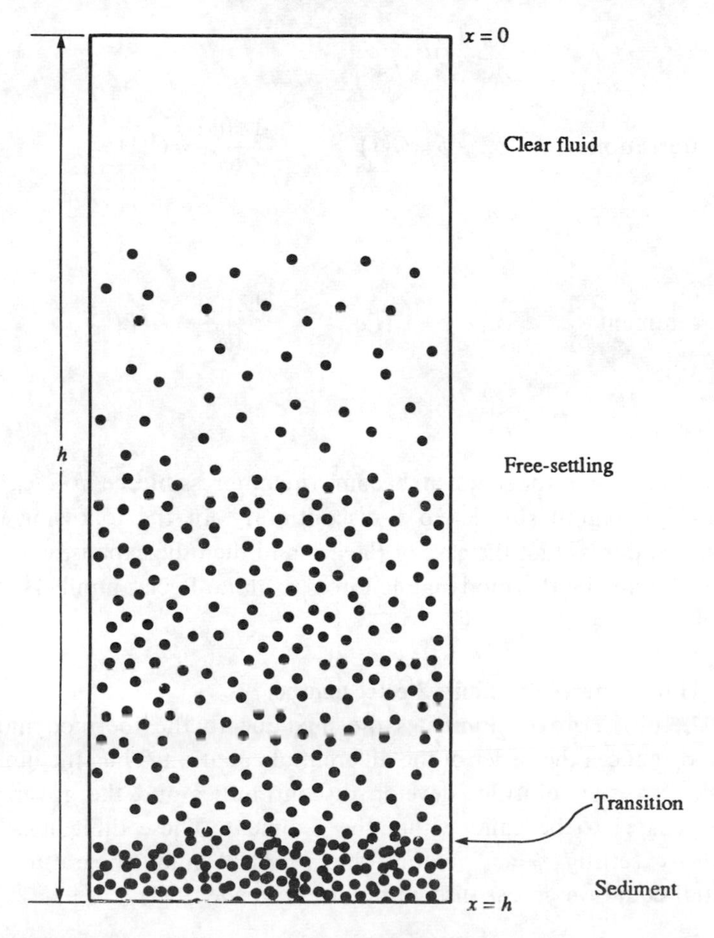

equation once, and applying the no-flux boundary condition at the bottom simply leaves

$$\frac{4\pi a^3}{3}\mathbf{g}=\frac{1}{n}\nabla\Pi. \tag{12.5.10}$$

Thus the thermodynamic force supports particles. In the bulk of the sediment this occurs because the osmotic pressure is large. But at the top of the sediment, the volume fraction changes from near closest packing to the value in the free-settling region. This produces a thin transition layer in which the thermodynamic forces are significant because $\partial\phi/\partial x = O(Pe)$.

These arguments identify three regions within the settling dispersion (Fig. 12.7):

(i) free-settling $\quad \phi_m-\phi=O(1) \quad \frac{dZ(\phi)}{d\phi}=O(1)$

$$\frac{\partial\phi}{\partial x}=O(1);$$

(ii) transition $\quad \phi_m-\phi=O(1) \quad \frac{dZ(\phi)}{d\phi}=O(1)$

$$\frac{\partial\phi}{\partial x}=O(Pe).$$

(iii) sediment $\quad \phi_m-\phi=O(Pe^{-1}) \quad \frac{dZ(\phi)}{d\phi}=O(Pe^2)$

$$\frac{\partial\phi}{\partial x}=O(Pe^{-1})$$

In the limit of $Pe\to\infty$ the sediment becomes incompressible, i.e. $\phi\to\phi_m$, and the transition region shrinks to a discontinuity. In the following two sections, we address first the case of $Pe=\infty$ and then the expansion of the sediment due to the thermodynamic forces at finite Peclet numbers.

12.6 Hard spheres at infinite Peclet number

Letting $Pe\to\infty$ eliminates the flux due to the thermodynamic forces and reduces the order of the differential equation. Thus the limit is singular, since the solutions develop discontinuities and the governing equation ceases to be valid within the sediment. These difficulties are resolved by setting $\phi=\phi_m$ within the sediment and integrating the differential equation across discontinuities (Kynch, 1952).

For a discontinuity at $x=L(t)$, the coordinate transformation $y=(x-L(t))Pe$ converts the dimensionless version of (12.5.7) into

$$\frac{1}{Pe}\frac{\partial\phi}{\partial t}+\frac{\partial}{\partial y}\phi(K(\phi)-L')=\frac{\partial}{\partial y}K(\phi)\frac{\mathrm{d}}{\mathrm{d}\phi}[\phi Z(\phi)]\frac{\partial\phi}{\partial y}. \tag{12.6.1}$$

Integrating the pseudosteady $O(1)$ equation with $\phi\to\phi^+$ and $\partial\phi/\partial y\to 0$ as $y\to-\infty$ yields

$$\phi(K(\phi)-L')-\phi^+(K(\phi^+)-L')=K(\phi)\frac{\mathrm{d}}{\mathrm{d}\phi}[\phi Z(\phi)] \tag{12.6.2}$$

Evaluating (12.6.2) as $y\to\infty$ then determines $L'\equiv \mathrm{d}L/\mathrm{d}t$ as follows:

(i) $\phi\to\phi_m$, so that

$$\lim_{y\to\infty}K(\phi)\left[\phi-\frac{\mathrm{d}}{\mathrm{d}\phi}[\phi Z(\phi)]\frac{\partial\phi}{\partial y}\right]=0,$$

leaving

$$L'=-\frac{\phi^+K(\phi^+)}{\phi_m-\phi^+} \tag{12.6.3}$$

as the rate of rise of the sediment.

(ii) $\phi\to\phi^-$ and $\partial\phi/\partial y\to 0$, so that

$$L'=\frac{\phi^+K(\phi^+)-\phi^-K(\phi^-)}{\phi^+-\phi^-} \tag{12.6.4}$$

for a discontinuity in the free-settling region.

The nature of permissible discontinuities follows from substituting (12.6.3) or (12.6.4) into (12.6.2). For the latter, this produces

$$K(\phi)\frac{\mathrm{d}}{\mathrm{d}\phi}[\phi Z(\phi)]\frac{\partial\phi}{\partial y}=\phi K(\phi)-\frac{(\phi-\phi^-)\phi^+K(\phi^+)-(\phi-\phi^+)\phi^-K(\phi^-)}{\phi^+-\phi^-}$$
$$=\Delta\phi K(\phi). \tag{12.6.5}$$

The right-hand side is simply the vertical distance between points at a particular ϕ on the flux curve and the line connecting points above and below the discontinuity, i.e. $(\phi^+, K(\phi^+))$ and $(\phi^-, K(\phi^-))$ in Fig. 12.8. Since $\partial\phi/\partial y\to 0$ as $y\to\pm\infty$ but $\partial\phi/\partial y\neq 0$ for $-\infty<y<+\infty$, (12.6.5) proves that the line cannot cross the flux curve. Hence the discontinuity depicted in Fig. 12.8 is impossible; the line must lie entirely below (above) the flux curve for $\partial\phi\partial y>(<)0$, in accord with the entropy condition of Lax (e.g., Rhee, Aris, & Amundson, 1987, §5.4).

To demonstrate the features of the settling process, we consider disordered hard-sphere dispersions for which $\phi_m = 0.64$ and

$$K(\phi) = (1-\phi)^{6.55}, \tag{12.6.6}$$

so the flux curve takes the form shown in Fig. 12.9. Tangents to this curve then determine the slope of the characteristics, $dx/dt = u$, and the velocity of the rising sediment corresponds to the slope of straight lines intersecting the horizontal axis at ϕ_m and the flux curve at ϕ_+. Two noteworthy features of the flux curve are:

(i) the inflection point at $\phi = 0.265$, separating Regions I and II, where $du/d\phi < 0$, from Regions III and IV, where $du/d\phi > 0$; and
(ii) the line from ϕ_m that is tangent to the flux curve at $\phi = 0.565$ and intersects the curve at $\phi = 0.024$.

Note that for $K(\phi)$ of the form (12.3.5), Regions II and III disappear for

$$-K_2 < \frac{2}{\phi_m}[1 + (1-\phi_m)^{1/2}] - 1,$$

Fig. 12.8 Discontinuity between regions settling freely with volume fractions ϕ^+ and ϕ^-.

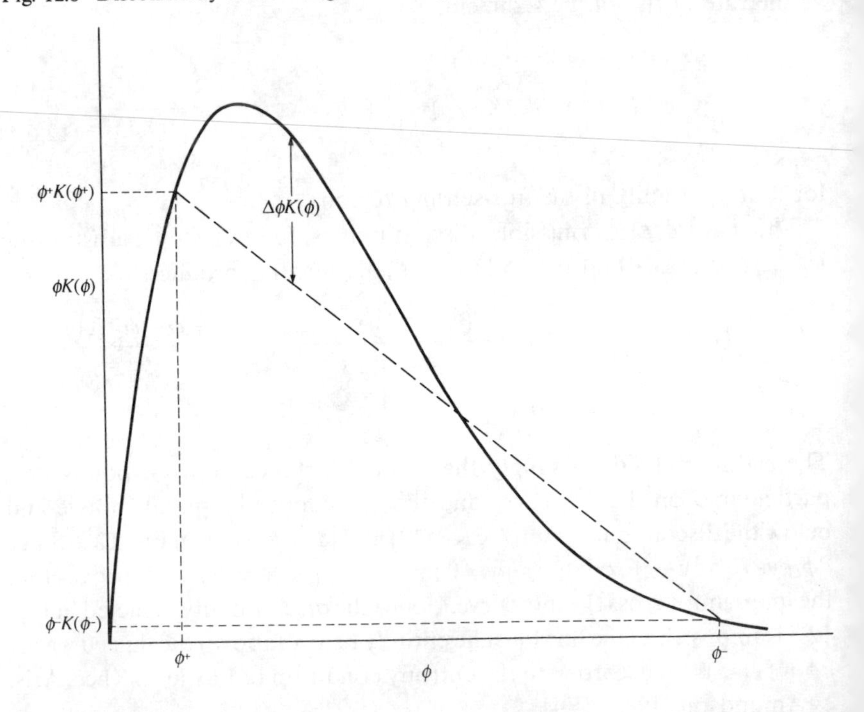

since the point of tangency vanishes. The significance of these four regions of the flux curve will be explained below.

The simplest and most familiar case is for a uniform volume fraction in each of the three regions – clear fluid, freely settling suspension, and sediment – such that

$$\phi = \begin{cases} 0 & 0 < x < U_{top}t \\ \phi_0 & U_{top}t < x < U_{sed}t \\ \phi_m & U_{sed}t < x < h \end{cases} \tag{12.6.7}$$

The discontinuities between the clear fluid and the settling suspension and the settling suspension and the sediment move with velocities

$$U_{top} = U_0 \frac{\phi_0 K(\phi_0)}{\phi_0} = U_0 K(\phi_0) \tag{12.6.8}$$

from (12.6.4) and

$$U_{sed} = -U_0 \frac{\phi_0 K(\phi_0)}{\phi_m - \phi_0} \tag{12.6.9}$$

Fig. 12.9. Flux curve for hard spheres constructed from (12.3.5), with $K_2 = -6.55$ and $\phi_m = 0.64$. The four regions defined by the broken lines correspond to: I, $0 \le \phi \le 0.024$; II, $0.024 \le \phi \le 0.264$; III, $0.264 \le \phi \le 0.565$; IV, $0.565 \le \phi \le 0.64$.

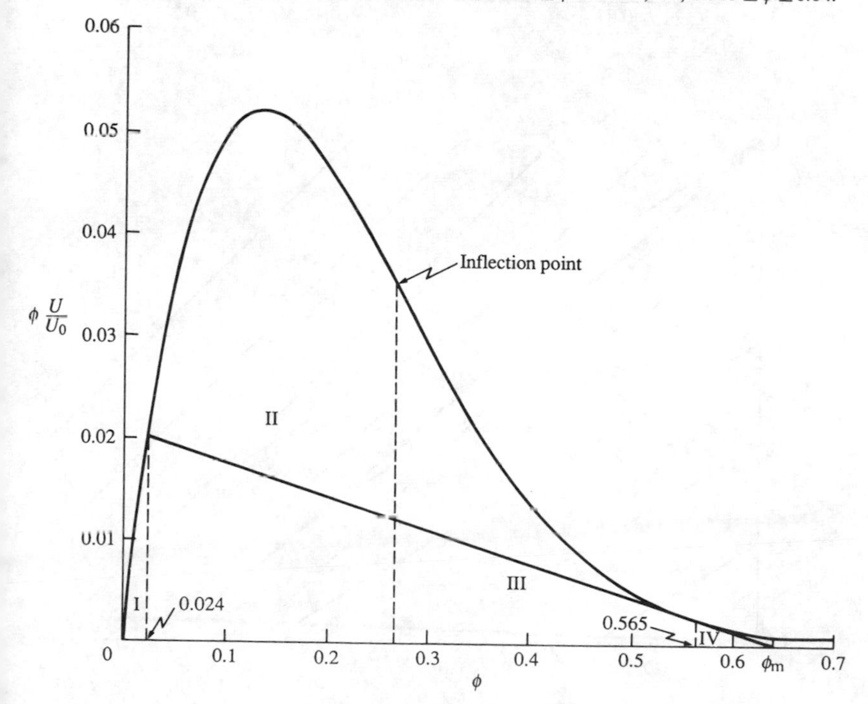

Fig. 12.10. Solutions for batch settling of hard spheres with $Pe = \infty$ and $\phi_0 = 0.024$: (*a*) Volume fraction as a function of position at several times. (*b*) Contour plot, showing: heavy unbroken line, discontinuities; light unbroken line, characteristics along which ϕ = const.; broken line, characteristics eliminated by discontinuities.

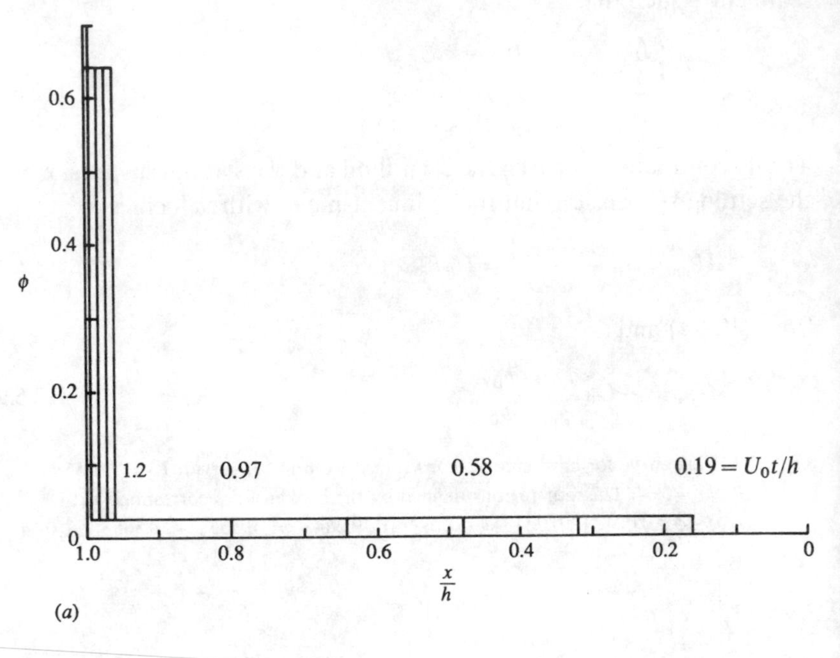

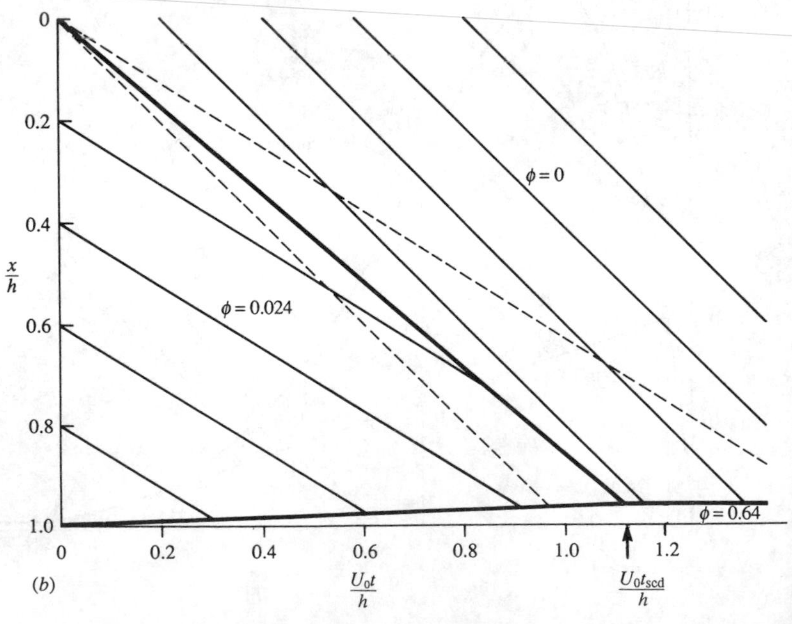

from (12.6.3), with $\phi^+ = \phi_0$, respectively. Thus the volume fraction varies as shown in Fig. 12.10(*a*) for $\phi_0 = 0.024$ and the sediment reaches its final height, $h\phi_0/\phi_m$, at time $h(1-\phi_0/\phi_m)/U_0K(\phi_0)$.

The construction of this solution via the method of characteristics proceeds as follows. The first step is to plot the characteristic curves originating at the boundaries of the domain where the initial and boundary conditions set ϕ (Fig. 12.10(*b*)). If these characteristics fill the entire $x-t$ plane without intersecting then the solution is complete and single-valued. Generally, however, the characteristics intersect as shown, implying an unacceptable multiplicity of solutions. The second step then is to eliminate this multiplicity by constructing one or more discontinuities, separating regions in which the solution varies continuously and propagating at velocities determined by (12.6.2). The corresponding chord on Fig. 12.9 lies below the flux curve for $\phi_0 \le 0.024$.

For values of ϕ_0 in Regions II and III, though, the chord connecting ϕ_0 on the flux curve with ϕ_m on the horizontal axis crosses the flux curve (Fig. 12.11(*a*)). So the jump in volume fraction from ϕ_0 to ϕ_m at the top of the sediment is inadmissable. Instead, solutions with discontinuities represented by chords lying below the flux curve must be constructed by recognizing that additional characteristics emanate from the origin with slopes corresponding to $\phi_0 \le \phi \le \phi_m$.

For example, in Region II a fan of characteristics from the origin (Fig. 12.11(*b*)) is bounded above by a discontinuity with the same velocity, i.e.

$$U_0 \frac{\phi_1 K(\phi_1) - \phi_0 K(\phi_0)}{\phi_1 - \phi_0} = U_0 \left. \frac{\mathrm{d}[\phi K(\phi)]}{\mathrm{d}\phi} \right|_{\phi=\phi_1}. \tag{12.6.10}$$

For $\phi_0 = 0.1$, this tangency condition determines $\phi_1 = 0.40$. The volume fraction at the bottom of the fan, $\phi^+ = 0.565$, is determined by the analogous condition

$$-U_0 \frac{\phi^+ K(\phi^+)}{\phi_m - \phi_+} = U_0 \left. \frac{\mathrm{d}[\phi K(\phi)]}{\mathrm{d}\phi} \right|_{\phi=\phi^+}. \tag{12.6.11}$$

Now the solution takes the form shown in Fig. 12.11(*c*), with a region of smoothly varying volume fraction determined by

$$\frac{x}{t} = -\frac{\mathrm{d}[\phi K(\phi)]}{\mathrm{d}\phi} \tag{12.6.12}$$

with $\phi_1 \le \phi \le \phi^+$, separating the uniform region from the sediment. Since the volume fraction increases through the fan, U_{top} decreases in the final stage of the process, termed the falling-rate period. The behavior in Region

III is similar, except that the discontinuity at the top of the fan vanishes so that $\phi_1 = \phi_0$.

Table 12.2 summarizes the key features of these solutions for incompressible sediments: the volume fractions at the top (ϕ_1) and bottom (ϕ^+) of the fan, the beginning of the falling-rate period (t_1), and the time for complete sedimentation (t_{sed}). The solutions are unique, provided all chords representing discontinuities lie below the flux curve and no characteristics emerge from the upper surface of the sediment, which bounds the domain of the solution. The validity of these assumptions is proven by the asymptotic arguments above, and is supported by the numerical solutions for finite Peclet numbers described in the next section.

Fig. 12.11. Solutions for batch settling of hard spheres, with $Pe = \infty$ and $\phi_0 = 0.10$: (*a*) Flux curve with chords corresponding to the inadmissable discontinuity $\phi_0 \rightarrow \phi_m$ and the three discontinuities required for the correct solution: $0 \rightarrow \phi_0$ at the top of the free-settling region; $\phi_0 \rightarrow \phi_1$ at the top of the fan; $\phi^+ \rightarrow \phi_m$ at the top of the sediment. (*b*) Contour plot, showing: heavy line, discontinuities; lighter line; characteristics; the beginning of the falling-rate period at $U_0 t_1/h$, and the end of the process at $U_0 t_{sed}/h$. (*c*) Volume fraction as a function of position and time.

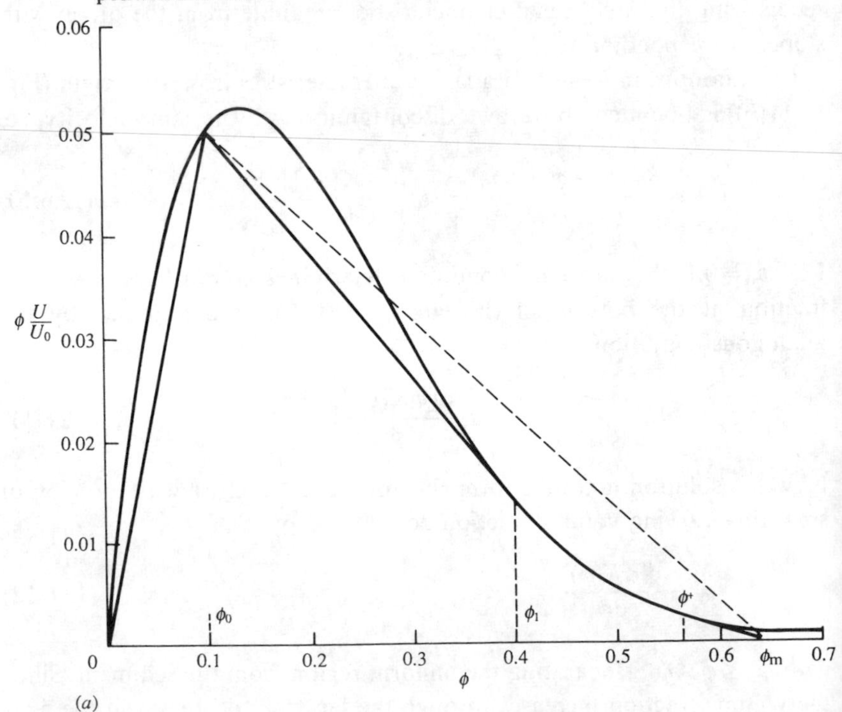

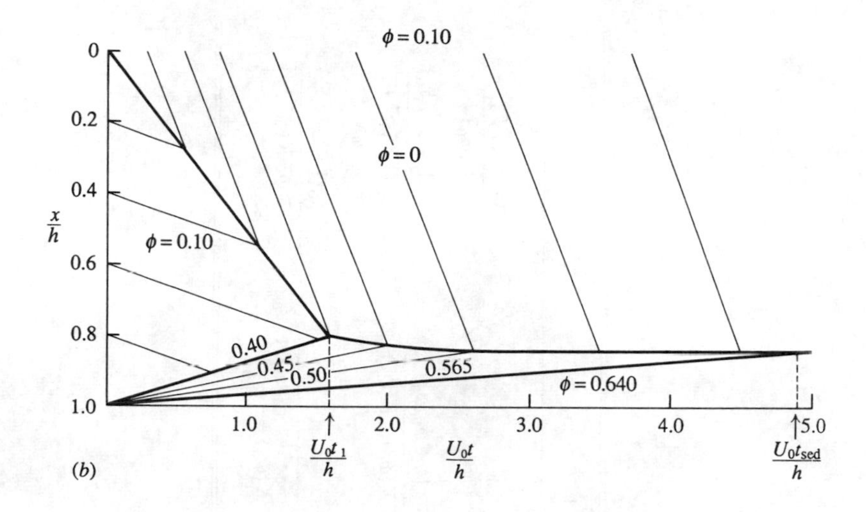

$\phi = 0.10$
$\phi = 0$
$\phi = 0.10$
0.40
0.45
0.50
0.565
$\phi = 0.640$
$\frac{x}{h}$
0
0.2
0.4
0.6
0.8
1.0
1.0
2.0
3.0
4.0
5.0
$\frac{U_0 t_1}{h}$
$\frac{U_0 t}{h}$
$\frac{U_0 t_{sed}}{h}$

(b)

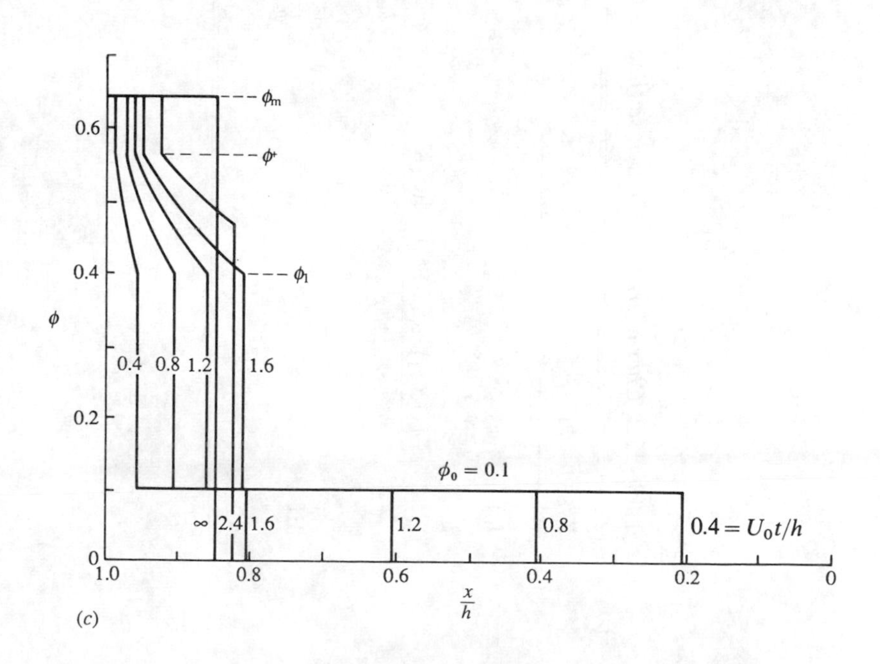

ϕ_m
ϕ^+
ϕ_1
ϕ
0.6
0.4
0.2
0
0.4
0.8
1.2
1.6
$\phi_0 = 0.1$
∞
2.4
1.6
1.2
0.8
$0.4 = U_0 t/h$
1.0
0.8
0.6
0.4
0.2
0
$\frac{x}{h}$

(c)

Table 12.2. *Characteristics of batch settling at* $Pe=\infty$

Region	ϕ_1	ϕ^+	t_1	t_{sed}
I, IV	ϕ_0	ϕ_m	t_{sed}	$\frac{(1-\phi_0/\phi_m)}{K(\phi_0)}$
II	$\frac{\phi_1 K(\phi_1)-\phi_0 K(\phi_0)}{\phi_1-\phi_0}=\frac{d(\phi_1 K)}{d\phi_1}$	$-\frac{\phi^+ K(\phi^+)}{\phi_m-\phi^+}=\frac{d(\phi^+ K)}{d\phi^+}$	$\frac{(1-\phi_0/\phi_1)}{K(\phi_0)-K(\phi_1)}$	$\frac{\phi_0(1-\phi^+/\phi_m)}{\phi^+ K(\phi^+)}$
III	ϕ_0	$-\frac{\phi^+ K(\phi^+)}{\phi_m-\phi^+}=\frac{d(\phi^+ K)}{d\phi^+}$	$\frac{1}{K(\phi_0)-\frac{d(\phi_0 K)}{d\phi_0}}$	$\frac{\phi_0(1-\phi^+/\phi_m)}{\phi^+ K(\phi^+)}$

12.7 Hard spheres at finite Peclet number

Direct numerical solutions of the full equation (12.5.7), with suitable forms for $K(\phi)$ and $Z(\phi)$, serve to illustrate the effects of the thermodynamic forces at finite Peclet number and confirm the arguments above as $Pe \rightarrow \infty$. Since large Peclet numbers are of primary interest, the thermodynamic forces become important only as $\phi \rightarrow \phi_m$. Although hard spheres at equilibrium exist in a face-centered-cubic solid phase for $\phi > 0.55$, computer simulations (Woodcock, 1981) and experiments (Pusey & van Megen, 1987) reveal a persistent disordered state up to random close packing at $\phi \approx 0.64$. So, the calculations described below pertain to a disordered fluid with (12.6.4) for $K(\phi)$ and

$$Z(\phi) = \frac{1.85}{\phi_m - \phi}, \tag{12.7.1}$$

with $\phi_m = 0.64$ in accord with the results of Woodcock (1981).

Results for $\phi_0 = 0.1$ (Auzerais, Jackson, & Russel, 1988) illustrate several points clearly:

(i) For $Pe = 850$, the thermodynamic forces smooth the discontinuities at the top of the uniform settling region only slightly but completely obliterate the distinction between the sediment and the fan (Fig. 12.12(*a*)). Indeed, the volume fraction within the sediment approaches ϕ_m only at long times, and then only at the very bottom.

(ii) Nonetheless, the contour plot of the solution (Fig. 12.12(*b*)) maintains the same qualitative form as at infinite Peclet number, with the top of the suspension, defined by the contour for $\phi = \phi_0/2 = 0.05$, falling at the velocity $U(\phi_0)$ independent of Pe.

(iii) For $Pe = 175\,000$, the solution (Fig. 12.13) does approach the analytical solution for $Pe = \infty$, though the discontinuity at the top of the sediment remains somewhat expanded.

The gradual transition from the fan to the sediment, even for $Pe \approx 10^5$, reflects the difference between the thermodynamic force at ϕ^+ and that at infinite dilution. The actual ratio of thermodynamic to gravitational forces at the top of the sediment,

$$\frac{D_0}{hU_0} \frac{\mathrm{d}[\phi Z(\phi)]}{\mathrm{d}\phi}\bigg|_{\phi+} = \frac{210}{Pe},$$

explains the considerably thicker transition layer. The approach of the numerical solutions to the analytical solution as $Pe \rightarrow \infty$, together with the asymptotic arguments mentioned above, affirm the validity of the results of §12.6.

Fig. 12.12. Numerical solutions for batch settling of hard spheres, with $Pe = 850$ and $\phi_0 = 0.10$ (Auzerais, Jackson, & Russel, 1988): (*a*) Volume fraction as a function of position and time. (*b*) Contour plot.

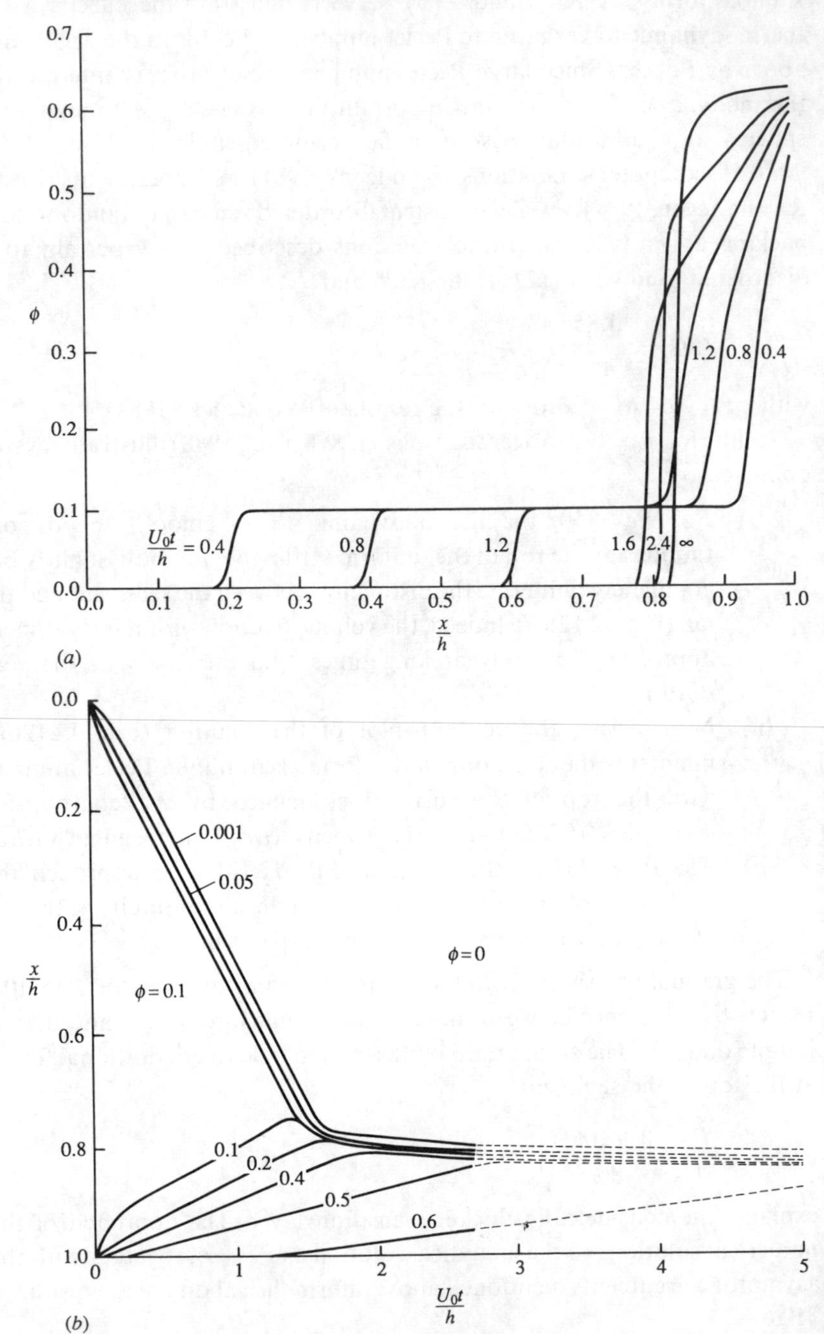

12.8 Summary

This chapter has highlighted the effects of interactions, both hydrodynamic and thermodynamic, on the sedimentation velocity and the transient settling process. The pair interaction theory for the velocity is exact at dilute concentrations and provides a basis for understanding and correlating the behavior of concentrated systems. In particular, it identifies the effects of attractive and repulsive potentials and the non-equilibrium structure responsible for the distinction between the velocities of colloidal and macroscopic particles. The treatment of batch settling demonstrates the classical theory for incompressible sediments to be the $Pe \rightarrow \infty$ limit for stable dispersions.

The basic approach has the potential for addressing several other interesting and important problems:

(i) the volume fraction dependence of the velocity in concentrated dispersions with stronger interparticle potentials or multimodal size distributions,

Fig. 12.13. Volume fraction as a function of position and time from numerical solutions for batch settling of hard spheres, with $Pe = 175\,000$ and $\phi_0 = 0.10$ (Auzerais, Jackson, & Russel, 1988).

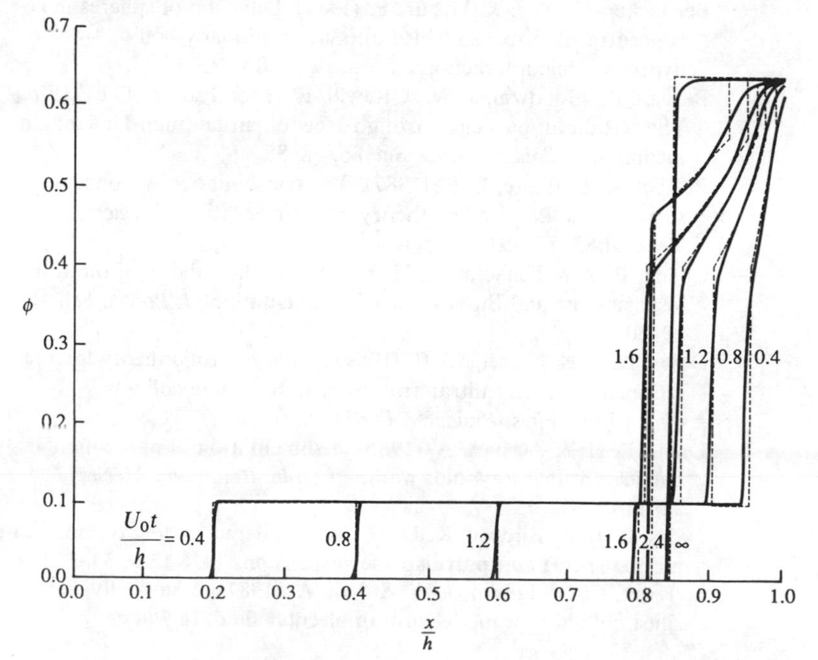

(ii) the behavior of non-spherical particles,
(iii) the settling of flocculated or phase separated systems (Buscall & White, 1987).

Further progress requires descriptions of the structure and the hydrodynamic interactions under more complex conditions than considered here.

References

Auzerais, F. M., Jackson, R. & Russel, W. B. (1988). The resolution of shocks and the effects of compressible sediments in transient settling. *J. Fluid Mech.* **195**, 437–62.

Batchelor, G. K. (1972). Sedimentation in a dilute dispersion of spheres. *J. Fluid Mech.* **52**, 245–68.

Batchelor, G. K. (1976). Brownian diffusion of particles with hydrodynamic interactions. *J. Fluid Mech.* **74**, 1–29.

Batchelor, G. K. (1982). Sedimentation of a dilute polydisperse system of interacting spheres. 1. General theory. *J. Fluid Mech.* **119**, 379–408.

Batchelor, G. K. & Wen, C. S. (1982). Sedimentation of a dilute polydisperse system of interacting spheres. 2. Numerical results. *J. Fluid Mech.* **124**, 495–528.

Beenacker, C. W. J. & Mazur, P. (1983). Diffusion of spheres in concentrated suspension: Resummation of many-body hydrodynamic interactions. *Phys. Lett.* **98A**, 22–4.

Buscall, R., Goodwin, J. W., Ottewill, R. H. & Tadros, T. F. (1982). The settling of particles through Newtonian and non-Newtonian media. *J. Colloid Interface Sci.* **85**, 78–86.

Buscall, R. & White, L. R. (1987). The consolidation of concentrated suspensions. Part 1. The theory of sedimentation. *J. Chem. Soc. Far. Trans.* **I 83**, 873–91.

Cheng, P. Y. & Schachman, H. K. (1955). The validity of the Einstein viscosity law and Stokes law of sedimentation. *J. Polym. Sci.* **16**, 19–30.

Davis, K. E. & Russel, W. B. (1989). An asymptotic theory for the sedimentation and ultrafiltration of hard sphere colloidal dispersions. *Phys. Fluids* **A, 1**, 82–100.

Davis, R. H. & Acrivos, A. (1985). Sedimentation of non-colloidal particles at low Reynolds numbers. *Ann. Rev. Fluid Mech.* **17**, 91–118.

Davis, R. H. & Birdsell, K. H. (1988). Hindered settling of semidilute monodisperse and polydisperse suspensions. *AIChE J.* **34**, 123–9.

de Kruif, C. G., Jansen, J. W. & Vrij, A. (1987). A sterically stabilized silica colloid as a model supramolecular fluid. In *Physics of*

Complex and Supramolecular Fluids (eds S. A. Safran and N. A. Clark), pp. 315–46. Wiley-Interscience.

Glendinning, A. B. & Russel, W. B. (1982). A pairwise additive description of sedimentation and diffusion in concentrated suspensions of hard spheres. *J. Colloid Interface Sci.* **89**, 124–43.

Hanratty, T. J. & Bandukwala, A. (1957). Fluidization and sedimentation of spherical particles. *AIChE J.* **3**, 293–6.

Jansen, J. W., de Kruif, C. G. & Vrij, A. (1986). Attractions in sterically stabilized silica dispersions. III. Second virial coefficients as a function of temperature as measured by means of turbidity. *J. Colloid Interface Sci.* **114**, 492–500. IV. Sedimentation. *J. Colloid Interface Sci.* **114**, 501–4.

Kops-Werkhoven, M. M., Pathmamanoharan, C., Vrij, A. & Fijnaut, H. M. (1982). Concentration dependence of the self-diffusion coefficient of hard spherical particles measured with photon correlation spectroscopy. *J. Chem. Phys.* **77**, 5913–22.

Kynch, G. J. (1952). A theory of sedimentation. *Trans. Far. Soc.* **48**, 166–76.

Mondy, L. A., Graham, A. L. & Jensen, J. L. (1986). Continuum approximations and particle interactions in concentrated suspensions. *J. Rheol.* **30**, 1031–51.

Newman, J., Swinney, H. L., Berkowitz, S. & Day, L. A. (1974). Hydrodynamic properties and molecular weight of fd bacteriofage DNA. *Biochemistry*. **13**, 4832–8.

O'Brien, R. W. (1979). A method for the calculation of the effective transport properties of suspensions of interacting particles. *J. Fluid Mech.* **91**, 17–39.

Pusey, P. N. & van Megen, W. (1987). Properties of concentrated suspensions of slightly soft colloidal spheres. In *Physics of Complex and Supramolecular Fluids* (eds S. A. Safran and N. A. Clark), pp. 673–98. Wiley-Interscience.

Reed, C. C. & Anderson, J. L. (1980). Hindered diffusion of a suspension at low Reynolds number. *AIChE J.* **26**, 816–27.

Rhee, H. K., Aris, R. & Amundson, N. R. (1987). *First Order Partial Differential Equations: Volume I Theory and Applications of Single Equations.* Prentice-Hall.

Woodcock, L. V. (1981). Glass transition in the hard sphere model and Kauzmann's paradox. *Ann. N.Y. Acad. Sci.* **371**, 274–98.

Problems

1 Approximate K_2 for hard spheres by evaluating (12.3.1) with the analytical far-field forms for the hydrodynamic functions.

2 Use the results for the adhesive excluded-shell potential (12.3.4) to approximate K_2 for polystyrene latices with $a = 1.0\,\mu\text{m}$ and $\psi_s = 50\,\text{mV}$ for ionic strengths of 10^{-3}–10^{-1} M.

3 Combine the results (12.4.4), (12.4.5), and (12.4.7) into approximate expressions for the settling velocities in a bimodal mixture concentrated in large spheres but dilute in small ones. Identify the nature of the dominate interactions.

4 Formulate the equations governing the transient settling of a bimodal mixture at infinite Peclet number. Then construct solutions via the method of characteristics for $\lambda=0.1$ and $\gamma=1.0$, with the initial concentrations $\phi=0.05$ and $\phi_2=0.005$. Assume a sediment volume fraction of 0.55 for the larger spheres, independent of the volume fraction of the smaller spheres.

5 Construct the solution for monodisperse spheres with the flux curve given below and $\phi_m=0.64$ settling at infinite Peclet number from an initial dispersion with $\phi_0=0.15$.

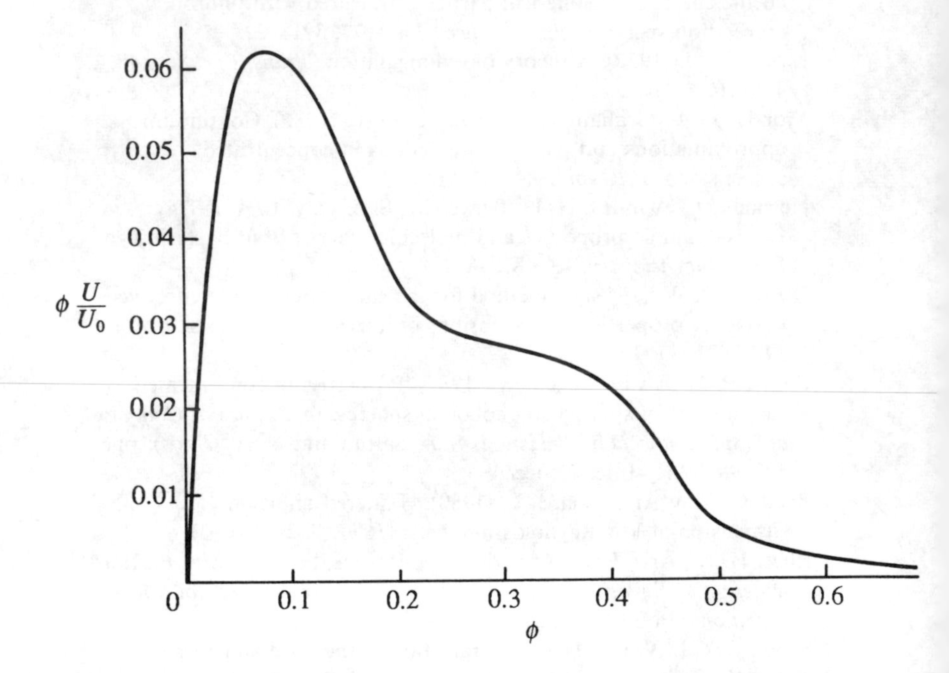

6 Derive the solution for the volume fraction within an equilibrium sediment, i.e. $\phi(x, \infty)$, from (12.5.7). Plot the profiles resulting from $Pe=10$, 10^2, and 10^3 for hard spheres initially at $\phi=0.1$.

13

DIFFUSION

13.1 Introduction

The analysis of the random Brownian motion of isolated spheres in a uniform suspension in §3.3 led to a mean-square displacement increasing linearly with time. The proportionality constant included the Stokes–Einstein diffusion coefficient, $D_0 = kT/6\pi\mu a$. Generalizing this concept to concentrated dispersions as

$$D_s = \frac{1}{6}\frac{\mathrm{d}\langle r^2\rangle}{\mathrm{d}t} \tag{13.1.1}$$

defines the self-diffusion coefficient D_s. At finite volume fractions, D_s differs significantly from D_0 because the random walk of an individual particle is hindered by interactions with neighboring particles. At short times the Brownian motion of a particle is affected only by hydrodynamic interactions with its neighbors. Consequently, the mean-square displacement initially increases linearly with time and (13.11) produces a well-defined short-time self-diffusion coefficient $D_s^0(\phi)$. At longer times, though, the particle must move past the surrounding particles, thereby generating an anisotropic distribution of neighbors and thermodynamic forces which slow its progress. At sufficiently long times a steady-state distribution evolves, so that the mean-square displacement again increases linearly with time and (13.1.1) yields a well-defined long-time self-diffusion coefficient $D_s^\infty(\phi)$. Measurement of the self-diffusion coefficient requires tagging a minority of the particles and detecting their mean-square displacements directly (Fig. 13.1(*a*)), as done by Perrin in 1910 for dilute suspensions (§1.2), or the decay of a spatial gradient in the number density of the tagged particles, e.g. by forced Rayleigh scattering. Whether the short- or long-time limit is observed depends on the spatial resolution of the technique.

In the presence of a gradient in the total number density of particles, the same Brownian motion produces a macroscopic flux of particles, generally termed gradient diffusion. The Langevin approach described in §3.3 can relate gradient diffusion to the microscopic dynamics (e.g. Pusey & Tough, 1985), but easier routes exist. The simplest is that of §12.5, since the term attributed to the thermodynamic force in (12.5.6) represents exactly this flux. Writing the diffusion flux in the conventional form of Fick's law, $-D(\phi)\nabla\phi$, then identifies the gradient diffusion coefficient from (12.5.6) as (Batchelor, 1976)

$$D(\phi)=D_0K(\phi)\frac{\mathrm{d}[\phi Z(\phi)]}{\mathrm{d}\phi}. \tag{13.1.2}$$

This generalized Stokes–Einstein equation, valid for arbitrary concentrations and volume fractions, accounts for hydrodynamic and thermodynamic interactions through the sedimentation coefficient $K(\phi)$ and the compressibility factor $Z(\phi)=\Pi/nkT$, respectively. As illustrated in Chapters 10 and 12, both K and Z depend strongly on volume fraction for concentrated dispersions. This does not invalidate Fick's law, though, since

Fig. 13.1. (*a*) Self-diffusion of a black sphere in uniform environment of white spheres. (*b*) Gradient diffusion of dispersion after removal of confining membrane.

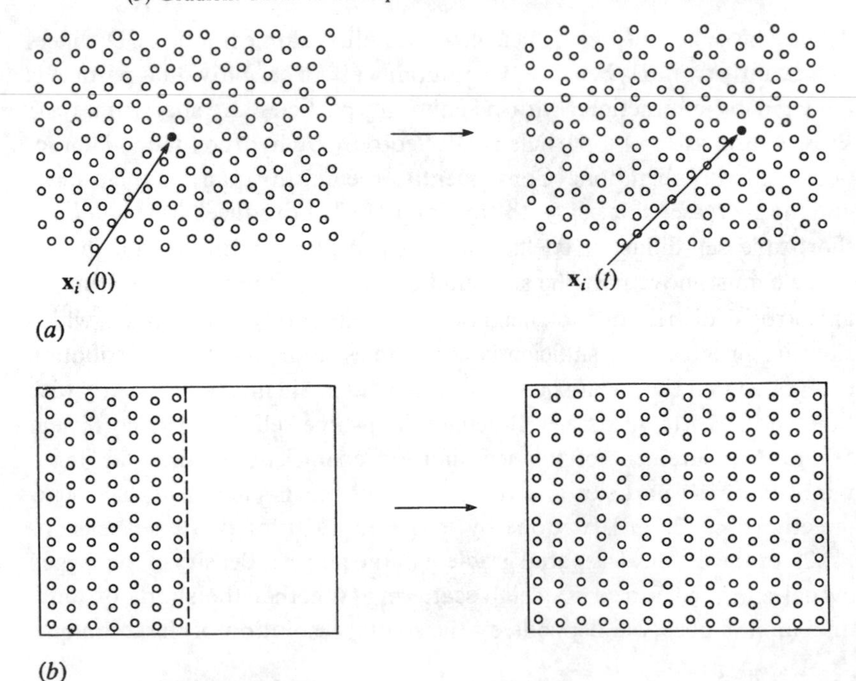

the flux remains linearly proportional to the gradient in the volume fraction.

The form of (13.1.2) permits us to understand how interactions between colloidal particles affect the gradient diffusion coefficient through their separate effects on the sedimentation velocity and the osmotic pressure. At dilute concentrations, repulsive interactions decrease $K(\phi)$ (§12.3), but increase the osmotic compressibility (§10.3). Conversely, attractive interactions increase the sedimentation velocity but decrease the osmotic compressibility. As an illustration of the behavior at higher concentrations, consider polystyrene latices at a low ionic strength separated by a membrane from a reservoir containing deionized water (Fig. 13.1(*b*)). Upon removal of the membrane, the dispersion will quickly expand into the reservoir, driven by the strong electrostatic repulsions responsible for the high osmotic pressure within the dispersion. Since the volume is fixed, this requires an equal but opposite volumetric flow of water which hinders the process. Thus the gradient diffusion coefficient, which characterizes this transient diffusion process, reflects the relative magnitudes of these opposing effects.

Gradient diffusion of colloidal particles to be discussed in §13.2 is important in several contexts similar to that depicted in Chapter 12 for transient settling. Another example is ultrafiltration, used to recover colloidal species such as proteins and to fabricate ceramics (Fig. 13.2). Both of these processes concentrate particles against a boundary at a rate proportional to the external driving force. In batch ultrafiltration, a

Fig. 13.2. Schematic of filtration and sedimentation processes.

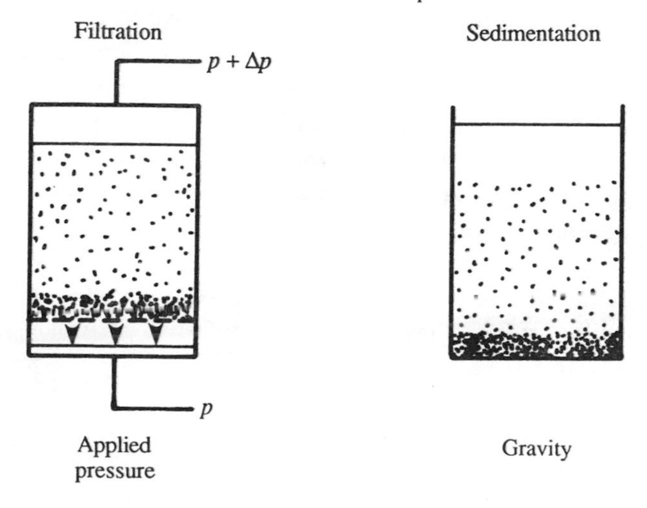

constant pressure difference might be applied, driving fluid through a membrane that rejects the suspended species. As particles accumulate, the resulting increase in the local osmotic pressure reduces the driving force for the fluid flow. Hence the rate decreases with time to a degree that depends on the thickness and structure of the dense layer of particles. Inability to rationalize experimental results for concentrated solutions of globular proteins with models based on $D = D_0$ spurred efforts to quantify the dependence of the diffusion coefficient on volume fraction (e.g. Kozinski & Lightfoot, 1971; Vilker, Colton, & Smith, 1981).

The latter portion of this chapter, §§13.4 to 13.6, deals with a characterization technique, photon correlation spectroscopy, that detects temporal fluctuations in the intensity of light scattered from a dispersion. The wavelength detected depends on the scattering angle and corresponds to a particular Fourier component of the fluctuations in concentration produced by Brownian motion. For long wavelengths, i.e. small scattering angles, these fluctuations decay at a rate governed by the gradient diffusion coefficient. But at larger scattering angles the wavelength can become comparable to the particle size. Ultimately, for wavelengths short relative to the particle radius, the self-diffusion coefficient controls the rate of decay.

At infinite dilution the gradient and self-diffusion coefficients both reduce to D_0. But at finite volume fractions the two differ significantly, because of the different weightings of the hydrodynamic and thermodynamic interactions between the particles. Hence photon correlation spectroscopy, which in principle can detect both, has assumed an important role in the characterization of colloidal dispersions, yielding valuable information about the particle size at dilute concentrations and the interparticle potential and viscous interactions in concentrated systems.

This chapter treats the gradient and self diffusion coefficients within the same context as the diffusion coefficients measured by photon correlation spectroscopy. The statistical mechanical derivation sketched in §13.3 provides the proper basis by generalizing (13.1.2) to apply for concentration gradients with a variable length scale such as those detected in light-scattering experiments.

13.2 Gradient diffusion of monodisperse spheres

For dilute suspensions the gradient diffusion coefficient follows from (13.1.2), with (10.3.3) and (12.3.4), as

$$D(\phi) = \frac{kT}{6\pi\mu a}[1 + (2A_2 + K_2)\phi + O(\phi^2)] \qquad (13.2.1)$$

$$\equiv D_0[1 + D_2\phi + O(\phi^2)].$$

Table 13.1

	Silica[a]	fd bacteriophage DNA[b]
A_2	4.0	4.0 ± 2.0
K_2	-6.6 ± 0.6	-6.7 ± 0.8
D_2	1.3 ± 0.2	1.2 ± 0.4

[a]Kops-Werkhoven & Fijnaut (1981); Kops-Werkhoven *et al.* (1982).
[b]Newman *et al.* (1974).

The expressions in §10.3 for the second virial coefficient A_2 and in §12.3 for K_2 characterizing the $O(\phi)$-correction to the sedimentation velocity illustrate the effect of the interaction potential on $D(\phi)$ as described below.

Hard spheres represent the base case, with $D_2 = 8 - 6.55 = 1.45$ (Batchelor, 1976), indicating that the enhancement due to the thermodynamic excluded volume just outweighs the hydrodynamic retardation, so that the diffusivity increases slightly with volume fraction. Data for the fd bacteriophage DNA in water and silica spheres in cyclohexane, which conform with the hard sphere predictions for A_2 and K_2, give values of D_2 of 1.2 ± 0.4 and 1.3 ± 0.2, respectively (Table 13.1).

Results for A_2 and K_2 for the adhesive excluded shell potential in (10.3.7) and (12.3.4) exemplify the effects of short-range attractions and repulsions. The corresponding values of D_2 vary with the range of the repulsion s_0 and the strength of the attraction $1/\tau$. Alternatively, one can express D_2 as a function of A_2 and s_0 with $1/\tau = 2s_0 - 4A_2/s_0^2$, yielding the family of curves in Fig. 13.3. Since $1/\tau \geq 0$, each curve terminates at $A_2 = s_0^3/2$, and the uppermost curve represents a limiting locus, with $1/\tau = 0$. The plot demonstrates the fact that D_2 is relatively insensitive to the detailed form of the interparticle potential and increases monotonically with the thermodynamic excluded volume. Thus a repulsive (attractive) interaction potential, in addition to the hard sphere, increases (reduces) the driving force through $2A_2$ more than the retardation represented by $-K_2$.

Considerable data exist in the literature on D_2 for a variety of systems including microemulsions, globular proteins, silica spheres, and polystyrene latices. Here we cite only that of Anderson, Rauh, & Morales (1978) (Table 13.2), obtained with the Taylor diffusion column, to show the general validity of the theory and the excluded-shell approximation. Photon correlation spectroscopy measurements will be discussed in later sections.

With polystyrene latices of $a=45.5\,\text{nm}$ in 10^{-2} and 10^{-3} M KCl solutions, the values for D_2, 5 and 19 respectively, significantly exceed those for hard spheres. Approximating electrostatic repulsions via linear superposition (Table 4.3) with a surface potential of 100 mV and the dispersion potential as discussed in §§5.9 and 5.10 leads to the values for s_0 and $1/\tau$ from (10.3.8) in Table 13.2. The corresponding predictions for D_2 from Fig. 13.3 agree reasonably well with the measurements and with values computed from the detailed potentials and the complete expression for K_2 (Anderson, Rauh, & Morales, 1978).

For the globular protein bovine serum albumin with $a=3.6\,\text{nm}$, the electrostatic repulsions enhance the diffusivity considerably more because $a\kappa<1$ at moderate ionic strengths (Table 13.2). Surface potentials can be derived from either the titrated charges or the electrophoretic mobilities; the dispersion potential is unknown but can be neglected at these ionic strengths. Again the observed trend in D_2 conforms to the predictions of the

Fig. 13.3. $D_2=K_2+2A_2$ versus A_2 obtained from the adhesive excluded-shell model (10.3.7 and 12.3.4).

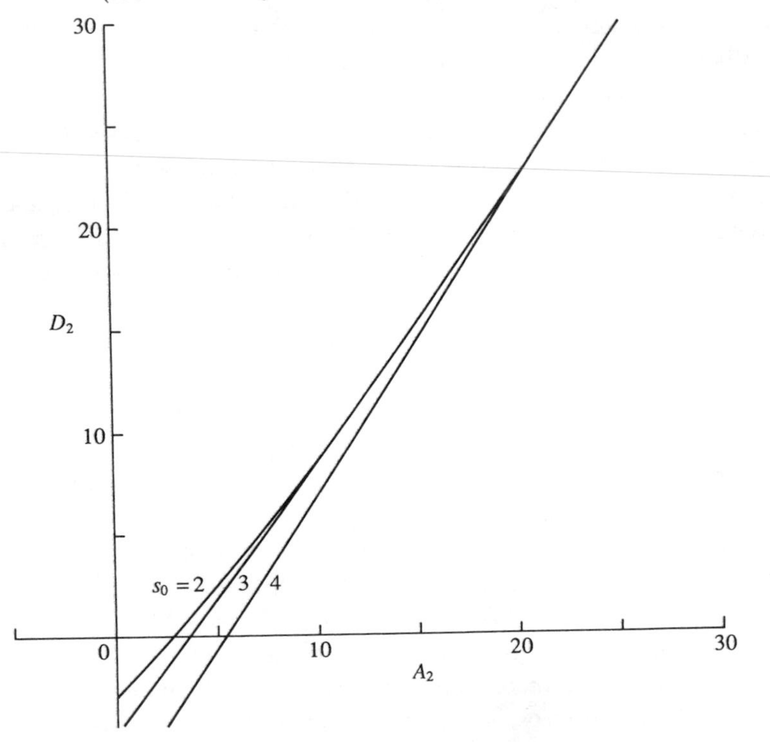

Table 13.2

(a) Polystyrene: $a = 45.5\,\text{nm}$; $\psi_0 = 100\,\text{mV}$ (assumed).

I/M	$a\kappa$	s_0	$1/\tau$	A_2	K_2	D_2 Theory	D_2 Expt.
10^{-2}	14.4	2.45	0.05	7.28	−9.2	5.4	5
10^{-3}	4.6	3.40	0.00	19.7	−17.7	21.7	19

(b) Bovine serum albumin: $a = 3.6\,\text{nm}$; $A_{\text{eff}} = 0$.

I/M	ψ_0/mV	$a\kappa$	s_0	A_2	K_2	D_2 Theory	D_2 Expt.
10^{-1}	15.8	3.6	2.21	5.4	−7.8	3	0
10^{-2}	21.5	1.14	3.04	14.1	−14	4	11
2.5×10^{-3}	21.0	0.57	3.93	30.3	−23	38	8
10^{-3}	22.5	0.36	5.11	66.7	−39	95	72
5×10^{-4}	23.3	0.25	6.31	126	−59	190	152

Data from Anderson, Rauh, & Morales (1978).

adhesive excluded-shell model (Table 13.2) and from the detailed form of the dilute theory (Fig. 13.4).

These results support the validity of the dilute theory and show the substantial effects of the interaction potential on the gradient diffusion coefficient. Although the theories remain strictly valid only for $D_2\phi \ll 1$, the suggestion that long-range repulsions increase the diffusivity well above the value at infinite dilution is borne out experimentally. One example is the data in Fig. 13.5 for bovine serum albumin at 10^{-3} and 0.15 M. At the higher ionic strength the diffusivity varies little at volume fractions up to 0.3, but the lower ionic strength generates a fivefold increase in the diffusivity at low volume fractions, owing to the long-range electrostatic repulsions.

Detailed predictions of the gradient diffusion coefficient at finite concentrations await an accurate theory for the sedimentation coefficient. However, a suggestion of the behavior for hard spheres follows from combining the Carnahan–Starling result (10.5.2) for the osmotic pressure with the semi-empirical expression (12.6.6) for $K(\phi)$ to obtain

$$D(\phi) = D_0(1-\phi)^{2.55}(1 + 4\phi + 4\phi^2 - 4\phi^3 + \phi^4). \qquad (13.2.2)$$

Fig. 13.4. Values of D_2 for bovine serum albumin in aqueous solutions at a range of ionic strengths (Anderson, Rauh, & Morales, 1978): ○, experimental results; ——, theoretical predictions from the detailed theory (10.3.4), (12.3.1), and (13.2.1).

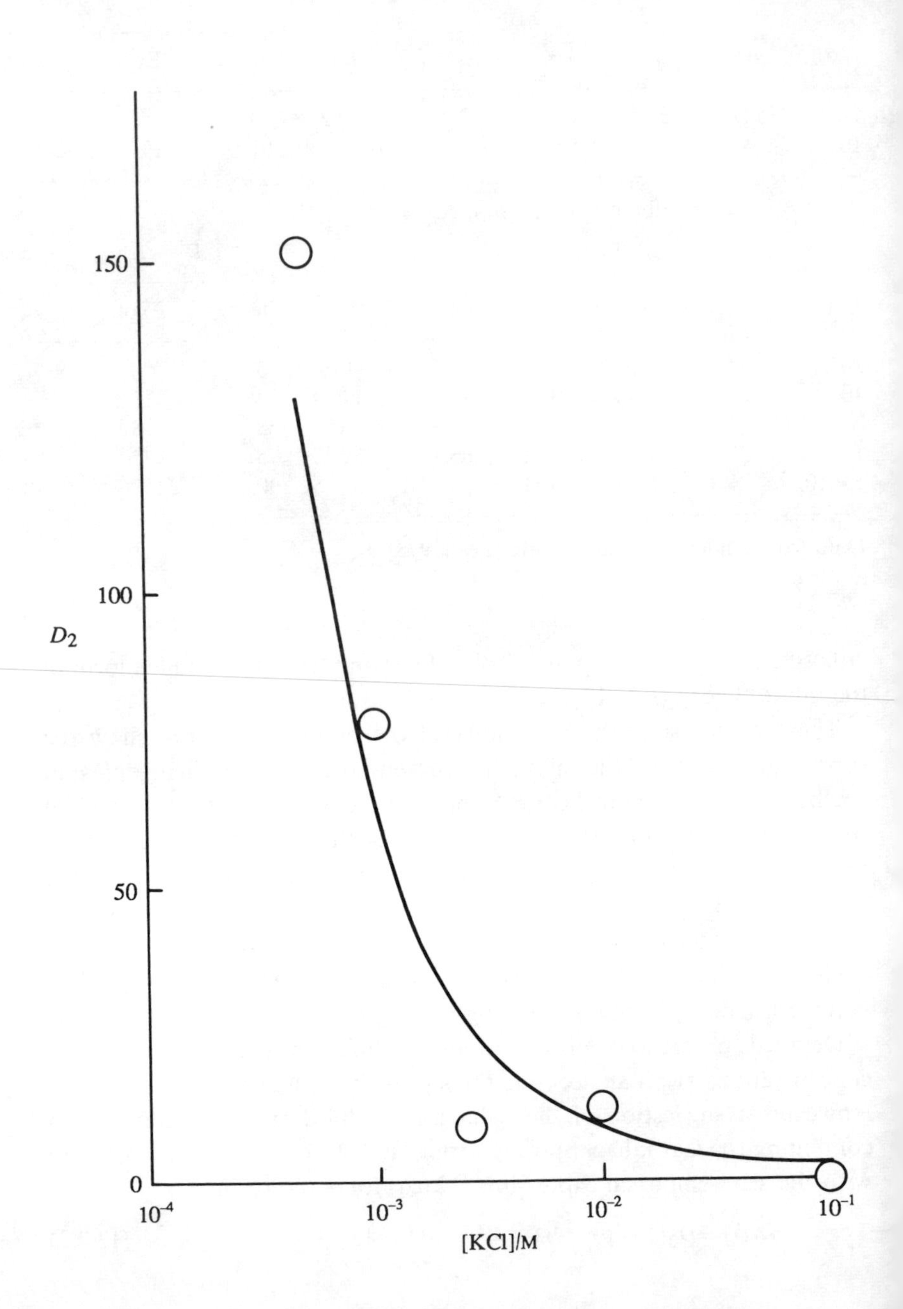

The result is compared in Fig. 13.5 with the measurements at the higher ionic strength where electrostatic repulsion is minimal. The discrepancy between the curve and the data may arise from the failure of (12.6.6) to represent accurately $K(\phi)$ for hard spheres (Fig. 12.3). Nonetheless, one sees clearly that the insensitivity to volume fraction reflects a near cancellation of the thermodynamic and hydrodynamic effects for $0.0 \leqslant \phi \leqslant 0.5$. Since $K \approx 10^{-2}$ and $\mathrm{d}(\phi Z(\phi))/\mathrm{d}\phi \approx 10^2$ at $\phi = 0.5$, and a priori theory must predict the effect of interactions on K within 1 per cent to yield $O(1)$ predictions of $D(\phi)$. Thus the theoretical task is a difficult one.

13.3 Equilibrium in the presence of an external potential

Consider a suspension of N identical spheres in a volume V, subjected to a weak, spatially periodic external potential with wave vector $\mathbf{q}$,

$$\Phi_{\mathrm{ext}} = \Phi(q)\exp(-i\mathbf{q}\cdot\mathbf{x}), \tag{13.3.1}$$

where $q^2 = \mathbf{q}\cdot\mathbf{q}$, $\Phi(q)/kT \ll 1$, and the wavelength is $2\pi/q$. The length scale characteristic of the variations in the radial distribution function (10.2.5) provides an appropriate dimensionless wave number, e.g. aq for hard spheres. The variation in the number density of particles produced by (13.3.1) takes the form

$$n[1 + \hat{n}(q, \phi)\exp(-i\mathbf{q}\cdot\mathbf{x})],$$

Fig. 13.5. Gradient diffusion coefficients as a function of volume fraction for bovine serum albumin in aqueous solutions: ●, 10^{-3} M (Anderson, Rauh, & Morales, 1978); ○, 0.15 M (Phillies, Benedek, & Mazur, 1976); ——, approximate theory for hard spheres (13.2.2).

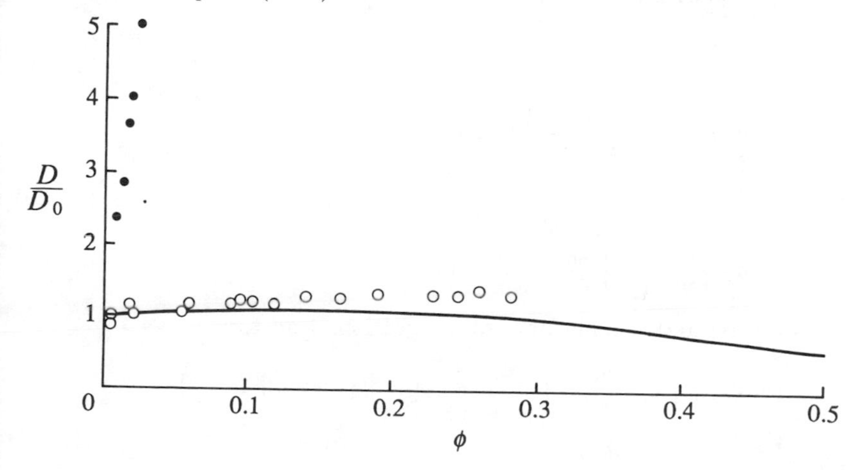

with the variation about the average number density n being small, i.e. $\hat{n}(q, \phi) \ll 1$.

At equilibrium, an equal but opposite diffusion flux must balance the flux generated by the external potential, so that

$$-nD(q, \phi)\nabla[\hat{n}(q, \phi)\exp(-i\mathbf{q}\cdot\mathbf{x})] - \frac{K(q, \phi)}{6\pi\mu a} n\nabla\Phi_{\text{ext}} = 0, \qquad (13.3.2)$$

or, with (13.3.1),

$$D(q, \phi) = -\frac{K(q, \phi)}{6\pi\mu a}\frac{\Phi(q)}{\hat{n}(q, \phi)}. \qquad (13.3.3)$$

Thus, determining the wavenumber-dependent diffusion coefficient requires deriving the generalized sedimentation coefficient, $K(q, \phi)$, and the relationship between the magnitude of the external potential and the variation in the number density.

At equilibrium, the N-particle distribution function conforms to the Boltzmann distribution

$$\begin{aligned} P_N &= P_N^0 \exp\left(-\sum_{i=1}^{N} \frac{\Phi_{\text{ext}}(\mathbf{x}_i)}{kT}\right) \\ &= P_N^0\left(1 - \sum_{i=1}^{N} \frac{\Phi_{\text{ext}}(\mathbf{x}_i)}{kT} + \ldots\right), \end{aligned} \qquad (13.3.4)$$

where P_N^0 is the distribution function in the absence of the potential. The Fourier transform of the variation in the number density is

$$\hat{n}(q, \phi) = \int\left(\frac{1}{N!}\int P_N \, \mathrm{d}\mathbf{x}_2 \ldots \mathrm{d}\mathbf{x}_N - \frac{n}{N}\right)\exp(i\mathbf{q}\cdot\mathbf{x}_1)\,\mathrm{d}\mathbf{x}_1. \qquad (13.3.5)$$

Since

$$n^2 g(r_{12}) = \frac{1}{(N-2)!}\int P_N^0 \, \mathrm{d}\mathbf{x}_3 \ldots \mathrm{d}\mathbf{x}_N$$

and $\qquad (13.3.6)$

$$n = \frac{1}{(N-1)!}\int P_N^0 \, \mathrm{d}\mathbf{x}_2 \ldots \mathrm{d}\mathbf{x}_N,$$

substitution and straightforward integrations lead to

$$\begin{aligned} \hat{n}(q, \phi) &= -\frac{\Phi(q)}{kT}\left(1 + \frac{4\pi n}{q}\int_0^\infty r(g-1)\sin qr \, \mathrm{d}r + \ldots\right) \\ &\equiv -\frac{\Phi(q)}{kT} S(q, \phi) + \ldots. \end{aligned} \qquad (13.3.7)$$

Thus $S(q, \phi)$, known as the static structure factor, characterizes the response of the dispersion to an external potential of wave number q and magnitude kT. Figure 13.6 illustrates the form for hard spheres. Small values of $S(q, \phi)$ indicate that thermal fluctuations produce small variations in number density, i.e. the dispersion is not very compressible. Indeed, in the limit of long wavelengths or small wavenumbers, $aq \ll 1$, the static structure factor reduces to the osmotic compressibility, $S(0, \phi) = \mathrm{d}\phi/\mathrm{d}[\phi Z(\phi)]$. Hence, additional long-range repulsions would reduce $S(0, \phi)$, while attractive potentials would have the opposite effect. In the short-wavelength limit, $aq \gg 1$, correlations between particles vanish and $S(\infty, \phi) = 1$.

Owing to the difficulty in representing many-body hydrodynamic interactions the corresponding hydrodynamic coefficient, $K(q, \phi)$, can be

Fig. 13.6. Static structure factor for dispersion of hard spheres calculated from (13.5.5), with the Percus–Yevick approximation (10.5.1) for the radial distribution function.

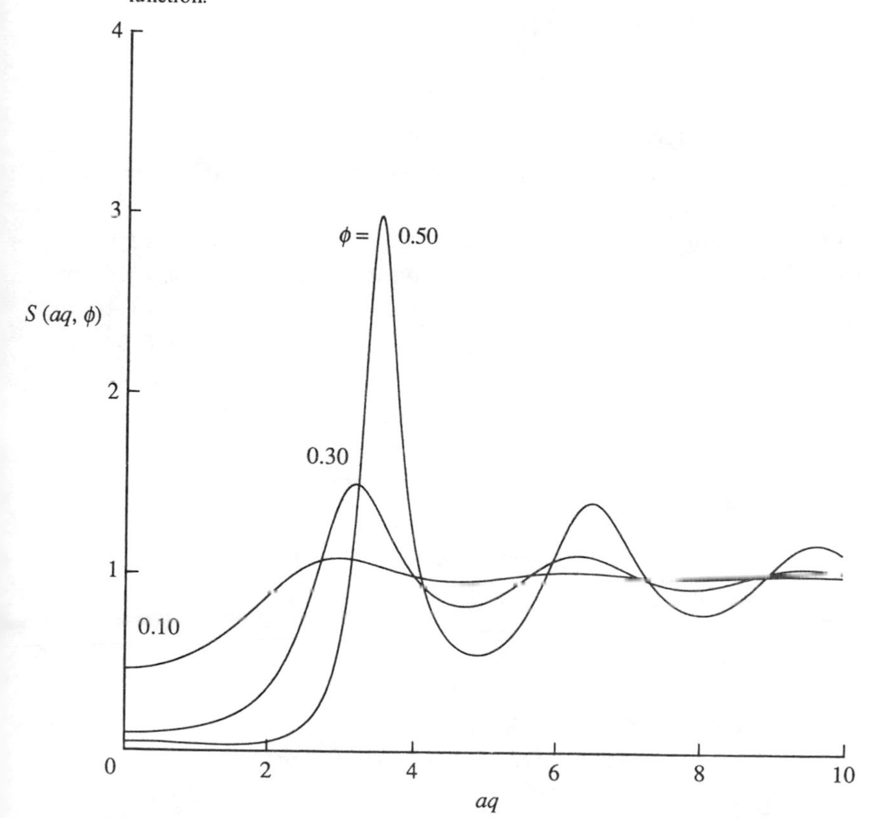

calculated rigorously only in the dilute limit. Generalizing (12.2.12) to account for the spatial variation in the external force on the ith particle,

$$\begin{aligned} \mathbf{F}_i &= -\nabla\Phi_{\text{ext}}(\mathbf{x}_i) \\ &= i\mathbf{q}\Phi(q)\exp(-i\mathbf{q}\cdot\mathbf{x}_i), \end{aligned} \tag{13.3.8}$$

leads to (Russel & Glendinning, 1981)

$$\begin{aligned} K(q,\phi) = 1 + \phi\Bigg(& \frac{15}{4(aq)^2}\left\{\cos 2aq - \frac{\sin 2aq}{2aq}\right\} \\ & + \frac{9}{2aq}\int_2^\infty \left\{\left(1-\frac{2}{s^2}\right)\left(\frac{\cos aqs}{aqs} - \frac{\sin aqs}{(aqs)^2}\right) + \left(1 - \frac{2}{3s^2}\right)\sin aqs\right\}(g-1)\,\mathrm{d}s \\ & + \int_2^\infty (A_{11} + 2B_{11} - 3)gs^2\,\mathrm{d}s \\ & + \frac{3}{(aq)^2}\int_2^\infty \left[2\left\{A_{12} - B_{12} - \frac{3}{4s}\left(1-\frac{2}{s^2}\right)\right\}\left(\cos aqs - \frac{\sin aqs}{aqs}\right)\right. \\ & \left. + \left\{A_{12} - \frac{3}{2s}\left(1 - \frac{2}{3s^2}\right)\right\} aqs \sin aqs\right] g\,\mathrm{d}s\Bigg) + O(\phi^2) \\ \equiv 1 + {} & K_2(q)\phi + O(\phi^2). \end{aligned} \tag{13.3.9}$$

The individual terms correspond to those in (12.3.1), with the near-field hydrodynamic interactions divided between the last two integrals in (13.3.9); these account for the hindrance of force-free neighbors and the convection from neighbors translating owing to the external force, respectively. For hard spheres in the long-wavelength limit, $aq \to 0$, the contributions are -5.0ϕ from the combined far-field effects of the backflow and the pressure gradient, -1.83ϕ from the passive hindrance, and 0.28ϕ from the translation of the neighbors. As aq increases from zero, the bulk motion and pressure gradient, and the velocities of neighboring particles, shift out of phase with respect to the test sphere, eventually becoming completely uncorrelated as $aq \to \infty$. Consequently, all interactions disappear except the passive hindrance.

Now the thermodynamic and hydrodynamic factors in (13.3.3) can be expressed in the form

$$D(q,\phi) = \frac{kT}{6\pi\mu a}\frac{K(q,\phi)}{S(q,\phi)}, \tag{13.3.10}$$

which is valid for arbitrary volume fractions and interparticle potentials. Indeed, the static structure factor encompassing the thermodynamic non-idealities can be evaluated over the full range of volume fractions for hard spheres and other simple potentials (cf. Chapter 10). Rigorous results for $K(q, \phi)$ are limited to the dilute limit noted above, however.

In the next section we turn our attention to photon correlation spectroscopy, showing that the technique measures the wavenumber-dependent diffusion coefficients (13.3.10) and discussing how the magnitude and the wavenumber dependence reflects the nature of the interparticle potential.

13.4 Principles of photon correlation spectroscopy

The results of §13.2 establish that the volume fraction dependence of the gradient diffusion coefficient arises from the combination of thermodynamic excluded volume, far-field effects of backflow and pressure gradients, and near-field hydrodynamic interactions. The arguments in §13.1 suggest that the corresponding variations in the short-time self-diffusion coefficient stem only from near-field hydrodynamic interactions. Hence, a technique capable of measuring both diffusion coefficients over a range of concentrations provides a powerful probe into the detailed dynamics of colloidal systems. In this and the following two sections we strive to show that photon correlation spectroscopy has this capability.

As noted in §3.4, temporal fluctuations in the intensity of light scattered from a dispersion derive from the Brownian motion of the particles. Experiments typically yield autocorrelations of the amplitude of the electric field. The principal difficulty lies in relating the decay of the autocorrelation function to the diffusion processes within the dispersion. For monodisperse spheres at infinite dilution, the decay is exponential and uniquely determines the diffusion coefficient D_0 through (3.4.6). However, at finite volume fractions, interactions between particles cause a variety of complications:

(i) a complex dependence of the initial decay rate on the wavenumber of the scattered light;
(ii) sensitivity of the initial decay rate to the volume fraction and the interparticle potential;
(iii) non-exponential decay of the autocorrelation function at longer times.

Our physical description and mathematical treatment below follow those of Pusey & Tough (1985) and Rallison & Hinch (1986), but address only the initial decay rate, which provides a sensitive and unambiguous measure of the dynamics of concentrated dispersions. These papers illustrate the complex behavior at longer times and thoroughly review earlier work on the subject (e.g. Pusey, 1975; Ackerson, 1976, 1978).

One must recognize at the outset that several different time scales affect the dynamics. As noted in §3.4, the diffusive processes of interest occur at times long with respect to the inertial time scale,

$$t_{\mathrm{i}} = \frac{m}{6\pi\mu a},$$

with m the mass of the sphere. The relevant time scale for diffusion, however, depends on the distance which the particles must travel. The light-scattering experiment detects fluctuations in number density with wavelengths equal to that of the scattered light, $2\pi/q$, with q the wavenumber. Hence the diffusion time characterizing the exponential decay of the autocorrelation function varies with wavenumber as

$$t_q = \frac{1}{q^2 D_0}.$$

In addition, there exists a microscopic diffusion time,

$$t_{\mathrm{a}} = \frac{a^2}{D_0},$$

associated with relaxation of the microstructure on the scale of the particle.

As stated in §3.4, the amplitude at time t of the electric field scattered at an angle θ relative to the incident beam is proportional to

$$E(q,t) = \sum_{j=1}^{N} I_j \exp(i\mathbf{q}\cdot\mathbf{x}_j(t)). \tag{13.4.1}$$

Here $\mathbf{x}_j(t)$ and I_j are the positions and scattering amplitudes of the individual particles and the wave vector $\mathbf{q}$ (Fig. 13.7) has magnitude $q = (4\pi/\lambda)\sin(\theta/2)$, with λ the wavelength of light in the medium. The autocorrelation function of the electric field amplitudes with time delay t is

$$F(q,t) = E(q,t)E^*(q,0)$$

$$= \frac{\sum_{k,j=1}^{N} I_k I_j \exp[i\mathbf{q}\cdot(\mathbf{x}_k(t) - \mathbf{x}_j(0))]}{\sum_{k,j=1}^{N} I_k I_j}. \tag{13.4.2}$$

The sum of exponentials accounts for the differences in the phases between the light scattering by the individual particles. Though the autocorrelation of the intensity of the scattered light is actually measured, the signal can be processed to extract $E(q,0)E^*(q,t)$ unambiguously, provided $N \gg 1$.

Now we must distinguish two different light-scattering experiments, one detecting the self-diffusion coefficient (13.1.1) and the other measuring the wavenumber-dependent diffusion coefficient defined by (13.3.10). The first experiment capitalizes on optical polydispersity, either naturally present or introduced intentionally. Suppose, for example, that all but a few ($N_1 \ll N$) of the particles have refractive indices identical to that of the fluid, so that their scattering amplitudes vanish. The remaining scatterers serve as identical, but non-interacting tracers. Setting $I_j = 0$ for $N_1 + 1 \le j \le N$ and $I_j = I_0$ for $1 \le j \le N_1$ in (13.4.2), and recognizing that the positions are uncorrelated with one another, produces the *self-intermediate scattering function*

$$F_s(q,t) = \frac{1}{N_1} \sum_{j=1}^{N_1} \exp[i\mathbf{q}\cdot(\mathbf{x}_j(t) - \mathbf{x}_j(0))]$$
$$= \langle \exp[i\mathbf{q}\cdot(\mathbf{x}_1(t) - \mathbf{x}_1(0))] \rangle. \tag{13.4.3}$$

The brackets indicate ensemble averaging over both the initial and current configurations, i.e.

$$\langle\ \rangle = \frac{1}{N!} \int\!\!\int (\ ) P_N(t) \mathrm{d}\mathbf{x}_1(t) \ldots \mathrm{d}\mathbf{x}_N(t) \mathrm{d}\mathbf{x}_1(0) \ldots \mathrm{d}\mathbf{x}_N(0). \tag{13.4.4}$$

Comparison with (3.4.3) and (3.4.6) suggests defining the self-diffusion coefficient by

$$D_s = -\frac{1}{q^2} \frac{\mathrm{d} \ln F_s}{\mathrm{d}t}. \tag{13.4.5}$$

Fig. 13.7. Schematic of light-scattering experiment.

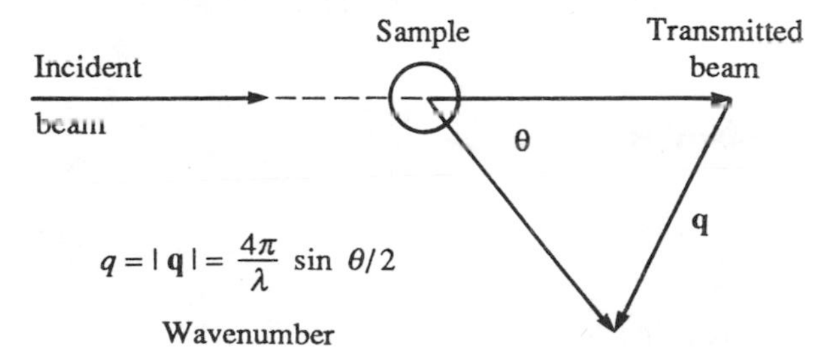

This experiment can be performed by matching the refractive indices of the particles and fluid as closely as possible, and then either adding a minority of a second species of similar size or depending on non-uniformities to provide a small population of scatterers (Vrij *et al.*, 1983).

The second experiment pertains to monodisperse particles with identical scattering amplitudes such that $I_k = I_0$ for $1 \leq k \leq N$, leading to the *intermediate scattering function* $F(q,t)$ with

$$\begin{aligned} F(q,t) &= \frac{1}{N} \sum_{k,j=1}^{N} \exp[i\mathbf{q} \cdot (\mathbf{x}_k(t) - \mathbf{x}_j(0))] \\ &= F_s(q,t) + (N-1)\langle \exp[i\mathbf{q} \cdot (\mathbf{x}_2(t) - \mathbf{x}_1(0))] \rangle. \end{aligned} \tag{13.4.6}$$

An expression analogous to (13.4.5) serves to define the corresponding diffusion coefficient. This is the experiment normally performed with colloidal systems. However, F_s still can be detected at long times if the second term, arising from correlations, decays faster.

For dilute systems, correlations between particles are negligible, so the second term in (13.4.6) vanishes and $F = F_s$. In this limit, $D_0 = -(\mathrm{d} \ln F/\mathrm{d}t)/q^2$, independent of time and wavenumber (e.g. §3.4). But interactions at finite concentrations alter both F and F_s, such that $(\mathrm{d} \ln F/\mathrm{d}t)/q^2$ and $(\mathrm{d} \ln F_s/\mathrm{d}t)/q^2$ often vary with wavenumber and time. The following sections define situations in which measurements of F and F_s yield meaningful wavenumber-dependent diffusion coefficients.

13.5 Initial decay of the autocorrelation functions

Deducing the dependence of the autocorrelation functions on the delay time t requires addressing the microscopic dynamics of the dispersion. This section proceeds only as far as the initial decay, thereby relating (i) $F(q,t)$ to the wavenumber-dependent diffusivity $D(q,\phi)$, which converges to $D(\phi)$ for $aq \to 0$ and to $D_s^0(\phi)$ for $aq \to \infty$, and (ii) $F_s(q,t)$ to $D_s^0(\phi)$ independent of aq. The more complicated question of the long-time behavior is discussed elsewhere (e.g. Rallison & Hinch, 1986).

The decay of the fluctuations is derived from the N-particle conservation equation (8.3.4)

$$\frac{\partial P_N}{\partial t} = -\sum_{i=1}^{N} \nabla_i \cdot \mathbf{j}_i,$$

with

$$\mathbf{j}_i = -\sum_{j=1}^{N} \boldsymbol{\omega}_{ij} \cdot (kT\nabla_j P_N + P_N \nabla_j \Phi) \tag{13.5.1}$$

and

$$P_N = P_N^0 \prod_{i=1}^{N} \delta(\mathbf{x}_i - \mathbf{x}_i(0)) \qquad \text{at } t=0$$

(Pusey & Tough, 1985; Rallison & Hinch, 1986). The initial condition simply specifies a discrete distribution of particles weighted by the equilibrium probability P_N^0. Multiplying by $\exp(-i\mathbf{q}\cdot\mathbf{x}_i(0))$ and integrating over $d\mathbf{x}_1(0) \ldots d\mathbf{x}_N(0)$ produces the transformed distribution function

$$\hat{P}_N = \int P_N \exp(-i\mathbf{q}\cdot\mathbf{x}_1(0))\, d\mathbf{x}_1(0) \ldots d\mathbf{x}_N(0), \tag{13.5.2}$$

which satisfies the same conservation equation but with

$$\hat{P}_N = P_N^0 \exp(-i\mathbf{q}\cdot\mathbf{x}_1) \qquad \text{at } t=0. \tag{13.5.3}$$

Substitution into the expressions for the scattering functions gives

$$F_s(q,t) = \frac{1}{N!}\int \hat{P}_N \exp(i\mathbf{q}\cdot\mathbf{x}_1)\, d\mathbf{x}_1 \ldots d\mathbf{x}_N$$

and (13.5.4)

$$F(q,t) = F_s(q,t) + \frac{(N-1)}{N!}\int \hat{P}_N \exp(i\mathbf{q}\cdot\mathbf{x}_2)\, d\mathbf{x}_1 \ldots d\mathbf{x}_N.$$

Now the initial values for the scattering functions and their derivatives can be found without solving the differential equation explicitly. The initial condition (13.5.3) immediately yields

$$F_s(q,0) = 1$$

$$F(q,0) = 1 + \frac{4\pi n}{q}\int_0^\infty r(g-1)\sin qr\, dr \tag{13.5.5}$$

$$= S(q,\phi).$$

The appearance of the static structure factor indicates that $F(q,0)$ measures the Fourier components of the fluctuating number density caused by thermal motion.

The initial slopes are obtained by differentiating (13.5.4) with respect to time, substituting (13.5.1) for $\partial\hat{P}_N/\partial t$, and integrating by parts. The results are

$$\frac{dF}{dt}(q,0) = -q^2kT\frac{1}{N!}\int \frac{\mathbf{qq}}{q^2} : [\boldsymbol{\omega}_{11}$$

$$+ (N-1)\boldsymbol{\omega}_{12}\exp\{i\mathbf{q}\cdot(\mathbf{x}_2 - \mathbf{x}_1)\}]P_N^0\, d\mathbf{x}_1 \ldots d\mathbf{x}_N$$

and (13.5.6)

$$\frac{dF_s}{dt}(q,0) = -q^2kT\frac{1}{N!}\int\frac{\mathbf{qq}}{q^2}:\boldsymbol{\omega}_{11}P_N^0\,d\mathbf{x}_1\,\ldots\,d\mathbf{x}_N.$$

The first is actually not suitable for calculations in its present form, since the long range of the hydrodynamic interactions embodied in $\boldsymbol{\omega}_{12}$ render the integral non-convergent. However, comparison with the expressions leading to $K(q,\phi)$ in §13.3 establishes that when renormalized and combined with (13.5.5),

$$\frac{dF}{dt}(q,0) = -q^2D_0K(q,\phi) \tag{13.5.7}$$

$$= -q^2D_0\frac{K(q,\phi)}{S(q,\phi)}F(q,0).$$

Noting (13.3.10) then gives

$$D(q,\phi) = -\frac{1}{q^2}\frac{d\ln F}{dt}(q,0). \tag{13.5.8}$$

Thus $D(q,\phi)$ characterizes the initial decay of concentration fluctuations with wavenumber q caused by the thermal motion.

Finally, recall from the earlier discussion that $S(\infty,\phi)=1$ and only the effect of the passive resistance of force-free neighboring spheres affects $K(\infty,\phi)$. Thus (13.5.6) and (13.5.7) together establish

$$D_s^0(\phi) = -\frac{1}{q^2}\frac{d\ln F_s}{dt}(q,0)$$

$$= \lim_{aq\to\infty} -\frac{1}{q^2}\frac{d\ln F}{dt}(q,0). \tag{13.5.9}$$

This short-time self-diffusion coefficient governs the initial decay of both short-wavelength fluctuations in a monodisperse system and fluctuations of an optical tracer at all wavenumbers.

During the decay of concentration fluctuations with finite wavenumbers, unequal driving forces, analogous to the external forces (13.3.8), act on neighboring particles. Hence the particles move at different velocities, thereby perturbing the microstructure from equilibrium and engendering a restoring force which retards their motion. The fact, that for $aq=O(1)$ the relaxation time for the microstructure, a^2D_0, compares with the diffusion time characterizing the decay of the fluctuation in number density, $1/q^2D_0$, makes the time dependence quite subtle. Indeed, the decay is exponential only at the short times treated above, and, for the self intermediate scattering function, at long times and small wavenumbers for which

$$D_s^\infty(\phi) = -\frac{1}{q^2}\frac{d\ln F_s}{dt} \quad \text{for} \quad t_a \ll t \ll t_q \quad \text{if} \quad aq \ll 1. \tag{13.5.10}$$

The difference between the two limits of the self diffusion coefficient is small for hard spheres, but should become substantial for strong long-range interaction potentials (e.g. Batchelor, 1983; Rallison & Hinch, 1986).

In summary, photon correlation spectroscopy produces either the intermediate scattering function, F, or the self-intermediate scattering function, F_s, depending on the mode of the experiment. Both decay with increasing delay time at a rate that depends on the wavenumber of the scattered light. Interpretation in terms of conventional diffusion processes is possible only for the initial decay at short times, $t \ll t_a$, for F and at short, $t \ll t_a$, or long times, $t_a \ll t \ll t_q$, for F_s. The diffusion coefficients corresponding to these situations depend on the volume fraction owing to both hydrodynamic and thermodynamic interactions.

13.6 Wavenumber-dependent diffusion coefficient

This section includes theoretical and experimental results for the wavenumber-dependent diffusion coefficient, $D(q, \phi)$, derived from the initial decay of F, and to a lesser extent, the short- and long-time self-diffusion coefficients, $D_s^0(\phi)$ and $D_s^\infty(\phi)$, derived from F_s. At dilute volume fractions, both $K(q, \phi)$ and $S(q, \phi)$ can be expanded in powers of the volume fraction, so that (13.3.10) yields

$$D(q, \phi) = D_0[1 + D_2(q)\phi + \ldots], \tag{13.6.1}$$

with $D_2(q) = 2A_2(q) + K_2(q)$ and $A_2(q)$ representing the corresponding coefficient in the expansion of the inverse of the static structure factor. Similar expansions for the self-diffusion coefficients would have $O(\phi)$-coefficients of D_{2s}^0 and D_{2s}^∞. As for previous topics, we begin with hard spheres that comprise a useful and realizable base case for assessing the effects of longer-range repulsions and short-range attractions.

The values of D_2 for hard spheres (Fig. 13.8), obtained by evaluating (13.3.9) and (13.5.5) with $g(r) = H(r - 2a)$, approach the limits of 1.45 for gradient diffusion when $aq < 0.2$ and of -1.83 for short-time self-diffusion when $aq > 4$. The latter simply represents the hydrodynamic resistance of force-free neighbors. The transition arises from the disappearance of the contributions to K_2 from the backflow and pressure gradient, -5.0ϕ, and to A_2 from the thermodynamic excluded volume, $+8.0\phi$, when the forces

(13.3.8) acting on interacting particles shift out of phase for $aq \geqslant O(1)$. The oscillations stem from the periodicity of these forces, i.e. (13.3.8), coupled with the different dependence on separation of the hydrodynamic and thermodynamic interactions.

Experiments with the organophilic silica spheres ($a=33\pm1$ nm), summarized by de Kruif, Jansen, & Vrij (1987) determined $D_2(0)$ and D_{2s}^{∞} from measurements of $F(q,t)$ and $F_s(q,t)$ for $aq \ll 1$. The data (Fig. 13.9) conform reasonably well with the theoretical lines for $D_2(0)=1.45$ and $D_{2s}^{\infty}=-2.1$ (Batchelor, 1983).

Measurements of the wavenumber dependence for $aq \geq 1$ require larger particles, which generally scatter more strongly, introducing the possibility of multiple scattering. Van Megen *et al.* (1985) obtained the requisite refractive index matching with poly(methyl methacrylate) spheres ($a=190\pm9$ nm) stabilized by poly(12-hydroxystearic acid) in carbon disulphide–hexane mixtures. The volume fractions studied lie outside the dilute regime. Nonetheless, we can compare the results for $D_0/D(q,\phi)$ with the dilute theory expressed as

$$\frac{D_0}{D(q,\phi)}=\frac{S(q,\phi)}{K(q,\phi)}$$
$$=\frac{1}{1+(2A_2(q)+K_2(q))\phi} \tag{13.6.2}$$

Fig. 13.8. Variation of D_2 with dimensionless wavenumber (Russel & Glendinning, 1981; Fijnaut, 1981): ——, hard spheres; - - -, excluded-shell potential with axes rescaled as D_2/s_0^3 and aqs_0.

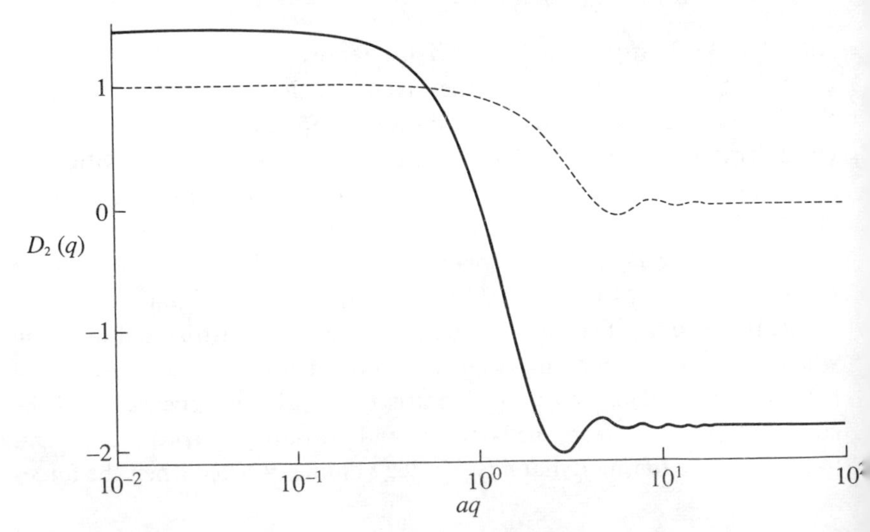

Fig. 13.9. Comparison of data from photon correlation spectroscopy measurements for dispersions of silica spheres with $a = 33 \pm 1$ nm in cyclohexane (de Kruif, Jansen, & Vrij, 1987) with predictions of dilute theories. (*a*) Gradient diffusion coefficient: ○, data; ——, prediction of $D_2(0) = 1.45$ for hard spheres (Batchelor, 1976). (*b*) Long-time self-diffusion: ○, data; ——, prediction of $D_{2s}^{\infty} = -2.11$ for hard spheres (Batchelor, 1983).

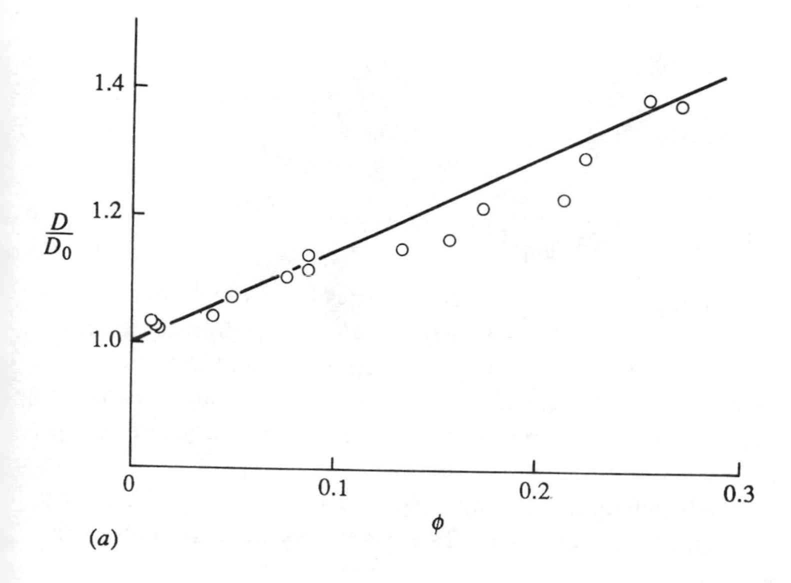

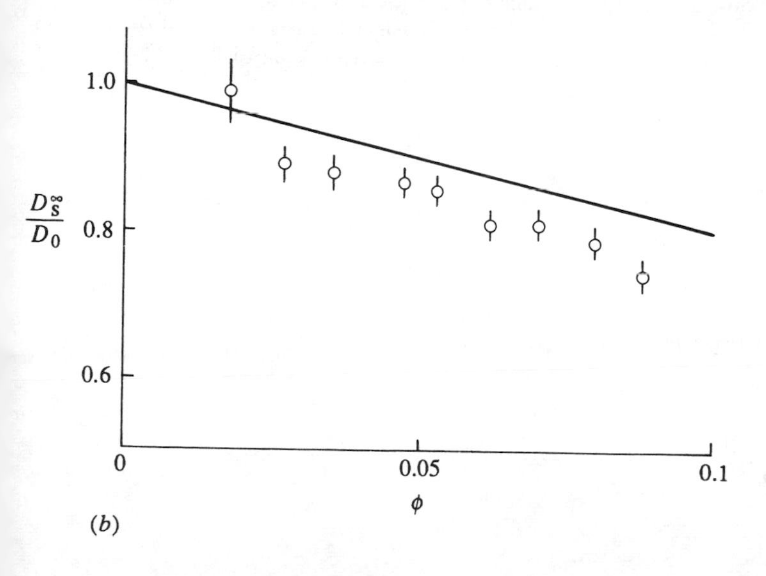

(Fig. 13.10). The oscillatory forms of the theory and the data are similar, but differ in detail owing to the approximate nature of (13.6.2).

To probe the effect of longer range attractive and repulsive potentials, we again turn to the adhesive excluded-shell model (§10.6). For adhesive hard spheres, i.e. $s_0=2$ but $1/\tau \neq 0$, the additional contributions to the static structure factor and the sedimentation coefficient from (13.3.9) and (13.5.5) expressed in the form of (12.3.4) are

$$\begin{aligned} A_a(q) &= -\frac{\sin 2aq}{2aq} \\ K_a(q) &= -0.295 + 0.286\frac{\cos 2aq}{(aq)^2} \\ &\quad -0.143\frac{\sin 2aq}{(aq)^3}[1-5.40(aq)^2]. \end{aligned} \tag{13.6.3}$$

The corresponding contribution to $D_2(q)$ varies in an oscillatory manner between $-1.13/\tau$, for $aq \ll 1$ to $-0.295/\tau$ for $aq \gg 1$. For gradient diffusion, the attraction reduces the osmotic compressibility more than it increases the sedimentation coefficient, as noted in §13.2. For self-diffusion, the production of doublets simply increases the hydrodynamic resistance. Hence diffusion is retarded in both limits.

The excluded-shell potential, with $s_0 \gg 2$ and $1/\tau = 0$, models long-range repulsions such as electrostatic interactions at low ionic strengths. This

Fig. 13.10. Wavenumber-dependent diffusion coefficient for poly(methylmethacrylate) spheres ($a = 190 \pm 9$ nm) in a carbon disulphide-hexane mixture at $\phi = 0.26$ (van Megen *et al.*, 1985): ●, $\lambda = 476.2$ nm; ×, $\lambda = 520.8$ nm; ○, $\lambda = 568.2$ nm; – – –, prediction from (13.6.2).

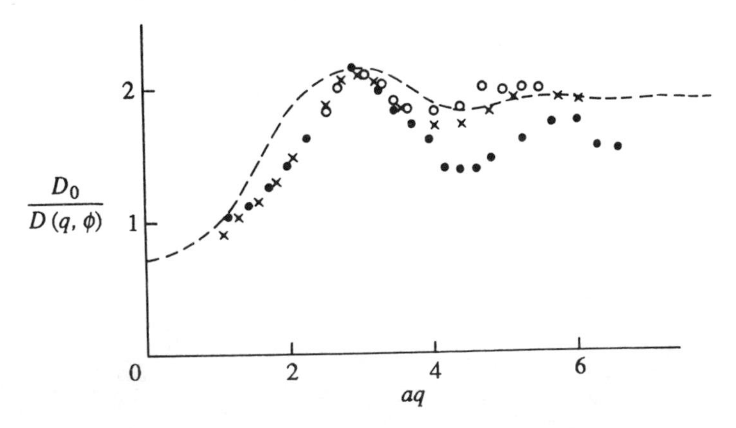

renders near-field hydrodynamic interactions insignificant, leaving (Altenberger, 1979)

$$D_2(q)=\frac{3s_0}{(aq)^2}\left(1-\frac{3}{2s_0}\right)\left(\frac{\sin aqs_0}{aqs_0}-\cos aqs_0\right), \tag{13.6.4}$$

which (broken line in Fig. 13.8) varies with wavenumber much as that for hard spheres. Note, however, that the magnitude for $aqs_0 \ll 1$, $D_2(0) \approx s_0^3$, substantially exceeds that for hard spheres, but for $aqs_0 \gg 1$ $D_2 \to 0$, since all interactions become negligible.

The wavenumber-dependent diffusion coefficients were actually first observed in experiments with polystyrene latices at low ionic strengths (e.g. Schaefer & Berne, 1974). The data of Brown *et al.* (1975) for deionized polystyrene latices with $a = 23.1$ nm at $0.75 \times 10^{-4} < \phi < 5.08 \times 10^{-4}$ in Fig. 13.11 exemplify the structure in $D(q, \phi)$. The solid curves demonstrate the validity of the approximation $D_0/D(q, \phi) \approx S(q, \phi)$ (Pusey, 1975) for conditions in which hydrodynamic effects are minimal, i.e. $K(q, \phi) \approx 1$.

Several points can be drawn from these results. First, the short-time self-diffusion coefficient, $D_s^0(\phi) = D(\infty, \phi)$ measures near-field hydrodynamic

Fig. 13.11. Inverse of wavenumber-dependent diffusion coefficients (●) compared with static structure factors (——) for deionized polystyrene latices ($a = 23.1$ nm) in water (Brown *et al.*, 1975): (i), $\phi = 6.4 \times 10^{-5}$; (ii), $\phi = 8.6 \times 10^{-5}$; (iii), $\phi = 1.5 \times 10^{-4}$; (iv), $\phi = 2.9 \times 10^{-4}$; (v), $\phi = 4.4 \times 10^{-4}$.

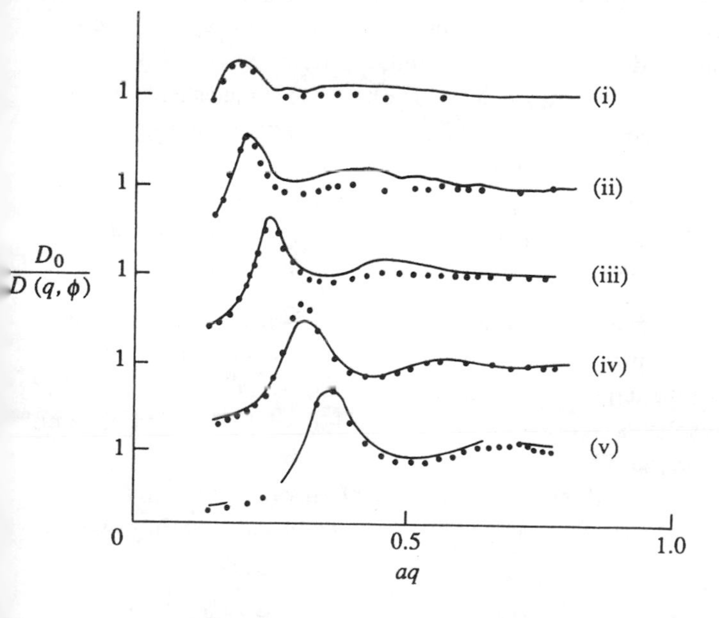

interactions only and decreases monotonically with increasing volume fractions. The interparticle potential affects the value only through the radial distribution function. Second, thermodynamic and far-field hydrodynamic interactions become important with decreasing wavenumber. The value of $D(q, \phi)$ then varies periodically with wavenumber with the period determined by the range of the repulsive potential. The magnitude is increased by long-range repulsions and decreased by short-range attractions, as for $D(\phi)=D(0, \phi)$. Thus, measurements by photon correlation spectroscopy of $D(q, \phi)$ over a suitable range of wavelengths probe in considerable detail both the hydrodynamic and the thermodynamic interactions.

13.7 Summary

In this chapter we have presented an expression for a diffusion coefficient, dependent on the wavenumber of the concentration gradient, valid for arbitrary volume fractions and interparticle potentials, provided the microstructure remains at equilibrium. Hydrodynamic interactions enter through a generalized sedimentation coefficient and thermodynamic interactions through the static structure factor. In the limit of small wavenumbers, the expression reduces to the familiar gradient diffusion coefficient. In the opposite limit, it conforms to the short-time self-diffusion coefficient. An analysis of the autocorrelation functions obtained from photon correlation spectroscopy establishes that the wavenumber-dependent diffusion coefficient governs the initial decay of intensity fluctuations detected at the corresponding scattering angle.

Explicit results for the hard sphere and adhesive excluded shell potentials in dilute systems demonstrate the effects of thermodynamic interactions. Experimental data for silica dispersions and aqueous and non-aqueous latices generally support the dilute theory and illustrate the behavior at higher volume fractions. These interactions have several consequences for the diffusion coefficient:

(i) The dependence on volume fraction and, hence, the magnitude at high concentrations depends on the mode of diffusion.

(ii) For gradient diffusion the weak dependence on volume fraction for hard spheres represents a cancellation of hydrodynamic and thermodynamic effects.

(iii) An increase with volume fraction generally indicates a long-range repulsion and a decrease, an attractive component to the potential.

The full effect of potentials more complex than the hard sphere on diffusion in concentrated dispersions has yet to be defined, but photon correlation spectroscopy clearly provides a powerful means of probing these interactions and has already provided insight into diverse colloidal systems.

References

Ackerson, B. J. (1976). Correlations for interacting Brownian particles. *J. Chem. Phys.* **64**, 243–6.

Ackerson, B. J. (1978). Correlations for interacting Brownian particles II. *J. Chem. Phys.* **69**, 684–90.

Altenberger, A. R. (1979). On the wave vector dependent mutual diffusion of interacting Brownian particles. *J. Chem. Phys.* **70**, 1994–2002.

Anderson, J. L., Rauh, F. & Morales, A. (1978). Particle diffusion as a function of concentration and ionic strength. *J. Phys. Chem.* **82**, 608–16.

Batchelor, G. K. (1976). Brownian diffusion of particles with hydrodynamic interactions. *J. Fluid Mech.* **74**, 1–29.

Batchelor, G. K. (1983). Diffusion in a dilute polydisperse system of interacting spheres. *J. Fluid Mech.* **131**, 155–75.

Brown, J. C., Pusey, P. N., Goodwin, J. W. & Ottewill, R. H. (1975). Light scattering study of dynamic and time averaged correlations in dispersions of charged particles. *J. Phys. A: Math. Gen.* **8**, 664–82.

de Kruif, C. G., Jansen, J. W. & Vrij, A. (1987). A sterically stabilized silica colloid as a model supramolecular fluid. In *Physics of Complex and Supramolecular Fluids* (ed. S. A. Safran and N. A. Clark), pp. 315–46. Wiley Interscience.

Fijnaut, H. M. (1981). Wave vector dependence of the effective diffusion coefficient of Brownian particles. *J. Chem. Phys.* **74**, 6857–63.

Kops-Werkhoven, M. M. & Fijnaut, H. M. (1981). Dynamic light scattering and sedimentation experiments on silica dispersions at finite concentrations. *J. Chem. Phys.* **74**, 1618–25.

Kops-Werkhoven, M. M., Pathmamanoharan, C., Vrij, A. & Fijnaut, H. M. (1982). Concentration dependence of the self-diffusion coefficient of hard spherical particles measured with photon correlation spectroscopy. *J. Chem. Phys.* **77**, 5913–22.

Kozinski, A. A. & Lightfoot, E. N. (1971). Ultrafiltration of proteins in stagnation flow. *AIChE J.* **17**, 81–5.

Newman, J., Swinney, H. L., Berkowitz, S. & Day, L. A. (1974). Hydrodynamic properties and molecular weight of fd bacteriofage DNA. *Biochemistry* **13**, 4832–8.

Phillies, G. D. J., Benedek, G. B. & Mazur, N. A. (1976). Diffusion in protein solutions at high concentration: a study by quasielastic light scattering spectroscopy. *J. Chem. Phys.* **65**, 1883–92.

Pusey, P. N. (1975). Dynamics of interacting Brownian particles. *J. Phys. A.: Math. Gen.* **8**, 1433–9.

Pusey, P. N. & Tough, R. J. A. (1985). Particle interactions. In *Dynamic Light Scattering: Applications of Photon Correlation Spectroscopy* (ed. R. Pecora), pp. 85–179. Plenum Press.

Rallison, J. M. & Hinch, E. J. (1986). The effect of particle interactions on dynamic light scattering from a dilute suspension. *J. Fluid Mech.* **167**, 131–68.

Russel, W. B. & Glendinning, A. B. (1981). The effective diffusion coefficient detected by dynamic light scattering. *J. Chem. Phys.* **74**, 948–52.

Schaefer, D. W. & Berne, B. J. (1974). Dynamics of charged macromolecules in solution. *Phys. Rev. Lett.* **32**, 1110–13.

van Megen, W., Ottewill, R. H., Owens, S. M. & Pusey, P. N. (1985). Measurement of the wave-vector dependent diffusion coefficient in concentrated particle dispersions. *J. Chem. Phys.* **82**, 508–15.

Vilker, V. L., Colton, C. K. & Smith, K. L. (1981). Concentration polarization in protein ultrafiltration: II. Theoretical and experimental study of albumin ultrafiltration in an unstirred cell. *AIChE J.* **27**, 632–45.

Vrij, A., Jansen, J. W., Dhont, J. K. G., Pathmamanoharan, C., Kops-Werkhoven, M. M. & Fijnaut, H. M. (1983). Light scattering of colloidal dispersions in non-polar solvents at finite concentration. Silica spheres as model particles for hard-sphere interactions. *Far. Dis.* **76**, 19–36.

Problems

1 The data in Fig. 12.3 indicate some uncertainty in $K(\phi)$ for hard spheres. Assess as convincingly as possible whether this might explain the difference between the theoretical curve and the experimental data in Fig. 13.5.

2 In ultrafiltration, particles accumulate at a semi-permeable membrane through which water passes. The resulting increase in volume fraction at the membrane drives diffusion of particles away from the membrane and reduces the driving force for water flow through the membrane. Since the density inevitably approaches close packing, the volume fraction dependence of the diffusion coefficient is important.
(i) Develop the appropriate conservation equations and boundary conditions for both momentum and the particle number density, analogous to (12.2.10) and (12.5.7).

(ii) Integrate the momentum equation to show that the actual pressure difference across the membrane equals the applied pressure difference minus the osmotic pressure of the dispersion at the membrane.
(iii) Integrate the particle conservation equation to find the steady-state volume fraction profile, $\phi(x, \infty)$, for $\phi(x,0)=\phi_0$ and $\phi(\infty,t)=0$. Compare with the corresponding sediment profiles from Problem 12.6 for the same Peclet numbers.

3 Derive from (13.4.2) the self-intermediate scattering function, $F_s(t)$, for a dilute bimodal mixture of spheres and calculate the corresponding first cumulant. Plot F_s and the line representing the truncated cumulant expansion for several size ratios to illustrate the difficulty in characterizing multimodal distributions.

4 Derive (13.3.7) and evaluate $A_2(q)$ and $K_2(q)$ for the excluded-shell potential.

5 Derive the forms (13.6.3) for the adhesive sphere potential, then evaluate and plot $D_2(q)$ for $1/\tau=0$, 1, and 2. Explain the variations with increasing stickiness in terms of the individual contributions to the hydrodynamic and thermodynamic coefficients.

6 Given a scattering instrument only operating for $45° < \theta < 135°$, identify the particle sizes necessary to detect $D_2(q)$ from $D_2(0)$ to $D_2(\infty)$ for silica spheres in cyclohexane.

7 The difference between D_{2s}^0 and D_{2s}^∞ arises from relaxation effects due to the perturbation of the configuration of surrounding particles by the motion of the tracer particle. To illustrate the effect, consider hard spheres with no hydrodynamic interactions such that $D_{2s}^0=0$.
(i) Solve (8.3.8) with $G=H=1$ through a regular perturbation expansion for $Pe \ll 1$ of the form

$$P_2(\mathbf{r})=n^2\left(1-\frac{\mathbf{r}\cdot\mathbf{U}_0}{r}f(r)\right)$$

to determine the nonequilibrium microstructure caused by a relative velocity $\mathbf{U}_0$.
(ii) Calculate the net interparticle force opposing the motion generated by the perturbed structure.
(iii) From the total force acting on the particle deduce the long time self diffusion coefficient, D_{2s}^∞.

14

RHEOLOGY

14.1 Introduction

In the preceding chapters we have examined the response of colloidal particles to interactions with one another in a quiescent fluid, to interactions with large collectors while being convected by the fluid, and to imposed forces due to electric fields, gravity, or concentration gradients. In each case, equilibrium or non-equilibrium, static or dynamic, the interparticle forces and the resulting suspension microstructure play key roles. Now we consider the stresses and the non-equilibrium microstructure generated in a flowing suspension when the velocity varies spatially on a scale large with respect to the size of the particles.

A Newtonian incompressible liquid is characterized by a linear relation between the stress tensor and the rate-of-strain tensor, with the constant of proportionality being the viscosity. Polymeric liquids are well known for their non-Newtonian behavior including shear-rate-dependent viscosities, elasticity manifested in recoil upon the cessation of flow, solid-like fracture during extrusion, and a variety of secondary flow phenomena. Colloidal suspensions also depart from Newtonian behavior. They often behave as solids requiring a finite stress, the yield stress, before deforming continuously as a liquid. The contrast with the polymeric liquids reflects the fundamentally different microstructures. Both microstructures deform under stress, but macromolecular systems can recover from strains of several hundred per cent because the restoring force increases with the degree of deformation. The interparticle forces governing the microstructure in colloidal dispersions generally have a short range and the magnitude decreases with increasing separation, providing no mechanism for recovery beyond strains of a few per cent. Hence the emphasis shifts to the linear viscoelastic properties, the yield stress, the viscosity, and the time dependence of the rheology.

The qualitative nature of the rheological response, i.e. the non-Newtonian behavior, is intimately related to the stability or phase behavior discussed in Chapters 8 to 10. A stable suspension at moderate volume fractions normally flows like a modestly shear-thinning, low-viscosity fluid. The dramatic effects appear when one interparticle force dominates. For example, adding electrolyte to an electrostatically stabilized suspension or introducing a poor solvent into a polymerically stabilized system, permits short-range attractions to create an amorphous solid. Imposition of a steady shear causes the material to fracture or yield and then flow plastically. Upon cessation of flow, the structure may take many hours or days to recover. On the other hand, removal of electrolyte from an aqueous dispersion can generate a crystalline material that responds to linear viscoelastic tests as a solid but creeps slowly under a constant stress. Alternatively, concentrating the original dispersion to close packing also forms a solid at rest. These flow at moderate stresses, acquiring two-dimensional order, but at a critical stress undergo an order–disorder transition accompanied by a substantial increase in viscosity (e.g. Hoffman, 1972, 1974).

The key to understanding and controlling these phenomena lies in characterizing the microstructure and its response to flow. This chapter begins with a description of the standard rheological measurements and the associated response of both solids and fluids. We then classify systems through dimensional analysis and use experimental data for model systems to characterize the rheology for particular interparticle potentials, e.g. hard sphere, electrostatic repulsion, polymeric repulsion, and weak attractions. After the phenomena are clearly defined, theoretical approaches are described. A successful theory must treat correctly both the interactions between particles and the resulting microstructure. We limit the discussion to exact descriptions for pair interactions in dilute dispersions. The results have been confirmed experimentally for dilute systems, and provide the basis for understanding several features of concentrated dispersions.

The theme throughout is the direct connection between macroscopic rheology and microstructure. The colloidal forces responsible for thermodynamic non-idealities at rest often dominate the response in flows that only slightly perturb the equilibrium microstructure. Non-ideal, e.g. non-Newtonian or non-Hookean, mechanical behavior accompanies substantial deformation from equilibrium.

14.2 Characterization of rheological behavior

Much of the diversity in the rheology of colloidal suspensions can be illustrated through a few relatively simple flows. In this section we confine our attention to three specific types of incompressible flows, one being time-independent, one periodic in time, and one transient. More complete treatments are readily available (Schowalter, 1978, Ch. 8; Tanner, 1985, Ch. 3; Bird, Armstrong, & Hassager, 1987, Ch. 4), but the material presented here should suffice for the purpose of this book.

In our discussion of fluids we assume that there exists an isotropic rest state and that the stress at a point in the fluid depends only on the history of the deformation of that material element. These assumptions comprise the notion of a 'simple' fluid central to most interpretations of rheological measurements (Schowalter, 1978, Ch. 8).

Consider now the steady, simple shear flow

$$u_x = \gamma y \tag{14.2.1}$$

with shear rate γ (Fig. 14.1). This flow belongs to a general class of viscometric flows for which each material element experiences a steady shear for all times. Consequently, for a simple fluid the stress depends only on the local rate of strain. Furthermore, if the fluid response is to be independent of the reference frame of an observer, then $\sigma_{xy}(\gamma) = -\sigma_{xy}(-\gamma)$, $\sigma_{yz} = \sigma_{xz} = 0$, and the normal stresses σ_{xx}, σ_{yy}, and σ_{zz} must be even functions of γ. Then the behavior of simple fluids can be described, to within an arbitrary isotropic pressure, by three independent functions of the material and the shear rate. Indeterminacy of the pressure is a consequence of the assumption of incompressibility, which means that no thermodynamic equation of state can be defined for the fluid.

The most common material function is the effective viscosity, $\eta(\gamma)$, defined for the simple shear flow by

$$\sigma_{xy} = \eta(\gamma)\gamma. \tag{14.2.2}$$

Fig. 14.1. Schematic of simple shear flow with $\mathbf{u} = (\gamma y, 0, 0)$, produced by translation of parallel plates.

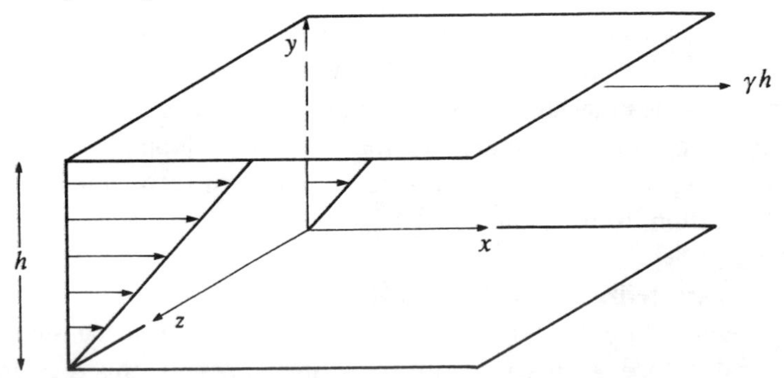

Because of the indeterminacy and isotropy of the pressure, the stress tensor generally is divided into isotropic and traceless portions as

$$\boldsymbol{\sigma} = -p\boldsymbol{\delta} + \boldsymbol{\tau}, \tag{14.2.3}$$

with tr $\boldsymbol{\tau}=0$ and tr $\boldsymbol{\sigma} = -3p$. Two additional material functions, the first and second normal stress differences, conventionally defined by

$$\begin{aligned} N_1(\gamma) &= \sigma_{xx} - \sigma_{yy} \\ N_2(\gamma) &= \sigma_{yy} - \sigma_{zz}, \end{aligned} \tag{14.2.4}$$

along with $\eta(\gamma)$ and the requirement tr $\boldsymbol{\tau}=0$, then determine completely the rheological behavior of a simple fluid in viscometric flows. For a Newtonian fluid, $\eta(\gamma)=\mu$ and $N_1(\gamma)=N_2(\gamma)=0$, suggesting that the normal stress differences arise from elasticity or memory of a material.

For the second prototypical flow we choose a sinusoidal oscillation

$$u_x = \gamma y \sin \omega t, \tag{14.2.5}$$

which can be obtained merely by oscillating one of the planes of Fig. 14.1 relative to the other. For sufficiently small frequencies, e.g., $\omega < 1/t_{\text{vis}} = \mu/\rho h^2$ for a Newtonian fluid with viscosity μ and density ρ (§2.7), inertial effects are negligible and the velocity profile is spatially linear. Much of the early work on viscoelasticity was based on the notion that a material generally responds as a combination of a Newtonian fluid (linearity between stress and rate of strain) and a Hookean solid (linearity between stress and strain). Then the stress resulting from (14.2.5) is

$$\sigma_{xy} = \eta'(\omega)\gamma \sin \omega t - \frac{G'(\omega)}{\omega}\gamma \cos \omega t. \tag{14.2.6}$$

Note that the viscous contribution to the stress, characterized by the dynamic viscosity $\eta'(\omega)$, is in phase with the velocity gradient, while the elastic contribution, characterized by a shear modulus, $G'(\omega)$, is in phase with the strain but 90° out of phase with the velocity gradient.

Other common measurements detect the transient response of a material to sudden loading or unloading of stress or strain. For example, a 'creep' experiment consists of applying the shear stress

$$\sigma_{xy} = \begin{cases} 0 & t \leq 0 \\ \sigma & 0 < t \end{cases} \tag{14.2.7}$$

and monitoring the shear strain $\varepsilon_{xy}(t)$. A modulus for this experiment is the creep compliance, defined by

$$J(t) = \frac{2\epsilon_{xy}(t)}{\sigma} \tag{14.2.8}$$

Table 14.1

	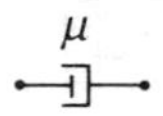 μ Viscous fluid	 G Elastic solid	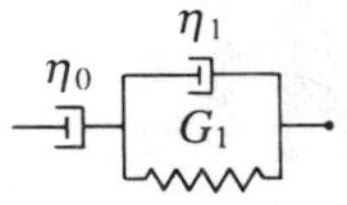 η_0, η_1, G_1 Viscoelastic fluid	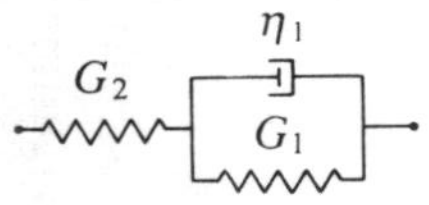 G_2, η_1, G_1 Viscoelastic solid
Steady shear	$\eta=\mu$	$\eta=\infty$	$\eta=\eta_0$	$\eta=\infty$
Small-amplitude oscillations	$\eta'=\mu$ $G'=0$	$\eta'=0$ $G'=G$	$\dfrac{\eta'}{\eta_0}=\dfrac{1+\dfrac{\eta'_\infty}{\eta_0}(\lambda\omega)^2}{1+(\lambda\omega)^2}$ $\dfrac{G'}{G'_\infty}=\dfrac{(\lambda\omega)^2}{1+(\lambda\omega)^2}$	$\dfrac{\eta'}{\eta'_0}=\dfrac{1}{1+(\lambda\omega)^2}$ $\dfrac{G'}{G'_\infty}=\dfrac{G_0/G'_\infty+(\lambda\omega)^2}{1+(\lambda\omega)^2}$
Creep	$J=\dfrac{t}{\mu}$	$J=\dfrac{1}{G}$	$J=\dfrac{1}{G'_\infty}\left(1-\dfrac{\eta'_\infty}{\eta_0}\right)^2\left(1-e^{-\frac{\eta_0}{\eta'_\infty}\frac{t}{\lambda}}\right)+\dfrac{t}{\eta_0}$	$J=\dfrac{1}{G_0}-\left(\dfrac{1}{G_0}-\dfrac{1}{G'_\infty}\right)\exp\left(-\dfrac{G_0 t}{G'_\infty\lambda}\right)$

$\eta_1 = \dfrac{\eta_0 \eta'_\infty}{\eta_0 - \eta'_\infty}$	$\eta_1 = \dfrac{\eta'_0}{(1 - G_0/G'_\infty)^2}$
$G_1 = \dfrac{G'_\infty}{(1 - \eta'_\infty/\eta_0)^2}$	$G_1 = \dfrac{G_0}{1 - G_0/G'_\infty}$
	$G_2 = G'_\infty$
$\lambda = \dfrac{\eta_0 - \eta'_\infty}{G'_\infty}$	$\lambda = \dfrac{\eta'_0}{G'_\infty - G_0}$

The moduli from these three experiments for a Newtonian fluid with

$$\boldsymbol{\sigma} = -p\boldsymbol{\delta} + 2\mu\mathbf{E} \tag{14.2.9}$$

and a Hookean solid with

$$\boldsymbol{\sigma} = -p\boldsymbol{\delta} + 2G\boldsymbol{\varepsilon} \tag{14.2.10}$$

provide convenient limits for comparison with the responses of more complex materials. Here, $\boldsymbol{\varepsilon}$ is the strain tensor and $\mathbf{E} = \mathrm{d}\boldsymbol{\varepsilon}/\mathrm{d}t$ the rate-of-strain tensor. The derivations are straightforward. For example, creep of a Newtonian fluid is described by substituting (14.2.7) into (14.2.9) to obtain $\sigma H(t) = 2\mu(\mathrm{d}\varepsilon_{xy}/\mathrm{d}t)$ and then integrating with $\varepsilon_{xy}(0) = 0$. The results for $J(t)$ and the other moduli are listed in Table 14.1. For linearly viscoelastic materials, mechanical analogs consisting of combinations of Hookean springs and viscous dashpots have been useful aids for interpreting experiments such as those described above. For general time-dependent flows with these more complex models, the variation of the stress or strain can readily be obtained via Laplace transforms, as long as the assumption of linear viscoelasticity is valid (Tanner, 1985, §2.7).

Two particular models and the results for the moduli are shown in Table 14.1 and Fig. 14.2. In this context, a 'fluid' is a material that continues to deform as long as a non-zero stress is applied, while a 'solid' undergoes only a finite deformation for a finite stress. Note that a single characteristic time, λ, scales the time and the frequency in the creep and oscillatory experiments, respectively. At low dimensionless frequencies, the shear modulus goes to zero for fluids as the stresses relax, but for solids the modulus attains a finite low-frequency plateau (Fig. 14.2(*a*)). As the frequency increases, the dynamic viscosity decreases and the shear modulus increases, since the material relaxes less during the period of the oscillation.

In the creep experiment, the solid deforms instantaneously as the load is applied, but the fluid does not (Fig. 14.2(*b*)). At long times, the deformation of the solid reaches a finite asymptote, while that of the fluid increases linearly. For each the combination of the initial slope or displacement and the slope and intercept of the linear portion of the curve determines the high and low frequency limits of the dynamic viscosity and the shear modulus.

More general treatments of viscoelastic behavior show that linear laws such as those just described apply for motions that are, in an appropriate sense, slow and/or small. However, for many features of interest with colloidal dispersions, the linearity hypothesis is too limiting. For a start, colloidal dispersions exhibit a shear-dependent viscosity in the steady flow of (14.2.1). Also, flows with varying degrees of extension and rotation can

Fig. 14.2. Rheometric functions for the viscoelastic fluid (——) and solid (- - -) models in Table 14.1, with $\eta_0/\eta'_\infty = 10$ for the fluid and $G_0/G'_\infty = 10^{-1}$ for the solid. (*a*) Shear modulus and dynamic viscosity as functions of frequency. (*b*) Creep compliance as function of time.

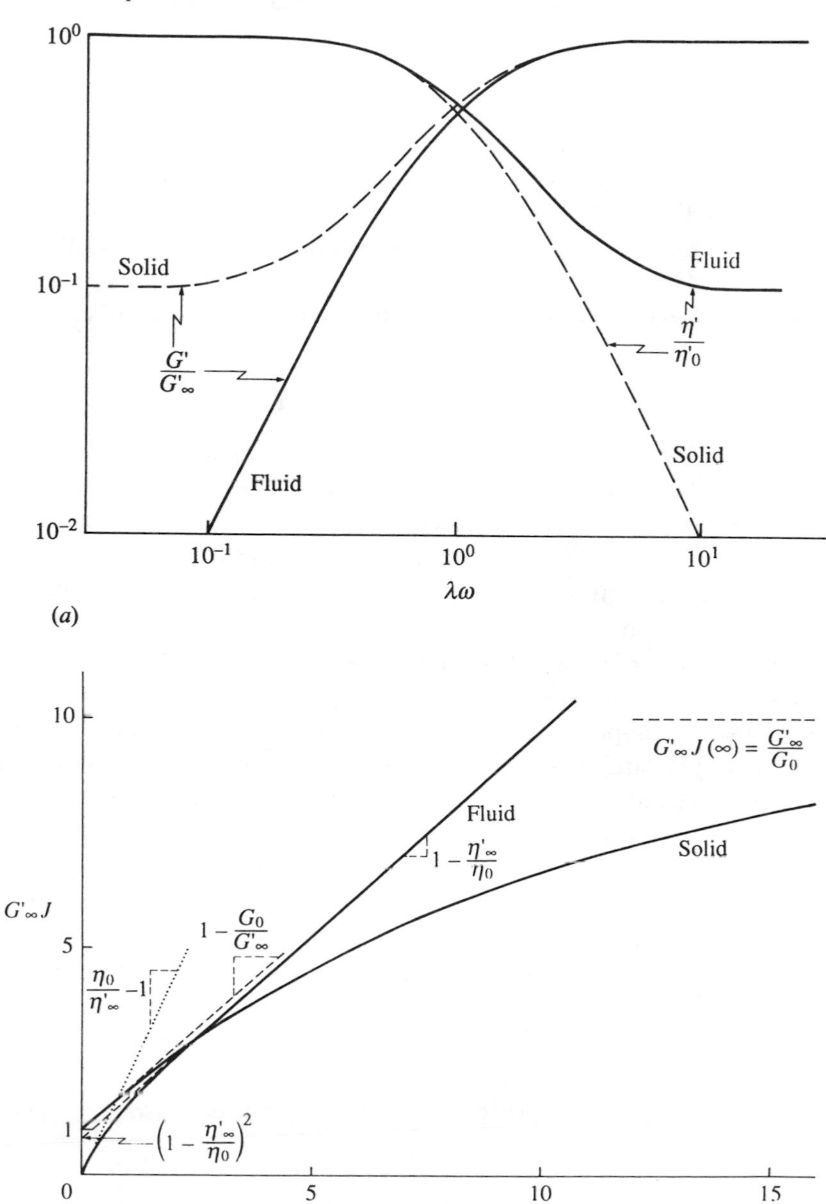

cause different alterations to the suspension microstructure. These differences will be reflected in the constitutive behavior of the dispersion. Nevertheless, the simple models and flows described above serve to illustrate a number of qualitatively important features of dispersions, such as the high and low frequency limits for the dynamic viscosity and the high and low shear limits for the steady shear viscosity.

14.3 Dimensional analysis

Dimensional analysis involves identifying parameters that affect the properties of interest and constructing an appropriate number of dimensionless groups. Choosing these groups as ratios of physical quantities provides a basis for assessing the relative importance of individual forces (Krieger, 1972; Krieger and Eguiluz, 1976; Chaffey, 1977; Russel, 1980).

This discussion pertains only to flows for which the characteristic length scale substantially exceeds the microscale, normally the particle size. Then the suspension behaves on the macroscale as a continuum whose stresses, normalized either on the solvent viscosity μ and the magnitude of the rate of strain γ as $\boldsymbol{\sigma}/\mu\gamma$ or on the thermal energy density as $a^3\boldsymbol{\sigma}/kT$, should depend on the dimensionless parameters characterizing the local velocity gradient and the microstructure of the suspension. For a steady, simple shear flow the dimensionless shear stress reduces to the relative viscosity, η/μ. For an oscillatory shear flow with frequency ω the component of the shear stress in phase with the velocity gradient yields the relative dynamic viscosity, η'/μ, and the out-of-phase component the dimensionless shear modulus, a^3G'/kT. The latter scaling is appropriate, since the elasticity arises from non-hydrodynamic forces. Table 14.2 summarizes the relevant physical quantities and a meaningful set of dimensionless groups.

Since viscous forces perturb the microstructure against the restoring effect of Brownian motion, the Peclet number, defined in terms of the relative velocity $a\gamma$ as $Pe = a^2\gamma/D_0$, gauges the magnitude of the departure from equilibrium. Simulations (e.g. Evans, Hanley, & Hess, 1984; Barnes, Edwards, & Woodcock, 1987) predict significant non-equilibrium effects and non-Newtonian rheology, even in molecular fluids, when $Pe > 1$, but the shear rates required when $a = 0.1$–1.0 nm ($\geq 10^{12}\,\mathrm{s}^{-1}$) are unattainable. Colloidal dispersions, however, depart from equilibrium at $\gamma = 1$–$1000\,\mathrm{s}^{-1}$. In fact, for particles larger than 10 μm in water the difficulty lies in attaining the equilibrium state, since $Pe = 300$ for $\gamma = 1\,\mathrm{s}^{-1}$.

The characteristic magnitudes of the interparticle potentials divided by the thermal energy provide the dimensionless groups encountered in §8.3 and 10.7:

Table 14.2

Functions	*Dimensional*	*Dimensionless*
stress	$\boldsymbol{\sigma}$	$a^3\boldsymbol{\sigma}/kT$
viscosity	η	η/μ
modulus	G'	a^3G'/kT
Time scales		
diffusion	a^2/D_0	
convection	γ^{-1}	$Pe=a^2\gamma/D_0$
Energies and length scales		
thermal	kT	
	a	
electrostatic	$\varepsilon_0\varepsilon\psi_s^2a$	$\varepsilon\varepsilon_0\psi_s^2a/kT$
	κ^{-1}	$a\kappa$
dispersion	A_{eff}	A_{eff}/kT
	a	
polymeric	$N^{1/2}avkT/l^4$	$N^{1/2}av/l^4$
	r_g	a/r_g
interparticle	$n^{-1/3}$	$\phi=(4\pi/3)a^3n$

$$\frac{A_{eff}}{kT} \qquad \frac{\text{dispersion}}{\text{thermal}},$$

$$\frac{\varepsilon_0\varepsilon\psi_s^2 a}{kT} \qquad \frac{\text{electrostatic}}{\text{thermal}},$$

$$\frac{ar_g^2 n_b}{N} \qquad \frac{\text{polymeric}}{\text{thermal}}$$

with A_{eff} the effective Hamaker constant, $\varepsilon_0\varepsilon$ the dielectic permittivity of the fluid, ψ_s the surface potential, N the number of segments in the polymer chain, r_g the radius of gyration and n_b the segment density. The ratio of the range of each to the sphere radius determines the additional groups:

$$a\kappa \qquad \frac{\text{sphere radius}}{\text{double-layer thickness}},$$

$$\frac{a}{r_g} \qquad \frac{\text{sphere radius}}{\text{radius of gyration}}.$$

Finally, the volume fraction, ϕ, which measures the ratio of the radius to the mean separation $n^{-1/3}$, indicates the importance of interactions in general.

The next four sections illustrate the roles of these groups in the rheology of dispersions through several examples.

14.4 Hard spheres

The simplest system, hard spheres, is affected by viscous forces, Brownian motion, and the excluded volume of the particles. Thus the relative steady shear and dynamic viscosities and the dimensionless shear modulus a^3G'/kT should depend on the volume fraction and the Peclet number or, equivalently, the dimensionless frequency $a^2\omega/D_0$. Three sets of data, for systems with characteristics listed in Table 14.3, confirm these expectations and define, to some extent, the functional relationships.

The data for the shear viscosity in Fig. 14.3 demonstrate the shear thinning generally observed between the well-defined low and high shear-limiting viscosities, η_0 and η_∞, respectively. At a fixed volume fraction, the relative viscosities for a variety of solvent viscosities and radii of the polystyrene latices superimpose when plotted against the Peclet number, as required by the dimensional analysis. Both the latices and the silica systems are stabilized by soluble layers with thicknesses of $L = 2$–5 nm. As noted in §9.2, dispersion forces become significant for large ratios of the sphere radius to the layer thickness, corresponding roughly to $a \geq 0.05$–$0.10\,\mu$m. For these systems, Krieger (1972) observed superposition for $a \leq 0.25\,\mu$m with the aqueous latices. So only these data are included below.

Both the low and high shear-limiting viscosities increase monotonically with volume fraction (Fig. 14.4(*a*)), with shear thinning only detectable for $\phi \geq 0.25$–0.30. Only for $\phi \geq 0.5$ do either the viscosity or the extent of shear thinning become substantial. De Kruif *et al.* (1986) extracted the expansions

$$\begin{aligned} \eta_0/\mu &= 1 + 2.5\phi + (4 \pm 2)\phi^2 + (42 \pm 10)\phi^3 + \ldots, \\ \eta_\infty/\mu &= 1 + 2.5\phi + (4 \pm 2)\phi^2 + (25 \pm 7)\phi^3 + \ldots \end{aligned} \tag{14.4.1}$$

in the dilute limit and correlated the data over the full range of volume fractions via

$$\begin{aligned} \frac{\eta_0}{\mu} &= \left(1 - \frac{\phi}{0.63}\right)^{-2}, \\ \frac{\eta_\infty}{\mu} &= \left(1 - \frac{\phi}{0.71}\right)^{-2}. \end{aligned} \tag{14.4.2}$$

The stress characterizing the shear thinning, i.e. the stress σ_c at which $\eta = (\eta_0 + \eta_\infty)/2$, scales roughly as kT/a^3 in accord with the dimensional

Table 14.3. *Model hard-sphere dispersions*

	$2a/nm$	*Solvent*
Krieger (1972)	108	
polystyrene	140	water
	180	
	77	benzyl alcohol
	110	meta-cresol
de Kruif *et al.* (1986) silica	76 ± 2	cyclohexane
Mellema *et al.* (1987) silica	76 ± 2	cyclohexane
	28 ± 2	
van der Werff and de Kruif (1988) silica	76 ± 2	cyclohexane
	46 ± 2	

Fig. 14.3. Relative steady shear viscosity as function of Peclet number for polystyrene latices of radii listed in Table 14.3 dispersed in water (——) and benzyl alcohol or meta-cresol (○ ●) (Krieger, 1972).

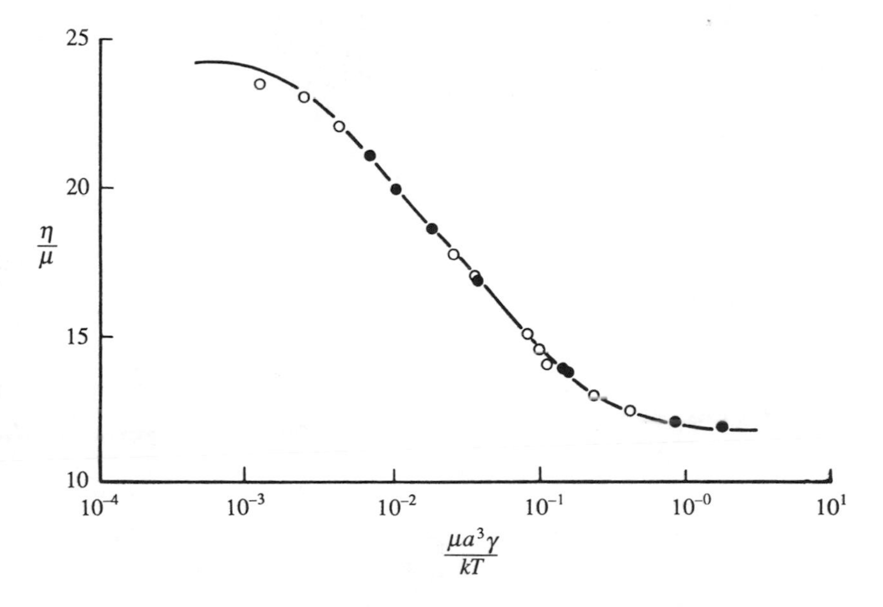

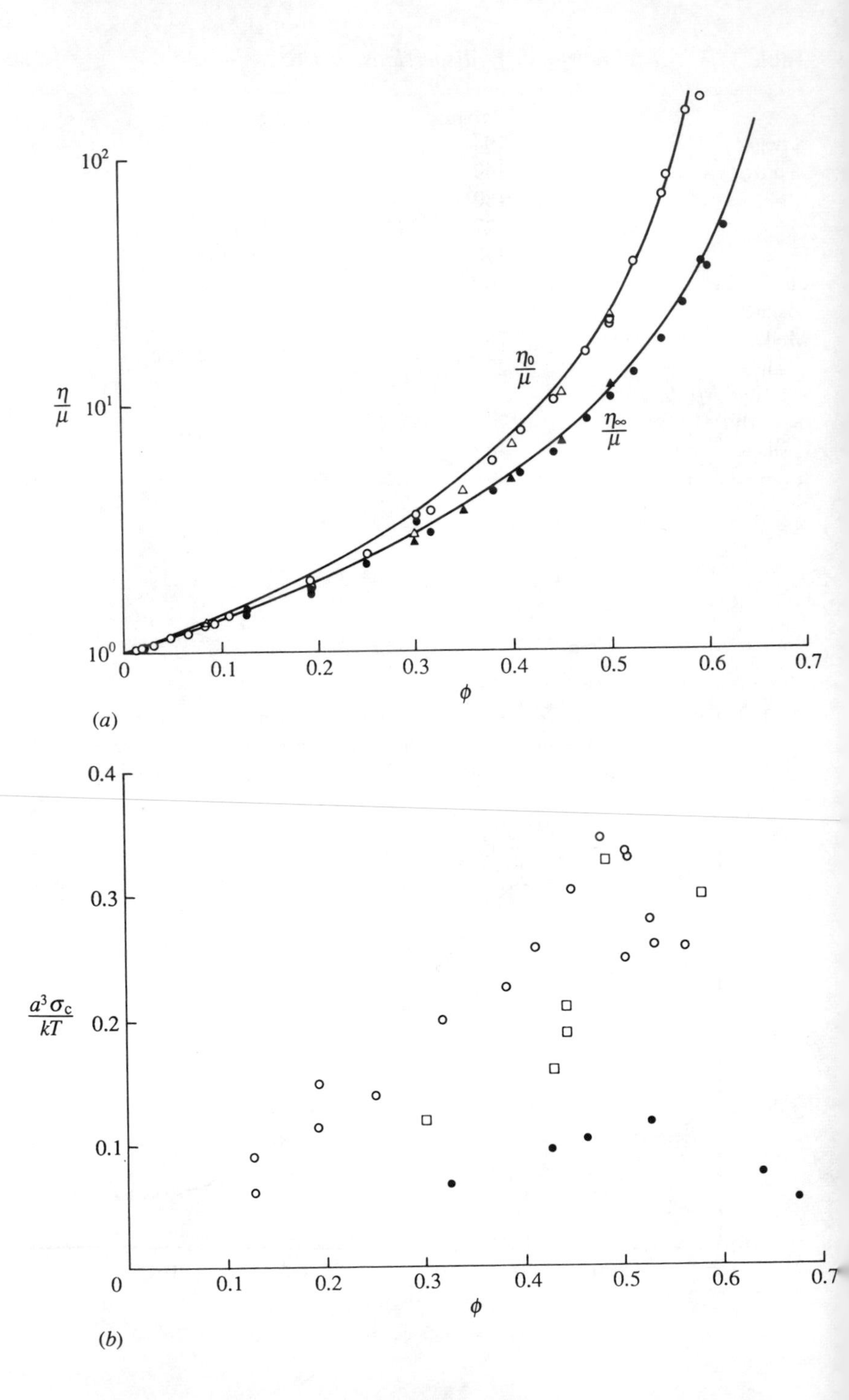

10²
10¹
10⁰
η/μ
η₀/μ
η∞/μ
0
0.1
0.2
0.3
0.4
0.5
0.6
0.7
φ
(a)
0.4
0.3
0.2
0.1
a³σc/kT
φ
(b)

Fig. 14.4. (*a*) Low- (open symbols) and high- (filled symbols) shear limiting viscosities for dispersions of hard spheres with characteristics listed in Table 14.3: ○, silica spheres in cyclohexane (de Kruif *et al.*, 1986); △, polystyrene latices in water (Krieger, 1972). (*b*) Dimensionless characteristic stresses for dispersions of hard spheres (silica with $a=78\pm2$ nm ○ and 46 ± 2 nm □ in cyclohexane, (de Kruif *et al.*, 1986; van der Werff and de Kruif, 1989)) and soft spheres (●, poly(methyl methacrylate) with $a=42$ nm stabilized by poly(12-hydroxystearic acid) in decanol (Frith, 1986)).

analysis and varies with volume fraction, as shown in Fig. 14.4(*b*). These results identify several interesting features of the structure and rheology of the hard-sphere systems.

First, the expansions (14.4.1) point out that shear thinning arises from three-body and higher-order interactions. This reflects the lack of long-range structure in dilute suspensions of hard spheres, for which three-body interactions are required to impart significant structure in $g(r)$ (e.g. Fig. 10.2). This structure generates thermodynamic stresses at low shear, i.e. $Pe \ll 1$. But at high Peclet numbers, viscous forces dominate both the structure and the stresses. Hence $\eta_0 - \eta_\infty$ gauges the degree of structure and the importance of non-hydrodynamic forces.

Second, the data and the correlations (14.4.2) suggest that dispersions cease to flow, i.e. the viscosity diverges, at a lower volume fraction in the low shear limit than in the high shear limit. The former roughly coincides with random close packing, where the osmotic compressibility diverges for a disordered fluid (§10.5). The fact that dispersions flow at high shear rates for $\phi > 0.63$ establishes that shear must orient and, perhaps, increase to some degree the ordered domains that must exist in the microstructure.

Third, at intermediate stresses the data correlate reasonably well with

$$\frac{\eta - \eta_\infty}{\eta_0 - \eta_\infty} = \frac{1}{1 + \left(\dfrac{\sigma}{\sigma_c}\right)^n} \tag{14.4.3}$$

with $1 \leq n \leq 2$. The stress σ_c (Fig. 14.4(*b*) increases with volume fraction in dilute dispersions, then appears to pass through a maximum and decreases to zero at $\phi = 0.63$, where the low shear viscosity diverges. The peak at $\phi \approx 0.5$ is the only indication that the hard-sphere disorder–order transition either occurs or is rheologically significant in these systems. Otherwise they behave as disordered fluids or glasses.

Mellema *et al.* (1987) measured the linear viscoelastic response of dispersions of silica spheres of two radii, 28 ± 2 nm and 76 ± 2 nm, at

Fig. 14.5. Shear moduli and dynamic viscosities measured for silica spheres at $\phi=0.46$: □∗, $a=28\pm2$ nm; ○+, $a=76\pm2$ nm (Mellema *et al.*, 1987). The broken lines correspond to the zero and infinite shear viscosities (de Kruif *et al.*, 1985) and the solid curves to the frequency dependence predicted by the viscoelastic fluid model of Table 14.1 with the measured values of η_0, η'_∞, and G'_∞.

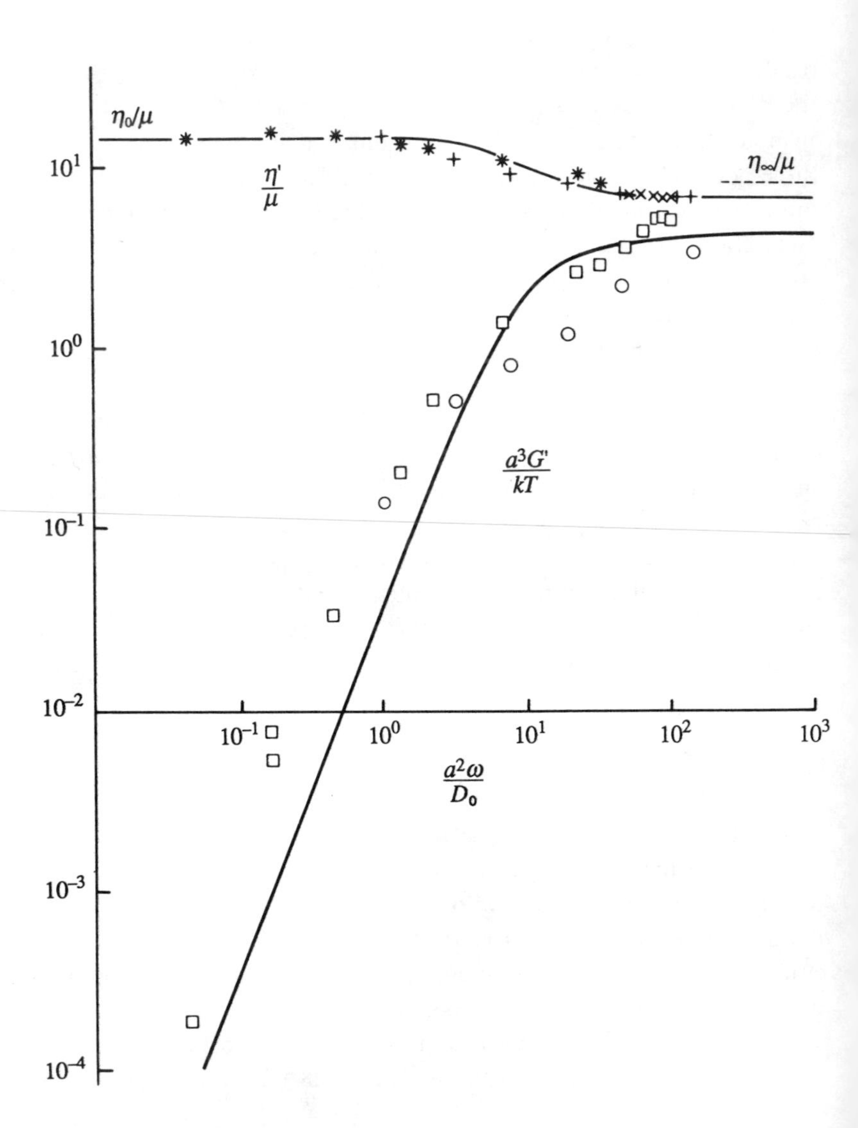

$\phi = 0.45$. The results (Fig. 14.5) reveal well-defined low and high frequency limits for the dynamic viscosity and, perhaps, the shear modulus, i.e.,

$$\lim_{\omega \to \infty} G' = G'_\infty,$$

$$\lim_{\omega \to \infty} \eta' = \eta'_\infty,$$

$$\lim_{\omega \to 0} G'/\omega^2 = \lambda^2 G'_\infty,$$

$$\lim_{\omega \to 0} \eta' = \eta_0.$$

The curves demonstrate that the expressions for η' and G' in Table 14.1 for the viscoelastic fluid model with a single relaxation time, $\lambda = (\eta_0 - \eta'_\infty)/G'_\infty$, correlate the data. The superposition of the two sets of data supports the scaling of the shear modulus with kT/a^3 and the frequency with D_0/a^2, as expected from the dimensional analysis. Finally, note that $\eta'_\infty \neq \eta_\infty$. Both correspond to situations in which Brownian motion is negligible, but the high-frequency oscillations perturb the equilibrium microstructure only slightly, while steady shear produces a highly non-equilibrium, hydrodynamically dominated structure.

In subsequent sections we reiterate several important points demonstrated by these three sets of data:

(i) The difference between the low and high shear viscosities (Fig. 14.4(*a*)), and hence the magnitude of the shear thinning, reflect the importance of non-hydrodynamic forces.
(ii) The viscous stress σ_c, characterizing the shear thinning, and the shear modulus G' scale on the interparticle potential, or kT for hard spheres.
(iii) Dispersions become solids at rest beyond random close packing, but a finite stress can induce flow and produce significant order.

The results also provide a baseline for assessing the importance of long-range interaction potentials in more complex systems.

14.5 Charged spheres

Aqueous latices inevitably bear a surface charge, introducing two additional dimensionless groups: $\varepsilon_0 \varepsilon \psi_s^2 a/kT$ and $a\kappa$. At moderate ionic strengths, for which $a\kappa \gg 1$ but the suspensions remain stable, the rheology resembles that for hard spheres as evidenced by the data of Krieger (1972) cited in the previous section. Reducing the ionic strength produces dramatic changes in the rheology, as in the phase behavior (e.g. §10.6), when the Debye length becomes comparable to the interparticle spacing. For example, the viscosity increases by orders of magnitude, a yield stress appears, and, under some conditions, shear disrupts the crystalline structure. To illustrate these effects quantitatively, we draw on three of many sets of experimental data.

Krieger & Eguiluz (1976) measured the steady shear viscosities for aqueous latices at a range of ionic strengths (Fig. 14.6). They observed that decreasing the ionic strength from 1.8×10^{-2} M to that of deionized water increases the viscosity by many orders of magnitude above that for uncharged spheres (broken line). Two points are worth noting:

(i) The high shear limit appears to vary little with ionic strength since hydrodynamic forces dominate both the structure and the stresses.
(ii) The divergence of the viscosity at a critical stress indicates a solid rest state that can sustain a finite stress before yielding and flowing.

The conditions at which the low-shear rheology suggests a change from a fluid to a solid correspond to those associated with the disorder–order transition for charged spheres (§10.6). If we assume that $\varepsilon_0 \varepsilon \psi_s^2 a/kT \approx 150$

Fig. 14.6. Steady shear viscosities for polystyrene latices ($a = 110$ nm) at $\phi = 0.40$ in water (Krieger & Eguiluz, 1976): ○, deionized; □, 1.9×10^{-4} M HCl; ■, 1.9×10^{-3} M HCl; ●, 1.9×10^{-2} M HCl; – – –, hard spheres.

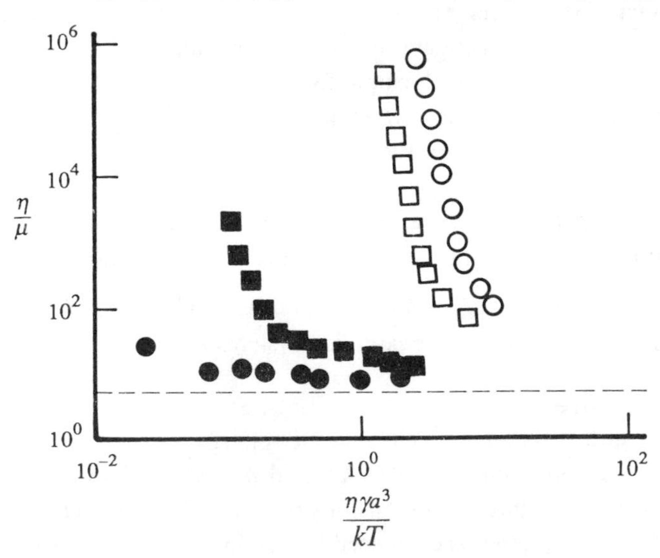

and $e\psi_s/kT \approx 4$, as in Fig. 10.8, the transition for the data in Fig. 14.6 should occur for $16 < a\kappa_0 < 51$ or $1.9 \times 10^{-3} < [\mathrm{HCl}] < 1.9 \times 10^{-2}$ M, exactly where the yield stress appears. Here κ_0^{-1} is the Debye length based on the added electrolyte alone.

The more sensitive steady shear and linear viscoelastic measurements of Buscall *et al.* (1982) with slightly smaller latices probe the dependence on volume fraction, ionic strength, and size in more detail. Low shear viscosities η_0 and static moduli G_0 were detected by applying very low, constant stresses ($< 1\ \mathrm{N\,m^{-2}}$) and monitoring the creep of the sample.

The data in Fig. 14.7 illustrate the correlation between the apparent divergence of the low shear viscosity and the appearance of a measurable shear modulus. This elastic response suggests that the transition from a disordered fluid to either an ordered or a glassy solid occurs at

Fig. 14.7. Low-shear limiting viscosities and static shear moduli for polystyrene latices ($a = 34$ nm) in 5×10^{-4} M NaCl: ●, η_0, ○, G_0 (Buscall *et al.*, 1982); ——, prediction from (14.5.2) with $\phi_m = 0.144$; – – –, prediction from (14.5.6) with $\frac{l_b G_0}{\kappa_0^2 kT} f\left(a\kappa_0, \frac{\phi}{\phi_m}\right) = \phi_{red}$.

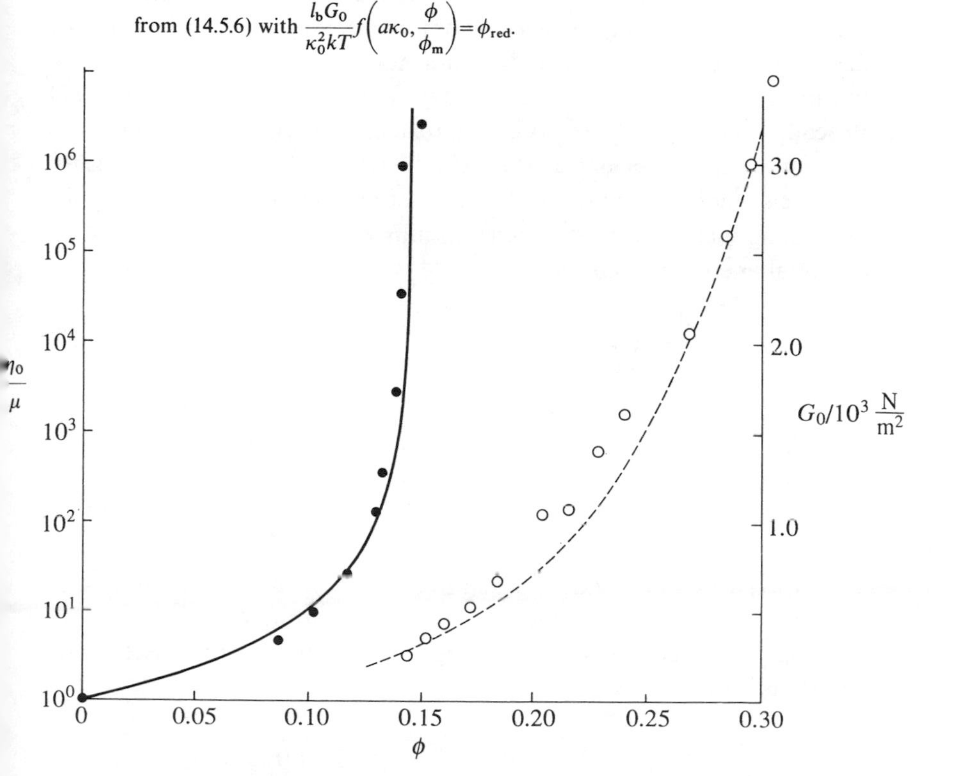

$\phi_m \approx 0.14$–0.15. The analysis in §10.6 predicts the disorder–order transition at $\phi = 0.50(2a/d)^3$, with

$$d = \frac{1}{\kappa} \ln\{\alpha/\ln[\alpha/\ln(\alpha/\ldots)]\}, \tag{14.5.1}$$

$\alpha = 4\pi\varepsilon_0\varepsilon\psi_s^2 a^2 \kappa \exp(2a\kappa)/kT$, and the Debye length $\kappa^{-1} < \kappa_0^{-1}$ accounting for both the added electrolyte and the counterions. Even without the phase transition, the low shear viscosity might be expected to diverge at random close packing of the equivalent hard spheres, i.e. $\phi = 0.63(2a/d)^3$. In either case, values of the surface potential, which were not measured, in the range of 25–75 mV provide the appropriate volume fraction at the transition from fluid to solid. Furthermore, the simple correlation

$$\frac{\eta_0}{\mu} = \left(1 - \frac{\phi}{\phi_m}\right)^{-2} \tag{14.5.2}$$

suggested by (14.4.2) conforms to the data reasonably well (Fig. 14.7).

Above the threshold volume fraction, the dispersion behaves as a viscoelastic solid. The shear moduli increase with increasing volume fraction and decreasing particle size or ionic strength. A correlation of the data can be constructed through arguments presented in a more quantitative form by several authors (e.g. Goodwin & Khidher, 1976; Joanny, 1979; Buscall *et al.*, 1982). They note that a small-amplitude static strain with amplitude ε produces an elastic energy density $G_0\varepsilon^2/2$. At the microscopic level this energy resides in the interaction potentials between pairs. Displacing pairs from their equilibrium separation r_m increases the potential energy of each pair by

$$\tfrac{1}{2}(\varepsilon r_m)^2 \frac{d^2\Phi}{dr^2}(r_m). \tag{14.5.3}$$

If each particle has N neighbors, the number of pairs per unit volume is $Nn/2$. Then comparison of (14.5.3) with the stored energy yields

$$G_0 \approx \tfrac{1}{2} N n r_m^2 \frac{d^2\Phi}{dr^2}(r_m). \tag{14.5.4}$$

Since the particles are close packed in an electrostatic sense $\pi n r_m^3/6 \approx \phi_m$ with ϕ_m and N depending on the details of the structure.

Using the linear superposition approximation for the electrostatic repulsion leads to

$$\frac{d^2\Phi}{dr^2} = 4\pi\varepsilon_0\varepsilon\psi_s^2 a^2 \frac{(\kappa r_m)^2 + 2\kappa r_m + 2}{r_m^3} \exp[-\kappa(r_m - 2a)] \tag{14.5.5}$$

with $r_m = 2a(\phi_m/\phi)^{1/3}$. The combination of (14.5.5) and (14.5.4) suggests plotting the scaled shear modulus

$$\frac{l_b G_0}{\kappa_0^2 kT} f\left(a\kappa_0, \frac{\phi}{\phi_m}\right)$$

with l_b the Bjerrum length (§8.2) and (14.5.6)

$$f(a\kappa_0, x^3) = \frac{1}{x^2\left(1 + \dfrac{x}{a\kappa_0} + \dfrac{x^2}{2(a\kappa_0)^2}\right)}$$

as a function of the reduced volume fraction

$$\phi_{red} = \frac{\phi}{\{a\kappa_0[1-(\phi/\phi_m)^{1/3}]\}^3}. \quad (14.5.7)$$

This leaves the dimensionless surface potential, $e\psi_s/kT$, as an independent variable.

Figure 14.8 illustrates the correlation between the scaled modulus and ϕ_{red} with $\phi_m = 0.63$ for $26 \le a \le 98$ nm, $0.14 \le \phi \le 0.41$, and $5 \times 10^{-5} \le I \le 2 \times 10^{-3}$ M. The deviations from a single curve could indicate

Fig. 14.8. Correlation of static shear moduli measured by Buscall *et al.* (1982), according to scaling of (14.5.6) and (14.5.7). ○, $a = 26$ nm; □, $a = 34$ nm; △, $a = 39$ nm; hexagons, $a = 98$ nm. The shadings indicate different ionic strengths.

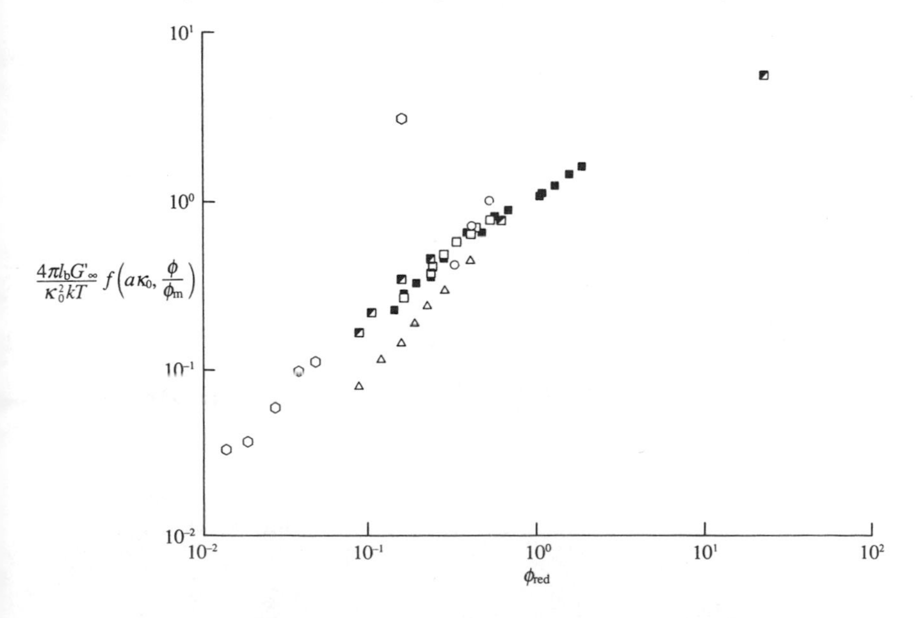

variations in $e\psi_s/kT$, which was not measured, the effects of counterions (cf. §10.6), or the approximate description of the interaction potential. Nonetheless, the correlation provides a basis for estimates; indeed, a simple proportionality between the dimensionless modulus and the reduced volume fraction conforms well to the data in Fig. 14.7.

In experiments at lower volume fractions, the effects of the structural transitions on the viscosity appear more vividly. For example, the data in Fig. 14.9 obtained with a sensitive Zimm viscometer pertain to polystyrene latices with $a=46$ nm and $\phi=0.04$ at several electrolyte concentrations. At 0–4 × 10^{-5} M HCl, the samples lie in the ordered region of the phase diagram and appear iridescent, but the order disappears at 5 × 10^{-5} M. The ordered phases exhibit an apparent yield stress on this plot, while the disordered phase clearly has a Newtonian low shear viscosity.

The new feature of these data is the transition evident at intermediate shear rates and electrolyte levels. For this system at 4 × 10^{-5} M HCl and $\dot{\gamma}=50$–$55\,s^{-1}$, the viscosity abruptly increases and the iridescence disappears. Lindsay & Chaikin (1985) offer the following interpretation. At rest, the colloidal crystal possesses three-dimensional symmetry, either bcc or fcc. Steady shear flow rearranges the crystal structure to permit an

Fig. 14.9. Shear stress as a function of shear rate for polystyrene latices ($a=45$ nm) at $\phi=0.04$ in □, deionized water; +, 1 × 10^{-5} M HCl; hexagons, 4 × 10^{5} M HCl; Δ, 5 × 10^{-5} M HCl; (Lindsay & Chaikin, 1985).

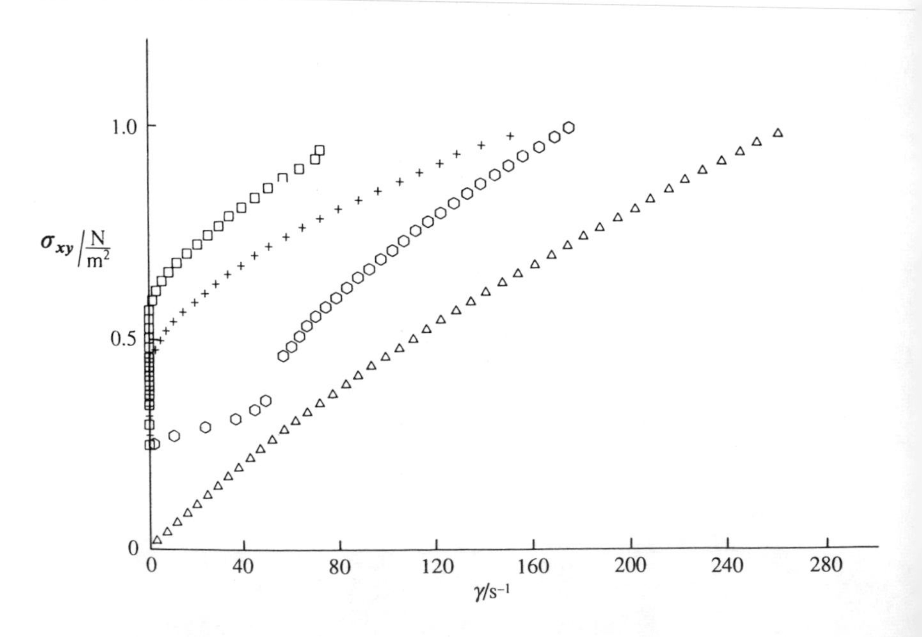

ordered flow of crystal planes translating perpendicular to the plane of shear. At a critical stress, however, this structure becomes unstable, producing a disordered state with a higher viscosity. The phenomenon requires that the viscous stresses overcome the elasticity of the crystal, so the critical stress increases with the shear modulus.

The data described above illustrate that long-range electrostatic repulsions dramatically alter the rheology of dispersions. The primary effect is to produce a solid state at relatively low volume fractions. The solid can assume either a metastable glassy or a crystalline structure. The low shear viscosity either diverges or becomes exceedingly large and a measurable static shear modulus appears. At finite stresses, the solids yield and flow; the crystals rearrange and ultimately become disordered at a critical stress.

14.6 Polymerically stabilized spheres

With dispersions stabilized by terminally anchored polymer, interactions between the polymer layers also increase the viscosity substantially. In this situation, however, the layer thickness fixes the range of the interaction and varies only modestly with the quality of the solvent for the anchored polymer. For example, Fig. 6.5 shows that L changes by only $O(1)$ amounts over a broad range of v/l^3, in contrast to the order of magnitude changes in the Debye length discussed in the previous section. With the following data, we demonstrate the correlation of the results through an equivalent hydrodynamic diameter for the sphere plus polymer layer and identify the conditions under which this fails due to the compressibility or deformability of the polymer.

A study of poly(methyl methacrylate) spheres stabilized by poly(12-hydroxystearic acid) and dispersed in decalin (Frith, 1986; Mewis *et al.*, 1989) exemplifies the behavior of this class of dispersions. Measurements for dilute systems of spheres with three different radii determined the coefficients in the expansion for the low shear limiting viscosity,

$$\eta_0/\mu = 1 + [\eta]n + k_h[\eta]^2 n^2 + \ldots, \tag{14.6.1}$$

listed in Table 14.4. Comparison with the dilute limit for spheres, (2.6.13), suggests defining an equivalent hydrodynamic volume fraction as

$$\begin{aligned} \phi &= 0.4[\eta]n \\ &= \left(1 + \frac{L}{a}\right)^3 \frac{4\pi a^3}{3} n. \end{aligned} \tag{14.6.2}$$

The values for $[\eta]$ imply that the stabilizing layer has a hydrodynamic thickness of $L \approx 8$–10 nm, so that $a/L \approx 5, 30, 60$ for the three particle sizes.

Table 14.4. *Poly(methyl methacrylate) spheres stabilized by poly(12-hydroxystearic acid) and dispersed in decalin* (Frith, 1986)

$2a$/nm	$3[\eta]/4\pi a^3$	k_h	L/nm	a/L
84	4.4	8.2	9	5
475	2.73	9.5	8	30
1220	2.6	10.0	10	60

Fig. 14.10. Low-shear limiting viscosities (○) and static moduli (●) for poly(methyl methacrylate) spheres ($a=42$ nm) stabilized with poly(12-hydroxystearic acid) ($L \approx 9$ nm) in decalin (Frith, 1986). The broken line represents the corresponding viscosities from (14.4.2).

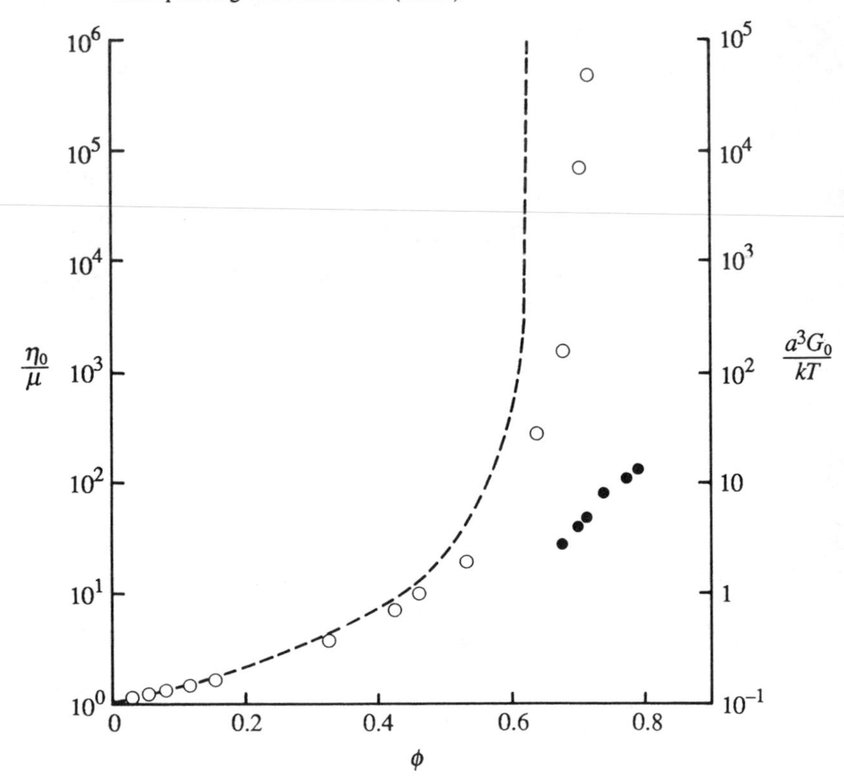

Since the earlier discussions (e.g. §9.2, §14.4) indicate that dispersion forces could affect the behavior of the latter two, we do not consider the low shear data from these systems.

Measurements of the viscosity at finite volume fractions reveal significant shear thinning, with plots of the relative viscosity as a function of the Peclet number resembling that for hard spheres (Fig. 14.3). However, the curves for different particle sizes at the same ϕ do not superimpose, reflecting the importance of the additional dimensionless group, a/L. Plotting the low- and high-shear relative viscosities, the shear modulus, and the dimensionless characteristic stress as functions of the equivalent volume fraction produces, in principle, a family of curves converging to the hard-sphere

Fig. 14.11. High shear-limiting viscosities for poly(methyl methacrylate) spheres (○, $a = 42$ nm; $*$, $a = 242$ nm; $+$, $a = 610$ nm) stabilized with poly(12-hydroxystearic acid ($L \approx 9$ nm) in decalin (Frith, 1986). The dashed line indicates the corresponding viscosities for hard spheres from (14.4.2).

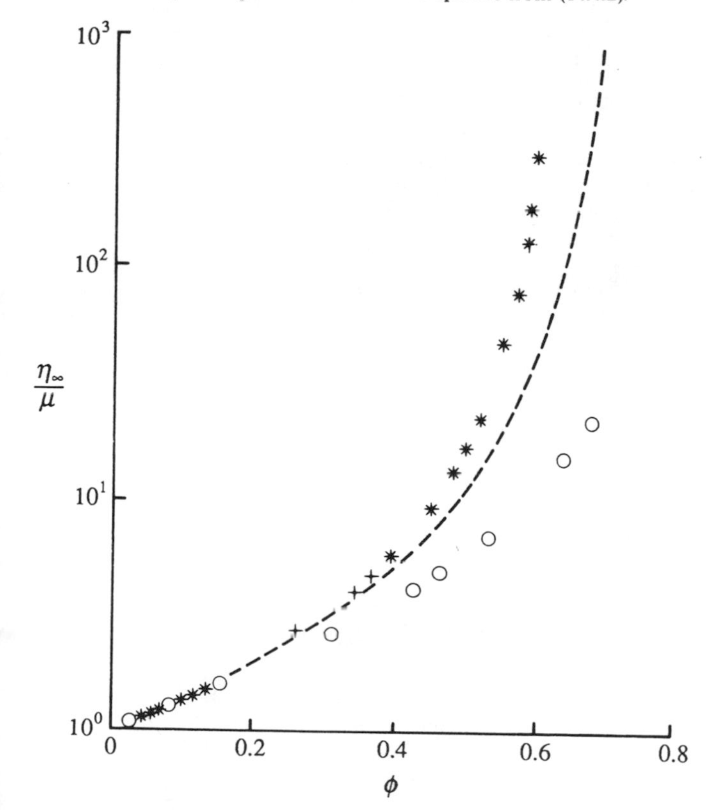

limit as $a/L \to \infty$ (Figs 14.4(*b*), 10, 11). Although the limited data do not define fully the dependence on a/L, several features deserve comment.

In the low shear limit (Fig. 14.10), the viscosities for $a/L \approx 5$ fall below the hard-sphere values and a measurable low-frequency plateau in the shear modulus appears for $\phi \geq 0.63$. This marks random close packing where the osmotic compressibility and the low shear viscosity diverge for the hard spheres. For these polymerically stabilized spheres, though, the repulsion is soft, as with charged spheres, permitting the osmotic compressibility, the shear modulus, and the viscosity to increase smoothly.

In the high shear limit (Fig. 14.11), the data for the two larger particles superimpose but lie somewhat above that for the hard spheres, suggesting a residual effect of the dispersion forces or a small error ($\approx$ 10 per cent) in the estimate of $[\eta]$. The viscosities for the smallest particles again fall below the hard-sphere curve. In this limit, the viscous stresses force the polymer layers to interpenetrate even at the lower volume fractions. As a consequence, σ_c is also smaller than for hard spheres (Fig. 14.4(*b*)).

Fig. 14.12. Equivalent hard-sphere volume fractions ϕ_{eff} for the polymerically stabilized spheres ($a = 42$ nm) extracted from data of Figs 14.11 and 14.12 by equating the measured viscosity to that for hard spheres: ○, low shear limit; ●, high shear limit.

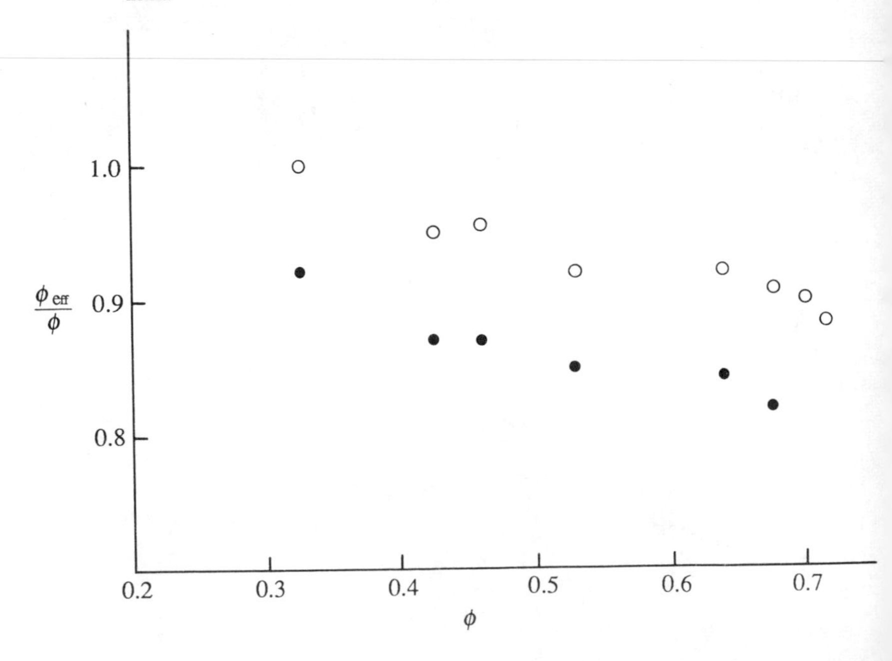

These effects become clearer if one defines an equivalent hard-sphere volume at each volume fraction by comparing the measured viscosities with those expected for hard spheres. The ratios of these equivalent volumes to the hydrodynamic volume, ϕ_{eff}/ϕ, decrease monotonically with increasing ϕ (Fig. 14.12), due to interpenetration of the polymer layers driven by the thermodynamic excluded volume in the low shear limit and the viscous forces at high shear rates. Although the deviations from unity are small the consequences are large due to the sensitivity of the viscosity to volume fraction.

14.7 Weakly flocculated dispersions

Flocculated suspensions respond elastically to small deformations with moduli that depend strongly on the volume fraction of particles. As with the stable systems that are close-packed, i.e. ϕ or $\phi_{eff} > 0.63$, the elasticity arises solely from the non-hydrodynamic interactions. In the flocculated case, however, the particles must form a volume-spanning network. For aggregation resulting from long-range attractions, e.g. dispersion- or polymer-induced forces acting across electrical double layers, deformation of the network changes the relative positions of the particles, thereby increasing the total potential energy of the system and producing an interparticle force tending to restore the equilibrium structure.

The primary difficulty in dealing with flocculated dispersions, both experimentally and theoretically, is the non-equilibrium nature of the structure. Consequently, the rheology is often history-dependent over very long time scales. However, for weakly flocculated dispersions with $-\Phi_{min}/kT \leqslant 20$ (c.f. Fig. 14.13), the structure recovers to a reproducible rest state after shear. Such dispersions would macroscopically phase-separate given sufficient time, but rheological studies performed relatively soon after formulation probe a metastable structure that changes negligibly during the experiment.

This section presents the results of two recent studies on well-characterized systems to illustrate some features of the rheology of weakly flocculated dispersions. First, the nature of the systems and the types of rheological measurements are described. Then the rheological properties are related to the interparticle potentials through scaling arguments similar to those in §14.5. Since the data are recent, our treatment is intended to be suggestive, not definitive, proposing ideas to be tested by additional data on these and other systems.

Goodwin *et al.* (1986) examined three monodisperse polystyrene latices with an adsorbed layer of a non-ionic surfactant in 0.5 M NaCl. The characteristics of the particles are listed in Table 14.5. The interaction potentials calculated with the Derjaguin approximation (4.10.10) for the electrostatic repulsion and the non-retarded Hamaker form (5.2.15) for the dispersion attraction have minima with magnitudes varying from 7 to 16 kT. Only the shear modulus was reported, and at a single frequency, 1200 s^{-1}. The corresponding dimensionless frequency, $a^2\omega/D_0 = 600$–4800, places these in the high-frequency limit. The high-frequency moduli proved to be independent of particle size, but very sensitive to volume fraction.

Patel & Russel (1987) studied two monodisperse polystyrene latices, stabilized electrostatically in 0.06 M NaCl, but flocculated or phase-separated by the addition of non-adsorbing polymer (Table 14.5). The interaction potentials, calculated with the Derjaguin approximation (4.10.10) for the electrostatic repulsion and the geometrical approximation

Fig. 14.13. Schematic of pair potential with soft repulsion and attractive minimum identifying the following: r_m, Φ_{min}, position and value of minimum; Φ''_{min}, curvature of potential at minimum; Φ'_{max} maximum attractive force.

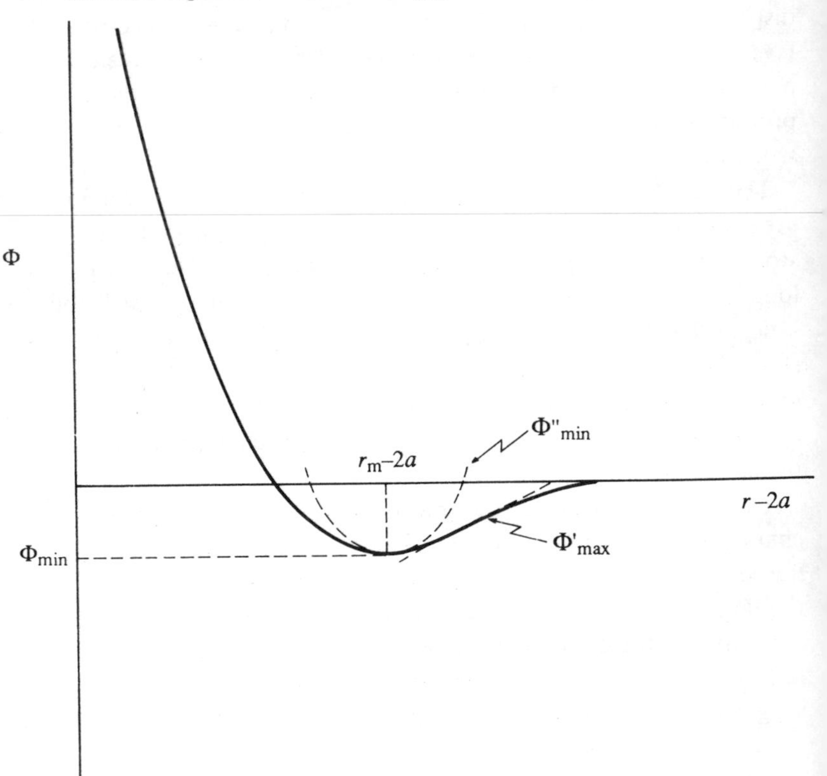

Table 14.5 *Characteristics of weakly flocculated dispersions*

(a) Goodwin *et al.* (1986)

a/nm	$\frac{\Phi_{min}}{kT}$	$\frac{a^2\Phi''_{min}}{kT}/10^6$	$\frac{kT}{a^2}\left(-\frac{\Phi_{min}}{kT}\right)^{3/2}\left(\frac{\Phi''_{min}}{kT}\right)^{1/2}$
487	−7.4	1.6	900
706	−11.2	4.7	950
958	−15.6	13.7	1030

(b) Patel & Russel (1987)

a/nm	$\frac{\Phi_{min}}{kT}$	$\frac{a^2\Phi''_{min}}{kT}/10^3$	$\frac{a\Phi'_{max}}{kT}$	Scale factor $/\frac{\mathrm{N}}{\mathrm{m}^2}$: $\frac{kT}{a^2}\left(-\frac{\Phi_{min}}{kT}\right)^{3/2}\left(\frac{\Phi''_{min}}{kT}\right)^{1/2}$	Scale factor $/\frac{\mathrm{N}}{\mathrm{m}^2}$: $\frac{\Phi'_{max}}{a^2}$
30	−11.9	0.6	23	1.5×10^5	3.6×10^3
	−17.9	1.1	35	3.8×10^5	5.3
110	−3.7	4.0	32	1.4×10^3	1.0×10^2
	−7.8	8.8	64	6.4×10^3	2.0
	−13.4	13.2	107	1.8×10^4	3.3
	−20.5	14.5	160	4.0×10^4	5.0

(6.4.13) for the volume restriction attraction, have attractive minima of 3 to 20 kT for samples in the fluid–solid region of the phase diagram (§10.7). Steady-shear and small-amplitude oscillatory measurements (Fig. 14.14) detected the yield stress σ_0 and the static modulus G_0. The dependence of these parameters on particle size and volume fraction and the polymer concentration, i.e. depth of the attractive minimum, were also measured.

These two sets of data can be correlated through a modified version of the arguments presented in §14.5. The form (14.5.4) for the modulus remains appropriate, but in weakly flocculated systems the number of nearest neighbors depends on the strength of the attractive potential and the volume fraction of particles as

$$N \approx n\int(g-1)\,\mathrm{d}\mathbf{r}. \tag{14.7.1}$$

Since at low concentrations $g=\exp(-\Phi/kT)\approx 1-\Phi/kT$, while

$$\Phi(r)=\Phi(r_m)+\tfrac{1}{2}(r-r_m)^2\frac{\mathrm{d}^2\Phi}{\mathrm{d}r^2}(r_m)+\ldots \tag{14.7.2}$$

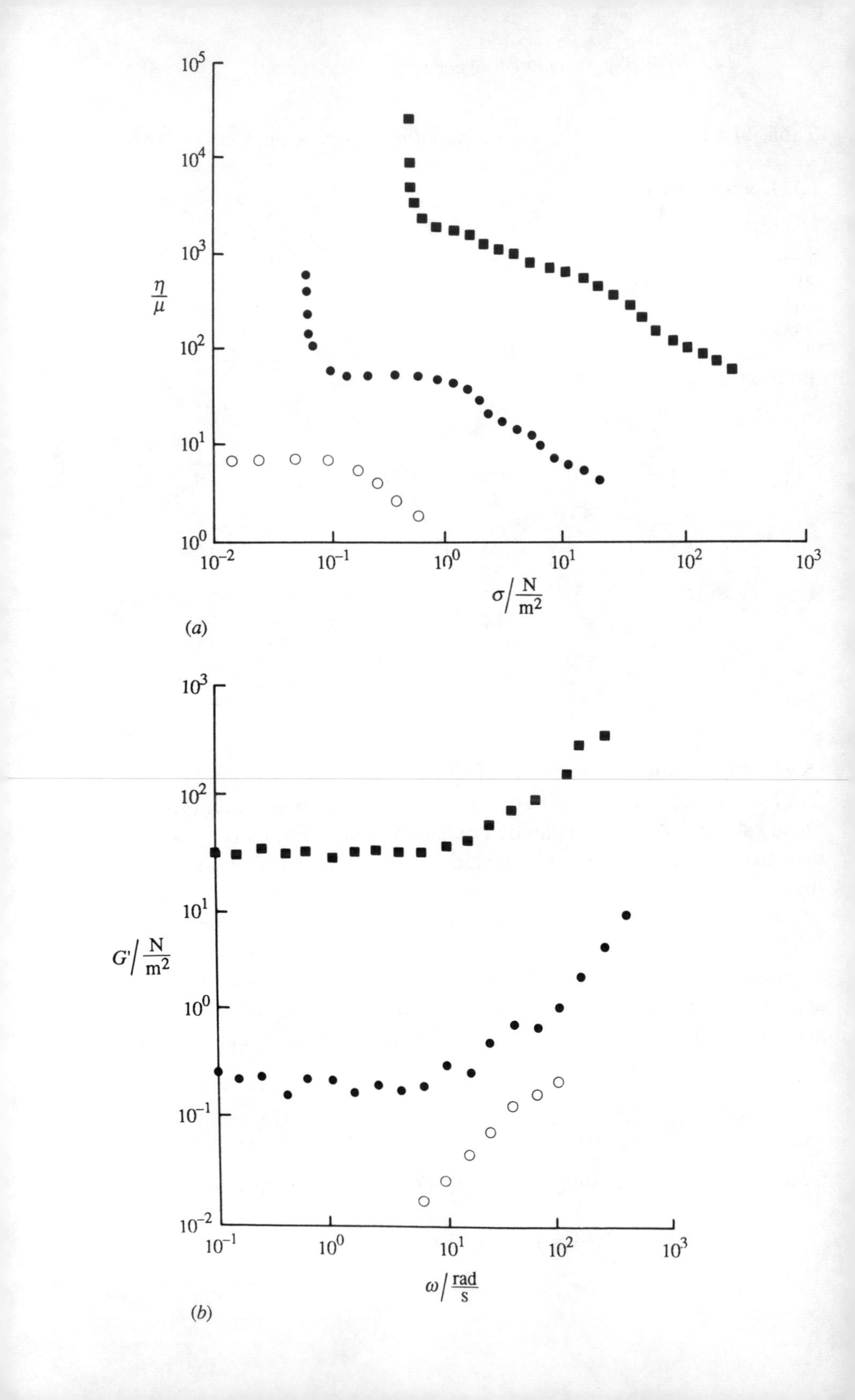

10^5
10^4
10^3
$\frac{\eta}{\mu}$
10^2
10^1
10^0
10^{-2}
10^{-1}
10^0
10^1
10^2
10^3
$\sigma / \frac{N}{m^2}$
(a)
10^3
10^2
10^1
$G' / \frac{N}{m^2}$
10^0
10^{-1}
10^{-2}
10^{-1}
10^0
10^1
10^2
10^3
$\omega / \frac{rad}{s}$
(b)

Fig. 14.14. Data for polystyrene latices ($a = 220$ nm) in water at 0.06 M NaCl and 1.5 per cent (by weight) Triton X-405 with soluble dextran ($M_w = 600$ kg/mol, $r_g = 33$ nm) added (Patel & Russel, 1987): ○, $\phi = 0.20$, $\Phi_{min} = -1.5kT$; ●, $\phi = 0.20$, $\Phi_{min} = -20kT$; ■, $\phi = 0.30$, $\Phi_{min} = -20kT$. (*a*) steady shear viscosity; (*b*) shear modulus.

in the vicinity of the minimum $\Phi(r_m) \equiv \Phi_{min}$, the number of nearest neighbors can be approximated as

$$N \approx -4\pi a^2 \left(-\frac{2\Phi_{min}}{\Phi''_{min}} \right)^{1/2} n \frac{\Phi_{min}}{kT} \tag{14.7.3}$$

with $\Phi''_{min} \equiv d^2\Phi/dr^2\,(r_m)$. Then, for $r_m \approx 2a$,

$$G_0 \approx \phi^2 \frac{kT}{a^2} \left(-\frac{\Phi_{min}}{kT} \right)^{3/2} \left(\frac{\Phi''_{min}}{kT} \right)^{1/2}. \tag{14.7.4}$$

Similarly, the yield stress corresponds to the maximum force per unit area that the network can withstand before rupturing. This is proportional to the maximum force acting between each pair multiplied by the number of bonds per unit area. In this case, though, we assume that yielding requires the rupture of only one bond per particle, so that

$$\sigma_0 \approx \frac{\phi^2}{a^2} \Phi'_{max}, \tag{14.7.5}$$

with Φ'_{max} denoting the maximum force (Fig. 14.13).

This simplistic argument does not account for the variation of the pair distribution function with volume fraction, and, therefore, anticipates a much weaker ϕ-dependence than observed. Nonetheless, the effects of particle size and the characteristics of the interaction potential embodied in (14.7.4) and (14.7.5) provide appropriate dimensionless forms for comparing these diverse sets of data (Fig. 14.15). For the polymer-induced attraction, the scaling correlates to some degree static moduli and yield stresses that vary over five orders of magnitude. For the dispersion attraction (5.2.15), the potential at a fixed separation increases approximately linearly with particle size. Hence the scale factor for the modulus becomes independent of radius, explaining the absence of the particle-size dependence for the high-frequency modulus. The difference of orders of magnitude between the dimensionless low- and high-frequency limits for the shear modulus suggests substantial relaxations in these weakly flocculated structures.

Significant scatter remains, reflecting the imperfect nature of the

Fig. 14.15. Correlation of (*a*) static and high frequency shear moduli and (*b*) yield stress according to the scalings suggested by (14.7.4) and (14.7.5):

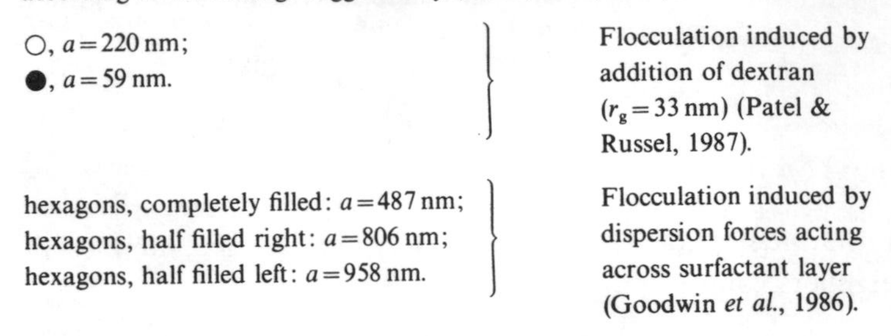

○, $a = 220$ nm; ●, $a = 59$ nm. — Flocculation induced by addition of dextran ($r_g = 33$ nm) (Patel & Russel, 1987).

hexagons, completely filled: $a = 487$ nm; hexagons, half filled right: $a = 806$ nm; hexagons, half filled left: $a = 958$ nm. — Flocculation induced by dispersion forces acting across surfactant layer (Goodwin *et al.*, 1986).

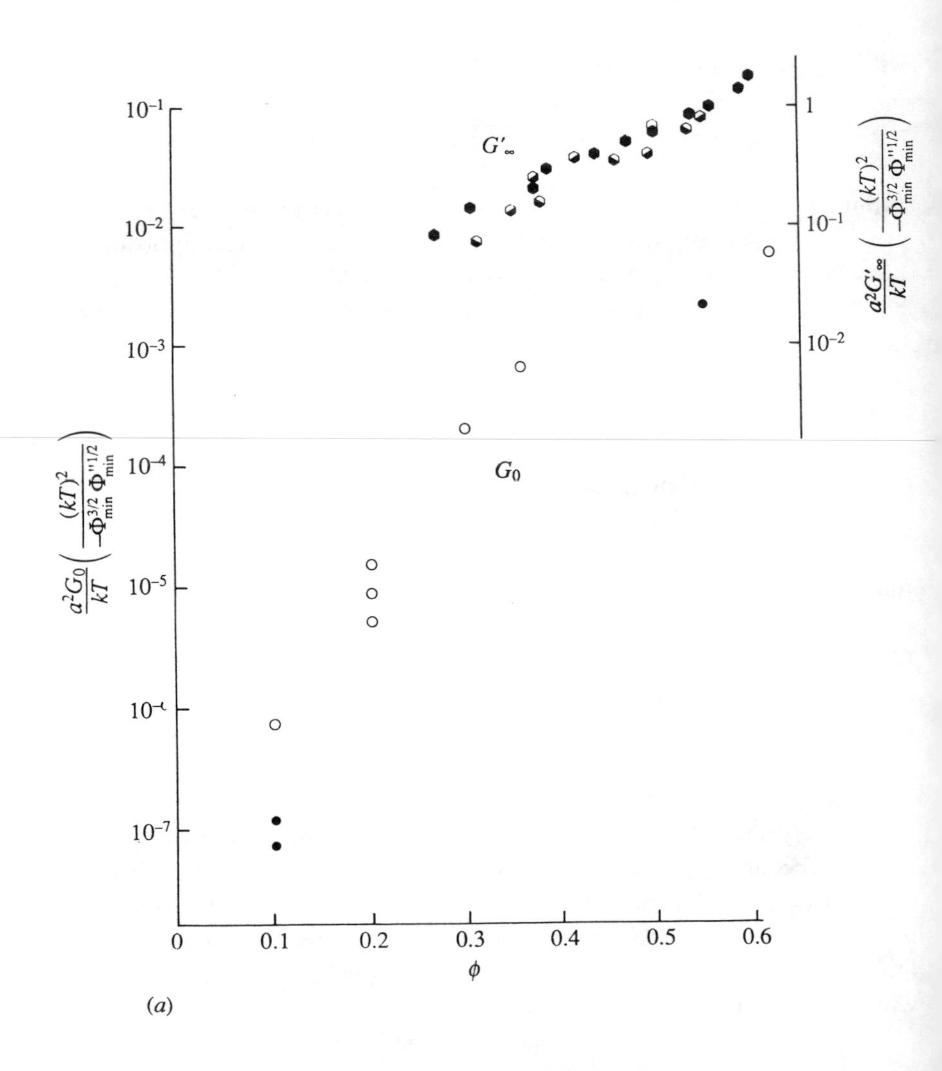

expressions for the interparticle potentials and the existence of independent dimensionless groups, e.g. ratios of the individual length scales, not entirely accounted for by the scaling. However, the exercise demonstrates at least qualitatively the relationship between interparticle forces, dispersion microstructure, and rheological properties responsible for the complex behavior of weakly flocculated dispersions. Similar attempts for strongly flocculated systems suggest somewhat different relationships but remain even sketchier.

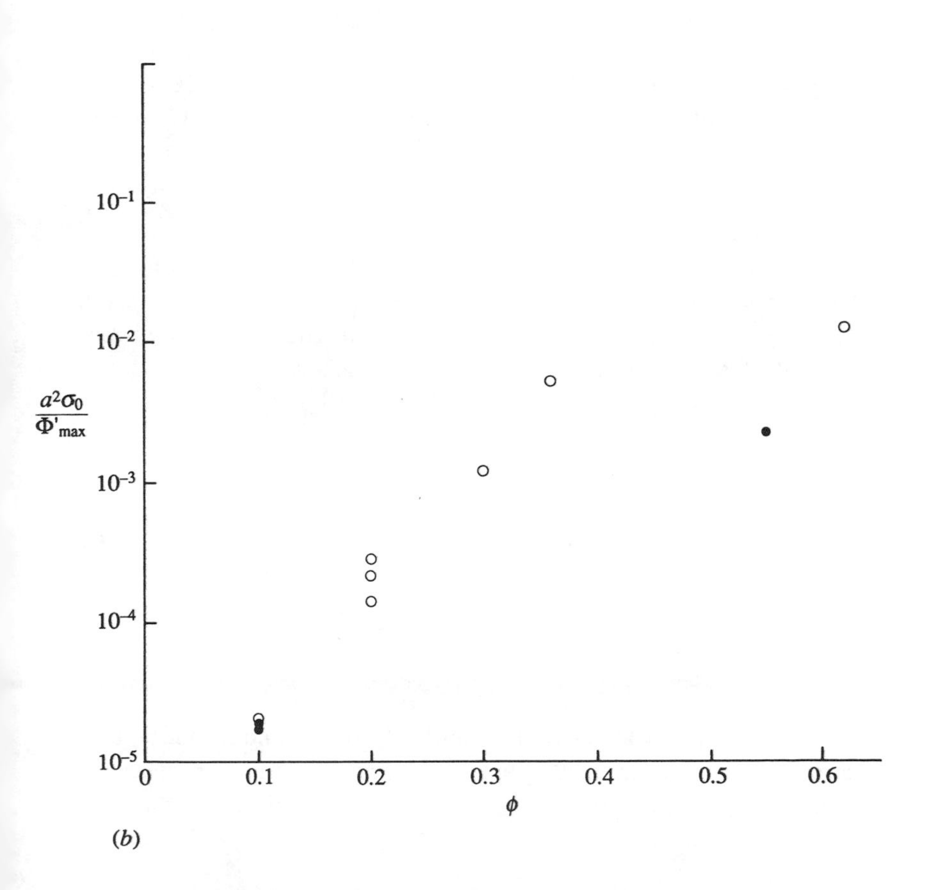

(*b*)

14.8 Motivation for pair interaction theories

Prediction of the rheological properties of a dispersion requires three elements:

(i) expressions for the hydrodynamic interactions and interparticle potentials as functions of the configuration;
(ii) a description of the non-equilibrium microstructure resulting from these forces; and
(iii) a valid method for averaging the interactions to obtain the bulk stress.

As for flocculation, sedimentation, and diffusion, a rigorous theory can be constructed only for dilute suspensions in which pair interactions dominate. Then the hydrodynamic functions presented in §2.9 and the interparticle potentials in Chapters 4–6 suffice and the microstructure, characterized by the pair distribution function P_2 (§8.3), is governed by (8.3.8). Proper averaging of these interactions, weighted by P_2, to obtain the macroscopic stresses is discussed in §14.10. The resulting prediction for the stresses, correct to $O(\phi^2)$, suggests not only the order of magnitude variations arising from the different interparticle potentials but also some of the non-Newtonian behavior seen in §§14.4 – 14.7.

14.9 Non-equilibrium microstructure

For dilute dispersions, the pair probability P_2 completely specifies the microstructure. The conservation equation, e.g. (8.3.8) for equal spheres, is composed of a balance among transient effects, convection by the applied flow, translation due to the interparticle forces, and Brownian motion as (Batchelor & Green, 1972)

$$\frac{\partial P_2}{\partial t}+\nabla_r\cdot P_2(\mathbf{U}-\boldsymbol{\omega}\cdot\nabla_r\Phi)=kT\,\nabla_r\cdot\boldsymbol{\omega}\cdot\nabla_r P_2,$$

with the boundary conditions (14.9.1)

$$\lim_{r\to 2a}\left\{\frac{\mathbf{r}}{r}\cdot[(\mathbf{U}-\boldsymbol{\omega}\cdot\nabla_r\Phi)P_2-kT\boldsymbol{\omega}\cdot\nabla_r P_2]\right\}=0,$$

$$P_2\to n^2 \qquad \text{as} \qquad r\to\infty.$$

The relative velocity $\mathbf{U}$ due to the applied shear field and the mobility $\boldsymbol{\omega}$ are described in §§2.8 and 2.9. The first boundary condition specifies impenetrable spheres and the second corresponds to a microstructure without long-range order.

Without flow, the balance between diffusion and the interparticle potential yields the Boltzmann distribution for pairs. An imposed shear flow convects the spheres relative to one another, perturbing the microstructure from equilibrium. A general solution to (14.9.1) would lead to the full constitutive equation governing the rheology of the dispersion for arbitrary flow type and strength. Unfortunately, the variable coefficients, $\mathbf{U}$, $\boldsymbol{\omega}$, and Φ, have precluded other than asymptotic solutions for large and small values of the Peclet number.

For $Pe \ll 1$, shear acts on the spherically symmetric equilibrium distribution to produce an $O(Pe)$-perturbation with the symmetry of the radial component of the relative velocity (Batchelor, 1977; Russel, 1978), i.e.,

$$P_2 = n^2 g(r)\left[1 - \frac{3\pi\mu a^3}{kT}\frac{\mathbf{r}\cdot\mathbf{E}\cdot\mathbf{r}}{r^2} f(r)\right]. \qquad (14.9.2)$$

Substitution into (14.9.1) determines the dimensionless equation governing the radial dependence as

$$\frac{1}{s^2}\frac{\mathrm{d}}{\mathrm{d}s}\left(s^2 G\frac{\mathrm{d}f}{\mathrm{d}s}\right) - G\frac{\mathrm{d}}{\mathrm{d}s}\left(\frac{\Phi}{kT}\right)\frac{\mathrm{d}f}{\mathrm{d}s} - \frac{6H}{s^2}f = -S + s(1-A)\frac{\mathrm{d}}{\mathrm{d}s}\frac{\Phi}{kT},$$

$$\lim_{s\to 2}\left[s(1-A) + G\frac{\mathrm{d}f}{\mathrm{d}s}\right] = 0, \qquad (14.9.3)$$

$$f \to 0 \quad \text{as } s \to \infty$$

with $S = 3B - (1/s^2)[\mathrm{d}(s^3 A)/\mathrm{d}s]$ and $s = r/a$. The terms on the left-hand side of the equation account for diffusion and translation due to the pair potential; the forcing terms on the right-hand side arise from the divergence of the convective velocity and convection across gradients in the equilibrium radial distribution function. Solutions for f are available for several interaction potentials, demonstrating the effects of both attraction and repulsion on the distribution function.

For hard spheres with $g = 1$ for $r > 2a$, numerical solutions are necessary to capture the effects of near-field hydrodynamic interactions (Batchelor, 1977). As shown in Fig. 14.16(a), $f(r)$ is positive with a maximum at contact; thus shear causes pairs to accumulate where the relative velocity brings particles together, i.e., $\mathbf{r}\cdot\mathbf{E}\cdot\mathbf{r} < 0$. While the clustering illustrated here arises from hydrodynamic interactions at small separations, the hard-sphere excluded volume would produce a similar result in their absence. For example, setting $G = H = 1$ and $A = S = 0$ in (14.9.3) leads to the solution $f = 32/3s^3$ (broken curve), which resembles the exact solution at least qualitatively.

For charged spheres with $\alpha = 4\pi\varepsilon_0\varepsilon\psi_s^2 a^2\kappa \exp(2a\kappa)/kT \gg 1$, g varies smoothly from zero to one at separations $\kappa d \approx \ln\{\alpha/\ln\{\alpha/\ln(\alpha. \ . \ .)]\}$. Complete analyses with, for example, the linear superposition approximation (4.10.12) for the electrostatic repulsion are straightforward (Russel, 1978), with the results shown in Fig. 14.16(*b*). However, an excluded shell potential (§10.3) with $s_0 = d/a$ satisfactorily mimics the electrostatic repulsion. For $s_0 = 2-4$, numerical solutions to (14.9.3) with the no-flux boundary condition at $s = s_0$ are again necessary (Russel, 1984). If $s_0 > 4$,

Fig. 14.16. Equilibrium and non-equilibrium components of radial distribution function defined by (14.9.2) for $Pe \ll 1$ and (14.9.4) or (14.9.6) for $Pe \gg 1$: (*a*) hard spheres with $Pe \ll 1$ (Batchelor, 1977), with broken line indicating far-field approximation; (*b*) charged spheres, with $\kappa d = 10$ and $Pe \ll 1$ (Russel, 1978). (*c*) excluded shell with $s_0 = 3$ and $Pe \ll 1$; (*d*) hard spheres in extensional flow with $Pe \gg 1$ (Batchelor & Green, 1972); (*e*) excluded shell with $d = 2a$ and $Pe \gg 1$ (Russel, 1980), with solid lines indicating singular layers.

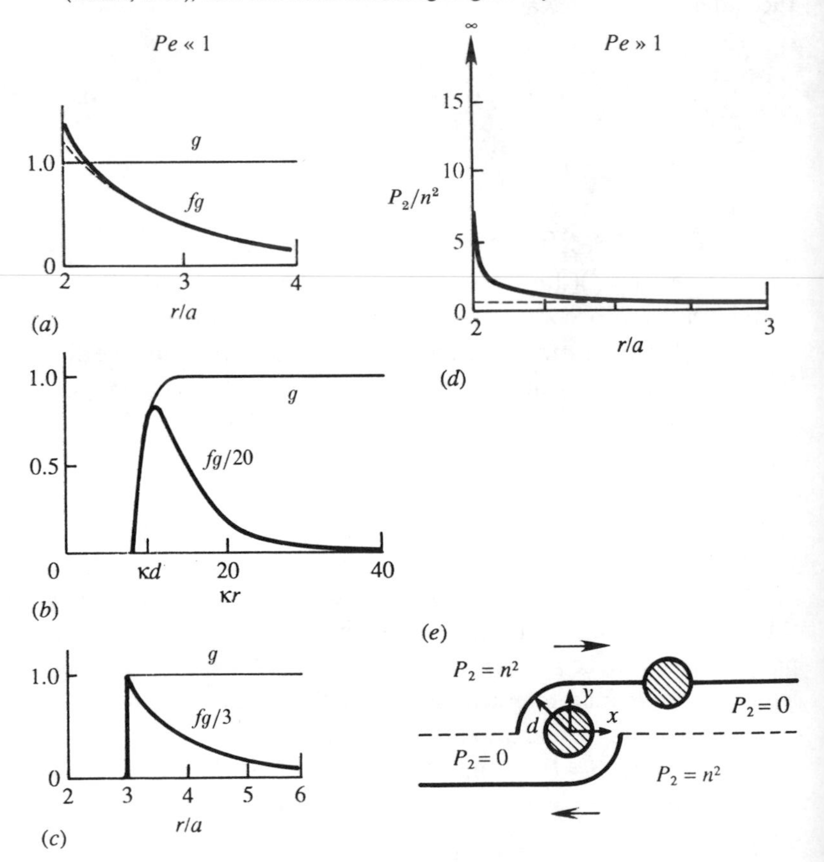

hydrodynamic interactions become insignificant, leaving $f \approx s_0^5/3s^3$. The radial dependence of the perturbation (Fig. 14.16(*c*)) resembles that for hard spheres but is displaced outward and has a larger magnitude.

The adhesive sphere (10.3.5) models attractions through a potential of zero range; the magnitude is infinite, but adds only a finite population of doublets to the hard sphere g. Since $\mathrm{d}\Phi/\mathrm{d}s$ is non-zero only at contact where $G=0$ and $A=1$, the stickiness does not enter (14.9.3), and, hence, has no effect on f (Russel, 1984). The doublets respond to the flow exactly as the other pairs near contact, preserving the symmetry of the non-equilibrium distribution function.

When $Pe \gg 1$, the diffusion term becomes negligible, reducing (14.9.3) to a first-order partial differential equation. The solution can be constructed by integrating the characteristic equations, (8.8.10) and (8.8.11). The first three determine the trajectory of interacting particles and the fourth the corresponding variation in P_2 (Batchelor & Green, 1972). Both the trajectories and P_2 depend on the dimensionless groups characterizing the ratio of the interparticle force to the viscous force. For P_2 to be determined uniquely, integration must begin upstream at large separations where $P_2 = n^2$.

For hard spheres, integrating the characteristic equations yields the spherically symmetric pair distribution function (Batchelor & Green, 1972)

$$P_2(r) = \frac{n^2}{1-A} \exp \int_r^{\infty} \frac{3(B-A)}{1-A} \frac{\mathrm{d}r}{r}. \tag{14.9.4}$$

This symmetry is characteristic of purely hydrodynamic interactions; the singularity at contact (Fig. 14.16(*d*)) reflects the lubrication stresses that cause interacting particles to accumulate at small separations. Recall, however, that the solution applies only along trajectories extending to upstream infinity. In a pure extension with zero vorticity, these trajectories cover the accessible volume, i.e., $r > 2a$. Flows such as simple shear, however, produce closed trajectories (§2.9) with no starting point for the integration, rendering the solution indeterminate in the absence of Brownian motion or three-body interactions.

For the excluded-shell potential with $s_0 = d/a \gg 2$, or for free-draining spheres, hydrodynamic interactions are unimportant. The particles follow the fluid streamlines; for example, for simple shear with $\mathbf{u} = (\gamma y, 0, 0)$, the trajectory equations reduce to

$$\frac{\mathrm{d}x}{\gamma y} = \frac{\mathrm{d}y}{0} = \frac{\mathrm{d}z}{0} \tag{14.9.5}$$

for $x^2+y^2+z^2>d^2$. Upon collision the pairs accumulate as doublets, rotate until the relative velocity $\mathbf{r}\cdot\mathbf{E}\cdot\mathbf{r}$ becomes positive for $x>0$, then separate and move downstream. As with adhesive spheres this population of doublets represents a discontinuity in the pair distribution; so the conservation equation does not apply at $r=\mathrm{d}$. However, integrating across the discontinuity to obtain a conservation equation for the doublet density leads to complete solutions of the simple form

$$P_2=n^2\begin{cases} H(s-s_0)+\dfrac{as_0}{3}\delta(s-s_0) & \begin{matrix} x<0, & y>0 \\ x>0, & y<0 \end{matrix} \\ H(\rho-s_0)+\dfrac{as_0}{3}\delta(\rho-s_0) & \begin{matrix} x>0, & y>0 \\ x<0, & y<0 \end{matrix} \end{cases} \tag{14.9.6}$$

with $(a\rho)^2=y^2+z^2$. Thus a repulsive interparticle force produces an asymmetric trajectory with a population of rotating doublets and a shadow zone downstream where $P_2=0$ (Fig. 14.16(*e*)).

For charged spheres at high Peclet number, the radius s_0 of the excluded shell corresponds to the separation at which the electrostatic repulsion balances the maximum viscous force due to the shearing flow (Blachford, Chan, & Goring, 1969; Russel, 1978).

$$3\pi\mu a^2\gamma s_0=4\pi\varepsilon\varepsilon_0\psi_s^2(a\kappa)^2 e^{2a\kappa}\frac{1+a\kappa s_0}{(a\kappa s_0)^2}e^{-a\kappa s_0}. \tag{14.9.7}$$

Consequently, s_0 decreases with increasing dimensionless shear rate Γ from

$$s_0\sim-\frac{\ln[\Gamma(\ln\Gamma)^2]}{a\kappa}$$

for

$$\frac{kT}{4\pi\varepsilon_0\varepsilon\psi_s^2 a}\frac{e^{-2a\kappa}}{a\kappa}\ll\Gamma\ll 1 \tag{14.9.8}$$

to

$$s_0\sim\frac{1}{a\kappa\Gamma^{1/3}}$$

for $1\ll\Gamma\ll(a\kappa)^{-3}$ and finally $s_0\sim 2$ for $(a\kappa)^{-3}\ll\Gamma$.

These results reveal several characteristic features of the non-equilibrium microstructure. For weak flows, the $O(Pe)$-perturbation is oriented in the principal direction of strain, with the magnitude depending on the range of the interparticle potential. In strong flows without closed trajectories, the asymptotic form is independent of Pe, but interparticle forces destroy the symmetry evident with purely hydrodynamic interactions.

14.10 Macroscopic stresses

The bulk stresses characterizing the macroscopic rheology of a suspension follow from averaging the interparticle forces, the hydrodynamic interactions, and the equilibrium and non-equilibrium microstructure. The development below parallels that of Batchelor and coworkers (Batchelor, 1970; Batchelor & Green, 1972; Batchelor, 1977; Russel, 1978) and addresses only the deviatoric stresses that arise from three sources: the mechanical stresses $\boldsymbol{\sigma}$ due to the viscous forces and the thermodynamic stresses $\boldsymbol{\tau}$ from Brownian motion and the interparticle potentials.

The mechanical stresses satisfy the local momentum equation

$$\nabla \cdot \boldsymbol{\sigma} = -\sum_{i=1}^{N} \mathbf{F}_i \delta(\mathbf{x}-\mathbf{x}_i), \tag{14.10.1}$$

with $\mathbf{F}_i$ denoting the total non-hydrodynamic force acting at the center of the ith sphere. Averaging over a homogeneous representative volume V containing N particles gives the contribution to the bulk stress as

$$\begin{aligned} \langle \boldsymbol{\sigma} \rangle &= \frac{1}{V} \int_V \boldsymbol{\sigma} \, \mathrm{d}V \\ &= 2\mu \mathbf{E} + \frac{1}{V} \sum_{i=1}^{N} \mathbf{S}_i, \end{aligned} \tag{14.10.2}$$

with

$$\mathbf{S}_i = \int_{V_i} (\boldsymbol{\sigma} - 2\mu \mathbf{e}) \, \mathrm{d}V.$$

The divergence theorem and (14.10.1) allow the stresslet $\mathbf{S}_i$, exerted on the fluid by each rigid particle, to be expressed in terms of the viscous tractions acting on the particle surface as

$$\mathbf{S}_i = \int_{A_i} \{\boldsymbol{\sigma} \cdot \mathbf{n}(\mathbf{x}-\mathbf{x}_i) - \tfrac{1}{3} \boldsymbol{\delta}(\mathbf{x}-\mathbf{x}_i) \cdot \boldsymbol{\sigma} \cdot \mathbf{n}\} \, \mathrm{d}A. \tag{14.10.3}$$

Further decomposition for pair interactions as

$$\mathbf{S}_i = \mathbf{S}_{i0} + \sum_{j \neq i} (\mathbf{S}'_{ij} - (\mathbf{x}_j - \mathbf{x}_i) \mathbf{C}_{ij} \cdot \mathbf{F}_i) \tag{14.10.4}$$

identifies explicitly the dipoles induced in force- and torque-free particles by the applied shear, $\mathbf{S}_{i0}$ for an isolated particle (2.6.11) plus $\mathbf{S}_i'$ due to hydrodynamic interactions from (2.9.4). The coupling tensor $\mathbf{C}_{ij}$ (2.9.3) relates the stresslet to the interparticle force, with $\mathbf{F}_i = -\mathbf{F}_j$. Since $\mathbf{C}_{ij}$ decays as r^{-3}, non-hydrodynamic forces contribute to the mechanical stress only through hydrodynamic interactions.

The thermodynamic stress is obtained from the change in the Helmholtz free energy induced by an arbitrary homogeneous deformation $\boldsymbol{\varepsilon}$. From (6.2.10) and (6.2.14), the free energy for a particular configuration can be written as

$$A = kT \ln P_N + \Phi, \tag{14.10.5}$$

with Φ denoting the sum of all the interparticle potentials. Then the effect of the deformation $\boldsymbol{\varepsilon}$ is

$$\mathrm{d}A = V\boldsymbol{\varepsilon} : \langle \boldsymbol{\tau} \rangle = kT\, \mathrm{d} \ln P_N + \mathrm{d}\Phi \tag{14.10.6}$$

for the usual representative volume V. With the chain rule for differentiation and the relation between $\boldsymbol{\varepsilon}$ and particle displacements, (14.10.6) becomes

$$\mathrm{d}A = \sum_{i=1}^{N} \mathbf{x}_i \cdot \boldsymbol{\varepsilon} \cdot \nabla_i (kT \ln P_N + \Phi). \tag{14.10.7}$$

For arbitrary deformations $\boldsymbol{\varepsilon}$ this means

$$\begin{aligned} \langle \boldsymbol{\tau} \rangle &= \frac{1}{V} \sum_{i=1}^{N} \mathbf{x}_i \nabla_i (kT \ln P_N + \Phi) \\ &= -\frac{1}{V} \sum_{i=1}^{N} \mathbf{x}_i \mathbf{F}_i \end{aligned} \tag{14.10.8}$$

where the total non-hydrodynamic force $\mathbf{F}_i$ includes the Brownian force, $-kT\nabla_i \ln P_N$, in addition to the interparticle force, $-\nabla_i \Phi = \nabla_i \sum_{j \neq i} \Phi(r_{ij})$.

Evaluation of these ensemble averages of both the mechanical and thermodynamic stresses is fraught with difficulties arising from non-convergent integrals. Several techniques are now available for renormalizing these integrals and extracting unambiguous expressions for the macroscopic stress. Here, we combine the approach of O'Brien (1979), described in the earlier treatment of sedimentation (§12.2), with that of Batchelor (1977).

Within the volume V, the velocity field can be represented through the integral solution (12.2.3) to the Stokes equation. The corresponding stress field, required by the expressions (14.10.3) for the dipoles, takes the form

$$\boldsymbol{\sigma}(\mathbf{x}) = -\int_{S+\Sigma A_i} \{\mathbf{J}\cdot\boldsymbol{\sigma}\cdot\mathbf{n} - \mathbf{n}\cdot(\mathbf{u}\cdot\mathbf{K})\}\mathbf{dx}', \tag{14.10.9}$$

with **K** a third rank tensor obtained from **I** and **J**. Recall that the integral over S, the surface bounding V, accounts for the motion of particles and fluid outside V and the integrals over the A_i for the particles within V. Substitution of (14.10.9) into (14.10.3), expansion of **J** and **K** about the center of the ith particle, and integration over A_i provides the requisite expressions for the dipoles.

Now the ensemble average of the mechanical contribution to the stress can be written as

$$\langle\boldsymbol{\sigma}\rangle = 2\mu\mathbf{E} + n\mathbf{S}_{10} + n\langle\mathbf{S}'_{12}\rangle - \frac{4\pi a^3}{3} n \int_S \mathbf{J}'\cdot\langle\boldsymbol{\sigma}'\rangle\cdot\mathbf{n}\,\mathbf{dx}' + O(\phi^3), \tag{14.10.10}$$

with $\mathbf{J}' = \mathbf{J} - \frac{1}{3}\boldsymbol{\delta}:\mathbf{J}\boldsymbol{\delta}$

$$\mathbf{S}_{10} = \frac{20\pi}{3} a^3 \mu \mathbf{E}.$$

Here the integrals over the A_i have been identified with $\mathbf{S}_{10}$ and $\mathbf{S}'_{12}$ for particles within V interacting in a pairwise fashion. Also $\boldsymbol{\sigma}'$, the disturbance due to the particles, replaces the full stress $\boldsymbol{\sigma}$ since $\mathbf{S}_{10}$, the dipole for an isolated sphere, fully accounts for the effect of the solvent stress $2\mu\mathbf{E}$. The ensemble average commutes with the integral over S as shown, because only $\boldsymbol{\sigma}'$ depends on the configuration of particles within V. The second term in the integrand of (14.10.9) disappears, since $\langle\mathbf{u}'\mathbf{n}\rangle \equiv \mathbf{0}$, and (14.10.10) shows that

$$\langle\boldsymbol{\sigma}'\rangle = n\mathbf{S}_{10} + O(\phi^2), \tag{14.10.11}$$

which suffices for the integral. Finally, one should note that the far-field form of the pair dipole (2.9.7) corresponds to

$$\lim_{r\to\infty} \mathbf{S}'_{12} = \frac{4\pi a^3}{3} \mathbf{S}_{10} : \nabla\mathbf{J}'. \tag{14.10.12}$$

Since $\mathbf{S}'_{12}$ decays as r^{-3} for $r\to\infty$, (14.10.10) contains a non-convergent integral. However, converting the integral over S to integrals over the fluid volume and A_1 via the divergence theorem produces

$$-\frac{4\pi a^3}{3}n\int_S \mathbf{J}'\cdot\langle\boldsymbol{\sigma}'\rangle\cdot\mathbf{n}\,\mathrm{d}\mathbf{x}' = \frac{4\pi a^3}{3}n^2\left\{\int_{A_1}\mathbf{J}'\cdot\mathbf{S}_{10}\cdot\mathbf{n}\,\mathrm{d}\mathbf{x}' - \int_{r\geq a}\mathbf{S}_{10}:\nabla\mathbf{J}'\,\mathrm{d}\mathbf{x}'\right\}. \tag{14.10.13}$$

Thus the volume integral over $r \geq a$ cancels the non-convergence in the ensemble average of $\mathbf{S}_{12}'$.

Similarly, the ensemble average of the Brownian portion of the thermodynamic stress fails to converge since $\int \nabla P_N \,\mathrm{d}\mathbf{x}_3 \ldots \mathrm{d}\mathbf{x}_N \sim r^{-3}$ as $r \to \infty$. However, applying the divergence theorem yields

$$\begin{aligned}\langle(\mathbf{x}_2-\mathbf{x}_1)kT\nabla_2 \ln P_N\rangle &= \frac{1}{(N-1)!}\int kT(\mathbf{x}_2-\mathbf{x}_1)\nabla_2 P_N\,\mathrm{d}\mathbf{x}_2\ldots\mathrm{d}\mathbf{x}_N\\ &= -\frac{1}{(N-1)!}\int kT\boldsymbol{\delta}P_N\,\mathrm{d}\mathbf{x}_2\ldots\mathrm{d}\mathbf{x}_N\\ &\quad+\frac{kT}{(N-1)!}\int\left\{\int_S(\mathbf{x}_2-\mathbf{x}_1)\mathbf{n}P_N\,\mathrm{d}\mathbf{x}_2\right.\\ &\quad\left.+\int_{r=2a}(\mathbf{x}_2-\mathbf{x}_1)\mathbf{n}P_N\,\mathrm{d}\mathbf{x}_2\right\}\mathrm{d}\mathbf{x}_3\ldots\mathrm{d}\mathbf{x}_N.\end{aligned} \tag{14.10.14}$$

But for $\mathbf{x}_2$ on S,

$$\frac{1}{(N-1)!}\int P_N\,\mathrm{d}\mathbf{x}_3\ldots\mathrm{d}\mathbf{x}_N = \frac{P_2}{N-1} = \frac{n}{V},$$

while (14.10.15)

$$\frac{1}{(N-1)!}\int P_N\,\mathrm{d}\mathbf{x}_2\ldots\mathrm{d}\mathbf{x}_N = n,$$

$$\int(\mathbf{x}_2-\mathbf{x}_1)\mathbf{n}\,\mathrm{d}\mathbf{x}_2 = V\boldsymbol{\delta},$$

so the first two terms cancel, leaving only the convergent integral over the surface $r=2a$ enclosing the hard-sphere excluded volume.

The final expression for the bulk stress consists of the mechanical and thermodynamic portions with contribution to each from both the Brownian and interparticle forces. For later purposes it is convenient to separate the terms due to hydrodynamic, interparticle, and Brownian forces as

$$\langle \boldsymbol{\sigma} \rangle^{\mathrm{H}} = 2\mu \mathbf{E}(1+\tfrac{5}{2}\phi) + \int \left(\mathbf{S}'_{12} - \frac{20\pi}{3} a^3 \mu \mathbf{e}'\right) P_2 \, \mathrm{d}\mathbf{r}$$
$$+ \frac{20\pi}{3} a^3 \mu \int \mathbf{e}'(P_2 - n^2) \, \mathrm{d}\mathbf{r}, \tag{14.10.16}$$

$$\langle \boldsymbol{\sigma} \rangle^{\mathrm{I}} = \tfrac{1}{2} \int (1-A) r \frac{\mathrm{d}\Phi}{\mathrm{d}r} \left(\frac{\mathbf{rr}}{r^2} - \tfrac{1}{3}\boldsymbol{\delta} \right) P_2 \, \mathrm{d}\mathbf{r}, \tag{14.10.17}$$

$$\langle \boldsymbol{\sigma} \rangle^{\mathrm{Br}} = -\tfrac{1}{2} kT \int S(r) \left(\frac{\mathbf{rr}}{r^2} - \tfrac{1}{3}\boldsymbol{\delta} \right) P_2 \, \mathrm{d}\mathbf{r}, \tag{14.10.18}$$

with $\mathbf{e}' = (4\pi a^3/3)\mathbf{E}:\nabla \mathbf{J}'$.

14.11 Results and comparison with experiment

Before proceeding it is instructive to examine the magnitude of the stress dipoles arising from the various forces. For the hydrodynamic interactions (14.10.10) indicates that $\langle \boldsymbol{\sigma} \rangle \approx \mu \phi^2 \mathbf{E}$, but for the Brownian and interparticle forces the scaling must account for the non-equilibrium microstructure. At low shear rates the $O(Pe)$-perturbation, together with $S = O(1)$ at separations of $O(2a)$, establishes that $\langle \boldsymbol{\sigma} \rangle^{\mathrm{Br}} \approx \mu \phi^2 \mathbf{E}$ also. High shear rates or long-range repulsions render the Brownian term entirely negligible as $S \sim (a/r)^6$. Since viscous forces cause the perturbation, the restoring interparticle force should scale as $F \approx \mu a^2 s_0 \gamma$, with s_0 representing the characteristic separation. Then

$\langle \boldsymbol{\sigma} \rangle^{\mathrm{I}} \approx \mu a^3 s_0^2 \mathbf{E}$	$\times$	$a^3 s_0^3$	$\times$	n^2	$\approx$	$\mu s_0^5 \phi^2 \mathbf{E}$.	(14.11.1)
dipole strength		volume		pair density			

For hard and adhesive spheres, $s_0 \approx 2$, leading to an $O(1)$ ϕ^2-coefficient in the expansion for the viscosity. But the direct contribution from repulsive potentials becomes large when $s_0 \gg 2$, as for charged spheres at low ionic strengths. Coupled with the estimates for s_0 discussed in the previous section, this explains at least qualitatively the dramatic dependence on ionic strength and shear rate of the steady shear viscosities measured for the deionized latices.

Substitution of the perturbation expansion for $Pe \ll 1$ (14.9.2) into the bulk stress identifies the low shear limiting viscosity as

$$\frac{\eta_0}{\mu}=1+\tfrac{5}{2}\phi+\tfrac{5}{2}\phi^2+\tfrac{15}{2}\phi^2\int_2^\infty Jgs^2\,\mathrm{d}s$$

near-field hydrodynamics

$$+\tfrac{9}{40}\phi^2\int_2^\infty s^3(1-A)\frac{\mathrm{d}g}{\mathrm{d}s}f\,\mathrm{d}s \tag{14.11.2}$$

interparticle force

$$+\tfrac{9}{40}\phi^2\int_2^\infty s^2Sgf\,\mathrm{d}s+O(\phi^3),$$

Brownian motion

with $J(s)$ representing the angular average of the renormalized pair dipole in (14.10.16).

For hard spheres, Batchelor (1977) calculated the low shear viscosity from the numerical solution for f as

$$\eta_0/\mu=1+2.5\phi+6.0\phi^2+O(\phi^3). \tag{14.11.3}$$

The bulk of the $O(\phi^2)$ coefficient comes from the viscous stresses associated with the uniform equilibrium state, leaving a smaller contribution (0.99) from the Brownian stresses generated by the non-equilibrium pair distribution function. The hard-sphere repulsion does not contribute, since $(1-A)\,\mathrm{d}g/\mathrm{d}s=0$ everywhere, as $A\rightarrow 1$ at contact and $g=1$ for $s>2$.

Data from careful experiments by Saunders (1961), with dilute polystyrene latices having $a=0.042$, 0.087 μm, agree reasonably well with the hard-sphere theory at the lowest volume fractions (Fig. 14.17), particularly in light of the difficulty in suppressing long-range potentials. Note, however, that the data depart from the asymptote, i.e., three-body interactions become significant, for $\phi\geq 0.1$.

For adhesive spheres with stickiness $1/\tau$, the potential contributes, since

$$\int_2^\infty (1-A)\frac{\mathrm{d}g}{\mathrm{d}s}f\,\mathrm{d}s=-\frac{1}{6\tau}\frac{\mathrm{d}}{\mathrm{d}s}[s^3(1-A)f]_{s=2}, \tag{14.11.4}$$

and the doublets also generate additional viscous, $2\mu\mathbf{E}(1.11/\tau)\phi^2$, and Brownian, $2\mu\mathbf{E}(-2.83/\tau)\phi^2$, stresses once one properly accounts for a subtle discontinuity in f at contact (Cichocki and Felderhof, 1990). Consequently, the low shear viscosity,

$$\frac{\eta_0}{\mu}=1+2.5\phi+\left(6.0+\frac{1.90}{\tau}\right)\phi^2+O(\phi^3), \tag{14.11.5}$$

exceeds that for hard spheres.

The excluded-shell repulsion with $s_0 \gg 2$ eliminates all but far-field hydrodynamic interactions. This suppresses the Brownian stress and reduces the viscous stress, but increases that from the interparticle force. The result of substituting the far-field form for f into (14.11.2) is (Russel, 1978)

$$\frac{\eta_0}{\mu} = 1 + 2.5\phi + (2.5 + \tfrac{3}{40}s_0^5)\phi^2 + O(\phi^3). \tag{14.11.6}$$

For smaller values of s_0, numerical solutions for f determine the values for the $O(\phi^2)$-coefficient plotted in Fig. 14.18.

Data for polystyrene latices (Stone-Masui & Watillon, 1968) and bovine serum albumin (Tanford & Buzzell, 1956) in Fig. 14.19 confirm the large magnitude of the coefficient at low ionic strengths and the accuracy of the asymptotic form of the theory with $a\kappa s_0 = \ln\{\alpha/\ln[\alpha/\ln(\alpha/\ldots)]\}$. Recall though that the dilute theory remains strictly valid only when $\eta_0/\mu - 1 \ll 1$.

Representing the predictions for the magnitude of the $O(\phi^2)$ coefficient as a function of the excluded volume (Fig. 14.18) emphasizes that the minimum in viscosity is associated with the hard spheres; both attractive, $A_2/A_2^{\text{HS}} < 1$, and repulsive, $A_2/A_2^{\text{HS}} > 1$, interactions increase the viscosity.

For $Pe \gg 1$, the viscous stresses and the interparticle potential govern the trajectories of the interacting particles. Brownian motion need be considered only if the resulting pair distribution function is indeterminate, as

Fig. 14.17. Low shear viscosity for dilute suspensions of hard spheres (Russel, 1980): ○, data for polystyrene latices (a=42, 87 nm) in water (Saunders, 1961); ——, theory of Batchelor (1977).

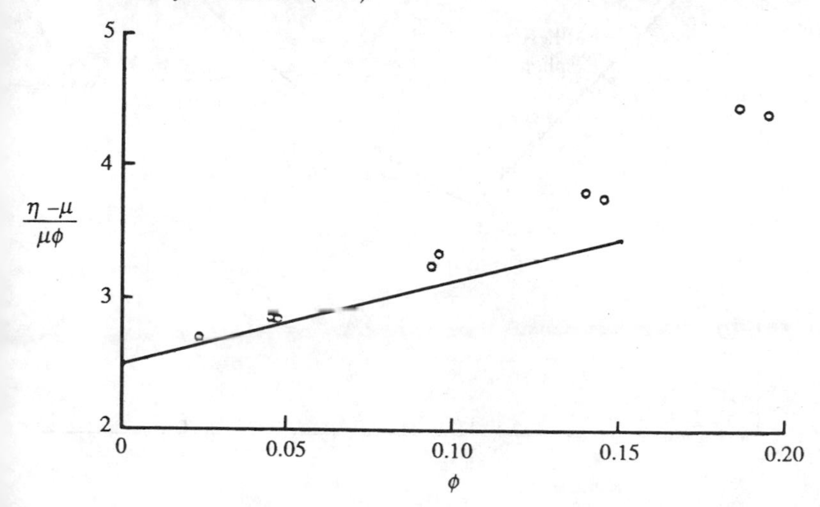

for hard spheres in a simple shear flow. For hard spheres in pure extension, the viscosity calculated with (14.9.4) and (14.10.16) (Batchelor & Green, 1972),

$$\eta_\infty/\mu = 1 + 2.5\phi + 7.6\phi^2 + O(\phi^3), \tag{14.11.7}$$

derives entirely from the viscous stress. The clustering induced by the near-field hydrodynamic interactions accounts for the difference between this result and the viscous term, $5.2\phi^2$, in the low shear limit. As a result, the suspensions are slightly strain-thickening at dilute concentrations.

For the excluded-shell potential with $s_0 \gg 2$ or for free-draining spheres, both affected only by far-field hydrodynamic interactions, the asymmetric trajectories in Fig. 14.16(*e*) translate into anisotropic stresses. From the viscous stress come only the single-particle and far-field effects,

$$\langle\sigma_{xy}\rangle^{\mathrm{H}} = 2.5\mu\gamma\phi(1+\phi).$$

Fig. 14.18. $O(\phi^2)$-coefficient for low shear viscosity as function of excluded volume from adhesive sphere and excluded-shell models (Russel, 1984).

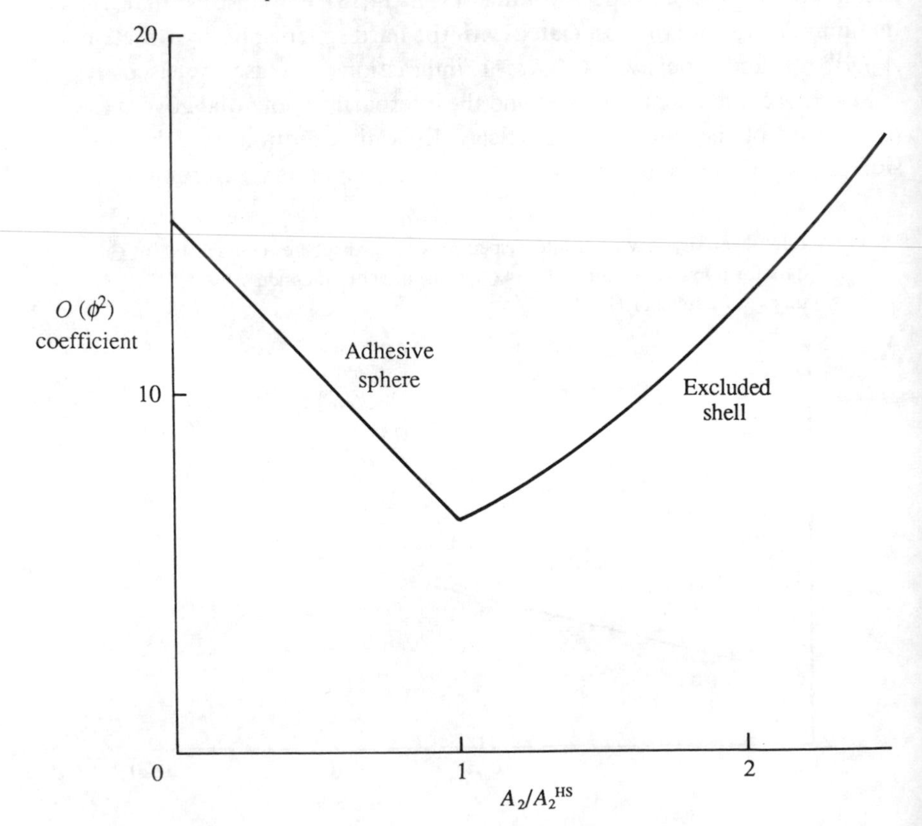

The remainder derives from the action of the repulsion on the population of doublets that accumulate owing to collisions. The magnitude of the repulsion at contact must balance the viscous force pushing the spheres together, i.e.

$$-\frac{d\Phi}{ds} = -\frac{3\pi\mu a^2 \mathbf{s}\cdot\mathbf{E}\cdot\mathbf{s}}{s} \quad \text{at} \quad s = s_0, \tag{14.11.8}$$

Fig. 14.19. $O(\phi^2)$-coefficient for low shear viscosity for charged spheres (Russel, 1980): ○, data for polystyrene latices (a=25–45 nm) (Stone-Masui & Watillon, 1968); △, data for bovine serum albumin (a=3.6 nm) (Tanford & Buzzell, 1954); ——, predictions from (14.11.6) with $a\kappa s_0 = \ln\{\alpha/\ln[\alpha/\ln\alpha\ldots]\}$

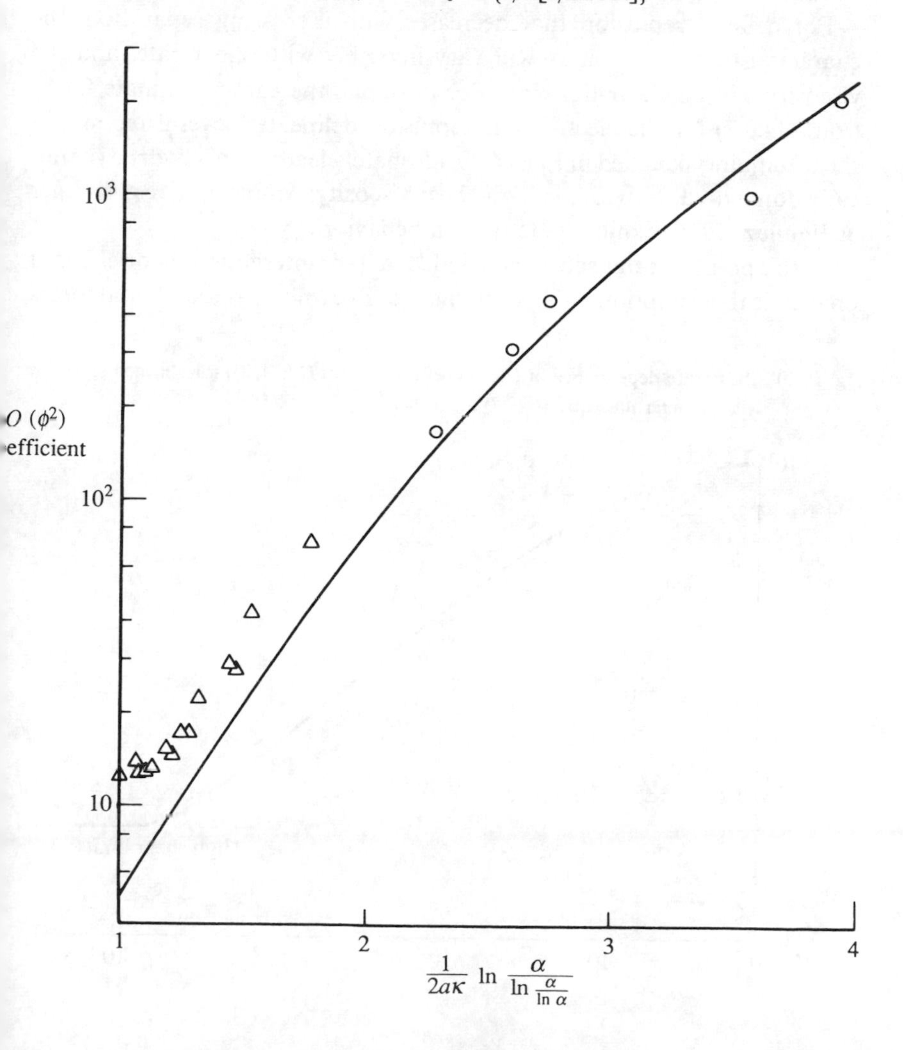

to maintain the no-flux boundary condition. The force is zero when $s > s_0$ or $\mathbf{s} \cdot \mathbf{E} \cdot \mathbf{s} > 0$. Substitution of (14.11.8) and (14.9.6) into the interparticle stress (14.10.17) and integration over the upstream quadrants of the collision surface produces a shear stress,

$$\langle \sigma_{xy} \rangle^{\mathrm{I}} = \tfrac{3}{40} \gamma s_0^5 \phi^2. \tag{14.11.9}$$

Summing the shear stresses determines the high shear viscosity as

$$\frac{\eta_\infty}{\mu} = 1 + \tfrac{5}{2}\phi + (\tfrac{5}{2} + \tfrac{3}{40} s_0^5)\phi^2 + \ldots, \tag{14.11.10}$$

which is identical to the low shear limit (14.11.6).

For a finite repulsion that decreases with increasing separation, the characteristic separation s_0 will vary inversely with shear rate, and the viscosity will be shear-thinning. For example, the earlier estimate for s_0 from (14.9.7) for the electrostatic repulsion delineates several regimes of shear thinning sketched in Fig. 14.20, ultimately leading to a hydrodynamically dominated, Newtonian high-shear viscosity. Ample data (e.g. Krieger & Eguiluz, 1976) exhibit exactly such behavior.

At this point it seems safe to conclude that pair interaction theories based on classical descriptions of hydrodynamic, Brownian, and colloidal forces

Fig. 14.20. Shear rate dependence of $O(\phi^2)$-coefficient from (14.11.10) for charged spheres, with s_0 determined by (14.9.7) for $a\kappa = 1$.

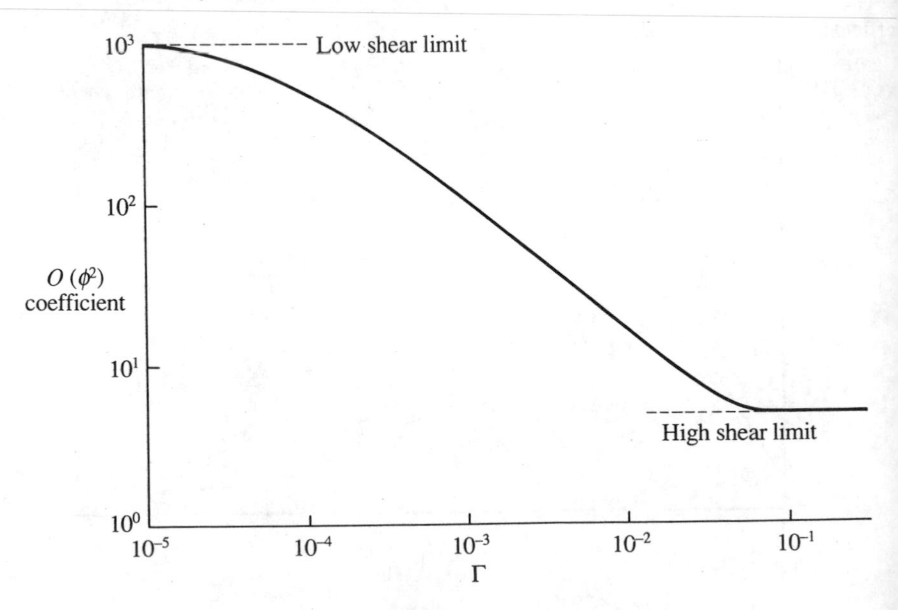

can predict quantitatively the rheological properties of dilute suspensions, at least at low shear rates. Furthermore, they provide physical interpretations for several of the phenomena described in earlier sections for concentrated systems.

14.12 Summary

The complex rheology of colloidal dispersions indicates the interplay of the full range of forces at the microstructural level. The shear viscosity, the shear modulus, and the extent of shear thinning all increase as interparticle potentials, either attractive or repulsive, become strong relative to Brownian motion. The low shear viscosity and static modulus, which derive from small perturbations of the structure at rest, provide the most sensitive probes of these particle potentials. The data show that η_0 diverges when the effective volume fraction, accounting for repulsive potentials, reaches random close packing or attractions cause the formation of a volume spanning network. At higher volume fractions, the yield stress and static modulus become measurable. The high-shear and high-frequency limiting viscosities are generally much lower and controlled by hydrodynamic interactions.

The non-equilibrium microstructures generated by shear become anisotropic when the viscous forces dominate Brownian motion. Indeed, diffraction and scattering techniques provide means of correlating structure with rheology. These and other complex phenomena, such as time-dependent effects and shear thickening, are quite important but lie beyond the scope of this treatment.

The analyses of pair interactions suggest the physics responsible for the behavior of concentrated dispersions and lay the basis for more general theories. Several approaches are currently being developed, attempting to achieve through various approximations suitable compromises between the intractability of rigorous descriptions of interactions among large ensembles of particles and the inadequacies of simplistic models that ignore essential features of either the structure or the forces.

References

Barnes, H. A., Edwards, M. F. & Woodcock, L. V. (1987). Applications of computer simulations to dense suspension rheology. *Chem. Eng. Sci.* **42**, 591–608.

Batchelor, G. K. (1970). The stress system in a suspension of force-free particles. *J. Fluid Mech.* **41**, 545–70.

Batchelor, G. K. (1977). The effect of Brownian motion on the bulk stress in a suspension of spherical particles. *J. Fluid Mech.* **83**, 97–117.

Batchelor, G. K. & Green, J. T. (1972). The determination of the bulk stress in a suspension of spherical particles to order c^2. *J. Fluid Mech.* **56**, 401–72.

Bird, R. B., Armstrong, R. C. & Hassager, O. (1987). *Dynamics of Polymeric Liquids. I. Fluid Mechanics.* Wiley.

Blachford, J., Chan, F. S. & Goring, D. A. I. (1969). Secondary electroviscous effect. Paths of approach of two charged spheres in a viscous medium. *J. Phys. Chem.* **73**, 1062–5.

Buscall, R., Goodwin, J. W., Hawkins, M. W. & Ottewill, R. H. (1982). Viscoelastic properties of concentrated latices: I. Methods of examination; II. Theoretical analysis. *J. Chem. Soc. Far. Trans. I* **78**, 2873–87, 2889–99.

Chaffey, C. E. (1977). Mechanisms and equations for shear thinning and thickening in dispersions. *Colloid and Poly. Sci.* **255**, 691–8.

Cichocki, B. & Felderhof, B. U. (1990). Diffusion coefficients and effective viscosity of suspensions of sticky hard spheres with hydrodynamic interactions. *J. Chem. Phys.* **93**, 4427–32.

de Kruif, C. G., van Iersel, E. M. F., Vrij, A. & Russel, W. B. (1986). Hard sphere colloidal dispersions: viscosity as a function of shear rate and volume fraction. *J. Chem. Phys.* **83**, 4717–25.

Evans, D., Hanley, H. J. M. & Hess, S. (1984). Non-Newtonian phenomena in simple fluids. *Physics Today* **37(1)**, 26–33.

Frith, W. J. (1986). The rheology of well characterized non-aqueous PMMA suspensions. Ph.D. thesis, Katholieke Universiteit Leuven.

Goodwin, J. W., Hughes, R. W., Partridge, S. J. & Zukoski, C. F. (1986). The elasticity of weakly flocculated suspensions. *J. Chem. Phys.* **85**, 559–66.

Goodwin, J. W. & Khidher, A. M. (1976). The effect of the addition of water soluble polymer on the elasticity of polymer latex gels. In *Colloid and Surface Science*, Vol. IV (ed. M. Kerker), pp. 529–47. Academic Press.

Hoffman, R. L. (1972). Discontinuous and dilatant viscosity behavior in concentrated suspensions. I. Observation of a flow instability. *Trans. Soc. Rheo.* **16**, 155–73. (1974) II. Theory and experimental tests. *J. Colloid Interface Sci.* **46**, 491–506.

Joanny, J. F. (1979). Acoustic shear waves in colloidal crystals. *J. Colloid Interface Sci.* **71**, 622–4.

Krieger, I. M. (1972). Rheology of monodisperse latices. *Adv. Colloid Interface Sci.* **3**, 111–36.

Krieger, I. M. & Eguiluz, M. (1976). The second electroviscous effect in polymer latices. *Trans. Soc. Rheol.* **20**, 29–45.

Lindsay, H. M. & Chaikin, P. M. (1985). Shear elasticity and viscosity in colloidal crystals and liquids. *J. Phys. (Paris) C3* **46**, 269–80.

Mellema, J., de Kruif, C. G., Blom, C. & Vrij, A. (1987). Hard sphere colloidal dispersions: mechanical relaxation pertaining to thermodynamic forces. *Rheologica Acta* **26**, 40–4.

Mewis, J., Frith, W. J., Strivens, T. A. & Russel, W. B. (1989). The rheology of suspensions containing polymerically stabilized particles. *AIChE J.* **35**, 415–22.

O'Brien, R. W. (1979). A method for the calculation of the effective transport properties of suspensions of interacting particles. *J. Fluid Mech.* **91**, 17–39.

Patel, P. D. & Russel, W. B. (1987). The rheology of polystyrene latices phase separated by dextran. *J. Rheology* **31**, 599–618.

Russel, W. B. (1978). The rheology of suspensions of charged rigid spheres. *J. Fluid Mech.* **85**, 209–32.

Russel, W. B. (1980). A review of the role of colloidal forces in the rheology of suspensions. *J. Rheology* **24**, 287–317.

Russel, W. B. (1984). The Huggins coefficient as a means for characterizing suspended particles. *J. Chem. Soc. Far. Trans. II* **80**, 31–41.

Saunders, F. L. (1961). Rheological properties of monodisperse latex systems, I. Concentration dependence of relative viscosity. *J. Colloid Sci.* **16**, 13–22.

Schowalter, W. R. (1978). *Mechanics of Non-Newtonian Fluids.* Pergamon.

Stone-Masui, J. & Watillon, A. (1968). Electroviscous effects in dispersions of monodisperse polystyrene latices. *J. Colloid Interface Sci.* **28**, 187–202.

Tanford, C. F. & Buzzell, J. G. (1956). Viscosity of aqueous solutions of bovine serum albumin between pH 4.3 and 10.5. *J. Phys. Chem.* **60**, 225–31.

Tanner, R. I. (1985). *Engineering Rheology*. Oxford University Press.

van der Werff, J. C. & de Kruif, C. G. (1989). Hard sphere colloidal dispersions: The scaling of rheological properties with particle size, volume fraction, and shear rate. *J. Rheology* **33**, 421–54.

Problems

1 Derive the expressions for the linear viscoelastic moduli in Table 14.1 by requiring that elements in series support equal stresses and those in parallel experience equal strains. The overall strain is the sum of the strains of elements in series.

2 Calculate the effective hard-sphere diameter as in §10.6 for charged spheres at the conditions of Fig. 14.7 and replot the data as a function of ϕ_{eff}.

3 A model fluid with a static modulus of 300 N/m^2 is needed to simulate cement to study the relationship between conventional field tests and

rheological properties. Such a fluid can be formulated from a dispersion of polystyrene latices with $a = 0.05\,\mu\text{m}$ and $\psi_s = 100\,\text{mV}$ in water, either by deionizing the dispersion or by adding a water-soluble polymer such as hydroxyethyl cellulose with $r_g = 25\,\text{nm}$ and $M = 115\,\text{kg/mol}$. Specify the appropriate conditions in both cases.

4 For a small static deformation, ε, of an ordered dispersion the affine approximation specifies that

$$P_N = P_N^0 + \sum_{i=1}^{N} \mathbf{x}_i \cdot \varepsilon \cdot \nabla_i P_N^0,$$

with

$$P_N^0 = \exp(-\Phi/kT)$$

and

$$\Phi = \tfrac{1}{2} \sum_{i=1}^{N} \sum_{j=1}^{N} \Phi(r_{ij}),$$

for pairwise additive potentials. Derive from these and the expression (14.10.8) for the thermodynamic stress a relationship between the static shear modulus G_0 and the pair potential.

5 (i) Obtain a formal solution to the time-dependent equivalent of (14.9.3) for small-amplitude, high-frequency oscillations such that $\mathbf{E} = \omega \mathbf{A} \exp(i\omega t)$, with $a^2\omega/D_0 \gg 1$ and $|\mathbf{A}| \ll 1$.
(ii) Develop from the expressions for the viscous and thermodynamic stresses the resulting forms for the high-frequency limiting moduli, G'_∞ and η'_∞.
(iii) Evaluate the $O(\phi^2)$ coefficients for charged spheres at low ionic strengths such that $a\kappa \leq 1$. The linear superposition approximation for Φ and far-field forms for the hydrodynamic functions should suffice.

6 The square-well potential

$$\Phi = \begin{cases} \infty & r < 2a \\ -\varepsilon & 2a < r < R \\ 0 & R < r \end{cases}$$

models rather crudely a combination of hard-sphere repulsion and long-range attraction. With this potential derive expressions for $g(r)$, $f(r)$ for weak steady-shear flows, and the low-shear limiting viscosity η_0. Either use far-field hydrodynamics or neglect hydrodynamic interactions altogether.

APPENDIX A. MEASURED PROPERTIES OF WATER AND SOME ELECTROLYTE IONS AT 298.16 K

Density: 997.1 kg/m^3
Shear viscosity: 8.91×10^{-4} kg/m s
Kinematic viscosity: 8.94×10^{-7} m^2/s
Dielectric constant: 78.3
Sound speed: 1.50×10^3 m/s:

Ion	Stokes radius/nm	Mobility/(μm/s)/(V/m)
H_3O^+	0.0256	3.60×10^{-1}
K^+	0.121	7.50×10^{-2}
Na^+	0.180	5.06×10^{-2}
Ca^{++}	0.153	5.96×10^{-2}
OH^-	0.0433	2.06×10^{-1}
Cl^-	0.203	7.71×10^{-2}
$SO_4^=$	0.115	7.94×10^{-2}

APPENDIX B. VECTOR AND TENSOR NOTATION

Vectors and tensors are denoted by boldface type. The operations follow the customary format. Some of the conventions are listed below; more extensive tabulations, including the equations of motion and the various operations in curvilinear coordinate systems are given by:

R. Aris (1962) *Vectors, Tensors, and the Basic Equations of Fluid Mechanics.* Prentice Hall

G. K. Batchelor (1967) *An Introduction to Fluid Mechanics.* Cambridge University Press

R. B. Bird, W. E. Stewart, and E. N. Lightfoot (1960) *Transport Phenomena.* John Wiley & Sons.

Operations on vectors (first order tensors) can be readily expressed in the cartesian system with **i**, **j**, and **k** denoting unit vectors in the x, y, and z directions, respectively. Relative to be the cartesian basis, the vector **A** has components A_x, A_y, A_z. With index notation, the components are denoted as $A_1, A_2,$ and A_3 and the unit vectors are $\mathbf{e}_1, \mathbf{e}_2,$ and $\mathbf{e}_3$. The *scalar product* is

$$\mathbf{A}\cdot\mathbf{B} \equiv A_xB_x + A_yB_y + A_zB_z \equiv A_1B_1 + A_2B_2 + A_3B_3 \equiv \sum_{i=1}^{3} A_iB_i.$$

The *cross* or *vector product* is

$$\mathbf{A}\times\mathbf{B} \equiv \begin{vmatrix} \mathbf{i} & \mathbf{j} & \mathbf{k} \\ \mathrm{A}_x & \mathrm{A}_y & \mathrm{A}_z \\ \mathrm{B}_x & \mathrm{B}_y & \mathrm{B}_z \end{vmatrix} \equiv (A_yB_z - A_zB_y)\mathbf{i} + (A_zB_x - A_xB_z)\mathbf{j} + (A_xB_y - A_yB_x)\mathbf{k}.$$

Similarly, the second order tensor **T** has components

$$\begin{pmatrix} T_{xx} & T_{xy} & T_{xz} \\ T_{yx} & T_{yy} & T_{yz} \\ T_{zx} & T_{zy} & T_{zz} \end{pmatrix} \text{ or } \begin{pmatrix} T_{11} & T_{12} & T_{13} \\ T_{21} & T_{22} & T_{23} \\ T_{31} & T_{32} & T_{33} \end{pmatrix}.$$

The *transpose* of **T**, $\mathbf{T}^{\mathrm{T}}$, is formed by interchanging the rows and columns of **T**. The *vector* or *dot product* of **T** and **A** is a vector, i.e.,

$$\begin{aligned}\mathbf{T}\cdot\mathbf{A} \equiv &(T_{xx}A_x + T_{xy}A_y + T_{xz}A_z)\,\mathbf{i} + \\ &(T_{yx}A_x + T_{yy}A_y + T_{yz}A_z)\,\mathbf{j} + \\ &(T_{zx}A_x + T_{zy}A_y + T_{zz}A_z)\,\mathbf{k}.\end{aligned}$$

More compactly,

$$\mathbf{T}\cdot\mathbf{A} \equiv \sum_{i=1}^{3}\sum_{j=1}^{3} T_{ij}A_j\mathbf{e}_i.$$

Similarly the *dot product* of two tensors **T** and **S** is a tensor

$$\mathbf{T}\cdot\mathbf{S} \equiv \sum_{i=1}^{3}\sum_{k=1}^{3}\sum_{j=1}^{3} T_{ij}S_{jk}\mathbf{e}_i\mathbf{e}_k.$$

The *scalar* or *double dot* product of two tensors is

$$\begin{aligned}\mathbf{T}:\mathbf{S} &\equiv T_{xx}S_{xx} + T_{xy}S_{yx} + T_{xz}S_{zx} + T_{yx}S_{xy} + T_{yy}S_{yy} + T_{yz}S_{zy} + T_{zx}S_{xz} + T_{zy}S_{yz} \\ &\quad + T_{zz}S_{zz} \\ &\equiv \sum_{i=1}^{3}\sum_{j=1}^{3} T_{ij}S_{ji}.\end{aligned}$$

A *dyadic product* of two vectors is a second order tensor, viz.,

$$\mathbf{AB} \equiv \begin{pmatrix} A_1B_1 & A_1B_2 & A_1B_3 \\ A_2B_1 & A_2B_2 & A_2B_3 \\ A_3B_1 & A_3B_2 & A_3B_3 \end{pmatrix},$$

i.e., the components of **AB** are A_iB_j. Accordingly, the unit tensor $\boldsymbol{\delta}$ has components δ_{ij} which stands for the Kronecker delta with $\delta_{ij} = 0$ unless $i = j$ and $\delta_{ii} = 1$. $\boldsymbol{\delta}$ can be written

$$\boldsymbol{\delta} = \mathbf{ii} + \mathbf{jj} + \mathbf{kk} = \sum_{j=1}^{3} \mathbf{e}_j\mathbf{e}_j.$$

Differentiation involving the *gradient* or 'del' operator is written as

$$\nabla(\ ) \equiv \mathbf{e}_1\frac{\partial(\ )}{\partial x} + \mathbf{e}_2\frac{\partial(\ )}{\partial y} + \mathbf{e}_3\frac{\partial(\ )}{\partial z}.$$

In this form the operator can be manipulated like a vector. Thus, the gradient of a scalar function of x, y, and z (or x_1, x_2, and x_3) yields

$$\nabla s \equiv \mathbf{i}\frac{\partial s}{\partial x} + \mathbf{j}\frac{\partial s}{\partial y} + \mathbf{k}\frac{\partial s}{\partial z} \equiv \sum_{i=1}^{3} \mathbf{e}_i \frac{\partial s}{\partial x_i}.$$

Similar operations can be employed with vectors, tensors and dyadics. For example, with a tensor $\mathbf{T}$,

$$\nabla \cdot \mathbf{T} \equiv \sum_{i=1}^{3} \mathbf{e}_i \sum_{j=1}^{3} \frac{\partial}{\partial x_j} T_{ji}.$$

The *curl* operator is written $\nabla \times (\ )$ and

$$\nabla \times (\ ) \equiv \begin{vmatrix} \mathbf{e}_1 & \mathbf{e}_2 & \mathbf{e}_3 \\ \dfrac{\partial}{\partial x_1} & \dfrac{\partial}{\partial x_2} & \dfrac{\partial}{\partial x_3} \\ (\ )_1 & (\ )_2 & (\ )_3 \end{vmatrix}.$$

AUTHOR INDEX

SUBJECT INDEX

-7878